CAMBRIDGE LIBRARY COLLECTION

Books of enduring scholarly value

Botany and Horticulture

Until the nineteenth century, the investigation of natural phenomena, plants and animals was considered either the preserve of elite scholars or a pastime for the leisured upper classes. As increasing academic rigour and systematisation was brought to the study of 'natural history', its subdisciplines were adopted into university curricula, and learned societies (such as the Royal Horticultural Society, founded in 1804) were established to support research in these areas. A related development was strong enthusiasm for exotic garden plants, which resulted in plant collecting expeditions to every corner of the globe, some-times with tragic consequences. This series includes accounts of some of those expeditions, detailed reference works on the flora of different regions, and practical advice for amateur and professional gardeners.

Flora Capensis

This seminal publication began life as a collaborative effort between the Irish botanist William Henry Harvey (1811–66) and his German counterpart Otto Wilhelm Sonder (1812–81). Relying on many contributors of specimens and descriptions from colonial South Africa – and building on the foundations laid by Carl Peter Thunberg, whose *Flora Capensis* (1823) is also reissued in this series – they published the first three volumes between 1860 and 1865. These were reprinted unchanged in 1894, and from 1896 the project was supervised by William Thiselton-Dyer (1843–1928), director of the Royal Botanic Gardens at Kew. A final supplement appeared in 1933. Reissued now in ten parts, this significant reference work catalogues more than 11,500 species of plant found in South Africa. Volume 5 appeared in three parts, the third comprising sections published between 1912 and 1913, covering Hydrocharideae to Scitamineae.

Cambridge University Press has long been a pioneer in the reissuing of out-of-print titles from its own backlist, producing digital reprints of books that are still sought after by scholars and students but could not be reprinted economically using traditional technology. The Cambridge Library Collection extends this activity to a wider range of books which are still of importance to researchers and professionals, either for the source material they contain, or as landmarks in the history of their academic discipline.

Drawing from the world-renowned collections in the Cambridge University Library and other partner libraries, and guided by the advice of experts in each subject area, Cambridge University Press is using state-of-the-art scanning machines in its own Printing House to capture the content of each book selected for inclusion. The files are processed to give a consistently clear, crisp image, and the books finished to the high quality standard for which the Press is recognised around the world. The latest print-on-demand technology ensures that the books will remain available indefinitely, and that orders for single or multiple copies can quickly be supplied.

The Cambridge Library Collection brings back to life books of enduring scholarly value (including out-of-copyright works originally issued by other publishers) across a wide range of disciplines in the humanities and social sciences and in science and technology.

Flora Capensis

*Being a Systematic Description
of the Plants of the Cape Colony,
Caffraria & Port Natal,
and Neighbouring Territories*

VOLUME 5: PART 3
HYDROCHARIDEAE TO SCITAMINEAE

WILLIAM H. HARVEY *ET AL.*

CAMBRIDGE
UNIVERSITY PRESS

CAMBRIDGE
UNIVERSITY PRESS

University Printing House, Cambridge, CB2 8BS, United Kingdom

Cambridge University Press is part of the University of Cambridge.

It furthers the University's mission by disseminating knowledge in the pursuit of
education, learning and research at the highest international levels of excellence.

www.cambridge.org
Information on this title: www.cambridge.org/9781108068130

© in this compilation Cambridge University Press 2014

This edition first published 1913
This digitally printed version 2014

ISBN 978-1-108-06813-0 Paperback

FLORA CAPENSIS.

VOL. V. SECT. 3.

DATES OF PUBLICATION OF THE SEVERAL PARTS OF THIS VOLUME.

PART I., pp. 1–192, was published *October*, 1912.

PART II., pp. 193–end, was published *March*, 1913.

FLORA CAPENSIS:

BEING A

Systematic Description of the Plants

OF THE

CAPE COLONY, CAFFRARIA, & PORT NATAL

(AND NEIGHBOURING TERRITORIES)

BY

VARIOUS BOTANISTS.

EDITED BY

SIR WILLIAM T. THISELTON-DYER, K.C.M.G.,
C.I.E., LL.D., D.Sc., F.R.S.

HONORARY STUDENT OF CHRIST CHURCH, OXFORD,
LATE DIRECTOR, ROYAL BOTANIC GARDENS, KEW.

*Published under the authority of the Governments of the
Cape of Good Hope, Natal and the Transvaal.*

VOLUME V. SECTION 3.
HYDROCHARIDEÆ TO SCITAMINEÆ.

LONDON:
L. REEVE & CO.,
6, HENRIETTA STREET, COVENT GARDEN.
Publishers to the Home, Colonial and Indian Governments,
1913.

LONDON:
PRINTED BY WILLIAM CLOWES AND SONS, LIMITED,
DUKE STREET, STAMFORD STREET, S.E., AND GREAT WINDMILL STREET, W.

PREFACE.

THE present section of Volume V. completes the description of the MONOCOTYLEDONS of South Africa to which Volumes VI. and VII. are devoted. Its preparation has been long postponed in the hope that the *Orchideæ* might be undertaken by Dr. BOLUS who had made a continuous study of the South African species from 1882 to the close of his life thirty years later. No botanist has ever succeeded in examining or in illustrating with his own hand so many of them in a living state. The "Icones Orchidearum Austro-Africanarum Extratropicarum" in two volumes, commenced in 1892, and the revision of the last proof-sheets of which occupied him "on the very eve of his death" on May 25th, 1911, are a worthy and permanent monument of his labours. Failing health had compelled him to abandon the elaboration of the *Ericaceæ* beyond the vast genus *Erica*, in which the late Professor GUTHRIE had collaborated with him, and it equally deterred him from attempting the study of the intricate problems which are involved in the synonymy of so many of the imperfectly described species of South African *Orchideæ*.

In the event I was reluctantly compelled to acquiesce, and I entrusted their elaboration to Mr. R. A. ROLFE, A.L.S., Assistant in the Herbarium of the Royal Botanic Gardens, Kew, who has long made the order his special study, and had previously contributed the *Orchideæ* to the seventh volume of the "Flora of Tropical Africa."

I am indebted for much friendly aid to Lieut.-Colonel Sir DAVID PRAIN, C.M.G., C.I.E., F.R.S., Director of the Royal Botanic Gardens, and my acknowledgments are also due to Mr. C. H. WRIGHT, A.L.S., and to Mr. N. E. BROWN, A.L.S.,

A 2

Assistant Keepers of the Herbarium, the former for reading the proofs and the latter for working out the localities and geographical distribution.

For the limits of the regions under which the localities are cited in which the species have been found to occur, reference may be made to the Preface to Volume VI.

Besides the maps already cited in the Preface to Volumes VI. and VII., the following have also been used :—Map of the Colony of the Cape of Good Hope and neighbouring territories. Compiled from the best available information. By JOHN TEMPLER HORNE, Surveyor-General, 1895. Stanford's map of the Orange Free State and the Southern parts of the South African Republic, etc., 1899. Carte du Théâtre de la Guerre Sud-Africaine. Par le Comte CAMILLE FAVRE, 1902.

A correct determination of the species represented by the names and descriptions of earlier writers is a necessary basis for an accurate nomenclature. With this object, in order to assist Dr. BOLUS and to the ultimate advantage of the Flora, the types of South African Orchids described by THUNBERG were generously entrusted to Kew in 1883 by Dr. TH. M. FRIES, the Director at the time of the Botanic Garden at Upsala, and in the following year the same kind service was rendered for those of SWARTZ by Dr. V. B. WITTROCK, Director of the Museum of Natural History at Stockholm. They were critically examined by Mr. N. E. BROWN and compared with the specimens of LINDLEY preserved at Kew. Errors in the correlation of the names of the earlier with those of later botanists were detected by him and communicated to Dr. BOLUS as well as to Mr. ROLFE.

Further acknowledgments for the loan of types are due to :—

Dr. J. I. BRIQUET, Director of the Botanic Gardens, Geneva ;

Geheimrath Dr. A. ENGLER, Director of the Botanic Garden, Dahlem ;

Professor C. A. M. LINDMAN, Curator of the Botanical Department of the Museum of Natural History, Stockholm ;

and for the same service, as well as for the gift of specimens, to :—

Dr. RUDOLPH SCHLECHTER, Berlin.

The general material worked out is due to a long series of

contributors of specimens of South African plants whose names are recorded in the previously published volumes.

The expenses of preparation and publication of the present section have been aided by a grant from the Government of the Transvaal.

W. T. T.-D.

WITCOMBE, 20*th February*, 1913.

Series 7.   GLUMACEÆ.—*Flowers* in the axils of scales (*glumes*) in
spikelets.   *Perianth* none, glumaceous or setose.   *Ovary* 1-ovuled
or divided into 1-ovuled cells.   *Seeds* albuminous.

SEQUENCE OF ORDERS CONTAINED IN VOL. V. SECT. 3, WITH BRIEF CHARACTERS.

Series I. MICROSPERMÆ. Ord. CXXIX.–CXXXI.

CXXIX. HYDROCHARIDEÆ (page 1). *Flowers* regular, small, unisexual. Outer *perianth* usually calycine, inner petaloid. *Stamens* 3–12 (*Aquatic herbs*).

CXXX. BURMANNIACEÆ (page 2). *Flowers* regular, hermaphrodite. *Perianth* petaloid. *Stamens* 3 or 6 (*Small terrestrial sometimes leafless herbs*).

CXXXI. ORCHIDEÆ (page 3). *Flowers* usually very irregular. *Perianth* petaloid. *Stamen* 1 (*Terrestrial or epiphytic herbs*).

Series II. EPIGYNÆ. Ord. CXXXII.

CXXXII. SCITAMINEÆ (page 313). *Flowers* irregular. Outer *perianth* usually calycine, inner petaloid. *Stamens* 1 or 5. *Ovary* inferior, 3-celled. *Seeds* with embryo in a central canal of the albumen (*Herbs usually perennial, rarely shrubs*).

ORDER CXXIX. **HYDROCHARIDEÆ**.

(By C. H. WRIGHT.)

Flowers usually unisexual. *Perianth* regular, 2-(rarely) 1-seriate, 3-merous, the inner whorl petaloid, white or yellow. *Stamens* 3–12, inserted on the base of the perianth, free or slightly united at the base ; anthers 2-celled, usually introrse; staminodes sometimes present. *Ovary* inferior, 1-celled with 3–6 parietal placentas, or 6–9-celled ; style short or very long ; stigmas 3 on the 1-celled or 6 on the several-celled ovaries, usually bifid ; ovules many or few, orthotropous or anatropous. *Fruit* various, often fleshy and indehiscent. *Seeds* exalbuminous ; embryo straight.

Floating or submerged, usually perennial, herbs ; leaves either all radical, or cauline and alternate, opposite or whorled, narrow in the South African genus ; flowers when in bud enclosed in a membranous or herbaceous, sessile or stalked spathe, the male usually numerous, the female solitary.

DISTRIB. Genera 14, only 1 in South Africa; species about 50, extending through the tropical and temperate regions of both hemispheres.

I. LAGAROSIPHON, Harv.

Flowers diœcious. *Male* : spathes ovate, bifid at the apex, many-flowered ; perianth 6-lobed, 3 outer lobes slightly larger than the inner ; stamens 3 ; staminodes 2–3. *Female* : spathes ovate or oblong, 1-flowered ; perianth-tube long, very slender ; limb of 6 lobes, the 3 outer slightly larger ; style as long as the perianth-tube ; stigmas 3, short, sometimes forked ; ovules few or many, orthotropous.

Aquatic herbs with long more or less branched stems, leafy throughout ; leaves alternate, subopposite or whorled, linear or linear-lanceolate; spathes axillary, solitary.

DISTRIB. Species about 12, chiefly in Tropical Africa, 1 in the Mascarene Islands and 1 in India.

1. **L. muscoides** (Harv. in Hook. Journ. Bot. iv. 1842, 230, t. 22) ; stem slender, 1 ft. or more long, sparingly branched ; leaves close together especially in the upper part of the stem, linear, acuminate, very minutely serrulate, glabrous, 1-nerved, about 9 lin. long, $\frac{1}{3}$ lin. wide; spathe compressed, 2-lobed, serrulate all round the margin ; male flower : perianth-lobes obovate, the outer slightly larger than the inner ; filaments subulate ; anthers dorsifixed, sagittate, mucronate ; staminodes 3, filiform, scabrous, twice as long as the stamens ; female flower : perianth-tube very slender, $1\frac{1}{2}$ in. long ; limb $1\frac{1}{2}$ lin. in diam. ; lobes obovate, the outer the larger ; staminodes

6, short, inserted in the throat of the perianth-tube; style shortly exserted; stigmas 3, short, forked; pericarp membranous. *Ridl. in Journ. Linn. Soc.* xxii. 233; *Wright in Dyer, Fl. Trop. Afr.* vii. 3. *Hydrilla dregeana, Presl, Bot. Bemerk.* 112. *H. muscoides, Planch. in Ann. Sci. Nat.* 3me *sér.* xi. 79. *Haloragea, Drège, Zwei Pfl. Documente,* 130. *Fluvialea, Drège, l.c.* 136, 140.

VAR. β, **major** (Ridl. in Journ. Linn. Soc. xxii. 233); a much more robust plant than the type, much resembling *Hydrilla verticillata,* Royle; leaves thick, opaque, dark green, 9 lin. long, 1–2 lin. wide, with short obtuse teeth on the pale margins; midrib obscure.

COAST REGION: Uitenhage Div.; swamp near the mouth of the Zwartkops River, *Drege,* 2276c! Bontjes River, Zuurberg Range, 2000 ft., *Drège,* 2276b! near Bethelsdorp, *Zeyher,* 1732! Bathurst Div.; in a valley between Devonshire Farm and Glenfilling, under 1000 ft., *Drège,* 2276a! Var. β: Victoria East Div.; Umdizine River, *Cooper,* 17!

KALAHARI REGION: Orange River Colony; Laai Spruit, *Burke*! and without precise locality, *Mrs. Barber*! Transvaal; Zwart River, New Scotland, *Nelson*! Var. β: Transvaal; Hoogevelt, Bronkerspruit, *Rehmann,* 6559! Trigards Fontein, *Rehmann,* 6678! Standerton, *Schlechter,* 3464!

EASTERN REGION: Var. β: Natal; near Greytown, *Wilms,* 2374!

Also in Tropical Africa.

ORDER CXXX. **BURMANNIACEÆ.**

(By C. H. WRIGHT.)

Flowers hermaphrodite. *Perianth* superior, petaloid; tube cylindrical or gibbous, sometimes winged; limb of 6 lobes in two series, the outer the larger, or 3 in one whorl. *Stamens* 3 or 6, inserted on the perianth; anthers with 2 widely separated cells; connective dilated. *Ovary* inferior, 1-celled with 3 parietal placentas or 3-celled with 3 axile placentas; style simple; stigmas 3, often 2–3-fid; ovules numerous. *Capsule* crowned by the marcescent perianth, terete or 3-angled or 3-winged, opening by apical valves, longitudinal slits or a lid. *Seeds* numerous, small; testa lax; embryo minute, undivided.

Small terrestrial herbs with green leaves or slender leafless parasites; cymes many-flowered, bracteate, more rarely flowers solitary.

DISTRIB. Genera 9, species about 50, dispersed through the tropics of both hemispheres.

I. **BURMANNIA,** Linn.

Perianth superior; tube prominently 3-angled or 3-winged; lobes 6, 3 inner smaller than the outer or absent. *Anthers* 3, subsessile in the perianth-tube; cells globose or clavate, dehiscing transversely; connective produced at the apex into an entire or 2-partite crest.

Ovary 3-celled; style shortly 3-lobed; ovules numerous. *Capsule* crowned by the persistent perianth, more or less strongly 3-winged. *Seeds* oblong or subglobose; testa very thin, reticulate or striate, usually adpressed, rarely lax and produced beyond the nucleus.

Erect unbranched herbs, sometimes very slender, coloured and with minute leaves or scales, sometimes stronger with larger basal leaves; flowers blue or white, rarely yellow, solitary and terminal or many and cymose.

DISTRIB. Species about 20, in the tropics of both hemispheres.

1. B. capensis (Mart. Nov. Gen. et Sp. i. 12); stem filiform, 4–5 in. long, naked or bearing 1 or 2 very minute leaves; flowers 3, terminal; calyx with 3 membranous angles or thin semicircular wings. *Harv. Gen. S. Afr. Pl. ed.* 2, 370; *Schlechter in Fedde, Repert. July,* 1912. *B. sp., Lam. Encycl.* i. 521.

SOUTH AFRICA : without locality, *Herb. Jussieu.*

There is doubt as to this plant being South African. Through the kindness of Prof. H. Lecomte, a photograph of this plant from the Paris Herbarium has been presented to Kew. It shows two very slender stems, respectively 2½ and 3¼ in. high, broken off above their bases and each bearing 3 flowers at the apex ; the whole bears a striking resemblance to *B. bicolor,* var. *africana,* Ridl. (Journ. Bot. 1887, 85), and *B. madagascariensis,* Baker (Journ. Linn. Soc. xx. 268). The specimen is labelled "Herb. Dr. Braguieres" and "M. Dupetit Thouars dit que celle-ci est son *Maburnia,*" a genus now united to *Burmannia* and which Mr. Baker believes to be the same as his *B. madagascariensis.*

ORDER CXXXI. **ORCHIDEÆ.**

(By R. A. ROLFE.)

Perianth superior, irregular, of six free or variously combined segments. *Sepals* equal, or the dorsal one (ventral if the flower is not reversed) different ; lateral ones sometimes united with each other or with the foot of the column, forming a sac- or spur-shaped base (*mentum*). *Petals* usually free, lateral ones more or less different from the sepals, ventral one, or dorsal if the flower is not reversed (*lip*), very different, variously lobed or entire, the so-called disc or central part usually bearing crests or appendages, and the base sometimes extended into a spur or sac. *Stamens* and *style* united into a central column, which faces the lip. *Anther* (in the African genera) solitary, on the top or back of the column, and more or less adnate to it, 2- or 4-celled. *Pollen grains* in 2–8 globose or club-shaped, waxy or granular masses (*pollinia*), which are free or cohere in pairs or fours or altogether by a viscid append-age, and are sometimes attached to a distinct stipes and gland. *Ovary* inferior, 1-celled, with parietal placentas, usually undeveloped at flowering time. *Stigma* either consisting of a viscid surface near the top of the concave side of the column or two-lobed and lateral ; in the former case it faces the lip, and is usually separated from the anther, below which it lies, by an appendage (*rostellum*). *Seeds* very numerous, minute, fusiform ; testa loose, reticulated, enclosing a homogeneous embryo.

Herbs of various habit, or rarely shrubby ; in many cases terrestrial, with tuberous roots or perennial creeping rhizomes, annual herbaceous stems, and solitary spicate or racemose flowers, or in many cases epiphytal, with perennial stems or branches, variously thickened and forming *pseudobulbs*, upon which the leaves and flowers are borne, or the latter sometimes produced below the leaves : leaves alternate, simple, entire, coriaceous or membranous ; inflorescence terminal, basal or axillary, spicate, racemose or paniculate, sometimes 1-flowered.

DISTRIB. The largest Order among monocotyledons, the species now known being estimated at over 5000, distributed into between 300 and 400 genera. They are found throughout the globe, except in the highest latitudes and altitudes, and the more remote oceanic islands, being very.rare, however, in very dry countries. The epiphytic species are for the most part confined to the intertropical zone, within which they are most numerous in the mountains of Tropical Asia and America. The great majority of African species are terrestrial.

Grammangis pardalina, Reichb. f., and *G. falcigera*, Reichb. f. in Flora, 1885, 541, are doubtfully included by Bolus (Journ. Linn. Soc. xxv. 185), with the locality S.E. Africa, but they were described in a paper on Comoro Island Orchids, and must be excluded. The genus is only Mascarene.

Tribe I. EPIDENDREÆ.—*Anther* 2-celled. *Pollinia* 2–8, waxy, united by a viscid appendage, free from the rostellum.

Sub-tribe 1. MALAXEÆ.—*Column* not produced into a foot at the base. *Pollinia* 4, sometimes cohering in pairs. *Lip* continuous with the base of the column. *Flowers* with spreading perianth-segments.

    I. **Liparis.**—Terrestrial herbs. *Leaves* membranous. *Flowers* in spikes, small, greenish.

Sub-tribe 2. DENDROBIEÆ.—*Column* produced into a foot at the base. *Pollinia* 4, sometimes cohering in pairs. *Lip* articulated to the foot of the column. *Flowers* with erect or subconnivent perianth-segments.

    II. **Megaclinium.**—Epiphytic herbs. *Pseudobulbs* ovoid. *Leaves* coriaceous. *Flowers* distichously arranged on either side of a dilated and often flattened rhachis.

Sub-tribe 3. BLETIEÆ.—*Column* footless. *Pollinia* 8. *Lip* inserted at the base of the column, and either convolute round it or more or less adnate to it ; its base produced into a slender spur.

    III. **Calanthe.**—*Lip* adnate to the column, its limb flat or reflexed and 3–4-lobed.

Tribe II. VANDEÆ.—*Anther-cells* usually confluent. *Pollinia* 2 or 4, attached singly or in pairs to a stipes and gland (a process of the rostellum), with with which they are carried away upon removal.

Sub-tribe 1. EULOPHIEÆ.—Usually terrestrial herbs. *Leaves* plicate and acute. *Lip* usually spurred or saccate at the base. *Column* with a distinct foot.

    IV. **Acrolophia.**—*Inflorescence* terminal. *Leaves* coriaceous and persistent. *Sepals* and *petals* equal or subequal, subconnivent ; petals similar to the sepals in colour or paler.

    V. **Eulophia.**—*Inflorescence* lateral. *Leaves* usually more or less membranous and not persistent. *Sepals* and *petals* equal or subequal, subconnivent or the sepals seldom reflexed ; petals usually very similar to the sepals in colour or paler.

VI. **Lissochilus.**—*Inflorescence* lateral. *Leaves* membranous and not persistent. *Sepals* generally strongly reflexed. *Petals* much larger, erect or suberect, usually differently coloured from the sepals.

Sub-tribe 2. CYMBIDIEÆ.—Epiphytic herbs. *Leaves* plicate and acute. *Lip* not spurred or saccate. *Column* without a foot.

VII. **Ansellia.**—*Sepals* and *petals* spreading, subequal. *Lip* inferior.

VIII. **Polystachya.**—*Sepals* and *petals* subconnivent, unequal. *Lip* superior.

Sub-tribe 3. SARCANTHEÆ.—Epiphytic herbs. *Leaves* distichous, coriaceous and variously bilobed, conduplicate at the base. *Lip* spurred or saccate at the base.

IX. **Angræcum.**—*Pollinia* with a single stipes and gland.

X. **Listrostachys.**—*Pollinia* with two distinct stipes attached to a single gland.

XI. **Mystacidium.**—*Pollinia* with two distinct stipes attached to two separate glands.

Tribe III. NEOTTIEÆ.—*Pollinia* granular or powdery, free or with apical caudicle and gland. *Anther-case* operculate or erect and persistent.

Sub-tribe 1. CORYMBIEÆ.—*Stems* tall and more or less woody. *Leaves* plicate or strongly ribbed. *Inflorescence* paniculate. *Floral-segments* and *column* much elongated. *Anther* and *rostellum* much as in the next sub-tribe.

XII. **Corymbis.**—*Sepals* and *petals* elongated, connivent and narrow below, somewhat spreading and dilated above. *Lip* erect, linear, dilated into a short recurved limb above.

Sub-tribe 2. SPIRANTHEÆ.—*Stems* dwarf, herbaceous. *Leaves* membranous, not plicately ribbed. *Floral-segments* and *column* usually short. *Anther* incumbent behind the somewhat elongated and beaked rostellum. *Pollinia* sectile, with a linear caudicle and gland.

XIII. **Zeuxine.**—Front lobe of the *lip* dilated and attached to the sac by a narrow claw.

XIV. **Platylepis.**—Front lobe of the *lip* sessile on the sac, not dilated in front into a distinct limb.

Sub-tribe 3. ARETHUSEÆ.—*Stems* from an underground tuber. *Leaves* radical, membranous, plicate. *Flowers* on leafless scapes, often borne before the leaves. *Column* clavate. *Pollinia* granular or powdery, without stipes or gland.

XV. **Pogonia.**—*Sepals* and *petals* free, more or less spreading. *Lip* without spur or sac.

Tribe IV. OPHRYDEÆ.—*Pollinia* granular or sectile, with a distinct caudicle and gland, sometimes with two distinct caudicles and glands. *Anther-cells* adnate to the column and persistent, often continuous with the rostellum.

Sub-tribe 1. GYMNADENIEÆ.—*Anther* erect. *Glands* of the pollinia not enclosed within a pouch. *Rostellum* without diverging side lobes. *Stigma* simple, situated on the front of the column.

XVI. **Brachycorythis.**—*Petals* about as long as the sepals, not broadly dilated at the apex. *Lip* not saccate or spurred at the base.

XVII. **Platanthera.**—*Petals* about as long as the sepals, not broadly dilated at the apex, forming with the dorsal sepal a hood over the column. *Lip* simple or 3-lobed, saccate or spurred at the base.

XVIII. **Schizochilus.**—*Petals* very small and membranous, not forming with the dorsal sepal a hood over the column. *Lip* 3-lobed, spurred or rarely obscurely saccate at the base.

XIX. **Bartholina.**—*Petals* about as long as the sepals, not broadly dilated at the apex. *Lip* broadly flabellate and multipartite, spurred at the base.

XX. **Holothrix.**—*Petals* longer than the sepals, narrow, sometimes fimbriate at the apex. *Lip* spurred at the base.

XXI. **Huttonæa.**—*Petals* broadly spathulate, concave and fimbriate. *Lip* broad, with fimbriate margin, not spurred at the base.

Sub-tribe 2. HABENARIEÆ.—*Anther* erect. *Glands* of the pollinia not enclosed within a pouch. *Rostellum* 3-lobed, with diverging, often much elongated side lobes. *Stigmas* 2, often elongate, clavate, and more or less lateral.

XXII. **Peristylus.**—*Lip* entire or 3-lobed, with short basal spur or sac. *Rostellum* triangular, minute, without diverging side lobes. *Stigmatic processes* short.

XXIII. **Stenoglottis.**—*Lip* 3–5-lobed, without basal spur. *Rostellum* triangular, small, without diverging side lobes. *Stigmatic processes* somewhat elongated.

XXIV. **Habenaria.**—*Lip* entire, 3-lobed or 3-partite, with an elongated or rarely a short or saccate basal spur. *Petals* simple or 2-lobed. *Rostellum* with small tooth-like front lobe and more or less elongated side lobes, which carry the anther channels. *Stigmatic processes* clavate, often much elongated.

XXV. **Bonatea.**—*Lip* 3-partite, with an elongated basal spur. *Petals* 2-partite, front lobes cohering at the base with the lip, lateral sepals and stigmatic processes. *Rostellum* with galeate front lobe and elongated side lobes, which carry the anther-channels. *Stigmatic processes* clavate and much elongated.

XXVI. **Cynorchis.**—*Lip* 3-lobed or rarely simple, with a short or elongated basal spur. *Rostellum* triangular, with cucullate front lobe, and more or less elongated side lobes, which carry the anther-channels. *Stigmatic processes* oblong, short.

Sub-tribe 3. DISEÆ.—*Anther* usually reclinate or more or less reflexed behind the column. Glands of the *pollinia* not enclosed within a pouch. *Rostellum* with more or less developed side lobes. *Stigma* simple or somewhat 2-lobed, concave or convex, often large, sometimes basal and somewhat adnate to the lip. *Lip* usually free from the column.

XXVII. **Satyridium.**—*Lip* superior, shortly 2-spurred at the base. *Sepals* and *petals* free. Glands of *pollinia* united.

XXVIII. **Satyrium.**—*Lip* superior, 2-spurred or 2-saccate at the base. *Sepals* and *petals* free. Glands of *pollinia* separate.

XXIX. **Aviceps.**—*Lip* superior, 2-saccate at the base. *Sepals* and *petals* united into an oblong limb. Glands of *pollinia* separate.

XXX. **Pachites.**—*Lip* superior, not spurred. *Sepals* and *petals* free and spreading or suberect. Glands of *pollinia* separate.

XXXI. **Orthopenthea.**—*Lip* superior, not spurred. Odd *sepal* anticous, concave or rarely subgaleate or shortly spurred near the base. *Petals* free and spreading. Glands of *pollinia* separate.

XXXII. **Monadenia.**—Odd *sepal* superior, galeate, spurred behind. *Petals* free. *Lip* narrow and entire. Glands of *pollinia* united.

XXXIII. **Amphigena.**—Odd *sepal* superior, galeate, ascending. *Petals* erect, included. *Lip* small and narrow, entire. *Rostellum* unappendaged. Glands of *pollinia* united.

XXXIV. **Herschelia.**—Odd *sepal* superior, galeate, shortly spurred behind. *Petals* free. *Lip* broad, concave and more or less fringed or sometimes long-spathulate. *Rostellum* 3-toothed at the apex. Glands of *pollinia* generally united.

XXXV. **Penthea.**—Odd *sepal* superior, flat, concave or not spurred. *Petals* free and spreading. Glands of *pollinia* separate. *Lip* narrowly linear.

XXXVI. **Disa.**—Odd *sepal* superior, galeate, with a short or sometimes much elongated spur behind. *Petals* free. *Lip* flat, entire, usually narrow. Glands of *pollinia* separate.

XXXVII. **Schizodium.**—Odd *sepal* superior, galeate, with an oblong or narrow spur behind. *Petals* free. *Lip* with a broad or concave base and an acuminate apex. Glands of *pollinia* separate.

XXXVIII. **Brownleea.**—Odd *sepal* superior, galeate, with an oblong or narrow and much elongated spur behind. *Petals* more or less adnate to the base of the dorsal sepal and sides of the column. *Lip* minute, dilated, more or less clasping the base of the column. Glands of *pollinia* separate.

Sub-tribe 4. CORYCIEÆ.—*Anther* usually reclinate or more or less reflexed behind the column, suberect in *Ommatodium*. Glands of *pollinia* not enclosed within a pouch. *Rostellum* with more or less developed side lobes. *Lip* adnate to the column throughout its length, and produced beyond the separated anther-cells into a variously-formed appendage. *Stigma* more or less 2-lobed.

*_Lateral sepals neither saccate nor spurred._

XXXIX. **Ceratandra.**—*Lip* superior, lunate, with a rather small obovate-oblong appendage. *Petals* cohering with the dorsal sepal into a concave limb. Lateral *sepals* free. Arms of *rostellum* narrow and suberect.

XL. **Ceratandropsis.**—Odd *sepal* superior, cohering with the petals into a concave limb. Lateral *sepals* free. *Lip* cordate-ovate, without appendage. Arms of *rostellum* elongated and ascending.

XLI. **Evota.**—Odd *sepal* superior, cohering with the ample petals into a spreading limb. Lateral *sepals* free. *Lip* lunate or hastate, with a large obcordate or bipartite appendage. Arms of the *rostellum* laterally dilated.

XLII. **Ommatodium.**—Odd *sepal* superior, cohering with the petals into a hood-shaped, somewhat concave limb. Lateral *sepals* free. *Anther* suberect, with the glands of the pollinia situated below.

XLIII. **Pterygodium.**—Odd *sepal* superior, cohering with the petals into a broad concave or spreading limb. Lateral *sepals* free. *Anther* inverted, with the glands of the pollinia uppermost.

XLIV. **Anochilus.**—*Lip* superior, with a large deflexed fleshy appendage. Odd *sepal* cohering with the petals into a large concave limb, sometimes ultimately free. Lateral *sepals* free.

XLV. **Corycium.**—Odd *sepal* superior, cohering with the saccate or very concave petals into a concave or much contracted limb. Lateral *sepals* united into a concave limb, or sometimes free. *Lip* with a large erect fleshy appendage.

**_Lateral sepals spurred or prominently saccate behind._

XLVI. **Disperis.**—Odd *sepal* superior, galeate or spurred, cohering with the petals into a galeate limb. Lateral *sepals* free. *Lip* produced into a variously-shaped erect appendage, sometimes larger than the reflexed limb.

I. LIPARIS, L. C. Rich.

Sepals spreading, free, or the lateral more or less connate, often falcate and broader than the dorsal. *Petals* usually very narrow. *Lip* inferior, adnate to the base of the column, usually deflexed or recurved above the erect base, entire, often bituberculate at the base. *Column* usually long and slender, incurved, usually margined or winged at the apex; anther terminal; pollinia 4, free or cohering by a viscid appendage.

Terrestrial or rarely epiphytic herbs, with or without pseudobulbs; leaves one or more, membranous and continuous with the sheath, or somewhat coriaceous and jointed on the sheath or pseudobulb; flowers small or medium-sized, in lax or dense racemes; bracts small or narrow; flowers small.

DISTRIB. Species about 120, found throughout the warm and temperate regions of the globe, but most numerous in Tropical Asia. All the South African species belong to the section *Mollifoliæ*.

Raceme dense ... ... (1) **capensis.**

Raceme lax:

Leaves ovate to elliptic ... ... (2) **Bowkeri.**

Leaves lanceolate ... ... (3) **Gerrardi.**

1. **L. capensis** (Lindl. in Ann. Nat. Hist. ser. i. ii. 314); pseudobulbs ovoid, $\frac{3}{4}$–1 in. long, usually 2-leaved; leaves spreading, ovate, subacute, rather thick and coriaceous, with depressed veins, shining above, $\frac{3}{4}$–$2\frac{1}{2}$ in. long, $\frac{1}{2}$–$1\frac{1}{2}$ in. broad; scapes 2–6 in. high; raceme dense, many-flowered; bracts lanceolate, acuminate, $\frac{1}{4}$–$\frac{1}{3}$ in. long; pedicels slightly longer than the bracts; dorsal sepal linear-oblong, obtuse, $\frac{1}{6}$–$\frac{1}{4}$ in. long, with revolute margins; lateral sepals ovate-oblong, subfalcate, subobtuse, $\frac{1}{8}$–$\frac{1}{6}$ in. long; petals linear, obtuse, about as long as the dorsal sepal, with revolute margins; lip recurved, elliptic-oblong, truncate or obscurely bilobed at the apex, $\frac{1}{8}$–$\frac{1}{6}$ in. long; sides inflexed towards the apex; column clavate, incurved, $\frac{1}{10}$–$\frac{1}{12}$ in. long. *Ridl. in Journ. Linn. Soc.* xxii. 272; *Bolus in Trans. S. Afr. Phil. Soc.* v. 103, *Orch. Cap. Penins.* 103, *Journ. Linn. Soc.* xxv. 180, *and Ic. Orch. Austr.-Afr.* i. *t.* 1; *Engl. Hochgebirgsfl. Trop. Afr.* 187; *Durand & Schinz, Conspect. Fl. Afr.* v. 8; *Rolfe in Dyer, Fl. Trop. Afr.* vii. 22. *L. Pappei, Lindl. ex Harv. Thes. Cap.* ii. 7. *Sturmia capensis, Sond. in Linnæa*, xix. 71.

SOUTH AFRICA: without locality, *Grey*!

COAST REGION: Cape Div.; near Cape Town and on the Flats, below 100 ft. *Bolus*, 4552! *MacOwan*, 2554! and in *Herb. Norm. Austr.-Afr.* 151! *Pappe*, 36! Wynberg Flats, *Prior*! near Rondebosch, *Bolus*, 4598! Steenberg, *Wolley-Dod*, 1098! Stellenbosch Div.; mountains of Hottentots Holland, *Zeyher*, 3887! Swellendam Div.; in woods and on rocks at Swellendam, *Bowie*! Komgha Div.; between Zandplaat and Komgha, *Drège*, 4560!

Also in Tropical Africa, if a fruiting specimen from the Cameroons is rightly referred here.

2. **L. Bowkeri** (Harv. Thes. Cap. ii. 6, t. 109); pseudobulbs ovoid or oblong, 1–$1\frac{1}{4}$ in. long, 3–4-leaved; leaves elliptic-oblong or ovate, acute, loosely sheathing at the base, limb $1\frac{1}{4}$–$4\frac{1}{2}$ in. long, $\frac{3}{4}$–$2\frac{1}{2}$ in.

broad, undulate, membranous, with 5 to 7 prominent nerves; scapes $2\frac{1}{2}$–8 in. high; racemes lax, about 4–12-flowered; bracts acuminate from a broad amplexicaul base, $\frac{1}{6}$–$\frac{1}{2}$ in. long; pedicels $\frac{1}{3}$–$\frac{1}{2}$ in. long; dorsal sepal narrowly linear-oblong, obtuse, $\frac{1}{3}$–$\frac{1}{2}$ in. long, with revolute margins; lateral sepals falcate oblong, obtuse, twisted, $\frac{1}{3}$ in. long; petals linear, obtuse, $\frac{1}{3}$–$\frac{1}{2}$ in. long, with revolute margins; lip subsessile or shortly clawed; limb recurved, ovate-orbicular, obtuse, concave, $\frac{1}{3}$ in. long, with a pair of erect calli near the base; column arcuate, $\frac{1}{5}$ in. long. *Ridl. in Journ. Linn. Soc.* xxii. 270; *Bolus in Journ. Linn. Soc.* xxv. 181, *and Ic. Orch. Austr.-Afr.* i. *t.* 2; *Durand & Schinz, Conspect. Fl. Afr.* v. 8; *Rolfe in Dyer, Fl. Trop. Afr.* vii. 21.

SOUTH AFRICA: without locality, *Mrs. Saunders*!

COAST REGION: Albany Div.; in mountain woods, Grahamstown, *MacOwan*, 1022! *Atherstone*! Mountains of Kaffraria, *Mrs. Barber*!

EASTERN REGION: Transkei; in crevices of rocks, in shade at Fort Bowker, *Bowker*, 454! Kreilis Country, *Bowker*! Natal, *Wood*, 5585! Zululand; Ingoma, on decayed trees, *Gerrard*, 1556!

Also in Tropical Africa.

3. **L. Gerrardi** (Reichb. f. in Flora, 1867, 118); stem very short, slightly swollen; leaves radical, 3 to 4, sheathing at the base, lanceolate, acute, membranous, 2–3 in. long, $\frac{1}{3}$–$\frac{1}{2}$ in. broad, with 3 prominent nerves below and 9 to 13 slender ones above; peduncle 5–7 in. long, with 2 or 3 lanceolate sheaths below the flowers; raceme lax, about 6–12-flowered; bracts lanceolate, acuminate, $\frac{1}{4}$ in. long; pedicels 3–5 lin. long; dorsal sepal oblong-lanceolate, acute, $\frac{1}{4}$ in. long; lateral sepals obliquely oblanceolate-oblong, obtuse, $\frac{1}{6}$ in. long; petals linear, subobtuse, $\frac{1}{4}$ in. long; lip ovate-suborbicular, bluntly apiculate, $\frac{1}{6}$ in. long, with a pair of blunt calli at the base; column stout, slightly curved, $\frac{1}{8}$ in. long. *Ridl. in Journ. Linn. Soc.* xxii. 275; *Bolus in Journ. Linn. Soc.* xxv. 181; *Durand & Schinz, Conspect. Fl. Afr.* v. 9.

EASTERN REGION: Natal, *Gerrard*!

Bolus (*Journ. Linn. Soc.* xxv. 181) includes *Liparis polycardia*, Reichb. f. in Flora, 1885, 543, as doubtfully South African, but it almost certainly occurs outside our limits. It was described in a paper on Comoro Orchids, and, although no habitat is given, it probably came from that region, for its affinity is with a small group of Mascarene species.

II. MEGACLINIUM, Lindl.

Dorsal sepal free, erect or spreading, longer than the triangular-ovate acuminate lateral ones, which are either falcate or reflexed about the middle, base of lateral adnate to the foot of the column. *Petals* smaller and narrower than the dorsal sepal, often very narrow. *Lip* articulated to the foot of the column, mobile, inflexed at the base, recurved above, entire or denticulate at the base, rarely somewhat 3-lobed. *Column* short, broadly dilated and winged at both sides, terminating above in acute or rounded teeth; base produced into a short foot; anther terminal, operculate, incumbent, 2- or rarely 1-celled; pollinia waxy, normally 4, but usually connate in pairs, without appendage.

Epiphytic herbs, with stout creeping rhizomes; pseudobulbs sessile in the axil of a sheath, mostly 3–5-angled, 2- or 1-leaved; scapes arising from the base of the pseudobulbs, simple; apex dilated into a flattened, often ensiform and almost foliaceous rhachis, along either side of which the flowers and bracts are distichously arranged; bracts usually ovate or triangular, ultimately much deflexed; flowers small, pedicelled, curved.

DISTRIB. Species about 30, mostly Tropical African, with 2 South African representatives, and 1 reported from Madagascar.

Pseudobulbs narrowly ovate-oblong, slightly 4-angled; leaves narrowly oblong ... ... (1) **Sandersoni**.

Pseudobulbs broadly ovate-oblong, strongly 4-angled; leaves elliptic-oblong ... ... (2) **scaberulum**.

1. **M. Sandersoni** (Oliv. in Bot. Mag. sub t. 5936); rhizome creeping, moderately stout; pseudobulbs narrowly ovoid-oblong, slightly 4-angled, 1–$1\frac{1}{4}$ in. long, 2-leaved; leaves narrowly oblong, obtuse, 2–$2\frac{1}{4}$ in. long, $\frac{1}{3}$–$\frac{1}{2}$ in. broad; scapes $3\frac{1}{2}$–4 in. long, with 3 or 4 oblong obtuse sheaths below the inflorescence; rhachis rather narrow, about 1–$1\frac{1}{4}$ in. long, $\frac{1}{6}$ in. broad, dark purple; bracts triangular-ovate, acute, about $\frac{1}{6}$ in. long; flowers not numerous, lurid purple; dorsal sepal linear-oblong, obtuse, somewhat recurved, concave, $\frac{1}{5}$–$\frac{1}{6}$ in. long; lateral sepals obliquely ovate, with falcate acute apex, $\frac{1}{8}$ in. long; petals linear, falcately curved, $\frac{1}{6}$ in. long; lip fleshy, acute, strongly recurved from a broad base, $\frac{1}{12}$ in. long; column short, with broad wings; capsule oblong, $\frac{1}{2}$ in. long. *Bulbophyllum Sandersoni, Reichb. f. in Flora*, 1878, 78; *Bolus in Journ. Linn. Soc.* xxv. 181, *and Ic. Orch. Austr.-Afr.* i. *t.* 3, *partly* (*excl. broad-leaved plant*); *Durand & Schinz, Conspect. Fl. Afr.* v. 16.

KALAHARI REGION: Transvaal; *Rendall*! on trees at Saddleback Kloof and Moodies, near Barberton, at 4100 ft., *Culver*, 6; Umzindine Creek, 4600 ft., *Galpin*, 688.
EASTERN REGION: Natal; *Sanderson*, 898! *Mrs. K. Saunders*!

2. **M. scaberulum** (Rolfe in Gard. Chron. 1888, iv. 6); rhizome creeping, stout; pseudobulbs ovoid, tetragonal, $\frac{3}{4}$–$1\frac{1}{4}$ in. long, 2-leaved; leaves elliptic-oblong, obtuse, coriaceous, $1\frac{1}{2}$–2 in. long, $\frac{1}{2}$–$\frac{3}{4}$ in. broad; scapes suberect, 4–6 in. long, with 3 or 4 oblong obtuse sheaths below the inflorescence; rhachis flattened, 2–4 in. long, $\frac{1}{3}$ in. broad, pale green, densely spotted and marbled with dusky-brown; triangular-ovate, acute, ultimately reflexed, $\frac{1}{6}$ in. long; flowers rather numerous, dull purple, becoming green spotted with purple at the base of the sepals; sepals papillose, dorsal narrowly lanceolate-linear, acute, $\frac{1}{3}$ in. long, lateral shorter, falcate, with broadly ovate base and narrowly acute apex; petals falcate, reflexed, narrowly linear, acute, $\frac{1}{6}$ in. long; lip sharply reflexed, base broadly orbicular, then rapidly narrowed into a broadly linear obtuse apex, grooved on the upper half, with two keels underneath; column short, with

broad wings, terminating upwards in a pair of short teeth. *Durand & Schinz, Conspect. Fl. Afr.* v. 16. *Bulbophyllum scaberulum, Bolus in Journ. Linn. Soc.* xxv. 181. *B. Sandersoni, Bolus, Ic. Orch. Austr.-Afr.* i. *t. 3, partly (broad-leaved plant only).*

EASTERN REGION : Pondoland ; *Tillett* ! Zululand, *Mrs. K. Saunders* !

Dr. Bolus has united this with the preceding species, but the two are quite distinct in numerous particulars.

III. CALANTHE, R. Br.

Sepals subequal, free, usually widely spreading. *Petals* like the sepals or narrower. *Lip* usually more or less adnate to the column ; limb spreading, 3-lobed, with the front lobe often 2-lobed ; disc variously crested or lamellate ; base mostly extended into a slender spur. *Column* short, without a foot ; wings usually united to the sides of the column. *Anther* subterminal, operculate, incumbent, 2-celled ; pollinia 8, oblong or obovate, somewhat compressed, the caudicles often somewhat attenuate and united by a viscid appendage. *Capsules* elliptic-oblong.

Terrestrial herbs, with short leafy stems, sometimes thickened at the base ; leaves 2 to several, petiolate ; limb elliptic- to oblong-lanceolate, acute or acuminate, plicate ; scapes erect from the rhizome, often tall, with a number of sheaths towards the base ; flowers in loose or dense racemes, usually showy ; bracts lanceolate or ovate-lanceolate.

DISTRIB. Species about 100, most numerous in Tropical Asia, extending to Japan and Australia, and sparingly represented in Africa, the West Indies and Central America.

1. **C. natalensis** (Reichb. f. in Bonplandia, 1856, 322) ; rhizome stout ; leaves tufted, petiolate, elliptic or elliptic-lanceolate, acuminate, plicate, 5–9-nerved, 8–15 in. long, 2¼–4 in. broad ; petioles 2–5 in. long ; scapes erect, 1–2 ft. high ; racemes lax, up to 9 in. long ; bracts linear-lanceolate, acute, ½–¾ in. long ; pedicels 1–1½ in. long ; flowers lilac, with purple lip which ultimately changes to buff ; sepals ovate-lanceolate, acute or acuminate, ¾ in. long ; petals slightly shorter than the sepals ; lip 3-lobed, ⅓ in. long ; side lobes oblong, obtuse ; front lobe obcordately 2-lobed ; disc with 3 rows of wart-like crests ; spur slender, curved, ¾–1 in. long ; column short and very stout. *N. E. Br. in Gard. Chron.* 1885, xxiv. 78, 136 ; *Bolus in Journ. Linn. Soc.* xxii. 65 ; xxv. 181, *and Ic. Orch. Austr.-Afr.* ii. *t. 7 ; Bot. Mag. t.* 6844 ; *Durand & Schinz, Conspect. Fl. Afr.* v. 18. *C. sylvatica, var. natalensis, Reichb. f. in Linnæa,* xix. 374.

COAST REGION : King Williamstown Div. ; Perie Forest, near King Williamstown, *D'Urban,* 108 ! *Bolus* ! *Walsh in Herb. Bolus,* 6101 !
EASTERN REGION : Natal ; at The Bluff, *Sanderson,* 1003 ! Kirkman's Cutting, Maritzburg Road, *Sanderson,* 1047 ! in Bush at 2400 ft., *Buchanan* ! Zululand ; Entumeni, *Wood,* 3949 ! Ingoma, *Gerrard,* 1538 ! Thlivia, *Gerrard,* 2173 !

IV. **ACROLOPHIA,** Pfitz.

Sepals subequal, free, subconnivent or spreading, the lateral not adnate to the foot of the column. *Petals* like the sepals or a little broader, similar in colour or sometimes paler. *Lip* continuous with the base of the column, usually narrow at the base, 3-lobed or sometimes entire above, generally saccate or spurred at the base, or without a spur in *A. ustulata*; side lobes suberect; front lobe expanded or sometimes concave; disc variously papillose or crested. *Column* short, clavate, not produced into a foot at the base; anther-bed oblique, erect, entire. *Anther* terminal, operculate, incumbent, semiglobose, conical or 2-horned above, imperfectly 2-celled; pollinia 4, ovoid, united in pairs, affixed to a broad stipes and gland. *Capsule* ovoid or oblong, with prominent thickened angles.

Terrestrial herbs; stems leafy at the base, not thickened into rhizomes; roots thick and fleshy; leaves in a more or less distichous basal tuft, narrow, coriaceous and persistent, plicate, conduplicate at the base; scapes or peduncles terminal, erect, usually loosely paniculate, racemose in *A. ustulata*; flowers small or medium-sized, usually lax; bracts small or narrow.

DISTRIB. Species 9, mostly limited to the south-west corner of the Cape Colony, and seldom far removed from the influence of the moist sea breezes.

Differs from *Eulophia* in the terminal inflorescence, leafy stems, which are not thickened into tubers; the very fleshy roots, and the coriaceous persistent leaves. It includes *Eulophia* section *Desciscentes*, Lindl. Gen. and Sp. Orch. 184.

Lip spurred or saccate at the base:
  Anther without horns: sepals $\frac{1}{6}$–$\frac{1}{4}$ in. long; front lobe
    of lip cochleate:
      Lip with ample side lobes   ...   .   (1) **micrantha.**

    Lip with obscure or no side lobes:
      Lip nearly erect with a constriction beneath near
        the base. .   ...   (2) **cochlearis.**

      Lip spreading without a constriction beneath   ...   (3) **Bolusii.**

  Anther 2-horned; sepals nearly $\frac{1}{2}$–$\frac{2}{3}$ in. long; front
    lobe of lip not cochleate:
  Sepals $\frac{1}{2}$ in. long or nearly so:
    Flowers rather stout in texture; nerves of lip
      strongly papillose or barbate:
      Capsule ellipsoid; nerves of disc with stout
        papillæ   ...   .   (4) **tristis.**

      Capsule spherical; nerves of disc with slender
        papillæ   ...   .   (5) **sphærocarpa.**

    Flowers submembranous; nerves of lip sparingly
      barbate ...   .   (6) **lunata.**

  Sepals $\frac{2}{3}$ in. long:
    Petals acute; disc of lip densely barbate; bracts
      not coloured   ...   .   (7) **lamellata.**

    Petals acuminate; disc of lip slightly barbate;
      bracts coloured ...   .   (8) **comosa.**

Lip neither spurred nor saccate   ...   .   (9) **ustulata.**

1. **A. micrantha** (Schlechter & Bolus in Journ. Bot. 1894, 332,
partly); rhizome stout; leaves tufted, 4–14, ensiform or linear,
acute, slightly curved or sometimes spreading, rigid, minutely
serrulate, 4–6 in. long, $\frac{1}{8}$–$\frac{1}{4}$ in. broad, dilated at the sheathing base;
inflorescence terminal, paniculate, 1–1$\frac{1}{2}$ ft. high, with numerous
lanceolate sheaths below; panicle $\frac{1}{2}$–1 ft. long, lax, many-flowered;
bracts lanceolate, acuminate, $\frac{1}{6}$–$\frac{1}{3}$ in. long; pedicels slender, $\frac{1}{4}$–$\frac{1}{3}$ in.
long; flowers small and numerous; sepals subconnivent, elliptic-
oblong, apiculate, $\frac{1}{8}$–$\frac{1}{6}$ in. long, greenish-brown; petals light yellow,
otherwise similar to the sepals; lip strongly 3-lobed, white, as long
as the sepals and petals; side lobes rounded, obtuse, with 5 or 6
thickened rose-coloured nerves; front lobe cochleate, suborbicular,
dotted with red, crisped and crenulate at the margin, rather larger
than the side lobes; disc with about 4 rows of cristate keels on the
front lobe; spur broadly oblong or globose, obtuse, $\frac{1}{3}$ as long as the
lip; column stout, more than $\frac{1}{2}$ as long as the segments; capsule
elliptic-oblong, over $\frac{1}{2}$ in. long. *Bolus, Ic. Orch. Austr.-Afr. ii. t. 8.
A. fimbriata, Schlechter in Engl. Jahrb. xxvi. 340. Eulophia
micrantha, Lindl. Gen. & Sp. Orch. 184, and in Hook. Comp. Bot.
Mag. ii. 202; Drège, Zwei Pfl. Documente, 128, 130, 131.*

SOUTH AFRICA : without locality or collector (*the type specimen*)!
COAST REGION : Bredasdorp Div. ; near Rietfontein Poort, *Schlechter*, 9703.
Sandy soils between the Cape Peninsula and Plettenbergs Bay, *Bowie*! Uitenhage
Div. ; between Coega and Sunday Rivers, under 1000 ft., *Drège*! between
Van Stadens Berg and Bethelsdorp, under 1000 ft., *Drège*! Bathurst Div. ;
between Kasuga River and Port Alfred, *Burchell*, 3962!
CENTRAL REGION : Somerset Div. ; Somerset East, *Bowker*!

See note under the following species.

2. **A. cochlearis** (Schlechter & Bolus in Journ. Bot. 1894, 332,
partly); rhizome stout; leaves tufted, 4–12, ensiform or linear,
acuminate or acute, slightly curved or sometimes spreading, rigid,
very minutely serrulate, 4–14 in. long, $\frac{1}{4}$–$\frac{1}{2}$ in. broad, dilated at the
sheathing base; inflorescence terminal, paniculate, 1–2 ft. high,
with numerous lanceolate sheaths below; panicle $\frac{1}{2}$–1 ft. long, lax,
many-flowered; bracts lanceolate, acuminate, $\frac{1}{8}$–$\frac{1}{3}$ in. long; pedicels
slender, $\frac{1}{3}$–$\frac{1}{2}$ in. long; flowers small, tawny brown, with a pale
yellow lip, and usually a deep red 3-lobed blotch near the base;
sepals and petals subconnivent, elliptic-oblong, subobtuse, nearly
$\frac{1}{4}$ in. long; lip cochleate, cuneate at the base, and with a sharp
constriction underneath below the middle, entire or with only very
minute side lobes, erect, irregularly crisped and crenulate at the
margin, as long as the sepals and petals; disc with 4 rows of small
filiform papillæ; spur broadly oblong, $\frac{1}{4}$ as long as the lip; column
stout, nearly $\frac{1}{2}$ as long as the segments; capsule elliptic-oblong, $\frac{3}{4}$ in.
long. *A. micrantha, Schlechter & Bolus in Journ. Bot. 1894, 332, partly.
Eulophia cochlearis, Lindl. in Hook. Comp. Bot. Mag. ii. 202 (not of
Bolus). E. micrantha, Drège, Zwei Pfl. Documente, 125, and in
Linnæa, xx. 217; Bolus in Journ. Linn. Soc. xxv. 183, and Ic. Orch.*

Austr.-Afr. i. *t.* 4; *Durand & Schinz, Conspect. Fl. Afr.* v. 23; *Kränzl. in Ann. Naturhist. Hofmus. Wien*, xx. 12.

SOUTH AFRICA : without locality, *Holland*, 14 !
COAST REGION : Mossel Bay Div.; Little Brak River, *Burchell*, 6177 ! between Little Brak River and Hartenbosch, *Burchell*, 6201 ! George Div.; near Silver River, *Penther*, 867. Knysna Div.; Dowkamma, under 500 ft., *Drège*. Uitenhage Div. ; Van Stadens Berg, *Zeyher*, 3897 ! Port Elizabeth Div. ; between Krakakamma Lake and the Leadmine River, *Burchell*, 4594 ! *Hutton*. Albany Div. ; near Grahams Town, *MacOwan*, 395 ! *Atherstone*! *Bolton*! *Miss Bowker*! near the coast, *Hutton*! Bathurst Div. ; Kowie sandhills, *MacOwan*, 395 ! King Williamstown Div., near King Williamstown, *D'Urban*, 129 ! Komgha Div. ; near the mouth of the Kei River, *Flanagan*, 1026 !
EASTERN REGION : Transkei ; Kreilis Country, *Bowker*, 11 !

This has been much confused with the preceding and following species, but an examination of the original types enables the matter to be cleared up. The plant called *Eulophia micrantha* by Bolus (Ic. Orch. Austr.-Afr. i. t. 4) is *E. cochlearis*, Lindl., while his *E. cochlearis* (l.c. t. 5) is a quite distinct species. The differences between the two were clearly pointed out by Bolus, but some of the specimens cited by him under the latter belong to the former. The true *E. micrantha*, Lindl., was at first misunderstood by Bolus, and was subsequently described as *Acrolophia fimbriata*, Schlechter.

3. **A. Bolusii** (Rolfe in Orch. Rev. 1911, 198); rhizome stout ; leaves tufted, 4–12, ensiform or linear, acuminate or acute, entire or very minutely serrulate, rigid, 4–14 in. long, $\frac{1}{4}$–$\frac{1}{3}$ in. broad, dilated at the sheathing base ; inflorescence terminal, paniculate, 1–2 ft. high, with numerous lanceolate sheaths below ; panicle $\frac{1}{2}$– 1 ft. long, lax, many-flowered ; bracts lanceolate, acuminate, $\frac{1}{6}$–$\frac{1}{3}$ in. long ; flowers small, whitish-yellow, striped with brown on the sepals and petals and suffused with the same on the base of the lip ; sepals and petals subconnivent, elliptic-oblong, subobtuse, nearly $\frac{1}{4}$ in. long ; lip cochleate, cuneate at the base, without a constriction beneath, reflexed from above the base, without side lobes, emarginate, longer than the sepals and petals, with crisped and crenulate margin ; disc with 3 rows of prominent filiform papillæ ; spur nearly globose, $\frac{1}{6}$ as long as the lip ; column stout, with prominent basal angles, $\frac{1}{2}$ as long as the sepals and petals. *A. cochlearis, Schlechter & Bolus in Journ. Bot.* 1894, 332, *partly. Eulophia cochlearis, Bolus in Journ. Linn. Soc.* xxv. 183, *in Trans. S. Afr. Phil. Soc.* v. 107, *Orch. Cap. Penins.* 107, *and Ic. Orch. Austr.-Afr.* i. *t.* 5; *Durand & Schinz, Conspect. Fl. Afr.* v. 20, *not of Lindl.*; *Kränzl. in Ann. Naturhist. Hofmus. Wien*, xx. 12.

SOUTH AFRICA : without locality, *Zeyher*, 1589 !
COAST REGION : Cape Div. ; Sandy downs east of Table Mountain, *Bolus*, 4561 ! Vyges Kraal, *Wolley-Dod*, 395 ! mountains near Cape Town, *Frænxel* ! Cape Flats, *Bunbury*! and without collector ! George Div. ; Zwart River, *Penther*, 408. Albany Div. ; Grahams Town, *Herb. Harvey* !

See note under the preceding species.

4. **A. tristis** (Schlechter & Bolus in Journ. Bot. 1894, 331); rhizome stout ; leaves tufted, 4–10, ensiform or linear, acuminate or acute, minutely serrulate, rigid, 5–12 in. long, $\frac{1}{4}$–$\frac{1}{2}$ in. broad,

dilated at the sheathing base; inflorescence terminal, paniculate, $\frac{3}{4}$–2$\frac{1}{4}$ ft. high, with numerous lanceolate sheaths below; panicle $\frac{1}{2}$–1$\frac{1}{4}$ ft. long, many-flowered, sometimes dense; bracts lanceolate, acuminate, $\frac{1}{4}$–$\frac{1}{2}$ in. long; pedicels $\frac{1}{3}$–1 in. long; flowers dull greenish-brown with a white lip, veined with pink on the side lobes; sepals and petals subconnivent, lanceolate-oblong, acute, $\frac{1}{2}$ in. long; lip 3-lobed, equalling or slightly longer than the sepals; side lobes oblong, obtuse, short; front lobe obovate, obtuse, undulate, $\frac{1}{4}$ in. long; disc with 5 crested keels; crests very prominent in front; spur clavate, obtuse, $\frac{1}{12}$ in. long; column clavate, $\frac{1}{4}$ in. long; anther-case with 2 stout diverging horns; capsule ellipsoid, nearly $\frac{3}{4}$ in. long. *Limodorum capense, Berg. Descr. Pl. Cap.* 347. *Satyrium capense, Houtt. Handl.* xii. 502, *t.* 86, *fig.* 3. *Satyrium triste, Linn. fil. Suppl.* 402. *Limodorum triste, Thunb. Prodr.* 4. *Eulophia tristis, Spreng. Syst.* iii. 720; *Lindl. Gen. & Sp. Orch.* 184; *Drège; Zwei Pfl. Documente,* 128, 136; *Sond. in Linnæa,* xix. 72; xx. 217; *Bolus in Trans. S. Afr. Phil. Soc.* v. 106, *Orch. Cap. Penins.* 106, *and Journ. Linn. Soc.* xxv. 181; *Durand & Schinz, Conspect. Fl. Afr.* v. 26; *Kränzl. in Ann. Naturhist. Hofmus. Wien,* xx. 12.

SOUTH AFRICA: without locality, *Brown!* *Masson!* *Reade!* *Thom,* 771! *Lehmann!* *Reeves!* *Mrs. Bowker!* *Mrs. Holland!*
COAST REGION: Cape Div.; Simons Bay, *MacGillivray,* 471! *Milne!* Hout Bay, *Alexander!* Table Mountain and above Klassenbosch, 1200–1400 ft., *Harvey!* *Prior!* *Bolus,* 4779! Orange Kloof, *Wolley-Dod,* 339! Klaver Vley, *Wolley-Dod,* 576! Constantia, *Schlechter,* 248! Devils Mountain, *Rehmann,* 958! Riversdale Div.; Garcias Pass, *Burchell,* 7038! George Div., *Bowie!* Knysna Div.; near Gokama (Gouwkamma) River, *Bowie!* Humansdorp Div.; near Clarkson, *Schlechter,* 6008! *Penther,* 85. Uitenhage Div.; near the Zwartkops River and on Van Stadens Berg, *Zeyher,* 285! *MacOwan,* 1220! Uitenhage, *Zeyher,* 600! between Van Stadens Berg and Bethelsdorp, *Drège,* 2205! Port Elizabeth Div.; Port Elizabeth, fide *Bolus.* Alexander Div.; Zuurberg Range, 2000–3000 ft., *Drège,* 2205! Bathhurst Div.; near Port Alfred, *Burchell,* 3984; Albany Div.; near Grahamstown, *Williamson!* *Bolton!* *MacOwan,* 105! between Rautenbachs Drift and Addo Drift, *Burchell,* 4213! Zuur Berg, between Cable Farm and Oakwell, *Atherstone,* 76! Eastern Frontier of Cape Colony, *Hutton!* *Hallack!* *Prior!*

5. **A. sphærocarpa** (Schlechter & Bolus in Journ. Bot. 1894, 331); plant robust, 2–3 ft. high; leaves equitant, ensiform, rigid, half as long as the scape, 1 in. broad; scape terminal, branched, with erect-spreading branches, bearing 10–20 flowers; flowers nodding, larger than in *E. tristis,* and the petals marked with two dark purple spots above the base; bracts acuminate, as long as the pedicel, which is about $\frac{2}{3}$ in. long; sepals and petals oblong-lanceolate, acute, over $\frac{1}{2}$ in. long; lip 3-lobed, cuneate at the base; side lobes short; front lobe oblong, obtuse, crisped at the margin; disc with fringed keels; spur short, inflated; capsule broadly ellipsoid or nearly spherical, about $\frac{3}{4}$ in. long. *Eulophia sphæro-carpa, Sond. in Linnæa,* xix. 74; *Bolus in Trans. S. Afr. Phil. Soc.* v. 106, *Orch. Cap. Penins.* 106, *and Journ. Linn. Soc.* xxv. 183; *Durand & Schinz, Conspect. Fl. Afr.* v. 25.

COAST REGION: Malmesbury Div. ; Saldanha Bay, *Ecklon & Zeyher* ! Cape Div. ; Sandy places on the Cape Flats, near Wynberg, *Ecklon & Zeyher.*

I have only seen a single leaf and flower with a couple of fruits, in the Lindley Herbarium.

6. **A. lunata** (Schlechter & Bolus in Journ. Bot. 1894, 332) ; rhizome stout ; leaves 6–10, distichous, mostly spreading, ensiform, acute, serrulate, rigid, conduplicate, 4–7 in. long, $\frac{1}{3}$–$\frac{1}{2}$ in. broad, dilated at the sheathing base ; inflorescence terminal, 1–2$\frac{1}{2}$ ft. high, racemose, with several lanceolate-oblong sheaths below ; raceme lax, many-flowered ; bracts linear-lanceolate, acute, $\frac{1}{3}$–$\frac{3}{4}$ in. long ; pedicels slender, $\frac{1}{2}$–$\frac{3}{4}$ in. long ; flowers rose ; sepals subconnivent, lanceolate-oblong, subacute, nearly $\frac{1}{2}$ in. long ; petals oblong, minutely apiculate, slightly longer than the sepals ; lip 3-lobed, rather longer than the petals ; side lobes broadly oblong, obtuse, short ; front lobe obovate-orbicular, obtuse, undulate, $\frac{1}{4}$ in. broad ; disc and front lobe bearing 7 closely crested keels ; spur oblong, obtuse, $\frac{1}{8}$ lin. long ; column clavate, nearly $\frac{1}{4}$ in. long ; anther-case with two stout diverging horns. *Eulophia barbata, Lindl. Gen. & Sp. Orch.* 184, *partly, not of Spreng. E. lunata, Schlechter in Verhandl. Bot. Ver. Brandenb.* xxxv. 45.

SOUTH AFRICA : without locality or collector in *Herb. Hook.* !
COAST REGION : Swellendam Div. ; Langeberg Range, near Zuurbraak, 3000 ft., *Schlechter*, 2169. Knysna Div. ; near Forest Hall, *Miss Newdigate in Herb. Bolus*, 6353 !

Lindley erroneously refers this to *E. barbata*, Spreng., and cites the collector as '' Thunberg,'' which is probably also erroneous.

7. **A. lamellata** (Schlechter & Bolus in Journ. Bot. 1894, 332) ; rhizome stout ; leaves tufted, 4–10, ensiform, acute, very minutely serrulate, rigid, 4–11 in. long, $\frac{1}{3}$–$\frac{1}{2}$ in. broad, dilated at the sheathing base ; inflorescence terminal, paniculate or rarely racemose, $\frac{3}{4}$–1$\frac{1}{4}$ ft. high, with several lanceolate sheaths below ; panicle 4–9 in. long, often rather dense ; bracts linear-lanceolate, acuminate, $\frac{1}{4}$–$\frac{3}{4}$ in. long ; pedicels slender, $\frac{1}{2}$–$\frac{3}{4}$ in. long ; flowers greenish lined with brown on the segments, and the lip white with pink side lobes ; sepals and petals subconnivent, lanceolate-oblong, acute, $\frac{1}{2}$–3 in. long ; lip 3-lobed, as long as the sepals and petals ; side lobes short, rounded ; front lobe broadly rounded, obtuse, undulate, with 7–9 very strongly crested keels ; spur clavate-oblong, $\frac{1}{8}$–$\frac{1}{6}$ in. long ; column clavate, nearly $\frac{1}{2}$ in. long ; anther-case with a pair of diverging horns. *Eulophia lamellata, Lindl. Gen. & Sp. Orch.* 184 ; *Sond. in Linnæa*, xx. 217 ; *Bolus in Trans. S. Afr. Phil. Soc.* v. 107, *t.* 22, *figs.* 4–7, *and Orch. Cape Penins.* 107, *t.* 22, *figs.* 4–7 ; *Durand & Schinz, Conspect. Fl. Afr.* v. 22.

SOUTH AFRICA : without locality, *Villett* ! *Hesse*, 9 ! *Harvey*, 142 !
COAST REGION : Cape Div. ; Zwart River, Cape Flats, *Zeyher* ! sea shore and on the dunes near Riet Valei, below 100 ft., *Zeyher*, 1590 ! Claremont and sand dunes near Cape Town, *Bolus*, 4558 ! near Vygeskraal Farm, *Wolley-Dod*, 396 !

sandy heaths, near Rondebosch, *MacOwan & Bolus, Herb. Norm. Austr.-Afr.* 152 !
Steenberg, 300 ft., *Dümmer*, 778 ! near Constantia, *Brown* ! between Cape Town
and George, *Rogers* ! Riversdale Div. ; Kampsche Berg, *Burchell*, 7072 !

8. **A. comosa** (Schlechter & Bolus in Journ. Bot. 1894, 331);
plant 1½ ft. high; leaves rigid, ensiform, very entire, inner as long
as the scape, the 2 or 3 outer shorter, ½ in. broad, contracted and
cucullate at the apex; spike oblong, 4 in. long; lower flowers
remote, upper approximate; bracts lanceolate, acuminate, rose-
coloured, the lower 2 in. long, upper 1 in. long, and crowded;
sepals linear-lanceolate, very acute, ⅔ in. long; petals lanceolate,
acute, rather shorter and broader than the sepals; lip 3-lobed,
obovate, cuneate at the base; side lobes short, somewhat obtuse;
front lobe obovate, somewhat crisped and crenulate, veined with
purple; disc with 7 somewhat obtuse parallel barbate keels; spur
cylindric, obtuse, ⅛ in. long. *Eulophia comosa, Sond. in Linnæa,*
xix. 72; *Bolus in Journ. Linn. Soc.* xxv. 183; *Durand & Schinz,*
Conspect. Fl. Afr. v. 20.

COAST REGION: Caledon Div.; region of the hot springs on the Zwart Berg,
1000–2000 ft., *Ecklon & Zeyher* !

I have only seen a single flower and bract in the Lindley Herbarium.

9. **A. ustulata** (Schlechter & Bolus in Journ. Bot. 1894, 332);
rhizome rather slender; roots very stout and fleshy; leaves tufted,
6–10, linear or linear-lanceolate, acuminate, entire, rigid, ¾–2 in.
long, scarcely 1/12 in. broad, except at the dilated sheathing base;
inflorescence terminal, shortly racemose, 2–4 in. high, with a few
narrowly lanceolate sheaths below; racemes 3–6-flowered; bracts
linear-lanceolate, acuminate, ⅓–½ in. long; pedicels ¼–⅓ in. long;
flowers deep chocolate-purple, drying nearly black; sepals and
petals subconnivent, oblong, subacute, fleshy, ¼–⅓ in. long; lip
3-lobed, rather shorter than the sepals and petals, slightly gibbous
at the base but without a spur; side lobes incurved, obtuse; front
lobe recurved, broadly oblong, obtuse, densely papillose above;
disc minutely papillose; column clavate, ⅛ in. long; anther-case
without horns. *Cymbidium ustulatum, Bolus in Journ. Linn. Soc.*
xx. 469. *Eulophia ustulata, Bolus in Trans. S. Afr. Phil. Soc.*
v. 110, *t.* 2, *Orch. Cap. Penins.* 110, *and in Journ. Linn. Soc.* xxv. 184;
Durand & Schinz, Conspect. Fl. Afr. v. 26.

COAST REGION : Cape Div.; in sandy soil in the valley opposite Farmer Peck's
Hotel on the Muizen Berg, 1200 ft., *Bolus*, 4848 ! *and in MacOwan & Bolus,*
Herb. Norm. Austr.-Afr. 153 !

V. **EULOPHIA,** R. Br.

Sepals subequal, free, subconnivent or spreading, the lateral ones
sometimes adnate to the foot of the column. *Petals* like the sepals
or a little broader, sometimes differently coloured. *Lip* continuous

with the base or foot of the column, sometimes a little contracted above the base, 3-lobed or entire; base usually variously saccate or spurred; side lobes erect or sometimes nearly obsolete; the middle one spreading or recurved; disc variously cristate or lamellate, rarely smooth. *Column* short, stout, the base sometimes produced into a more or less distinct foot, to which the lateral sepals are united, forming a *mentum*; anther-bed oblique, erect, entire. *Anther* terminal, operculate, incumbent, semiglobose, conical or rarely acuminate; apex more or less bilobed, imperfectly 2-celled; pollinia 4, ovoid, united in pairs, affixed to a broad stipes and gland. *Capsules* ovoid or oblong, rarely elongated, with prominent thickened angles.

Terrestrial herbs or rarely epiphytes; stems leafy at the base, creeping, often thickened into rhizomes, sometimes forming aërial pseudobulbs; leaves distichous, often narrow and elongated, usually plicate, rarely conduplicate and coriaceous; scapes or peduncles lateral, variously sheathed below, racemose or rarely paniculate above; flowers small, medium-sized or sometimes large, usually lax; bracts small or narrow.

DISTRIB. Species over 200, chiefly in the tropical and sub-tropical regions of the Old World, with the headquarters in Africa. About 70 South African species are known, and the Tropical African representatives are scarcely less numerous, very few being common to both.

The limits of the genus are difficult to define, but I have included *Cyrtopera*, Lindl., and *Eulophidium*, Pfitz. The first was characterised by having the column prolonged at the base into a distinct foot, to which the lateral sepals are united, forming a mentum, but there is such a marked gradation between the species with and those without a foot that the distinction between the two genera is now given up by common consent. *Eulophidium* was separated from *Eulophia*, and referred to *Maxillarieæ*, chiefly on account of its non-plicate leaves, but the floral structure is completely that of *Eulophia*. *Lissochilus*, R. Br., however, which is sometimes united with *Eulophia*, is here retained, as in the *Flora of Tropical Africa*, being in the main well characterised by the short reflexed sepals and much larger erect petals.

The South African species have been much neglected, and have never been the subject of any complete revision. The flowers are often very membranous, and the fleshy crests and appendages of the lip are changed so much during the process of drying, and are so difficult to restore by boiling, while the details of colour are lost, that it has been very difficult in some cases to define specific limits, and a study of living specimens would probably necessitate some re-arrangement. Several others are very imperfectly known, and a few of the old species seem to have been lost sight of. The genus is commended to those who have an opportunity of studying it in the field. *E. alismatophylla* (Reichb. f. in Flora, 1885, 543), and *E. sclerophylla* (Reichb. f. in Flora, 1885, 542), were included by Dr. Bolus as doubtfully South African (*Journ. Linn. Soc.* xxv. 184) but probably occur outside our limits. They were described in a paper on Comoro Island Orchids, and although no habitat is given they probably come from that region.

Leaves coriaceous, not plicate, variegated...   ...   ...   (1) **Mackenii.**

Leaves plicate or strongly veined, not variegated:
   Plants with fusiform, aërial pseudobulbs and fleshy
      rigid leaves:
      Sepals narrowly linear, strongly revolute or circinate
         at the apex; lip narrowly pandurate ...   ...   (2) **caffra.**
      Sepals linear-oblong, somewhat revolute at the apex;
         lip broadly pandurate   ...   .   (3) **circinata.**
   Plants with subterranean rhizomes and more or less
      membranous leaves:

†Spur slender, oblong, clavate or somewhat elongated :
‡Flowers ¼–½ in. long, arranged in lax or elongated
　racemes :
Sepals oblong or broadly oblong, ¼–⅓ in. long :
　Petals elliptic-oblong, obtuse or apiculate,
　　distinctly broader than the sepals :
　Lip distinctly longer than broad :
　　Front lobe of lip broadly oblong　...　(4) natalensis.
　　Front lobe of lip obcordate　...　...　(5) tenella.
　　Lip nearly as broad as long...　...　...　(6) nutans.
　Petals lanceolate or elliptic-lanceolate, scarcely
　　broader than the sepals :
　　Disc of lip with numerous strong papillæ...　(7) corallorrhizi-
　　　　　　　　　　　　　　　　　　　　　　　　formis.
　　Disc of lip with few slender papillæ　...　(8) flaccida.
Sepals lanceolate or oblong-lanceolate, ⅓–½ in.
　long :
　Petals not much broader than the sepals :
　Lip longer than broad :
　　Petals obtuse or apiculate, rather broader
　　　than the sepals :
　　Side lobes of the lip scarcely half as
　　　broad as the front lobe　...　...　(9) carunculifera.
　　Side lobes of the lip nearly as broad
　　　as the front lobe :
　　　Spur curved, spreading　...　...　(10) laxiflora.
　　　Spur straight, adpressed to the ovary　(11) Galpini.
　　Petals acute, scarcely broader than the
　　　sepals :
　　Leaves contemporaneous with the spike　(12) æmula.
　　Leaves not developed at flowering
　　　time　...　...　...　...　(13) longipes.
　　Lip broader than long　...　...　...　(14) chlorantha.
　Petals about twice as broad as the sepals :
　Lip strongly 3-lobed :
　　Front lobe of lip as broad as long　...　(15) Flanaganii.
　　Front lobe of lip longer than broad　...　(16) brachystyla.
　Lip obscurely 3-lobed　...　...　...　(17) zeyheriana.
‡‡Flowers over ½ in. long, arranged in lax or elongated
　racemes :
§Petals about as long as the sepals :
　*Spur slender, a third to half as long as
　　the lip :
　Leaves straight, contemporaneous with the
　　scape :
　　Side lobes of lip longer than broad　...　(18) purpurascens.
　　Side lobes of lip broader than long　...　(19) gladioloides.
　Leaves usually immature at flowering time :
　　Petals free :
　　Side lobes of lip not diverging :
　　　Spur of lip broader at the middle
　　　　than the apex　...　...　...　(20) ovatipetala.
　　　Spur of lip clavate or narrowed
　　　　about the middle :

Front lobe of lip broader than long :
Disc of lip with a dense mass of
papillæ   ...   .   (21) **Engleri.**
Disc of lip with few papillæ in
front ...   .   (22) **transvaalensis.**
Front lobe of lip longer than broad   (23) **aliwalensis.**
Side lobes of lip somewhat diverging...   (24) **hians.**
Petals adnate to the column for half their
length. .   ...   (25) **violacea.**
**Spur stouter, less than a third as long as
the lip :
‖Petals as broad as or broader than the sepals :
¶Leaves elongate-linear or ensiform and
narrow (unknown in 37, 43 and 48):
Sepals elliptic or oblong and sub-
obtuse :
Crests very prominent, chiefly on
front lobe of lip :
Petals suborbicular, twice as broad
as the sepals ...   .   (26) **platypetala.**
Petals elliptic-oblong, slightly
broader than the petals   ...   (27) **lissochiloides.**
Crests of lip not prominent and
central ...   .   (28) **Saundersiæ.**
Sepals lanceolate or oblong-lanceolate
and acute or acuminate :
Lip with ample rounded side lobes,
and under ¾ in. long :
Front lobe of lip rounded, not or
scarcely longer than broad :
Crests of lip rather slender and
scattered :
Sepals and petals ½–¾ in. long   (29) **barbata.**
Sepals and petals ¾–1 in. long :
Keels of lip slender and not
verrucose at the base   (30) **Thunbergii.**
Keels of lip slender and
distinctly verrucose at
the base   ...   ...   (31) **robusta.**
Crests of lip strong and chiefly
aggregated on the front
lobe :
Sepals and petals about ¾ in.
long   ...   .   (32) **dregeana.**
Sepals and petals about ½ in.
long   ...   .   (33) **Nelsoni.**
Front lobe of lip oblong, much
longer than broad   ...   ...   (34) **oblonga.**
Lip with obscure or small side lobes
and over ¾ in. long :
Petals broadly elliptic and sub-
obtuse :
Lip distinctly 3-lobed ...   ...   (35) **Haygarthii.**
Lip obscurely or shortly 3-lobed :
Leaves under ½ in. broad   ...   (36) **Macowani.**
Leaves over ½ in. broad   ...   (37) **parvilabris.**

Petals oblong or elliptic-oblong,
and usually subacute :
  Front lobe of lip narrowly
  oblong :
    Sepals acute or shortly acumi-
    nate   ...   . (38) **deflexa.**

    Sepals long-acuminate   ... (39) **acuminata.**

  Front lobe of lip broadly elliptic-
  oblong :
    Lip subentire or obscurely
    3-lobed ...   . (40) **Allisoni.**

    Lip pandurately 3-lobed   ... (41) **calanthoides.**

¶¶Leaves lanceolate, elliptic-lanceolate or
short and somewhat broadened about
the middle :
  Lip broadly 3-lobed :
    Lip shorter than the petals :
      Front lobe of lip obtuse ...   ... (42) **Bakeri.**

      Front lobe of lip acute   ...   ... (43) **Rehmanni.**

    Lip as long as the petals   ...   ... (44) **latipetala.**

    Lip subentire ...   . (45) **subintegra.**

‖‖Petals much narrower than the sepals   ... (46) **Woodii.**

§§Petals about half as long as the sepals :
  Leaves an inch or more broad ; sepals oblong (47) **Meleagris.**

  Leaves narrowly linear ; sepals narrowly
  linear ...   . (48) **antennata.**

‡‡‡Flowers over $\frac{1}{2}$ in. long and arranged in congested
racemes or short heads :
  Flowers about 1 in. long, arranged in abbrevi-
  ated racemes :
    Bracts elliptic-lanceolate   ...   . (49) **Cooperi.**

    Bracts linear-lanceolate :
      Side lobes of lip broader than long, blackish-
      purple ...   . (50) **Zeyheri.**

      Side lobes of lip longer than broad, yellow (51) **ensata.**
  Flowers $\frac{1}{2}$-$\frac{3}{4}$ in. long, arranged in congested
  heads :   ...   . (52) **leontoglossa.**

††Spur distinctly conical :
  Spike subcorymbose ...   . (53) **stenantha.**

  Spike more or less elongated :
    Leaves undeveloped or short at flowering time :
      Sepals acuminate and over $\frac{3}{4}$ in. long :
        Keels of lip closely and minutely papillose :

          Petals not much broader than the sepals (54) **fragrans.**

          Petals nearly twice as broad as the sepals (55) **Sankeyi.**

        Keels of lip strongly and more distantly
        papillose :
          Side lobes of lip membranous and
          nearly as broad as the front lobe   ... (56) **litoralis.**

          Side lobes of lip somewhat fleshy and
          only half as broad as the front lobe (57) **nigricans.**

Sepals acute or subacute, ½ to nearly ¾ in.
long :
Lip subentire... ... ... (58) **inandensis.**

Lip 3-lobed :
Side lobes of lip much shorter than the
front lobe :
Leaves linear :
Side lobes of lip rounded ; front lobe
convex :
Sepals over ½ in. long ... ... (59) **bilamellata.**

Sepals under ½ in. long ... ... (60) **oliveriana.**

Side lobes of lip quadrate ; front
lobe involute ... ... (61) **rupestris.**

Leaves oblong ... ... (62) **inæqualis.**

Side lobes of lip nearly as long as the
front lobe ... ... (63) **Pegleræ.**

Leaves developed at flowering time, elongate-
linear or grass-like, ¾–1¼ ft. long :
Spikes cylindrical, many-flowered ... ... (64) **papillosa.**

Spikes one-sided, few-flowered ... ... (65) **aurea.**

†††Spur didymous ... ... (66) **junodiana.**

††††Spur obsolete :
Flowers under ¾ in. long :.
Side lobes of lip small, much shorter than the
front lobe ... ... (67) **aculeata.**

Side lobes of lip ample or rounded, nearly as
long as the front lobe :
Lip ⅓ in. long, deeply 3-lobed :
Keels of lip verrucose from base to apex ;
front lobe not concave ... ... (68) **Huttoni.**

Keels of lip verrucose at the base of the
concave front lobe ... ... (69) **foliosa.**

Lip ¼ in. long, obscurely 3-lobed ... ... (70) **Boltoni.**

Flowers ¾–1 in. long ... ... (71) **tabularis.**

1. **E. Mackenii** (Rolfe in Gard. Chron. 1892, xii. 583) ; pseudobulbs
aërial, oblong or ovoid-oblong, covered with loose ovate acute imbricating sheaths, 1-leaved, about ¾ in. long ; leaves petioled, coriaceous, elliptic-oblong, subacute, irregularly banded with dark green
spots on a light grey ground, 3–5 in. long, ¾–1¼ in. broad ; petioles
½–1¼ in. long ; scapes erect, 6–8 in. long, with a few lanceolate
acuminate sheaths ; racemes lax, several-flowered ; bracts ovate-
lanceolate, acute, ⅛–¼ in. long ; flowers small, sepals and petals light
green, lip white, lined with reddish-purple on the front lobe ; sepals
elliptic-oblong, obtuse, over ¼ in. long ; petals elliptic, obtuse,
rather broader than the sepals ; lip longer than the petals, ⅓ in.
broad, 4-lobed ; side lobes broadly rounded, obtuse, diverging ;
front lobe nearly quadrate, truncate or crenulate, somewhat
diverging, not much exceeding the side lobes ; disc smooth, with
a 2-lobed callus at the base ; spur clavate-oblong, obtuse or

slightly didymous, ⅓ as long as the lip; column stout, ½ as long as the segments.

EASTERN REGION: Natal; Verulam and Coldmore, *McKen,* 11! and without precise locality, *Sanderson,* 1054! *Wood!*

2. **E. caffra** (Reichb. f. in Flora, 1865, 186); pseudobulbs ovoid-oblong or stoutly fusiform, about 6 or 7 in. long, 2-leaved at the apex; leaves linear-oblong, acute, rigid and more or less fleshy, about 10 in. long, ¾–½ in. broad; scapes basal, erect, stout, 2½–4 ft. long, loosely panicled above; bracts ovate-lanceolate, acuminate, ¼–⅓ in. long; pedicels 1–1¼ in. long; flowers lax, sepals and petals green with darker nerves, and the lip white with purple disc and keels; sepals spreading, linear-oblong, acute, revolute at the apex, 1–1¼ in. long; petals similar to the sepals but rather shorter and broader, and not spreading at the base; lip 3-lobed, oblong, 1–1¼ in. long; side lobes narrow, rounded at the apex, with about 6 radiating forked veins on each; front lobe elliptic-ovate, subacute, much crisped, with numerous radiating veins.; disc with 5 crenulate keels, the 3 inner longer and much elevated in front; spur clavate, much curved, ¼–⅓ in. long; column clavate, ⅓ in. long; anther 2-horned; capsule ellipsoid, 1¼ in. long. *Bolus in Journ. Linn. Soc.* xxv. 184; *Durand & Schinz, Conspect. Fl. Afr.* v. 19.

EASTERN REGION: Natal; Tugela River, *Gerrard,* 1814! and without precise locality, *Sanderson,* 1002! 1015! Zululand, *Warner.*

An imperfectly known species. It was described from living flowers sent by Mr. Warner of Chelmsford to Reichenbach as from Zulu territory, and Reichenbach has remarked that Sanderson's 1015 appears to be identical. It resembles the Tropical African *E. Petersii,* Reichb. f., in habit.

3. **E. circinata** (Rolfe in Kew Bulletin, 1910, 280); pseudobulbs ovoid-oblong or stoutly fusiform, about 4 in. long, 2–3-leaved at the apex; leaves linear-oblong, acute, rigid and more or less fleshy somewhat serrulate, about 6–8 in. long; scapes basal, erect, rather stout, 2–3 ft. high, slightly branched above; bracts subulate-lanceolate, acuminate, about ¼ in. long; pedicels 1–1¼ in. long; flowers numerous, lax; sepals spreading, narrowly linear-oblong, acute, strongly revolute or circinate at the apex, ¾ in. long; petals erect, rather shorter and broader than the sepals, somewhat revolute at the apex; lip broadly pandurate or 3-lobed, ¾ in. long; side lobes very narrow, with about 5 radiating forked veins on each; front lobe suborbicular, undulate, with numerous radiating veins; disc with 5 crenulate keels, the 3 inner longer and much elevated in front; spur subclavate, much curved, about ¼ in. long; column clavate, ¼–⅓ in. long; anther 2-horned; capsule ellipsoid, 1½ in. long.

KALAHARI REGION: Transvaal; Komati Poort, *Kirk,* 64!

Similar to *E. caffra,* Reichb. f., in habit, but with more slender inflorescence, shorter bracts and flowers, and a relatively broader lip. The circinate petals are remarkable.

Galpin, 965, from mountain ravines at Barberton, 3500 ft., represented at Kew by a single flower, belongs to the same group. The pseudobulbs are said to be 7 in. long, the leaves rigid and deeply channelled, 15 in. long, and the spike 4 ft. long. The flower is rather more like *E. caffra*, Reichb., in structure.

4. **E. natalensis** (Reichb. f. in Flora, 1865, 186); rhizome slender, moniliform; leaves linear, acuminate, 4–7 in. long, $\frac{1}{12}$–$\frac{1}{8}$ in. broad; scape slender, 8–12 in. long; sheaths lanceolate, acuminate, 1–2 in. long; raceme usually many-flowered, at length somewhat lax; bracts lanceolate, acuminate, $\frac{1}{8}$–$\frac{1}{2}$ in. long; flowers purple, rather smaller than in *E. violacea*, Reichb. f., brown and yellow (*Sanderson*); sepals triangular, acute; petals ovate, subacute; lip dilated from a broad cuneate base, trifid; sides lobes not divergent, obtusely triangular; front lobe quadrate; disc with 5 thickened, minutely crenulate veins, extending to the apex of the lip, and a rounded crest at the base; spur clavate-cylindric, half as long as the pedicelled ovary; column broadly oblong, $\frac{1}{12}$ in. long. *Bolus in Journ. Linn. Soc.* xxv. 184; *Durand & Schinz, Conspect. Fl. Afr.* v. 24. *E. collina, Schlechter in Engl. Jahrb.* xxvi. 336.

EASTERN REGION: Transkei; Krielis Country, *Bowker*! Natal; Clairmont Flat, *Sanderson*, 476! Up Park, 500 ft., *Wood*, 1417! Mount Edgecombe, 400 ft., *Wood*, 5783! Tugela, *Wood*, 11786! and without precise locality, *Gueinzius*! *Mrs. K. Saunders*!

I cannot separate *E. collina*, Schlechter.

5. **E. tenella** (Reichb. f. in Linnæa, xx. 681); rhizome creeping, thickened at the nodes; leaves narrowly linear, 7–11 in. long, about $\frac{1}{12}$ in. broad; scape 8 in. high, with 4 long acute sheaths; spike over 2 in. long, 8–20-flowered; bracts linear-lanceolate, acuminate, $\frac{1}{6}$–$\frac{1}{3}$ in. long; pedicel incurved, about $\frac{1}{3}$ in. long; flowers very small; sepals elliptic-oblong, shortly apiculate, $\frac{1}{4}$–$\frac{1}{3}$ in. long, greenish-brown; petals elliptic-ovate, obtuse, broader than the sepals, yellowish-green; lip trilobed, $\frac{1}{4}$–$\frac{1}{3}$ in. long by half as broad, yellowish-green; side lobes obtuse; front lobe obcordate, emarginate, very minutely crenulate; disc with 5–7 crenulate keels, the 3 middle much thickened and prominent; spur oblong, obtuse, $\frac{1}{3}$ in. long; column stout, broadened at the apex, $\frac{1}{12}$ in. long; capsule oblong, $\frac{1}{2}$ in. long. *Bolus in Journ. Linn. Soc.* xxv. 183; *Durand & Schinz, Conspect. Fl. Afr.* v. 26.

COAST REGION: Riversdale Div. ? Kleinfontein, *Mund*!

Only known from the original specimen in the Berlin Herbarium.

6. **E. nutans** (Sond. in Linnæa, xix. 73); leaves linear-ensiform, nearly as long as the scape; scape 1 ft. high, with 3 membranous sheaths; raceme lax, 6–8-flowered; bracts ovate, acute; pedicels as long as the bracts; flowers small, drooping; sepals lanceolate, acute, coloured outside; petals oblong, obtuse, very thin; lip ovate with a rounded base, 3-lobed; side lobes oblong, obtuse, veined; front lobe obovate; disc with a few scale-like crests; spur ascending, cylindric,

subemarginate, half as long as the lip ; capsule pendulous, clavate, with somewhat winged angles. *Bolus in Journ. Linn. Soc.* xxv. 183, *and Ic. Orch. Austr. Afr.* ii. *t.* 11 ; *Durand & Schinz, Conspect. Fl. Afr.* v. 24 ; *Kränzl. in Ann. Naturhist. Hofmus. Wien,* xx. 12.

COAST REGION : Uitenhage Div. ; Van Stadens Berg, *Zeyher,* 3900 ! Albany Div. ; Howisons Poort, near Grahamstown, at 1500 ft., *Schönland* ! Coldstream, 2300 ft., *Bolus,* 7366, *Glass.* Stockenstrom Div. ; Kat River Mountains, *Zeyher.* EASTERN REGION : Transkei ; Kentani, 1500 ft., *Miss Pegler,* 290. Griqualand East ; foot of Insizwa Range, *Krook, Penther,* 142. Ixopo District, Umkomanzi River, *Krook, Penther,* 118.

A very imperfectly known species. Zeyher's specimen is poor, and partly gone to fruit, but the one identified as such by Dr. Schönland agrees fairly well with it.

7. **E. corallorrhiziformis** (Schlechter in Engl. Jahrb. xx. Beibl. 50, 9, 26) ; rhizome creeping, thickened at the nodes ; leaves 1 or 2, radical, erect, narrowly linear, acute, 4–6 in. long ; scape appearing with the young leaves and from the same sheath, erect, slender, 9–12 in. long, with a few narrow spathaceous sheaths below ; raceme 1½–3 in. long, rather lax, many-flowered ; bracts linear-lanceolate, acuminate, ¼ to nearly ½ in. long ; pedicels about ¼ in. long ; flowers yellow ; sepals subconnivent, oblong-lanceolate, acute, ¼–⅓ in. long ; petals slightly broader than the sepals, otherwise similar ; lip 3-lobed, about as long as the sepals ; side lobes diverging, shortly and broadly oblong ; front lobe suborbicular, with about 7 rows of rather long erect papillæ, closely arranged ; spur oblong, obtuse, 1/10 in. long ; column oblong, about as long as the spur.

KALAHARI REGION: Transvaal ; Magaliesberg Range, *Burke & Zeyher,* 171 ! near Rustenberg, 4000 ft., *Pegler,* 1001 ! in marshes, Umlomati Valley, Barberton, 4000 ft., *Galpin,* 1221, *Culver,* 77 ! near Middleburg, *Schlechter* ; near Donkerhoek, 4900 ft., *Schlechter,* 3725 ; near Lydenburg, *Atherstone* !

8. **E. flaccida** (Schlechter in Engl. Jahrb. xx. Beibl. 50, 3) ; a slender erect herb, ¾–1¼ ft. high ; scape flexuous, with 3 or 4 narrow acute adpressed membranous sheaths ; leaves 2 or 3, narrowly elongate-linear, acute, flaccid, ½–¾ ft. long ; raceme lax, 10–12-flowered ; bracts spreading, lanceolate, acuminate, membranous ; flowers somewhat spreading, short ; sepals subequal, oblong, shortly acute, about ¼–⅓ in. long ; petals elliptic-oblong, obtuse, as broad as the sepals, but rather shorter ; lip broadly oblong, 3-lobed, as long as the petals ; side lobes erect, oblong, subobtuse ; front lobe broadly subquadrate, dilated towards the apex, truncate, somewhat excised, with 3 thickened keels, becoming verrucose in front ; spur cylindric, obtuse, ⅓ as long as the lip ; column ½ as long as the lip ; anther emarginate at the apex. *Bolus, Ic. Orch. Austr.-Afr.* ii. *t.* 20.

COAST REGION : Komgha Div. ; grassy valley near Komgha, 2000 ft., *Flanagan,* 1645 !
EASTERN REGION : Natal ; The Bluff, near Durban, 200 ft., *Schlechter,* 2860, marsh near Clairmont Flat, 50 ft., *Wood,* 1430 ! 4075 !

9. **E. carunculifera** (Reichb. f. in Flora, 1881, 329); leaves in fascicles of 4 to 6, ensiform-linear, acuminate or acute, strongly veined, conduplicate at the base, 3–4 in. long, about $\frac{1}{6}$ in. broad; scape 1–1$\frac{1}{4}$ in. long, rather slender, with 3 or 4 strongly veined spathaceous sheaths below; raceme lax, 6–8 in. long, many-flowered; bracts lanceolate, acuminate, $\frac{1}{4}$–$\frac{1}{3}$ in. long; pedicels slender, about $\frac{1}{2}$ in. long; flowers small; sepals oblong-lanceolate, acuminate, $\frac{1}{4}$–$\frac{1}{3}$ in. long; petals elliptic-oblong, subobtuse or apiculate, rather shorter and much broader than the sepals; lip elliptic-oblong, somewhat attenuate at the base, about as long as the petals, 3-lobed; side lobes very short, broadly rounded; front lobe ovate-orbicular or broadly rounded, obtuse, very short; disc with 3 to 5 thickened keels below and 5 to 7 strongly verrucose or papillose keels in front; spur clavate or oblong, about $\frac{1}{10}$ in. long; column broadly oblong, $\frac{1}{8}$ in. long. *Bolus in Journ. Linn. Soc.* xxv. 184; *Durand & Schinz, Conspect. Fl. Afr.* v. 20.

EASTERN REGION: Natal; without precise locality, *Buchanan*, 4! *and in Herb. Sanderson*, 1062!

A drawing in the Kew collection, localised as from Tongaat, Natal, *Sanderson*, may belong here, as the habit agrees, but the structural details are not shown.

10. **E. laxiflora** (Schlechter in Engl. Jahrb. xx. Beibl. 50, 4, 26); an erect glabrous herb, $\frac{3}{4}$–1$\frac{1}{4}$ ft. high; leaves in a fascicle of 4, suberect, narrowly linear, very acute, 6–10 in. long; scapes subflexuous, with numerous adpressed membranous acute sheaths; spikes lax, many-flowered; bracts spreading, oblong-lanceolate, acuminate, $\frac{1}{4}$–$\frac{1}{2}$ in. long; flowers small, spreading; sepals subequal, oblong-lanceolate, acute, nearly $\frac{1}{2}$ in. long; petals oblong, acute, $\frac{1}{3}$ in. long, by nearly half as broad; lip as long as the petals, oblong, 3-lobed; side lobes oblong, obtuse; front lobe subquadrate, obtuse, with 5 thickened densely papillose nerves, extending along the disc to near the base; spur incurved, cylindric, obtuse, one-third as long as the lip; column $\frac{1}{6}$ in. long, dilated at the apex; anther rounded, subemarginate at the apex. *Bolus, Ic. Orch. Austr.-Afr.* ii. *t.* 15.

COAST REGION: Queenstown Div.; on Lesseyton Nek, 4000 ft., *Galpin*, 1713. Komgha Div.; near Komgha, 2000 ft., *Flanagan*, 1699. Kei River Mouth, 100 ft., *Flanagan*, 1031.

CENTRAL REGION: Aliwal North Div.; near Aliwal North, 4200 ft., *Bolus in Herb. Bolus*, 10543.

KALAHARI REGION: Orange River Colony; Leeuw Spruit and Vredefort, *Barrett-Hamilton*! Transvaal; among grasses near Pretoria, *Schlechter*, 4143, between Rustenberg and Johannesberg, *Miss Pegler*; Springbok Flats, Waterberg District, *Burtt-Davy*, 1024, between Pietersburg and Houtbosch, *Bolus*, 11164.

EASTERN REGION: Pondoland; near Umtata, *Miss Pegler*! Natal; Inanda, *Wood*, 471! near Itafamasi, *Wood*, 725! near Blauw Krantz, *Wood*, 3430; Zaai Laager, near Estcourt, *Wood*, 3422! near Newcastle, 3000–4000 ft., *Wood*, 6748! Alexandria District, 2600 ft., *Rudatis*, 808! near Durban, 3000–4000 ft., *Sutherland*! Wessels Neck, 3500 ft., *Schlechter*, 3397.

11. **E. Galpini** (Schlechter in Engl. Jahrb. xx. Beibl. 50, 10); plant very slender, 14–18 in. high; leaves in a fascicle of 3 or 4,

narrowly linear, acute, veined, up to 1 ft. long; scape erect, slender, lateral, with adpressed membranous acute sheaths; spike lax, many-flowered; bracts erect, lanceolate, very acute, nearly as long as the slender pedicelled ovary; flowers small; sepals lanceolate, acute, ½ in. long; petals rather shorter and broader than the sepals; lip oblong, shorter than the petals; side lobes incurved, porrect, obliquely lanceolate, subacute; front lobe oblong-lanceolate, rounded at the apex, much longer than the side lobes; disc with 4 elevated lines from the mouth of the spur to the middle of the front lobe, long-papillose at the apex; spur short, obtuse, straight, adpressed to the ovary; column footless, slender, a third shorter than the lip.

KALAHARI REGION: Transvaal; Umlomati Valley, near Barberton, 3900 ft., *Galpin*, 1151; Musidora, near Barberton, 300 ft., *Culver*, 82.

Only known to me from description.

12. **E. æmula** (Schlechter in Engl. Jahrb. xx. Beibl. 50, 26); an erect herb, ¾–1¼ in. high; leaves in a fascicle of 4 or 5, erect or somewhat spreading, narrowly linear, acute, ½–1 ft. long; scape ¾–1¾ ft. long, with numerous lanceolate acuminate sheaths; raceme lax, many-flowered; bracts lanceolate or ovate-lanceolate, acuminate, usually spreading, ¼–¾ in. long; flowers generally spreading, purple; sepals subequal, lanceolate-oblong, acute, nearly ½ in. long; petals elliptic-oblong, apiculate, as long as the sepals; lip as long as the petals, 3-lobed; side lobes oblong, obtuse or subacute, short; front lobe obovate or rounded, obtuse; disc with 3–5 crested nerves; spur clavate or obtuse, slightly curved, ⅛ in. long; column stout, ⅙–¼ in. long. *Bolus, Ic. Orch. Austr.-Afr.* ii. *t.* 16.

SOUTH AFRICA: without locality, *Zeyher*, 1583!
COAST REGION: Komgha Div.; grassy slopes near Komgha, 2000 ft., *Flanagan*, 1117, 1699!
KALAHARI REGION: Orange River Colony; Caledon River, *Burke*! Rhinoster Kop, *Burke*, 508! Harrismith, *Sankey*, 257! Transvaal; Magaliesberg, *Burke*! *Rech*, 597! near Botsabelo, 5000 ft., *Schlechter*, 4057! near Elim, Spelonken, *Schlechter*; Moodies, near Barberton, 2000–3000 ft., *Culver*, 14! near Lydenburg, *Wilms*, 1403! Bronkhorst Spruit, *Wilms*, 1410! Wandewald, Pretoria, *Nelson*, 506! 507! Springbok Flats, Waterberg District, *Burtt - Davy*, 1024! near Nylstroom, *Burtt-Davy*, 2007! near Pretoria, 4500 ft., *Burtt-Davy*, 7487! Uit-gevallen, near Heidelberg, *Burtt-Davy*, 9144! and without precise locality, *Holub*!
EASTERN REGION: Transkei; Krielis Country, *Bowker*! 340! Tembuland; near Umtata, *Miss Pegler*! Natal; Sevenfontein, near Boston, 3000–4000 ft., *Wylie in Herb. Wood*, 5372! Swaziland; Dalriach, near Mbabane, 4500 ft., *Bolus*, 12302!

A specimen collected near Palapye, 3000 ft., in Tropical Bechuanaland, apparently belongs to the same species.

13. **E. longipes** (Rolfe in Kew Bulletin, 1910, 280); rhizome stout and woody, somewhat thickened at the nodes; leaves in a fascicle of 4 or 5, linear, acute, undeveloped at flowering time (2½ in. in the one seen), with 3 or 4 broad enveloping sheaths; scape over 1½ ft. long, with 2 or 3 lanceolate sheaths below; raceme lax,

about 6 in. long ; flowers medium-sized ; bracts ovate-lanceolate,
acute, strongly veined, $\frac{1}{3}-\frac{1}{2}$ in. long; pedicels about $\frac{3}{4}$ in. long ;
sepals broadly lanceolate, acute, $\frac{1}{2}$ in. long; petals similar to the
sepals in shape and size, but rather broader at the base ; lip 3-lobed,
rather shorter than the petals ; side lobes broadly oblong, subobtuse ; front lobe broadly elliptic, obtuse, with about 7 thickened
veins, bearing numerous stout papillæ, the three central extending
as papillose veins to near the base ; spur clavate, slender, about $\frac{1}{6}$ in.
long ; column clavate, $\frac{1}{4}$ in. long, with a short foot.

EASTERN REGION : Natal, *Buchanan*, 3 !

14. **E. chlorantha** (Schlechter in Engl. Jahrb. xx. Beibl. 50, 9) ;
leaves 2 or 3 in a fascicle, erect, narrowly linear, very acute, shorter
than the scape ; scape lateral, erect, slender, with several membranous
acute sheaths, 7–8 in. high ; raceme loosely many-flowered ; bracts
suberect, linear, setaceous-acuminate, about as long as the ovary ;
flowers medium-sized, green ; sepals lanceolate, acute, $\frac{1}{3}$ in. long ;
petals ovate-oblong, obtuse, rather shorter than the sepals, and
scarcely as broad ; lip 3-lobed from a cuneate base, as long as the
petals ; side lobes small, subtriangular, obtuse ; front lobe broadly
ovate, obtuse ; disc with 4 thickened lines from the apex of the lip
to the mouth of the spur, crested towards the apex ; spur subcylindric, obtuse, short, curved ; column half as long as the lip,
dilated at the apex ; anther produced into an obtuse appendage.

KALAHARI REGION : Transvaal ; mountain slopes, Berea, near Barberton,
3500 ft., *Culver*, 3 !
EASTERN REGION : Swazieland ; grassy slopes, Komassan Range, Havelock
Concession, 3500 ft., *Saltmarshe in Herb. Galpin*, 652 !

15. **E. Flanaganii** (Bolus in Trans. S. Afr. Phil. Soc. xvi. (1905)
143) ; a slender erect herb, $1\frac{1}{4}-1\frac{1}{2}$ ft. high ; leaves in a fascicle of
about 4, developed at flowering time, linear-ensiform, acuminate,
rigid, many-nerved, 10–13 in. long, $\frac{1}{8}-\frac{1}{4}$ in. broad ; scape $1\frac{1}{4}-1\frac{1}{2}$ ft.
high, straight or somewhat flexuous, with about 4 acute or acuminate
sheaths below ; racemes somewhat lax and one-sided, 12–20-
flowered ; bracts ovate, acuminate, the lower $\frac{1}{3}$ in. long ; sepals spreading, equal, shortly acute, about $\frac{1}{3}$ in. long, glaucous green or somewhat livid ; petals deflexed above the lip, oblong, apiculate, as long
as the sepals and nearly twice as broad, pale lilac with purple
apical margin ; lip porrect, cuneate, 3-lobed, as long as the petals,
pale lilac with purple margin ; side lobes short ; front lobe quadrate,
slightly retuse or nearly truncate; disc with a cluster of acuminate
setiform papillæ below the apex, and three rows of shorter papillæ
below ; spur ovate, $\frac{1}{12}$ in. long ; column oblong, subtetragonal,
without foot. *Bolus, Ic. Orch. Austr.-Afr.* ii. *t.* 14.

COAST REGION : Queenstown Div. ; mountain slopes near Queenstown, 4000 ft.,
Galpin, 1713. Komgha Div. ; grassy slopes near the mouth of the Kei River,
200 ft., *Flanagan*, 1029 !
CENTRAL REGION : Aliwal North Div. ; Elands Hoek, 4500 ft., *Bolus*, 10544.

16. E. brachystyla (Schlechter in Engl. Jahrb. xxvi. 336); erect, slender, much like *E. hians*, Spreng., in habit ; rhizomes rhomboid-oblong, depressed, small ; leaves radical, in fascicles of 2–4, erect or somewhat spreading, acute, rather rigid, not fully developed until after flowering time, usually shorter than the scapes, 6–10 in. long, about $\frac{1}{8}$ in. broad ; scapes erect, sometimes a little flexuous, 8–14 in. long, with 3 or 4 lanceolate sheaths below ; racemes 2–3 in. long, lax, many-flowered ; bracts lanceolate, acuminate, somewhat spreading, about half as long as the ovary ; flowers much like those of *E. violacea*, Reichb. f. ; sepals oblong, apiculate, 5-nerved, over $\frac{1}{3}$ in. long ; petals broadly obovate-oblong, obtuse, over twice as broad as the sepals, several-nerved ; lip 3-lobed, over $\frac{1}{3}$ in. long ; side lobes obliquely oblong, obtuse ; front lobe oblong, obtuse or subtruncate ; disc with 3–5 papillose keels ; spur subcylindric, obtuse, about $\frac{1}{8}$ in. long ; column about $\frac{1}{8}$ in. long.

EASTERN REGION : Griqualand East ; Insiswa Range, 6800 ft., *Schlechter*, 6489 !

17. E. zeyheriana (Sond. in Linnæa, xix. 73); leaves narrowly linear, as long as the scapes, scarcely $\frac{1}{6}$ in. broad ; scape slender, 1 ft. high, clothed with loose ovate acute membranous sheaths ; bracts ovate, acute, about $\frac{1}{2}$ in. long ; flowers densely corymbose, or ultimately somewhat racemose ; sepals lanceolate, acuminate, $\frac{1}{3}$ in. long ; petals obtuse, about as long as the sepals, nearly three times as broad ; lip hastately 3-lobed ; side lobes basal, horizontally spreading, ovate, acute, membranous ; front lobe obovate, obtuse, $\frac{1}{4}$ in. long ; disc somewhat fleshy ; spur cylindric, obtuse, nearly $\frac{1}{3}$ as long as the lip. *Bolus in Journ. Linn. Soc.* xxv. 183 ; *Durand & Schinz, Conspect. Fl. Afr.* v. 26.

COAST REGION : Queenstown Div. ; in marshy places on the Winter Berg, *Zeyher*.

I have only seen a single leaf and flower in Dr. Lindley's Herbarium.

18. E. purpurascens (Rolfe in Kew Bulletin, 1910, 281); rhizome stout, thickened and corm-like at the nodes ; leaves contemporaneous with the scape, 3 or 4 in a fascicle, straight and suberect, narrowly elongate-linear, subacute, closely veined, $\frac{1}{2}$–1 ft. long, with two or three lanceolate protecting sheaths at the base ; scape lateral, erect, straight, $1\frac{1}{4}$–$2\frac{1}{2}$ ft. long, with several narrow imbricate sheaths below ; racemes elongate, many-flowered, 3–7 in. long ; bracts lanceolate, acuminate, $\frac{1}{4}$–$\frac{1}{2}$ in. long, usually suberect ; pedicels slender, $\frac{1}{3}$–$\frac{1}{2}$ in. long ; flowers medium-sized ; sepals subconnivent or somewhat spreading, lanceolate or narrowly oblong-lanceolate, acute, $\frac{1}{2}$–$\frac{3}{4}$ in. long, brown or greenish-purple ; petals elliptic-lanceolate, subacute, nearly as long as the sepals, light purple, pink or whitish with purple spots ; lip deeply 3-lobed, nearly as long as the petals ; side lobes broadly oblong, obtuse or rounded ; front lobe obovate, obtuse or rounded at the apex, almost covered with strongly ramentaceous or crested veins, the 3 central but little crested below the middle ; spur clavate, slender, straight or somewhat curved, ⅙ to nearly ¼ in. long ; column clavate, ¼ in. long, footless.

SOUTH AFRICA : without locality, *Zeyher*, 3899 !
COAST REGION : Uitenhage Div. ; Van Stadens Berg, among grasses, *Zeyher*, 600 !
EASTERN REGION : Transkei ; Krielis Country, common in the eastern districts, *Mrs. Barber*, 340 ! Griqualand East ; damp places on mountains near Clydesdale, 3000 ft., *Tyson*, 1912 ! Natal ; Wentworth Flat and around Durban, 10–800 ft., *Sanderson*, 496 ! Inanda, *Wood*, 434 ! Howick, 3000–4000 ft., *Wood*, 11787 ! Umlalazi, *Wood*, 11790 ! Tongaat River, 500 ft., *Wood*, 700 ! *and in MacOwan & Bolus, Herb. Norm. Austr.-Afr.* 1367 ! Umlaas, *Mudd*, 132 ! Umogoye Flat, 2000 ft., *Rudatis*, 542 ! Fairfield, 1800 ft., *Rudatis*, 806 ! and without precise locality, *Buchanan*, 6 ! *Gerrard & McKen*, 10 !

A common and well-marked species, which exists under four different names in Herbaria, but none of them are correct, and I cannot find it enumerated under any of them.

19. **E. gladioloides** (Rolfe in Kew Bulletin, 1910, 281) ; rhizome not seen ; leaves in a fascicle of 2 or 3, straight, erect, narrowly elongate-linear, acute or attenuate at the apex, closely veined, with a strong midrib, 1–1½ ft. long, protected by 2 or 3 lanceolate acute imbricate sheaths at the base ; scape erect, straight, 2 ft. high, with numerous narrow imbricate sheaths to above the middle ; raceme about 6 in. long, many-flowered ; bracts lanceolate, very acuminate, strongly veined, ⅓–¾ in. long ; pedicels slender, about ¾ in. long ; flowers lilac (*Wood*) ; sepals linear-lanceolate, apiculate or acute, over ½ in. long ; petals oblong or elliptic-oblong, subobtuse, rather shorter than the sepals ; lip 3-lobed, nearly as long as the petals ; side lobes short, with rounded apex ; front lobe suborbicular, obtuse, with numerous very strongly crested veins, forming a dense central patch and extending to the base in 3 thickened and slightly verrucose keels ; spur clavate, slender, curved at the apex, ⅙ in. long ; column clavate, ⅓ in. long, footless.

EASTERN REGION : Natal ; Lidgetton, 3000–4000 ft., *Wood*, 7922 !

20. **E. ovatipetala** (Rolfe in Kew Bulletin, 1910, 281) ; rhizome and leaves not seen ; scapes rather stout, ¾–1¼ ft. high, with numerous narrow sheaths ; racemes lax, many-flowered, 3–8 in. long ; bracts ovate-lanceolate or elliptic-lanceolate, acuminate, strongly veined, ¼–½ in. long ; pedicels slender, ½ in. or more long ; flowers spreading or sometimes drooping, medium-sized ; sepals oblong or lanceolate-oblong, acute, over ½ in. long, rather membranous ; petals ovate, obtuse, rather longer than the sepals and more than twice as broad ; lip broadly ovate-elliptic, 3-lobed, obtuse or rounded at the apex, rather shorter than the sepals ; side lobes broadly rounded ; front lobe about as broad as long ; disc with 5–7 slightly thickened veins, bearing numerous slender filaments, chiefly about the middle, veins slightly verrucose behind ; spur slender and somewhat curved above, conical at the base, nearly ¼ in. long ; column oblong or slightly clavate, nearly ¼ in. long, with a short foot.

COAST REGION : Port Elizabeth Div. ; Emerald Hill, *Herb. Bolus,* 10674 !

Dr. Bolus remarks of this species : " Locality unknown, but found amongst a bundle of plants from Emerald Hill."

21. E. Engleri (Rolfe in Kew Bulletin, 1910, 282) ; rhizome stout, thickened at the nodes ; leaves immature at flowering time ; scapes lateral, erect, about 1 ft. high, with several lanceolate sheaths below ; raceme lax, about 9-flowered ; bracts ovate or ovate-lanceolate, acuminate, ⅓–½ in. long ; pedicels slender, about ½ in. long ; sepals oblong or elliptic-oblong, obtuse, about ½ in. long ; petals broadly elliptic, as long as the petals ; lip strongly 3-lobed, rather shorter than the petals ; side lobes broad, with somewhat rounded apex ; front lobe broadly orbicular, broader than long; disc covered with a mass of very prominent short fleshy papillæ, terminating behind in about five verrucose keels ; spur clavate, scarcely ¼ in. long ; column clavate, under ¼ in. long.

KALAHARI REGION : Transvaal, among stones on Klip River Mountains, 6000 ft., *Engler,* 2745 !

Described from a specimen in the Berlin Herbarium.

22. E. transvaalensis (Rolfe in Kew Bulletin, 1910, 282) ; rhizome and leaves not seen ; portion of scape seen over ½ ft high, with one lanceolate acute sheath ; raceme lax, 4 in. long, 6-flowered ; bracts elliptic-ovate, acute or shortly acuminate, spreading, ¼–⅓ in. long ; pedicels slender, ⅓ in. long ; sepals broadly oblong, subacute, nearly ½ in. long ; petals ovate-oblong, subobtuse, broader than the sepals and about as long ; lip broadly ovate, 3-lobed, nearly as long as the sepals ; side lobes broadly rounded ; front lobe nearly quadrate, apiculate, rather broader than long ; disc with 5–7 thickened veins, bearing several stout papillæ in front, extending to the base as 3 slightly verrucose keels ; spur clavate, nearly straight, ¼ in. long ; column broadly oblong, over ⅛ in. long, with a short foot.

KALAHARI REGION : Transvaal ; Vaal Bank, between the Devils Kantoor and Pretoria, 5000–6000 ft., *Bolus,* 10676 !

23. E. aliwalensis (Rolfe in Kew Bulletin, 1910, 368) ; rhizome not seen ; leaves undeveloped at flowering time, those seen very short and in a fascicle of 3 or 4, protected by 3 or 4 broad strongly veined sheaths ; scape lateral, about 1½ ft. high, with about 3 lanceolate acute sheaths ; raceme about 4 in. long, 7- or 8-flowered ; bracts elliptic-ovate, acute, purplish when dried, ¼–⅓ in. long ; pedicels rather stout, about ¾ in. long ; flowers medium-sized, " yellow-brown outside ; " sepals subconnivent, elliptic-oblong, sub-acute, over ½ in. long ; petals elliptic, subobtuse, much broader than the sepals and about as long ; lip broadly elliptic, 3-lobed, rather shorter than the sepals ; side lobes broadly rounded ; front lobe suborbicular, with 7–9 thickened crested veins, stoutly papillose in front, the 3 central extending to the base, but scarcely verrucose

below the middle; spur slender, subclavate, nearly straight, $\frac{1}{6}$ in.
long; column broadly oblong, $\frac{1}{4}$ in. long, footless.

CENTRAL REGION: Aliwal North Div.; Elands Hoek, near Aliwal North,
4550 ft., on dry flat grassy ground, *Bolus*, 10671 !

24. E. hians (Spreng. Syst. iii. 720); rhizomes very stout and
woody, irregularly thickened at the nodes; leaves in fascicles of
3 or 4, small or undeveloped at flowering time, narrowly ensiform,
acute, suberect or somewhat spreading, conduplicate, prominently
veined, up to 6 in. long on the specimens seen; scapes lateral, erect,
$\frac{1}{2}$-$1\frac{1}{4}$ ft. high, with several acute spathaceous sheaths below;
racemes lax, usually many-flowered; bracts ovate-oblong or ovate-
lanceolate, acute or acuminate, $\frac{1}{3}$-$\frac{1}{2}$ in. long; pedicels $\frac{1}{2}$-$\frac{3}{4}$ in. long;
flowers medium-sized, with brownish sepals and pink or lilac petals
and lip; sepals somewhat spreading, narrowly oblong, subobtuse,
nearly $\frac{1}{2}$ in. long; petals elliptic-oblong, obtuse, rather longer than and
twice as broad as the sepals; lip strongly 3-lobed, about as long as
the petals, cuneate at the base; side lobes diverging, oblong, subobtuse, small; front lobe broadly obovate or obcordate; disc with
3-5 thickened keels, crenulate or papillose in front; spur narrowly
clavate, somewhat curved, obtuse, $\frac{1}{6}$-$\frac{1}{4}$ in. long; column broadly-
clavate, $\frac{1}{4}$ in. long; anther somewhat 2-lobed. *Lindl. Gen. & Sp.
Orch.* 183; *Bolus in Journ. Linn. Soc.* xxv. 182, *and Ic. Orch. Austr.-
Afr.* ii. *t.* 18; *Durand & Schinz, Conspect. Fl. Afr.* v. 22; *Schlechter
in Engl. Jahrb.* xx. *Beibl.* 50, 25. *Satyrium hians, Linn. f. Suppl. Pl.*
401. *Limodorum hians, Thunb. Prodr. Pl. Cap.* 3; *Thunb. Fl. Cap.
ed. Schult.* 30. *Eulophia clavicornis, Lindl. Comp. Bot. Mag.* ii. 202.
E. emarginata, Lindl. Comp. Bot. Mag. ii. 202; *Drège, Zwei Pfl.
Documente,* 46; *Krauss in Flora,* 1845, 305; *Beitr. Fl. Cap- und
Natal.* 157. *E. platypetala, Krauss in Flora,* 1845, 305; *Beitr. Fl.
Cap- und Natal.* 157 (*not of Lindl.*).

SOUTH AFRICA: without precise locality, *Masson*! *Mrs. Bowker*!
COAST REGION: Cape Div.; between Cape Town and George, *Rogers*! George
Div.; between Malgaten River and Great Brak River, *Burchell*, 6147! Bathurst
Div.; between Blue Krantz and Kaffir Drift Military Post, *Burchell*, 3705!
Albany Div.; Grahamstown, *Bolton*! *MacOwan*! *Williamson*! Howisons Poort,
Hutton! Bedford Div.; Kaga Berg, *Hutton*! *Galpin*, 170. Fort Beaufort Div.;
Cooper, 413! Stockenstrom Div.; Kat River, *Ecklon and Zeyher*, 10! Kat Berg,
Drège, 2213c! *Hutton*! Queenstown Div.; *Cooper*, 1320! Cathcart Div.; near
Thomas River, *Cooper*, 322! between Shiloh and Windvogel Mountain, *Drège*,
2213d! Komgha Div.; grassy hills near the Kei River, *Flanagan*, 68! British
Kaffraria; *Hutton*! Eastern Frontier, *MacOwan*, 76! *Hutton*!
KALAHARI REGION: Orange River Colony; Bethlehem, *Richardson*! and with-
out precise locality, *Cooper*, 981! 1876! *Thomas*! Basutoland; near Leribe,
Dieterlen, 291! Transvaal; various localities, *McLea in Herb. Bolus*, 5817! *Bolus*,
10673! 10675! *Sanderson*! *Galpin*, 509! 550, *Miss Leendertz*, 282! 3131!
Wilkinson! *Wilms*, 1404! 1408! *Culver*, 2! 52, *Rand*, 793! 794! *Jenkins*, 7205!
8040! *Burtt-Davy*, 1048! 2156! 2378! *Rech*, 1002!
EASTERN REGION: Transkei; on grassy hills, *Mrs. Barber*, 432! Krielis
Country, *Bowker*, 16! 248! Kentani, 1200 ft., *Miss Pegler*, 370! Tembuland;
hill-sides at Bazeia, 2000-3000 ft., *Baur*, 263! 509! Pondoland; Egossa, *Sim*,
2453! Griqualand East; near Kokstad, 5100 ft., *Tyson*, 1601! 1602! Natal;

various localities, *Sutherland*! *Wood,* 205! 4264! 10823! 11821! *Wylie in Herb. Wood,* 11823! *Wilms,* 2281! 2200 ft., *Sanderson*! *Krauss,* 254! 406! *Buchanan,* 5! *Mrs. Fannin,* 62! *Molyneau*! *Schlechter,* 3431, *Rudatis,* 674! Swaziland; at Embabaan, *Burtt-Davy,* 3014; and without precise locality, *Miss Stewart,* 42!

25. **E. violacea** (Reichb. f. in Linnæa, xx. 683); leaves linear, acuminate, 1–4 in. long, ⅛ in. broad; scape stout, 1 ft. high, with about 4 sheaths; raceme lax, about 10-flowered, 3–4 in. long; flowers violet?; sepals oblong, acute, nearly ½ in. long, ⅛ in. broad; petals ovate, obtuse, agglutinated to the back of the column, rather shorter and broader than the sepals; lip quadrate, 3 lobed, as long as the sepals by about as broad; side lobes rounded, about a fourth as long as the lip; front lobe quadrate, obtuse-angled, emarginate; disc with the mid nerve and principal side nerves crested; spur cylindrical, subemarginate, about half as long as the lip; column clavate, over ⅛ in. long. *Bolus in Journ. Linn. Soc.* xxv. 183; *Durand & Schinz, Conspect. Fl. Afr.* v. 26.

COAST REGION: Knysna, Div.; Doukamma, *Mund*! Knysna Hills, *Bowie*! Albany Div.; Stones Hill, near Grahamstown, 5000 ft., *Schönland in Herb. Bolus,* 5980! Stockenstrom Div.; Philipton (Kat River), *Ecklon & Zeyher*!

26. **E. platypetala** (Lindl. in Hook. Comp. Bot. Mag. ii. 202); rhizome stout, thickened at the nodes; leaves in a fascicle of 4 to 6, ensiform, acuminate, conduplicate, closely veined, somewhat curved, 3–7 in. long, with 2 or 3 closely veined sheaths at the base; scapes lateral, rather stout, ¾–1¼ ft. high, with 2 or 3 broad spathaceous acute sheaths below; raceme loosely many-flowered; bracts ovate or ovate-lanceolate, shortly acuminate, ¼–½ in. long; pedicels ½–¾ in. long; flowers rather large; sepals somewhat spreading, elliptic-oblong, subobtuse, nearly ¾ in. long; petals suborbicular, obtuse, nearly as long as the sepals; lip strongly 3-lobed, rather longer than the sepals; side lobes somewhat diverging, broadly oblong, obtuse; front lobe broadly obovate from a cuneate base, obtuse or emarginate, twice as long as the side lobes; disc with 3 to 5 thickened keels from near the base to the apex, much elevated and papillose or fimbriate in front; spur clavate-oblong, ¼ in. long; column clavate, ⅓ in. long. *Bolus in Journ. Linn. Soc.* xxv. 183, *and Ic. Orch. Austr. Afr.* ii. *t.* 17; *Durand & Schinz, Conspect. Fl. Afr.* v. 24.

COAST REGION: Riversdale Div.; by the Zoetemelks River, *Burchell,* 6611! hills near the Zoetemelks River, *Burchell,* 6767! Knysna Div.; near Forest Hall, Plettenberg Bay, *Miss Newdigate in Herb. Bolus,* 6888!

27. **E. lissochiloides** (Lindl. in Hook. Comp. Bot. Mag. ii. 203); rhizome stout, much thickened at the nodes; leaves in a fascicle of 5 or 6, ensiform, acute, conduplicate, 3–4 in. long, with 2 or 3 reduced membranous sheaths at the base; scape lateral, about 1 ft. high, with 3 or 4 spathaceous acute membranous sheaths; raceme

short, with a few rather large flowers ; bracts oblong-lanceolate, acuminate, about $\frac{1}{2}$ in. long ; pedicels $\frac{3}{4}$ in. long ; sepals spreading, elliptic-oblong, subobtuse, nearly $\frac{3}{4}$ in. long ; petals rather broader than the sepals, otherwise similar ; lip 3-lobed, as long as the sepals ; side lobes broadly rounded-oblong ; front lobe obovate-oblong, obtuse, much longer than the side lobes ; disc with 3 to 5 fleshy crenulate keels from below the middle upwards, terminating in three high lamellæ in front ; spur oblong, obtuse, $\frac{1}{6}$ in. long ; column clavate, $\frac{1}{4}$ in. long, base prolonged into a short foot, forming with the side lobes of the lip a broad short mentum. *Bolus in Journ. Linn. Soc.* xxv. 183 ; *Reichb. f. in Walp. Ann.* vi. 644 ; *Durand & Schinz, Conspect. Fl. Afr.* v. 22.

COAST REGION : Riversdale Div. ; hills near the Zoetemelks River, *Burchell*, 6764 !

28. **E. Saundersiæ** (Rolfe in Kew Bulletin, 1910, 368) ; rhizome and leaves not seen ; upper part of scape with two oblong sheaths, lower part not seen ; raceme short, about 6-flowered ; bracts ovate, shortly acuminate, $\frac{1}{3}$–$\frac{1}{2}$ in. long ; pedicels $\frac{3}{4}$–1 in. long, rather stout ; flowers rather large, yellow and brown spotted (*Saunders*) ; sepals spreading, elliptic-oblong, obtuse or subapiculate, $\frac{3}{4}$–$\frac{7}{8}$ in. long ; petals broadly elliptic, obtuse, much shorter and broader than the sepals ; lip broadly elliptic, about as long as the petals, 3-lobed ; side lobes quadrate, obtuse or truncate, shorter than broad, very closely veined ; front lobe broadly elliptic oblong, obtuse or rounded at the apex, as broad as long, closely veined ; disc with about 7 approximate thickened verrucose or shortly papillose keels ; spur broadly oblong, obtuse, stout, $\frac{1}{6}$ in. long, straight or slightly curved ; column broadly oblong, $\frac{1}{4}$–$\frac{1}{3}$ in. long.

EASTERN REGION : Natal ; without precise locality, *Mrs. K. Saunders*, 1 !

29. **E. barbata** (Spreng. Syst. iii. 720) ; tubers subglobose, $\frac{3}{4}$–1 in. broad ; leaves 4 or 5 in a tuft, suberect or recurved, narrowly ensiform or linear, acute, 4–10 in. long, $\frac{1}{6}$–$\frac{1}{3}$ in. broad ; scape erect, compressed, 6–12 in. long, with numerous lanceolate acute imbricate sheaths ; racemes 3–4 in. long, many-flowered ; bracts lanceolate, acuminate, $\frac{1}{3}$–$\frac{1}{2}$ in. long ; flowers about $\frac{3}{4}$ in. long, green and white ; pedicels $\frac{1}{2}$–$\frac{3}{4}$ in. long ; sepals oblong-lanceolate, acute, $\frac{1}{2}$–$\frac{2}{3}$ in. long, green ; petals elliptic–oblong, subobtuse, about twice as broad as the sepals, white ; lip more or less distinctly 3-lobed, broadly elliptic, rather shorter than the petals, white ; side lobes short, broadly rounded ; front lobe broadly rounded, obtuse or truncate ; disc with 3 to 5 thickened much fimbriate keels extending to beyond the middle of the front lobe ; spur oblong, obtuse, $\frac{1}{6}$ in. or more long ; column clavate, nearly $\frac{1}{4}$ in. long ; anther-case obscurely 2-lobed. *Lindl. Gen. & Sp. Orch.* 184, *partly* ; *Bolus in Journ. Linn. Soc.* xxv. 182 ; *Durand & Schinz, Conspect. Fl. Afr.* v. 19 ; *Kränzl. in Ann. Naturhist. Hofmus. Wien*, xx. 11. *E. ovalis, Lindl. Comp. Bot.*

Mag. ii. 202 ; *Drège, Zwei Pfl. Documente,* 128, 132. *E. dregeana, Bolus, Ic. Orch. Austr.-Afr.* ii. *t.* 9, *partly, not of Lindl. Serapias capensis, Linn. Mant.* 293. *Limodorum barbatum, Thunb. Prodr.* 4 ; *Thunb. Fl. Cap. ed. Schult.* 29.

SOUTH AFRICA : without locality, *Bowie !* *Zeyher,* 1587 !

COAST REGION : Uitenhage Div. ; between Van Stadens Berg and Bethelsdorp, *Drège.* Addo, 1000–2000 ft., *Drège !* 2213b ! Uitenhage, *Pappe,* 75 ! Bathurst Div. ; Port Alfred, *Hutton in Herb. Schönland,* 303 ! Round Hill, 1000 ft., *Bolus,* 7368 ; Kowie Flats, *MacOwan,* 184 ! Linch's Post, near Kowie River, *Bowie,* 6 ! mouth of Kleinemund River, *MacOwan.* Albany Div. ; Cross Fountains Farm, *Atherstone !* and without precise locality, *Hutton !* Bedford Div. ; sandy flats near the mouth of the Fish River, *Atherstone,* 11 ! Komgha Div. ; sandy flats near the mouth of the Kei River, *Flanagan in MacOwan, Herb. Austr.-Afr.,* 1533 ! Stockenstrom Div. ; Chumie Peak, *Scully,* 173. Queenstown Div. ; Queenstown District, *Zeyher,* 6. King Williamstown Div. ; Flanagan Farm, near King Williamstown, *Krook, Penther,* 158, *Flanagan,* 2254. Eastern frontier of Cape Colony, *Hutton !* *Prior !*

CENTRAL REGION : Somerset Div. ; summit of Bosch Berg, 4500 ft., *MacOwan,* 184 ! and without precise locality, *Bowker !*

EASTERN REGION : Griqualand East ; near Nalogha, *Krook, Penther,* 200 ! foot of Insiswa Range, *Krook, Penther,* 137. Natal ; without precise locality, *Mrs. Fannin,* 61 ! *Wood !*

The only specimen cited by Lindley as *E. barbata* belongs to *E. lunata,* Schlechter, and the collector, "Thunberg," is believed to be erroneous.

30. **E. Thunbergii** (Rolfe in Kew Bulletin, 1910, 369) ; rhizome stout, thickened at the nodes ; leaves in a fascicle of 3 or 4, linear, subacute, somewhat recurved, 3–8 in. long, with 2 or 3 short imbricate sheaths below ; scapes lateral, $\frac{3}{4}$–$1\frac{1}{4}$ ft. long, with 2 or 3 lanceolate sheaths below ; racemes 3–5 in. long, somewhat lax ; bracts lanceolate, acuminate, $\frac{1}{2}$–$\frac{3}{4}$ in. long ; pedicels about $\frac{3}{4}$ in. long ; flowers rather large ; sepals lanceolate-attenuate from a broad base, acuminate, over $\frac{3}{4}$ in. long ; petals narrowly ovate, sub-obtuse, shorter than the sepals and about twice as broad ; lip broadly ovate, 3-lobed, nearly as long as the petals ; side lobes broad with a rounded apex ; front lobe broadly rounded, as broad as long ; disc with 5–7 somewhat thickened crested veins, veins somewhat verrucose in front and at the base of the side lobes ; spur slender, somewhat curved, about $\frac{1}{6}$ in. long ; column clavate, $\frac{1}{4}$ in. long.

SOUTH AFRICA : without locality, *Thunberg !*

Based on two specimens which Thunberg had referred to his *Limodorum barbatum,* but which differ in the lax inflorescence, larger flowers, etc. A drawing of the specimens and a single flower are preserved at Kew.

31. **E. robusta** (Rolfe in Kew Bulletin, 1910, 369) ; rhizome stout, woody, thickened at the nodes ; leaves in a fascicle of 4 to 6, ensiform or elongate-linear, acute, strongly veined, suberect or somewhat recurved, $\frac{1}{3}$–1 ft. long, with 2 or 3 short exterior sheaths at the base ; scape basal, erect, stout, $\frac{3}{4}$–$1\frac{1}{2}$ ft. high, with numerous lanceolate or spathaceous somewhat imbricate sheaths ; racemes many-flowered, somewhat dense, 3–6 in. long ; bracts lanceolate,

acuminate, ½–1 in. long; pedicels about ¾ in. long; flowers rather
large, yellow, or the petals and lip pink or white and the sepals
brown; sepals oblong-lanceolate or attenuate from a broader base,
acute or acuminate, over ¾ in. long; petals elliptic-oblong or ovate,
obtuse or subacute, as long as and sometimes longer than the sepals
and over twice as broad; lip broadly ovate, 3-lobed, rather shorter
than the petals; side lobes broadly rounded; front lobe broadly
elliptic or rounded; disc with 5–7 somewhat thickened veins,
verrucose behind and with scattered crests or papillæ in front; spur
oblong, rather stout, nearly ⅙ in. long; column broadly oblong,
about ¼ in. long.   *E. dregeana, Schlechter in Engl. Jahrb.* xx. *Beibl.*
50, 25, *not of Lindl.*; *Bolus, Ic. Orch. Austr. Afr.* ii. *t.* 9, *partly.*

KALAHARI REGION : Orange River Colony; Bethlehem, *Richardson*! Venters-
burg, *Low*! and without precise locality, *Cooper,* 976! 3595! Basutoland;
Drakensberg, 8000 ft., *Sanderson,* 628! near Leribe, *Dieterlen,* 134! Transvaal;
Magaliesberg, *Schlechter,* 3617, *Burke,* 407! Mooi River, *Burke*! near Heidelberg, 6000 ft., *McLea in Herb. Bolus,* 5618! *Wilms,* 1390! *Burtt-Davy,* 5627!
9131! near Barberton, 2000–4000 ft., *Culver,* 17! *Galpin,* 719; Pilgrims Rest,
Greenstock! Johannesburg, *Ommanney,* 5! 79! *Rand,* 1097! near Pretoria,
4600 ft., *Miss Leendertz,* 680, *Nelson,* 512! *Burtt-Davy,* 1044! 3853! Wonder-
fontein, *Nelson,* 263! *Bolus,* 12306; Little Olifants River, 3000 ft., *Schlechter,*
4049! Ermelo, *Wilkinson*! *Burtt-Davy,* 1005! 2159! 5476! Baastkraal, *Jenkins,*
6242! Great Olifants River, *Nilkerk,* 7507! Carolina, *Rademacher,* 7494! and
without precise locality, *Sanderson*! *Hallack*!
EASTERN REGION : Tembuland; Umtata, *Miss Pegler in Herb. Bolus,* 10672!
Griqualand East; on rough slopes near Kokstad, *Tyson,* 1086! 1596. Natal;
Oliviers Hoek, sources of the Tugela River, 4000 ft., *Allison,* B! near Camperdown, *Wood,* 1960, near Durban, *Miss Doidge,* 5883! and without precise locality,
3000 ft., *Allison,* Z 111!

This has been more or less confused with *E. dregeana,* Lindl., and may yet contain more than a single species. It is very variable, and the range of colour is
remarkable. Examination of living plants may show further differences. The
two specimens cited by Schlechter as *E. dregeana,* Lindl., I have not seen, but
geographically they should belong here, and I have included them.

32. **E. dregeana** (Lindl. in Comp. Bot. Mag. ii. 202); rhizome
stout, woody, thickened at the nodes; leaves in a fascicle of 4 to 6,
ensiform, acute, strongly nerved, more or less recurved, 4–9 in. long;
scape basal, erect, stout, ¾–1½ ft. high, with numerous loose somewhat imbricate spathaceous acute sheaths; racemes short, rather
dense, many-flowered; bracts lanceolate or oblong-lanceolate,
acuminate, ½–1 in. long; pedicels stout, ½–¾ in. long; flowers rather
large; sepals oblong-lanceolate, acute, ¾–1 in. long; petals elliptic-
oblong or subovate, obtuse, about twice as broad as the sepals; lip
deeply 3-lobed, rather shorter than the petals; side lobes broadly
rounded; front lobe obovate-orbicular, obtuse; disc with 3 strongly
barbate keels in front and 5 crenulate keels behind; spur oblong,
obtuse, ⅙–¼ in. long; column clavate, ¼ in. long.   *Drège, Zwei Pfl.*
Documente, 142, 146, 147, 152; *Bolus in Journ. Linn. Soc.* xxv. 183;
Durand & Schinz, Conspect. Fl. Afr. v. 21, *partly.*

COAST REGION : Peddie Div.; Fish River Hills, near Trumpeters Drift, *Drège*!
4573a!

EASTERN REGION: Transkei; between Gekau (Geua) River and Bashee River, 1000-2000 ft., *Drège*! Tembuland; Morley, among grasses, 1000-2000 ft., *Drège*. Pondoland; between Omsamwubo and Omsamcaba, *Drège*, 4573b! Natal; Fairfield, Alexandra District, 2300 ft., *Rudatis*, 568!

The plant cited by Schlechter as *E. dregeana,* Lindl., has been referred to the preceding species, which see. The var. *angustior*, Reichb. f. (*Linnæa*, xix. 75), from moist places near Shiloh, on the Klipplaat River, in Queenstown Div., I have not seen. It is said to have more distant sheaths and smaller flowers.

33. E. Nelsoni (Rolfe in Kew Bulletin, 1910, 369); rhizome not seen; leaves in a fascicle of 4 or 5, narrowly linear, acute, 3-9 in. long, closely veined; scape about 1½ ft. long, with several lanceolate sheaths below; raceme about 5 in. long, lax, about 8-flowered; bracts lanceolate, acuminate, ½-¾ in. long; pedicels ½-¾ in. long; flowers medium-sized; sepals lanceolate, acuminate, over ½ in. long; petals narrowly ovate-oblong, apiculate, rather shorter than the sepals and about twice as broad; lip broadly elliptic, about as long as the petals, strongly 3-lobed; side lobes broadly oblong, obtuse or rounded at the apex; front lobe suborbicular, apiculate, somewhat undulate; disc with several thickened veins, and bearing a tuft of prominent papillæ, chiefly on the lower half of the front lobe, veins below somewhat verrucose; spur oblong, obtuse, rather stout, about ⅛ in. long; column clavate, nearly ¼ in. long, with a very short foot.

KALAHARI REGION: Transvaal; Wanderwald, in Pretoria district, *Nelson*, 297!

34. E. oblonga (Rolfe in Kew Bulletin, 1910, 370); rhizome and leaves not seen; scape stout, with lanceolate imbricate sheaths, base not seen; raceme about 4 in. long, dense, many-flowered; bracts lanceolate, narrowly acuminate, ½-¾ in. long, strongly veined; pedicels ¾ in. long; flowers medium-sized, white; sepals oblong-lanceolate, acuminate, somewhat undulate, about ¾ in. long; petals oblong or elliptic-oblong, subobtuse, undulate, rather shorter than the sepals and about as broad; lip oblong, obtuse, 3-lobed below the middle, nearly ¾ in. long; side lobes oblong, obtuse, somewhat diverging, about ⅙ in. long; front lobe oblong, obtuse, ⅓ in. long, with several thickened veins and numerous slender filaments from the base of the side lobes upwards, veins somewhat verrucose below; spur slender, obtuse, straight, over ⅙ in. long; column clavate, over ¼ in. long.

COAST REGION: Albany Div.; Nazaar Hills, collector not stated!

Very distinct in the shape and details of the lip.

35. E. Haygarthii (Rolfe in Kew Bulletin, 1910, 370); rhizome not seen; leaves in a fascicle of 7 or 8, broadly ensiform or elongate-lanceolate, attenuate and acute above, recurved, 3-9 in. long, with 5-7 prominent veins, expanded into broad sheaths below; scapes lateral, 2-2½ ft. long, with numerous lanceolate imbricate sheaths up to the raceme; raceme 4-6 in. long, somewhat dense and many-flowered ; bracts lanceolate, acuminate, closely veined, ½–1 in. long ; pedicels about ¾ in. long ; flowers rather large ; sepals lanceolate or oblong-lanceolate, acute or acuminate, 1 in. long ; petals narrowly ovate or elliptic-ovate, acute, rather shorter than the sepals and about twice as broad ; lip broadly elliptic, about as long as the petals, 3-lobed ; side lobes broadly oblong, obtuse ; front lobe orbicular ; disc with 5–7 thickened veins, closely papillose or fringed on the lower half of the front lobe, verrucose behind ; spur oblong, obtuse, somewhat slender, ⅛ in. long ; column clavate, ¼ in. long.

EASTERN REGION : Natal; Camperdown, 2000 ft., *Wood, 469*! *Haygarth in Herb. Wood,* 1960 !

36. E. Macowani (Rolfe) ; rhizome stout and woody, much thickened at the nodes ; leaves in a fascicle of 5–7, elongate-linear or narrowly ensiform, acute, closely veined, recurved or some-times suberect, ½–1 ft. long, expanded into a broader sheath below ; scape lateral, stout, 1–1½ ft. long, with numerous lanceo-late sometimes imbricate sheaths ; racemes 4–6 in. long, often somewhat lax ; bracts lanceolate, acuminate, ½–1 in. long ; pedicels about ¾ in. long ; flowers large, yellow ; sepals oblong-lanceolate, acute, over ¾ in. long ; petals ovate or elliptic-ovate, acute or subacute, rather longer than the sepals and about three times as broad ; lip broadly elliptic, shorter than the petals, shortly or obscurely 3-lobed ; side lobes small, rounded or broadly oblong ; front lobe broadly elliptic or suborbicular ; disc with 3 somewhat thickened papillose or crested veins, sometimes with 2 or 3 additional papillose veins below the middle ; spur clavate or oblong, somewhat curved, ⅙–¼ in. long ; column broadly oblong, ¼ in. long ; capsule broadly elliptic, 1¾ in. long.

SOUTH AFRICA : without locality, *Zeyher*, 1586 !
COAST REGION : Bathurst Div. ; Kasouga River, *MacOwan*, 184 ! Kowie River *MacOwan* ! between Bathurst and Perie, *Hutton* ! sandy plains near the mouth of the Kowie River, *Hutton* ! *and in MacOwan & Bolus, Herb. Austr.-Afr.* 1215 ! Stockenstrom Div. ; Chumie (Tyumie) Mountain, on rocky slopes, *Scully*, 173 ! *and in Herb. Bolus*, 5918 ! *Tyson*, 1086 ! King Williamstown Div. ; Keiskamma, *Mrs. Hutton* !
CENTRAL REGION : Somerset Div. ; Somerset East (probably an error of locality), *Bowker* !

37. E. parvilabris (Lindl. in Hook. Comp. Bot. Mag. ii. 201) ; leaves in a fascicle of about 3, elongate, linear-lanceolate, acute, ¾–1¼ ft. long, ¾–1 in. broad, conduplicate at the base ; scape 1½–2 ft. long, stout, with numerous imbricate spathaceous sheaths ; raceme 6 in. or more long, many-flowered ; bracts elliptic-lanceolate, acuminate, strongly striate, ¾–1 in. long ; pedicels ¾ in. long ; flowers large ; sepals oblong-lanceolate, acute, over 1 in. long ; petals broadly elliptic, obtuse, rather longer than the sepals ; lip broadly elliptic, obtuse, obscurely 3-lobed, about ⅔ as long as the petals ; side lobes obtuse, not diverging ; front lobe as broad as

long ; disc with about 5 slender keels, more prominent and fimbriate near the base; spur clavate-oblong, obtuse, $\frac{1}{6}$ in. long ; column clavate-oblong, nearly $\frac{1}{4}$ in. long. *Bolus in Journ. Linn. Soc.* xxv. 183 ; *Drège, Zwei Pfl. Documente,* 151, 153 ; *Reichb f. in Walp. Ann.* vi. 645 ; *Durand & Schinz, Conspect. Fl. Afr.* v. 24.

EASTERN REGION : Pondoland ; between St. Johns River and Umsikaba River, 1000–2000 ft., *Drège,* 4575 ! between the great Waterfall and Umsikaba River, *Drège.*

Schlechter (*Engl. Jahrb.* xx. *Beibl.* 50, 4, 25) cites as *E. parvilabris,* Lindl., specimens collected in the Transvaal, near Johannesburg, *Endemann,* and on the Houtbosch Berg, 6500 ft., *Schlechter,* 4394, but I think they cannot belong here. I have not seen them.

38. E. deflexa (Rolfe in Kew Bulletin, 1895, 192) ; leaves lanceolate-linear, acute, 6–16 in. long, about $\frac{1}{2}$ in. broad ; scape elongate, 2 ft. high, with several long lanceolate-spathaceous sheaths below ; bracts ovate-lanceolate, acute, $\frac{1}{2}$–$\frac{3}{4}$ in. long ; pedicels nearly 1 in. long ; sepals spreading, lanceolate-oblong, acute, carinate, over $\frac{3}{4}$ in. long, about $\frac{1}{4}$ in. broad, light purple-brown ; petals somewhat deflexed, ovate, subobtuse or apiculate, over $\frac{3}{4}$ in. long, nearly $\frac{1}{2}$ in. broad, lined with lilac-purple on a lighter ground ; lip 3-lobed, $\frac{3}{4}$ in. long, under $\frac{1}{2}$ in. broad, veined with lilac-purple on a lighter ground, keels yellowish-white ; sides lobes oblong, obtuse ; front lobe orbicular-ovate, obtuse ; disc with 3 thickened strongly barbellate whitish-yellow keels ; spur oblong, $\frac{1}{6}$ in. long ; column clavate, $\frac{1}{4}$ in. long.

EASTERN REGION : Natal, *Allison* !

39. E. acuminata (Rolfe) ; rhizome and leaves not seen ; scapes stout, 1–1$\frac{3}{4}$ ft. high, with numerous lanceolate imbricate sheaths below ; racemes rather lax, 4–8 in. long, many-flowered ; bracts linear-lanceolate, acuminate, 1–1$\frac{1}{2}$ in. long ; pedicels about $\frac{3}{4}$ in. long ; flowers rather large, yellow ; sepals oblong-lanceolate, very acuminate, 1–1$\frac{1}{4}$ in. long ; petals elliptic-oblong, very acute, rather longer than the sepals by over twice as broad ; lip somewhat 3-lobed, shorter than the petals and about as broad ; side lobes very short, with rounded obtuse apex and numerous radiating nerves ; front lobe elliptic-oblong, apiculate ; disc with 3 thickened, somewhat verrucose keels below, becoming more slender towards the apex, and somewhat puberulous ; spur oblong, obtuse, curved, $\frac{1}{6}$ in. long ; column clavate, stout, $\frac{1}{6}$ in. long. *E. calanthoides, Bolus, Ic. Orch. Austr.-Afr.* i. *sub t.* 51, *partly, not of Schlechter.*

EASTERN REGION : Natal ; near Estcourt, *Wood,* 3428 !

Quite distinct from *E. calanthoides,* Schlechter, with which it has been confused.

40. E. Allisoni (Rolfe) ; leaves 5–7 in a fascicle, ensiform, acute, plicate, about 4 in. long at flowering time, $\frac{1}{3}$–$\frac{1}{2}$ in. broad ; scape rather stout, about 2½ ft. high, with a few loose sheaths below ; raceme 6 in. long, many-flowered ; bracts linear-lanceolate, acuminate, ¾–1 in. long ; pedicels about ½ in. long ; flowers rather large, light yellow, with numerous light red radiating lines on the side lobes of the lip ; sepals linear-lanceolate, very acuminate, ¾–1 in. long ; petals elliptic-oblong, subacute, shorter than the sepals by nearly three times as broad ; lip subentire or obscurely 3-lobed, elliptic, apiculate, as long as and rather broader than the petals ; side lobes reduced to a mere lateral notch ; disc with 5 slightly thickened keels at the base and numerous veins above, smooth or obscurely puberulous ; spur clavate, about ⅛ in. long ; column stout, clavate, nearly ¼ in. long. *E. calanthoides, Bolus, Ic. Orch. Austr.-Afr.* i. *t.* 51 (*as to figure and Allison specimen only*).

SOUTH AFRICA : without locality, *Allison* !
COAST REGION : Albany Div. ; near Grahamstown, *Todd* !

Distinct from *E. calanthoides,* Schlechter, with which it has been confused.

41. E. calanthoides (Schlechter in Engl. Jahrb. xx. Beibl. 50, 1) ; rhizome stout and woody, thickened at the nodes ; leaves about 7–8 in a fascicle, ensiform or elongate lanceolate, acute, conduplicate at the base, ½–1¼ ft. long, ¾–2¼ in. broad, somewhat recurved, with 5 prominent veins ; scapes lateral, 1–2 ft. high, with several lanceolate acuminate somewhat imbricate sheaths ; racemes ⅓–⅔ ft. long, usually somewhat dense and many-flowered ; bracts lanceolate, acuminate, ½–1 in. long ; pedicels ½–¾ in. long ; flowers large, cream-coloured ; sepals linear-lanceolate or oblong-lanceolate, very acuminate, 1–1¼ in. long ; petals nearly as long as the sepals and over twice as broad, elliptic-oblong, acute ; lip broadly elliptic, rather shorter than the petals, 3-lobed ; side lobes small, broadly rounded ; front lobe much larger, elliptic-oblong, subobtuse ; disc with 3–7 slightly thickened veins, somewhat crested towards the base ; spur slender, curved, obtuse, ⅙–¼ in. long ; column clavate, about ¼ in. long, with a short foot ; capsule oblong, 1 in. long.

KALAHARI REGION : Orange River Colony ; without precise locality, *Cooper,* 974 ! 3596 ! Transvaal ; near Nylstrom, *Burtt-Davy,* 2011 ! Potgieters Rust, *Rogers,* 2500 !
EASTERN REGION : Natal ; Sevenfontein, near Boston, 3000–4000 ft., *Wylie in Herb. Wood,* 5363 ! Oliviers Hoek, sources of Tugela River, 5000 ft., *Allison* ! Shafton, Howick, *Mrs. Hutton,* 200 ! Van Reenen, 5000–6000 ft., *Wood,* 5569 ! Fairfield, Alexandra District, 2300 ft., *Rudatis,* 567 ! and without precise locality, *Wood,* 4626.

This species was based upon *Wood,* 4626, which I have not seen, but *Wylie,* 5363, is marked by Wood himself as identical. The plant figured by Dr. Bolus is different, and the name seems to have been used in an aggregate sense. See *E. acuminata* and *E. Allisoni.*

42. E. Bakeri (Rolfe) ; rhizome not seen ; leaf elongate-lanceolate, acute, closely veined, attenuate at the base, 8 in. long, nearly ¾ in. broad ; scape about a foot high with 6 or 8 pinkish flowers ; sepals oblong-lanceolate, acute or shortly acuminate, over ¾ in. long ;

petals ovate-elliptic, minutely apiculate, rather longer than the sepals and over twice as broad ; lip ovate-elliptic, 3-lobed, rather shorter than the petals ; side lobes narrow, with rounded or obtuse apex; front lobe broadly elliptic or suborbicular; disc with 5–7 thickened keels, strongly fringed above the middle, verrucose below ; spur oblong, subobtuse, somewhat curved, over $\frac{1}{6}$ in long; column broadly oblong, over $\frac{1}{4}$ in. long.

KALAHARI REGION: Transvaal ; high ridge outside Johannesburg, 5000 ft., *Baker* !

This species has been described from rather imperfect material.

43. E. Rehmanni (Rolfe) ; leaves 6–8 in a spreading tuft, linear-oblong, acuminate, attenuate with 5–7 prominent veins, attenuate and conduplicate at the base, $\frac{1}{3}$–$\frac{1}{2}$ ft. long, $\frac{3}{4}$–$1\frac{1}{4}$ in. broad, with 2 or 3 short exterior sheaths; scape erect, $1\frac{1}{2}$ ft. high, with 3–4 distant sheaths ; raceme somewhat dense, 8–12-flowered ; bracts linear-lanceolate, acuminate, $\frac{3}{4}$–1 in. long ; pedicels $\frac{1}{2}$ in. long ; flowers large ; sepals lanceolate, acuminate, over $\frac{3}{4}$ in. long ; petals elliptic-ovate, subobtuse, rather longer than the sepals and about 3 times as broad ; lip shorter and narrower than the petals, ovate, 3-lobed ; side lobes broadly rounded, short; front lobe ovate, subacute ; disc with numerous approximate slightly thickened and crenulate keels at the base ; spur slender, clavate, $\frac{1}{4}$ in. long ; column stout, over $\frac{1}{6}$ in. long.

KALAHARI REGION : Transvaal ; Houtbosch (Woodbush), *Rehmann*, 5845 !

44. E. latipetala (Rolfe) ; leaves 5–7 in a spreading tuft, broadly linear-oblong, attenuate, with 5–7 prominent veins, conduplicate at the base, $\frac{1}{3}$–1 ft. long, 1–$1\frac{1}{2}$ in. broad, with 2 or 3 short exterior sheaths ; scape erect, stout, 1 ft. high, clothed with numerous broad imbricate sheaths ; raceme somewhat dense, 6–12-flowered ; bracts linear-lanceolate, acuminate, $\frac{3}{4}$–1 in. long; pedicels $\frac{1}{2}$ in. long ; flowers large ; sepals ovate-oblong, acute, nearly $\frac{3}{4}$ in. long ; petals broadly ovate, subobtuse, as long as the sepals, over $\frac{1}{2}$ in. broad ; lip as long and nearly as broad as the petals, 3-lobed ; side lobes broadly rounded, short; front lobe suborbicular-oblong, obtuse ; disc with 3–5 somewhat thickened keels at the base, smooth ; spur very short and stout, obtuse, subconical ; column very stout, over $\frac{1}{6}$ in. long.

KALAHARI REGION : Transvaal ; Houtbosch (Woodbush) Mountains, Pietersburg district, 5800 ft., *Bolus*, 10975 !

45. E. subintegra (Rolfe) ; leaves elongate-lanceolate, acute or acuminate, plicate, $\frac{3}{4}$–1 ft. long, $1\frac{1}{4}$–$1\frac{1}{2}$ in. broad (base not seen) ; scape slender, with a few sheaths below (the upper 10 in. only seen) ; raceme lax, 4 in. long ; bracts linear-lanceolate, acuminate, $\frac{1}{2}$–$\frac{3}{4}$ in. long; pedicels $\frac{1}{2}$ in. long ; flowers rather large ; sepals linear-lanceolate, acuminate, $\frac{3}{4}$ in. long, brownish ; petals oblong or elliptic-oblong, subacute, as long as the sepals by over twice as broad, yellow ; lip entire or subentire, elliptic, subobtuse, shorter and broader than the petals ; disc slightly puberulous, with 7–9 slightly thickened keels below, terminating in numerous nerves above ; spur oblong, obtuse, curved, about ⅙ in. long; column clavate, over ⅙ in. long.

EASTERN REGION : Natal ; damp places at Oliviers Hoek, sources of Tugela River, 5000 ft., *Allison*, S !

46. **E. Woodii** (Schlechter in Engl. Jahrb. xx. Beibl. 50, 5) ; a glabrous erect herb, 12–16 in. high ; leaves 3 or 4 in a fascicle, erect, linear, acute, 4–12 in. long ; scape subterete, with loose adpressed elongate acute membranous sheaths ; spikes lax, several-flowered ; bracts erect or somewhat reflexed, membranous, linear, acute, equalling or exceeding the ovary ; sepals subequal, ovate, subacute, 1 in. long, nearly ½ in. broad in the middle ; petals ovate-lanceolate, somewhat obtuse, rather shorter and much narrower than the sepals ; lip as long as the petals, oblong, 3-lobed, with two short calli from the mouth of the spur to the front lobe ; side lobes short, rounded ; front lobe oblong, obtuse or truncate ; spur very short, incurved, cylindrical ; column slender, carinate behind ; anther subglobose, shortly apiculate.

EASTERN REGION : Natal ; near Berlin Mission Station, *Wood*, 3577.

Only known to me from description.

47. **E. Meleagris** (Reichb. f. in Linnæa, xx. 683) ; rhizome stout ; leaves in a fascicle of 4 to 6, linear-oblong to elongate-lanceolate, acute, with 3 to 5 prominent veins, membranous, ½–1 ft. long, ¾–1¼ in. broad, with 2 or 3 exterior reduced sheaths at the base ; scapes stout, 1 to over 1½ ft. high, with several spathaceous lanceolate sheaths ; raceme about 6 in. long, loosely many-flowered ; bracts lanceolate or linear-oblong, acute, ⅓–¾ in. long ; pedicels ½–¾ in. long ; sepals more or less spreading, oblong, rather broader towards the apex, apiculate, rather fleshy, about ½ in. long ; petals suborbicular-oblong, apiculate, half as long as the sepals ; lip 3-lobed, rather longer than the petals ; side lobes obliquely semiovate, acute, ⅙ in. long ; front lobe oblong, acute, disc with 3 prominent keels, markedly crenulate in front ; spur cylindrical, rather stout, obtuse, somewhat curved, ⅙ in. long ; column stout, strongly angled, nearly as long as the petals. *Bolus in Journ. Linn. Soc.* xxv. 183 ; *Durand & Schinz, Conspect. Fl. Afr.* v. 23.

SOUTH AFRICA : without locality, *Krebs* !

COAST REGION : Stockenstrom Div. ; among shrubs at Benholm near Stockenstrom, *Scully*, 189 ! *and in Herb. Bolus*, 5916 !

EASTERN REGION : Griqualand East ; near Fort Donald, 5000 ft., *Tyson*, 1611 !

48. **E. antennata** (Schlechter in Engl. Jahrb. xxvi. 334) ; rhizome subterranean, oblique, ovoid-oblong, subcylindrical or slightly de-

pressed (*Schlechter*); leaves reduced to short linear acuminate
sheaths situated at the base of the stem, pallid (*Schlechter*); stems
generally fascicled (*Schlechter*), slender, somewhat flexuous, 1–2 ft.
high, with a few narrow sheaths below; racemes lax, 6–8 in. long,
8–12-flowered; bracts ovate-lanceolate, acute, $\frac{1}{6}$–$\frac{1}{4}$ in. long; pedicels
slender, $\frac{1}{3}$ to over $\frac{1}{2}$ in. long; flowers pallid (*Schlechter*); sepals linear,
acute, very narrow at the base, over $\frac{3}{4}$ in. long, all more or less
erect, 1-nerved; petals oblong, apiculate, subconnivent, 3-nerved,
under $\frac{1}{2}$ in. long; lip 3-lobed, over $\frac{1}{2}$ in. long; side lobes oblong,
obtuse or rounded at the apex; front lobe suborbicular or obovate-
oblong, undulate, over $\frac{1}{8}$ in. long; disc with 5 crenulate keels, more
prominent and verrucose or papillose in front; spur oblong, obtuse,
somewhat diverging, over $\frac{1}{6}$ in. long; column clavate, $\frac{1}{4}$ in. long.

EASTERN REGION: Delagoa Bay; near the Inkomati River, opposite Incanhini
Island, 50 ft., and near Massinga in the Inhambane region at 100 ft., *Schlechter*;
Inyamasan, *Schlechter*, 12075!

Schlechter describes this as allied to the Tropical African *E. galeoloides*, Kränzl.,
and *E. gastrodioides*, Schlechter, and remarks that all three are pallid and half-
saprophytic leafless herbs.

49. E. Cooperi (Reichb. f. in Flora, 1881, 330); leaves about
4 or 5 in a fascicle, ensiform, acuminate, immature at flowering time, with ample oblong imbricate acute sheaths at the
base; scapes lateral, $\frac{1}{2}$–$\frac{3}{4}$ ft. high, with several ample spathaceous
acuminate sheaths; racemes subcorymbose, dense; bracts ovate-
lanceolate, acuminate, about $\frac{3}{4}$ in. long; pedicels about $\frac{1}{2}$ lin. long;
flowers rather large, green and purple; sepals oblong-lanceolate,
acute, nearly 1 in. long; petals elliptic-lanceolate, acute, broader
than the sepals; lip broadly elliptic, 3-lobed, $\frac{3}{4}$ in. long; side lobes
broad, with rounded apex; front lobe with rounded apex, nearly as
broad as long; disc with 5 to 7 verrucose keels, slightly papillose in
front; spur broadly oblong, truncate, about $\frac{1}{10}$ in. long; column
clavate, $\frac{1}{3}$ in. long. *Bolus in Journ. Linn. Soc.* xxv. 184; *Durand
& Schinz, Conspect. Fl. Afr.* v. 21.

KALAHARI REGION: Orange River Colony; without precise locality, *Cooper*,
977!

50. E. Zeyheri (Hook. f. Bot. Mag. t. 7330); rhizomes stout,
woody, thickened at the nodes; leaves 2 to 3 in a fascicle, elongate-
lanceolate or linear-lanceolate, acuminate, strongly veined, $\frac{3}{4}$–$1\frac{1}{2}$ ft.
long, $\frac{1}{4}$–$1\frac{1}{4}$ in. broad, with 2 or 3 ovate-lanceolate sheaths at the
base; scapes lateral, stout, 1–$1\frac{3}{4}$ ft. high, with several loose
spathaceous sheaths; racemes very short and dense, subcorymbose
when young; bracts linear or lanceolate-linear, acuminate, $\frac{3}{4}$–$1\frac{1}{4}$ in.
long; pedicels $\frac{1}{2}$–$\frac{3}{4}$ in. long; flowers large, yellow, with a deeper
yellow disc to the lip and dark purple side lobes; sepals and petals
ovate-lanceolate or elliptic-oblong, shortly acuminate, 1–$1\frac{1}{4}$ in. long;
lip broadly elliptic, about as long as the petals, 3-lobed, side lobes
short, broadly rounded; front lobe suborbicular, obtuse, very broad;

disc with about 11 to 13 slightly thickened veins, bearing numerous
long slender filaments, and 2 crenulate keels at the base; spur
oblong, obtuse, ¼ in. long; column clavate-oblong, stout, ¼ in. long.
E. bicolor, Reichb. f. & Sond. in Flora, 1865, 186, *partly, not of
Blume*; *Bolus in Journ. Linn. Soc.* xxv. 184, *and Ic. Orch. Austr.-
Afr.* ii. *t.* 24; *Durand & Schinz, Conspect. Fl. Afr.* v. 19. *E. zeyheriana, Schlechter in Engl. Jahrb.* xx. *Beibl.* 50, 25, *not of Sond.*

KALAHARI REGION: Basutoland, *Bryce*! Transvaal; Magaliesberg Range,
Burke, 336! near Bronkhorst Spruit, *Wilms,* 1549! ridges near Johannesburg,
5000 ft., *Baker*! *Rand,* 1096! *Ommanney,* 75! 76! *Mainwaring,* 1042! near
Lydenburg, 4900 ft., *Schlechter,* 3940! Elandsfontein. near Johannesburg,
5500 ft., *Gilfillan in Herb. Galpin,* 1428! Rustenburg, 4500 ft., *Nation,* 34! near
Potchefstroom, *Nelson,* 331! Ermelo, *Collins,* 6339! *Burtt-Davy,* 5476a! Pilgrims
Rest, *Greenstock*! Kudus Poort, Pretoria, *Rehmann,* 4690! Heidelberg district,
Burtt-Davy, 5630! 9132! *Miss Leendertz,* 1038.

EASTERN REGION: Tembuland; Bazeia, 2000 ft., *Baur,* 809! near the
Chwenka River, between Maclear and Umtata, 3700 ft., *Bolus,* 8736. Griqualand
East, 2500-5000 ft., *Tyson,* 1591. Natal; near Maritzburg, *Sanderson,* 826!
Inanda, *Wood,* 1070! Lidgetton, 3000-4000 ft., *Wood,* 7921! Oliviers Hoek,
source of Tugela River, 4000 ft., *Allison*! Ginginklovu, up to 200 ft., *Haygarth
in Herb. Wood,* 11780! *Wylie in Herb. Wood,* 11782! and without precise
locality, *Sanderson*!

51. **E. ensata** (Lindl. in Bot. Reg. t. 1147); rhizome very stout
and woody, much swollen at the nodes; leaves in fascicles of 2 to 4,
linear or ensiform, acute or acuminate, strongly veined, conduplicate
at the base, 1-1½ ft. long, ¼-½ in. broad; scapes lateral, erect, stout,
with numerous loose imbricate spathaceous sheaths, 1¼-2 ft. long;
racemes short and very dense, corymbose when young; bracts linear
or linear-lanceolate, acuminate, ¾-1 in. long; pedicels rather shorter
than the bracts; flowers light yellow, with the disc and fringes of
the lip deep yellow; sepals and petals subconnivent, oblong-lanceolate or elliptic-lanceolate, acute or acuminate, about 1 in. long; lip
broadly elliptic, 3-lobed, rather shorter than the petals; side lobes
broadly rounded, front lobe broadly elliptic-oblong, obtuse; disc
with numerous strongly fringed keels; spur linear-oblong, obtuse,
somewhat curved, about ¼ in. long; column clavate-oblong, ¼ in.
long. *Lindl. Gen. & Sp. Orch.* 183; *Drège, Zwei Pfl. Documente,* 141,
145; *Krauss in Flora,* 1845, 305; *Beitr. Fl. Cap- und Natal.* 157;
Bolus in Journ. Linn. Soc. xxv. 183, *and Ic. Orch. Austr.-Afr.* ii.
t. 26; *Durand & Schinz, Conspect. Fl. Afr.* v. 21.

SOUTH AFRICA: without locality, *Prior*!
COAST REGION: Bathurst Div.; between Kowie and Kap River, *Drège*! near
Kowie, *Drège,* 4571! Between Bathurst and the Kowie River, *Atherstone*!
Kasouga River, *MacOwan*! 717! near the Karega River, 500 ft., *Zeyher,* 2,
Pappe! Albany Div.; Flats near Seven Mountains, *Hutton*! Komgha Div.; near
the mouth of the Kei River, 200 ft., *Flanagan,* 1030.
EASTERN REGION: Transkei; Krielis Country, *Bowker*! between Gekau (Geua)
River and Bashee River, 1000-2000 ft., *Drège.* Kentani, 1200 ft., *Miss Pegler,*
301! Pondoland; margins of woods near Egossa, 1000 ft., *Tyson,* 2842! Natal;
near Attercliffe and Verulam, 200-700 ft., *Sanderson,* 108! The Bluff, Wentworth,
and near Durban, *McKen,* 7! *Miss Doidge,* 5880! Attercliffe, *Sanderson,* 494!

Wood, 708 ! Inanda, *Buchanan*, 7 ! *Wood*, 275 ! 383 ! near Natal Bay, *Krauss*; near Durban, *Sanderson*, 25 ! Alexandra District, *Rudatis*, 242 ! and without precise locality, *Mrs. K. Saunders* ! *Wood*, 1020 ! *Sanderson*, 108 ! 494 ! 4000 ft., *Allison*, L iii ! Swaziland ; Hlakikulu, 4100 ft., *Miss Stewart*, 53 ! near Bremersdorp, 2300 ft., *Bolus*, 12304.

The locality "Sierra Leone," suggested by Lindley for this species when originally describing it, is clearly erroneous.

52. E. leontoglossa (Reichb. f. in Flora, 1881, 329) ; tubers sub-

globose, about $\frac{3}{4}$ in. broad; leaves 2 or 3 in a fascicle, linear or lanceolate-linear, acute or acuminate, 4–14 in. long, $\frac{1}{3}$–$\frac{1}{2}$ in. broad ; scapes erect, 4–12 in. long, with a few lanceolate acuminate sheaths below ; flower-heads congested or rarely oblong, 1–2 in. long ; bracts linear or linear-lanceolate, acuminate, $\frac{1}{2}$–$\frac{3}{4}$ in. long ; pedicels $\frac{1}{4}$–$\frac{1}{3}$ in. long ; flowers yellow, greenish-buff, pink or white ; sepals and petals subconnivent, lanceolate-oblong, subobtuse or apiculate, about $\frac{1}{3}$–$\frac{2}{3}$ in. long ; lip 3-lobed, elliptic-oblong, narrowed at the base, about as long as the petals ; side lobes somewhat divergent, oblong, obtuse or truncate, short ; front lobe elliptic-oblong, obtuse ; disc with 5 obscure keels below, papillose above, and with the surface of the front strongly papillose all over ; spur oblong or subclavate, obtuse, $\frac{1}{6}$ in. long ; column clavate, $\frac{1}{4}$ in. long. *Bolus in Journ. Linn. Soc.* xxv. 184 ; *Durand & Schinz, Conspect. Fl. Afr.* v. 22 ; *Schlechter in Engl. Jahrb.* xx. *Beibl.* 50, 25 ; *Kränzl. in Ann. Naturhist. Hofmus. Wien*, xx. 12. *E. lissochiloides, Krauss in Flora*, 1845, 305 ; *Beitr. Fl. Cap- und Natal.* 157, *not of Lindl.*

KALAHARI REGION : Orange River Colony ; Bethlehem, *Richardson* ! Witte Berg, *Mrs. Barber*, 647 ! Basutoland ; near Leribe, *Dieterlen*, 405 ! Transvaal ; Magaliesberg, *Burke* ! near Lydenburg, *Atherstone* ! *Schlechter*, 3971 ! Bronkhorst Spruit, *Wilms*, 1399 ! Mac Mac, *Mudd* ! Pretoria, *Crawley* ! *Burtt-Davy*, 3851 ! Saddleback Mountain, Barberton, 4000–5000 ft., *Galpin*, 720 ! Ivy Range, near Barberton, *Thorncroft*, 382 ! Crocodile River, Limpopo Sources, *Nelson*, 272 ! Ermelo, *Wilkinson* ! *Burtt-Davy*, 992 ! 2162 ! Johannesburg, *Ommanney*, 41 !

EASTERN REGION : Tembuland ; Umnyolo, 3000 ft., *Baur*, 811 ! Griqualand East ; near Kokstad, 5000 ft., *Tyson*, 1538 ! Natal ; near Durban, *Plant*, 53 ! *Sanderson* ! near Pietermaritzburg, *Krauss* ! 251 ! Inanda, *Wood*, 1085 ! in valley below Spout, *Wood*, 1069 ! between Greytown and Newcastle, *Wilms*, 2282 ! Klip River, *Sutherland* ! Tongaat, *McKen* ! Umbogintwini River and near Attercliffe, *Sanderson*, 173 ! hills near Pinetown, 1000 ft., *Wood*, 538 ! 5479 ! *Sanderson*, 483 ! *Schlechter*, 3171 ! near Van Reenen, Drakensberg Mountains, 5000–6000 ft., *Wood*, 7545 ! hills, Polela Division, *Mrs. Clarke*, 49 ! and without precise locality, *Buchanan* ! *Mrs. K. Saunders* ! *Wood* ! *Last* !

53. E. stenantha (Schlechter in Engl. Jahrb. xx. Beibl. 50, 27) ;

scape erect, about 10 in. high, with numerous erect acute sheaths ; spike corymbose, many-flowered ; bracts erect, narrow, acute, exceeding the flowers ; flowers spreading, long-pedicelled ; sepals lanceolate, acute, subequal, $\frac{3}{4}$ in. long ; petals obliquely oblong-lanceolate, shortly acute, as long as the sepals and rather broader ; lip as long as the sepals, 3-lobed, cuneate at the base ; side lobes very small ; front lobe ovate, obtuse, large, crenulate-undulate at the margin ; disc with the central nerves thickened from the base, papillose-verrucose in front; spur short, obtuse; column not half as long as the lip, with a rather long foot; anther shortly apiculate.

KALAHARI REGION : Transvaal ; stony cliffs on the Elandspruit Mountains, 7000 ft., *Schlechter*, 4004.

Only known to me from the description.

54. E. fragrans (Schlechter in Engl. Jahrb. xx. Beibl. 50, 27); leaves several in a fascicle, not developed until after flowering, much shorter than the scape, linear, acute, somewhat spreading; scape erect, 12–16 in. high, covered with short acute or acuminate cucullate sheaths; spikes dense, subcorymbose or oblong, several-flowered ; bracts ovate-lanceolate, acute, somewhat spreading, shorter than the pedicels ; flowers whitish-yellow, among the largest in the genus, somewhat spreading, deflexed after flowering ; sepals subequal, lanceolate, acute, over $\frac{3}{4}$ in. long, $\frac{1}{4}$ in. broad ; petals ovate, acute, as long as the sepals, $\frac{1}{2}$ in. broad ; lip with cuneate base, 3-lobed ; side lobes short, acute ; front lobe ovate, obtuse or subacute, twice as long as the side lobes ; disc with about 6 thickened subcristate nerves ; column over half as long as the lip ; foot produced and forming with the base of lip a short conical spur ; anther emarginate at the apex.

KALAHARI REGION : Transvaal, near Heidelberg, 5000 ft., *Schlechter*, 3531 !

55. E. Sankeyi (Rolfe) ; rhizome not seen ; leaves in a fascicle of 3 or 4, oblong-linear, acute, with about 5 prominent veins, 4–6 in. long, often somewhat recurved, with 1 or 2 short basal sheaths ; scapes about $\frac{3}{4}$ ft. long, stout, with several broadly oblong-lanceolate somewhat imbricate sheaths ; racemes about 4 in. long, somewhat lax, 10–12-flowered ; bracts ovate-lanceolate or elliptic-lanceolate, shortly acuminate, $\frac{1}{2}$–$\frac{3}{4}$ in. long ; pedicels about $\frac{3}{4}$ in. long; flowers cream-coloured, rather large ; sepals ovate-lanceolate, acute, $\frac{3}{4}$–1 in. long, dorsal rather broader than the lateral ; petals ovate, acute or abruptly and shortly acuminate, about as long as the sepals and twice as broad ; lip ovate, shortly 3-lobed, rather smaller than the petals ; side lobes short, rounded at the apex ; front lobe broadly ovate, apiculate or subacute, somewhat undulate ; disc with 5–7 thickened verrucose keels from the base to beyond the middle ; spur oblong, subobtuse, about $\frac{1}{4}$ in. long ; column clavate, over $\frac{1}{4}$ in. long, with a short broad foot ; anther minutely apiculate.

KALAHARI REGION : Orange River Colony ; Harrismith, *Sankey*, 306 !

56. E. litoralis (Schlechter in Engl. Jahrb. xxvi. 338); tubers oblique, thick, 2–3 in. long ; leaves absent at flowering time (not seen) ; scapes lateral, stout, 1–1$\frac{1}{2}$ ft. high, with several narrow spathaceous sheaths ; bracts oblong-lanceolate, acuminate, $\frac{3}{4}$–1 in. long ; pedicels about $\frac{3}{4}$ in. long ; flowers yellow, rather large ; sepals oblong-lanceolate, acuminate, about $\frac{3}{4}$ in. long ; petals elliptic-oblong, acute, about as long the sepals ; lip broadly elliptic, 3-lobed, rather broader than the petals ; side lobes broad, with acute apex ; front lobe obovate-orbicular, emarginate ; disc with about 9 rows of papillæ in front, and a pair of slightly thickened keels behind ; spur oblong, obtuse, $\frac{1}{8}$ in. long ; column clavate, nearly $\frac{1}{2}$ in. long ; anther apiculate.

COAST REGION : Caledon Div. ; sand dunes near Hawston, at the mouth of the Bot River, *Schlechter*, 9468 !

57. E. nigricans (Schlechter in Engl. Jahrb. xx. Beibl. 50, 5, 26) ; a glabrous erect herb, 1–1$\frac{1}{2}$ in. high ; leaves 2, erect, linear, acute ; scape subterete, densely covered with membranous cucullate acute sheaths ; spikes lax, many-flowered ; bracts suberect, linear-lanceolate, acute, $\frac{3}{4}$–1$\frac{1}{2}$ in. long ; sepals subequal, lanceolate, acute, $\frac{3}{4}$–1 in. long ; petals rather shorter and broader than the sepals ; lip nearly as long as the petals, 3-lobed ; side lobes short, subtriangular, usually acute ; front lobe ovate-oblong, obtuse ; disc verrucose or with very short papillæ ; spur very short, conical, obtuse ; column clavate, $\frac{1}{3}$ in. long, prolonged at the base into a horizontal foot $\frac{1}{4}$ in. long, to which the lateral sepals are partially attached at the base ; anther subglobose, apiculate ; pollinia with two distinct short caudicles.

KALAHARI REGION : Transvaal, 4700 ft., *Schlechter*, 4147 !
EASTERN REGION : Natal ; near Inanda, *Wood*, 335 !

58. E. inandensis (Rolfe) ; rhizome and leaves not seen ; scape nearly 1 ft. long, with several short oblong-lanceolate sheaths below ; raceme 3 in. long, 7–8-flowered ; bracts lanceolate, acuminate, $\frac{1}{3}$–$\frac{1}{2}$ in. long ; pedicels over $\frac{1}{2}$ in. long ; flowers medium-sized, yellow and brown ; sepals elliptic-oblong, apiculate, over $\frac{1}{2}$ in. long ; petals elliptic, subobtuse, rather shorter and much broader than the sepals ; lip ovate, obtuse, subentire, about as long as the petals ; disc with 3–5 thickened and slightly verrucose keels ; spur conical, obtuse, nearly $\frac{1}{4}$ in. long ; column clavate, over $\frac{1}{4}$ in. long ; anther obtuse.

EASTERN REGION : Natal ; Inanda, *Wood*, 976 !

59. E. bilamellata (Schlechter in Engl. Jahrb. xx. Beibl. 50, 1, 26) ; a glabrous terrestrial herb, about 1 ft. high ; leaves about 5, somewhat spreading, linear, acute, short ; scape erect, subterete, with loose broadly ovate acuminate sheaths ; racemes lax, many-flowered ; bracts membranous, somewhat spreading, ovate, acuminate, shorter than the pedicel ; dorsal sepal lanceolate, acute, $\frac{2}{3}$ in. long, $\frac{1}{4}$ in. broad ; lateral sepals somewhat spreading, obliquely ovate-lanceolate, subacute, about $\frac{3}{4}$ in. long, $\frac{1}{3}$ in. broad ; petals oval, subacute, narrowed at the base, thickened at the apex, about $\frac{1}{2}$ in. long, $\frac{1}{3}$ in. broad ; lip oblong, obscurely 3-lobed, concave, smooth, as long as the petals ; side lobes very short, obtusely truncate ; front lobe ovate, rounded at the apex, margin undulate ; disc with 2

short parallel keels ; spur pyramidal, incurved, obtuse at the apex,
half as long as the lip ; column short ; anther rounded.

VAR. β, **euryceras** (Schlechter in Engl. Jahrb. xx. Beibl. 50, 26) ; lip destitute
of keels ; mouth of spur broader.

KALAHARI REGION : Transvaal ; near Johannesburg, *Endemann* ! near Greyling-
stad, 6000 ft., *Vandeleur* ! rocky hill-slopes near Barberton, 3500 ft., *Culver*, 32 !
Saddleback Mountain, near Barberton, 4000–5000 ft., *Galpin*, 546 ! Warmbaths,
Miss Leendertz, 1300 !
EASTERN REGION : Natal ; among grasses near Inchanga, 700 ft., *Marloth*,
4170 ! Zululand ; near Dunns Coast, *Mrs. C. Saunders* ! Swaziland ; Embabaan,
Miller, 3012 ! Var. β : Natal ; near Emberton, among grasses, 1800 ft., *Schlechter*,
3232.

60. **E. oliveriana** (Bolus in Journ. Linn. Soc. xxv. 185) ; rhizome
stout, thickened at the nodes ; leaves 4–5 in a fascicle, small at
flowering time, ensiform-linear, acute, 6 in. long, closely veined, with
2 or 3 lanceolate exterior sheaths ; scape lateral, 1–2 ft. long, with
several oblong-lanceolate sheaths ; raceme 3–6 in. long, compact or
somewhat lax, many-flowered ; bracts oblong-lanceolate, acute or
acuminate, ½–¾ in. long ; pedicels ½–¾ in. long ; flowers medium-
sized, yellow and brown ; sepals elliptic or elliptic-oblong, obtuse,
about ¼ in. long ; petals broadly elliptic, rather shorter and broader
than the sepals ; lip about as long as the petals, strongly 3-lobed ;
side lobes with obtuse or rounded apex, narrowed behind ; front
lobe oblong or narrowly oblong, obtuse ; disc convex, with 3–5
thick fleshy verrucose keels ; spur very short, broadly conical ;
column broadly oblong, over ¼ in. long, with a distinct foot.
Durand & Schinz, Conspect. Fl. Afr. v. 24 ; *Bolus, Ic. Orch. Austr.-*
Afr. ii. *t.* 10. *Cyrtopera oliveriana, Reichb. f. in Flora,* 1881, 329.

COAST REGION : Stockenstrom Div. ; summit of Kat Berg, 5000 ft., *Galpin*,
1684. King Williamstown Div. ; Pirie, near King Williamstown, 4000 ft., *Sim*,
1268.
KALAHARI REGION : Orange River Colony ; Elands River Valley and Besters
Vallei, near Harrismith, 6000 ft., *Flanagan*, 1986, *Bolus*, 8300. Transvaal ;
Mount Sheba, near Barberton, 3700 ft., *Bolus*, 9787 ! Hooge Veld, between
Middelburg and Elands Spruit Mountains, 6800 ft., *Schlechter.*
EASTERN REGION : Tembuland ; Umnyolo, near Bazeia, 3000 ft., *Baur*, 754 !
Griqualand East ; Vaal Bank, *Haygarth in Herb. Wood*, 4202. Natal ; Inanda,
Wood, 279 ! 659 ! Liddesdale, 5000 ft., *Wood*, 4259 ; Van Reenens Pass, 5000–
6000 ft., *Wood*, 7545 ; Inswatzi, *Wood*, 11781 ! Gillits, 1950 ft., *Wood*, 11788 !
Niginya, *Wylie in Herb. Wood*, 11824 ! and without precise locality, *Buchanan*,
13 !

61. **E. rupestris** (Reichb. f. in Linnæa, xx. 682) ; leaves lanceolate,
acuminate, with 3 prominent veins, curved, distichous, 6 in. long ;
scape about 1½ ft. high, with about 5 lanceolate sheaths ; racemes
about 3 in. long, dense-flowered ; bracts ovate-lanceolate, acute, ¾ in.
long ; flowers brown, the inner segments paler ; sepals oblong, acute,
½ in. long by about half as broad ; petals ovate, acute, shorter than
the sepals ; lip 3-lobed, united to the foot of the column and produced
into a short mentum, ⅓ in. long ; side lobes quadrate, obtuse ; front
lobe oblong, acute, involute and crisped at the margin ; disc with

elevated papillose nerves on the upper half ; column semiterete, with inflexed wings and a pair of small basal angles. *Bolus in Journ. Linn. Soc.* xxv. 183 ; *Durand & Schinz, Conspect. Fl. Afr.* v. 25. *E. rupincola, Reichb. f. in Bonplandia,* 1857, 38.

COAST REGION : Div. ? ; Pardekop, *Mund & Maire.*

Only known to me from the description.

Not to be confounded with *E. rupestris,* Lindl., which is a synonym of the Indian *E. campestris,* Wall.

62. E. inæqualis (Schlechter in Engl. Jahrb. xx. Beibl. 50, 3, 26) ; leaves about 4 in a fascicle, somewhat spreading, linear-oblong, subacute, 2–4 in. long ; scape stout, 6–14 in. high, with about 3 short spathaceous acuminate sheaths ; raceme lax, few- or many-flowered ; bracts ovate-oblong, acuminate, nearly half as long as the pedicels ; flowers erect, yellow with dull purple sepals ; sepals subequal, oblong-lanceolate, acute, about $\frac{1}{2}$ in. long, $\frac{1}{8}$ in. broad ; petals ovate-oblong, subacute, rather shorter than the sepals and almost twice as broad ; lip as long as the petals, 3-lobed ; side lobes oblong, subobtuse, short ; front lobe rounded, concave, with crenulate-undulate margin ; disc with 2 crested keels at the base, and more numerous papillose crests in the centre ; spur clavate, short ; column slender, produced into a long foot ; anther rounded, emarginate.

EASTERN REGION : Natal ; near Pinetown, *Mrs. Button in Herb. Sanderson,* 1011 ! among grasses near Ladysmith, 4200 ft., *Schlechter,* 3431.

This species was based upon *Sanderson,* 1011, from Natal, of which the original drawing is at Kew, and from which the above description has been taken.　I have not seen the specimen afterwards added by Schlechter, but from the reference to *E. hians,* Spreng., I suspect that it may be different, for some specimens of the latter had been lubelled *E. inæqualis,* Schlechter.

63. E. Peglerse (Rolfe) ; rhizome and leaves not seen ; scape with a few lanceolate bracts ; raceme short, about 6-flowered ; bracts oblong-lanceolate, acuminate, about $\frac{1}{2}$ in. long ; pedicels rather longer than the bracts ; flowers medium-sized ; sepals oblong-lanceolate, acute, over $\frac{2}{3}$ in. long ; petals oblong-lanceolate, subacute, rather shorter than the sepals ; lip deeply 3-lobed, about $\frac{1}{2}$ in. long ; side lobes oblong, obtuse, somewhat diverging, nearly as long as the front lobe ; front lobe obovate or broadly obovate-oblong, obtuse, $\frac{1}{4}$ in. long ; disc puberulous, with a central thickened slightly verrucose keel, ending abruptly about the middle of the front lobe, more slender towards the base ; spur stout, obtuse, very short ; column clavate, over $\frac{1}{4}$ in. long, with a foot one-third its length.

EASTERN REGION : Transkei ; Kentani, *Miss Pegler in Herb. Bolus,* 10677 !

64. E. papillosa (Schlechter in Engl. Jahrb. xx. Beibl. 50, 25, in note) ; a slender erect very glabrous herb ; leaves in a fascicle of 2 to 3, erect, flaccid, linear, acute, 1–2 ft. long, $\frac{1}{4}$–$\frac{1}{2}$ in. broad ; scape lateral, erect, $1\frac{1}{2}$–$3\frac{1}{4}$ ft. high, with numerous membranous acute sheaths; spikes cylindric or pyramidal, dense, many-flowered; bracts somewhat spreading, lanceolate, aristate, rather shorter than the pedicels; flowers rather small, yellow, sometimes with purple side lobes to the lip; sepals somewhat spreading, subequal, ovate-oblong, acute, $\frac{1}{3}$–$\frac{1}{2}$ in. long; petals ovate-elliptic, acute, rather shorter than the sepals; lip as long as the petals, broadly oblong, concave, 3-lobed; side lobes short, semiovate, obtuse, erect; front lobe subquadrate, somewhat narrowed at the base, emarginate at the apex; disc with a few papillæ in front, and a pair of short parallel lamellæ at the base; column about half as long as the lip, base produced into a short foot, and forming with the base of the lip an obtuse mentum. *E. chrysantha, Schlechter in Engl. Jahrb.* xx. *Beibl.* 50, 2. *Cyrtopera papillosa, Rolfe in Kew Bulletin,* 1893, 336.

KALAHARI REGION: Transvaal; Barberton, *Thorncroft,* 4857!

EASTERN REGION: Transkei; swamps by the mouth of the Bashee River, *Bowker,* 453! Natal; marshes near Inanda, and near Verulam, *Wood,* 785! hill near Botha's, *Wood,* 942! Shafton, Howick, *Mrs. Hutton,* 239! Karkloof, 200 ft., *Wood in MacOwan Herb. Austr.-Afr.,* 1532! near Durban, *Gerrard & McKen,* 745! 2174! Dargle Farm, *Mrs. Fannin,* 130! Congella Flat, Umbilo River, *Sanderson,* 497! Dumisa, 2000 ft., *Rudatis,* 243! and without precise locality, *O'Brien! Sanderson,* 63! 829! *Buchanan! Mrs. K. Saunders!*

65. **E. aurea** (Kränzl. in Bull. Herb. Boiss. v. 635); tubers subterranean, thickened; leaves at flowering time immature, grass-like, acuminate, 10–12 in. long, about $\frac{1}{3}$ in. broad, with a few obtuse or acute sheaths at the base; scapes slender, 1$\frac{3}{4}$–2 ft. long, somewhat flexuous above; raceme short, one-sided, few-flowered; bracts rhomboid, acuminate or aristate, convolute, shorter than the ovary; flowers deep yellow; dorsal sepal ovate-oblong, obtuse, about $\frac{1}{3}$ in. long; lateral sepals broadly ovate, obtuse, produced into an obtuse mentum at the base; petals oblong, subacute, shorter than the sepals; lip 3-lobed; side lobes spreading, acute; front lobe spathulate, suborbicular in front, margin crenulate or dentate; disc with radiating keels from the middle almost to the margin; spur very short; column nearly half as long as the sepals.

EASTERN REGION: Delagoa Bay; marshy places, in white sand, *Junod,* 187.

Only known to me from the description.

66. **E. junodiana** (Kränzl. in Bull. Herb. Boiss. v. 634); stem leafy, with spathaceous basal sheaths, which are dotted with red or black; leaves linear-lanceolate, acuminate (immature), 6 in. long, over $\frac{1}{3}$ in. broad; scape 1$\frac{1}{4}$–2 ft. long, flexuous, with a few short sheaths; racemes few- to many-flowered; bracts lanceolate, acute, minute; sepals oblong, acute or obtuse, $\frac{1}{3}$ in. long, green; petals rather broader than the sepals, otherwise similar; lip purple with darker veins, obsoletely 3-lobed, compressed, cordate-ovate, recurved at the apex, margin crisped or undulate; side lobes nearly obsolete; disc with 2 thickened undulate keels extending nearly to the apex, and a smaller central and 2 outer keels; spur broad, compressed,

didymous, strongly incurved ; column slightly curved, acute, half as long as the sepals.

EASTERN REGION : Delagoa Bay, *Junod*, 122.

Only known to me from the description, in which it is compared with the Madagascar *E. pulchra*, Lindl.

67. **E. aculeata** (Spreng. Syst. iii. 720) ; rhizome creeping, rather stout, scaly, irregularly thickened between the nodes ; leaves 1 or 2, erect, linear or lanceolate-linear, acuminate, somewhat rigid, $\frac{1}{4}$–1$\frac{1}{4}$ ft. high, $\frac{1}{6}$–$\frac{1}{2}$ in. broad ; scapes erect, $\frac{1}{4}$–1$\frac{1}{2}$ ft. high, with several linear-lanceolate acuminate sheaths ; raceme subcapitate or oblong, up to 2$\frac{1}{2}$ in. long, dense, many-flowered ; bracts lanceolate or linear-lanceolate, acuminate, $\frac{1}{4}$–$\frac{1}{2}$ in. long ; pedicels $\frac{1}{8}$–$\frac{1}{3}$ in. long ; flowers cream-colour or yellow ; sepals subconnivent, oblong or oblong-lanceolate, subobtuse, $\frac{1}{3}$–$\frac{1}{2}$ in. long ; petals broadly elliptic, obtuse, about as long as the sepals ; lip 3-lobed, $\frac{1}{4}$–$\frac{1}{3}$ in. long by nearly as broad, spurless ; side lobes broadly oblong, obtuse, front lobe triangular-oblong, obtuse, slightly undulate, equalling or slightly exceeding the side lobes ; disc with 4 rows of papillæ extending to the apex of the lip ; spur obsolete ; column clavate, $\frac{1}{4}$–$\frac{1}{8}$ in. long. *Bolus in Trans. S. Afr. Phil. Soc.* v. 109 (*excl. syn. Reichb. f.*), *Orch. Cap. Penins.* 109, *partly, and in Journ. Linn. Soc.* xxv. 182. *Satyrium capense, Linn. Amœn. Acad.* vi. 110, *and Sp. Pl. ed.* 2, 1339. *S. aculeatum, Linn. f. Suppl.* 402. *S. pedicellatum, Linn. f. Suppl.* 402. *Serapias aculeata and S. pedicellata, Thunb. Prodr.* 3. *Cymbidium aculeatum, Sw. in Schrad. Journ.* ii. 225. *C. pedicellatum, Sw. in Schrad. Journ.* ii. 224. *C. plicatum, Harv. in Comp. Bot. Mag.* ii. 203 ; *Hook. Ic. Pl. t.* 104. *Cyrtopera pedicellata, Lindl. Gen. & Sp. Orch.* 190. *Eulophia pedicellata, Spreng. Syst. Veg.* iii. 720. *E. odontoglossa, Reichb. f. in Linnæa,* xix. 373, xx. 684 (*ex Reichb. f. in Flora,* 1883, 463). *E. plicata, Bolus in Journ. Linn. Soc.* xix. 336 (*excl. syn.* 3 *and* 4). *E. capensis, Bolus in Journ. Linn. Soc.* xix. 336, *and Ic. Orch. Austr.-Afr.* ii. *t.* 25 ; *Durand & Schinz, Conspect. Fl. Afr.* v. 20 ; *Kränzl. in Ann. Naturhist. Hofmus. Wien,* xx. 12.

SOUTH AFRICA : without locality, *Rogers* ! *Harvey* ! *Mrs. Holland,* 12 ! *Zeyher,* 1591 ! *Bunbury* !

COAST REGION : Cape Div. ; Cape Flats, *Pappe* ! near Rondebosch, *Burchell,* 222 ! *Bolus,* 3900 ! Table Mountain, *Zeyher* ! *Harvey* ! *Prior* ! 2200 ft., *Bolus,* 3900 ! *Burchell,* 653 ! lower plateau, *Wolley-Dod,* 2226 ! Stellenbosch Div. ; Hottentots Holland, near Sir Lowrys Pass, 900 ft., *Bolus,* 4207 ! Caledon Div. ; Little Houwhoek, *Zeyher* ! Riversdale Div. ; Kampsche Berg, *Burchell,* 7094 ! George Div. : George, *Penther,* 103. Knysna Div. ; Knysna Hills, *Bowie* ! Knysna, without collector ! Zitzikamma River, *Penther,* 298 ; Elands River, *Penther,* 299. Humansdorp Div. ; near Clarkson, *Penther,* 301. Uitenhage Div. : near Van Stadens River mountains and on the hills of Adow, *Zeyher,* 300 ! Zwartkops River, without collector, 257 ! foot of Witte Klip, *MacOwan,* 2129 ! Port Elizabeth Div. ; around Krakakamma, *Burchell,* 4571 ! Albany Div. ; Howisons Poort, *Mrs. Hutton* ! near Grahamstown, *Galpin,* 3087.

CENTRAL REGION : Somerset Div. ; Bosch Berg, 4800 ft., *MacOwan,* 1859 !

KALAHARI REGION: Transvaal: mountain tops, Saddleback, Barberton, 4000 ft., *Culver*, 86!

EASTERN REGION: Tembuland; Engcobo Mountain, 4500 ft., *Bolus*, 10294.

68. **E. Huttonii** (Rolfe); rhizome stout and woody, much thickened at the nodes; leaves 2 or 3 in a fascicle, erect, elongate-linear, acute or acuminate, closely veined, $\frac{1}{2}$–1 ft. long; scapes lateral, erect, $\frac{1}{2}$–1$\frac{1}{4}$ ft. long, with numerous lanceolate imbricate sheaths; raceme short, usually dense, 1–3 in. long; bracts lanceolate, acuminate, $\frac{1}{4}$ to over $\frac{1}{2}$ in. long; pedicels about $\frac{1}{2}$ in. long; flowers medium-sized, dull brown or red; sepals elliptic-lanceolate or oblong-lanceolate, acute or subacute, about $\frac{1}{2}$ in. long; petals elliptic-oblong, subobtuse, about as long as the petals and rather broader; lip as long as the petals, broader than long, subequally 3-lobed; side lobes somewhat diverging, broadly oblong, with rounded apex; front lobe suborbicular or broadly oblong, obtuse or rounded at the apex, with 3–5 prominent strongly crested keels from middle to apex, more slender behind; spur obsolete; column clavate, over $\frac{1}{4}$ in. long.

COAST REGION: Uitenhage Div.: Zwartkops River, without collector, 257! Stockenstrom Div.; Kat Berg, *Hutton*! Chumie (Tyumie) Peak, *Scully in Herb. Bolus*, 5917! Queenstown Div.; Winterberg Range, *Mrs. Barber*, 533! Stutterheim Div.; Dohne Peak, near Fort Cunynghame, *Bolus*, 10293!

KALAHARI REGION: Orange River Colony, without locality, *Cooper*!

EASTERN REGION: Griqualand East; Vaal Bank, *Haygarth in Herb. Wood*, 4202! near Kokstad, 5000 ft., *Tyson*, 1085. Natal; hill near Liddesdale, 5000 ft., *Wood*, 4259! Howick, 3000–4000 ft., *Wood*, 11818! 11819!

This has been more or less confused with the preceding and following species.

69. **E. foliosa** (Bolus in Journ. Linn. Soc. xix. 337); rhizome stout, irregularly thickened at the nodes; leaves 3 or 4 in a fascicle, erect, linear-lanceolate or elongate-lanceolate, acute, $\frac{1}{3}$–1$\frac{1}{2}$ ft. long; scape lateral, erect, $\frac{1}{2}$–1$\frac{1}{2}$ ft. long, with 3–4 lanceolate acute sheaths; racemes 2–3 in. long, dense, many-flowered; bracts linear or linear-lanceolate, acuminate, $\frac{1}{3}$–$\frac{3}{4}$ in. long; pedicels $\frac{1}{3}$–$\frac{1}{2}$ in. long; flowers medium-sized, green with purple lip; dorsal sepal oblong-lanceolate, acute, $\frac{1}{3}$–$\frac{1}{2}$ in. long, lateral sepals rather broader than the dorsal, attached to the foot of the column; petals elliptic or elliptic-lanceolate, rather broader than the sepals; lip shorter than the petals, strongly 3-lobed; side lobes diverging, broadly oblong, obtuse or rounded at the apex; front lobe suborbicular, rather larger than the lateral; disc with several rows of small papillæ; spur obsolete; column clavate, $\frac{1}{6}$–$\frac{1}{4}$ in. long. *Bolus in Journ. Linn. Soc.* xxv. 183; *Durand & Schinz, Conspect. Fl. Afr.* v. 21. *E. reichenbachiana, Bolus in Journ. Linn. Soc.* xxv. 185, *and Ic. Orch. Austr.-Afr.* ii. *t.* 19, *partly*; *Schlechter in Engl. Jahrb.* xx. *Beibl.* 50, 25. *E. aculeata, Bolus in Trans. S. Afr. Phil. Soc.* v. 109, *and Orch. Cape Penins.* 109, *partly, not of Spreng. E. Buchanani, Durand & Schinz, Conspect. Fl. Afr.* v. 19. *Cyrtopera foliosa, Lindl. in Hook. Comp. Bot. Mag.* ii. 203; *Drège, Zwei Pfl. Documente*, 147. *Cymbidium Buchanani, Reichb. f. in Flora*, 1881, 329.

KALAHARI REGION: Orange River Colony; without precise locality, *Cooper*, 975! 980! Transvaal; Houtbosch, *Rehmann*, 5841! near Heidelberg, *Wilms*, 1397a! west slope of Vaal Kop, *Scully*, 190! Carolina District, near Slackfontein Beacon, 6000 ft., *Burtt-Davy*, 2965! Elands Spruit Range, 6500 ft., *Schlechter*, 4001, near Belfast, 6500 ft., *Bolus*, 12305!

EASTERN REGION: Transkei; Kentani, 1200 ft., *Miss Pegler*, 204! Tembuland; mountain slopes near Bazeia, 2500 ft., *Baur*, 459! between Bashee River and Morley, 1000–2000 ft., *Drège*! Gatberg, 4000 ft., *Baur*, 1168! Griqualand East; near Kokstad, 5000 ft., *Tyson*, 1085! *and in MacOwan & Bolus, Herb. Norm. Austr.-Afr.* 548! Natal; Inanda, *Wood*, 263! Pinetown, 1000 ft., *Sanderson*, 484! Kelvĭn Grove, near Dundee, 4000–5000 ft., *Wood*, 5370! Van Reenens Pass, 5000–6000 ft., *Wood*, 5863. Bishopstowe, *Sanderson*, 1006! near Durban, *Miss Doidge*, 5928! Umgaye, Alexandra District, 2000 ft., *Rudatis*, 341! and without precise locality, *Mrs. K. Saunders*! *Buchanan*, 12! *Last*! Zululand; *Mrs. McKenzie*! Swaziland; Mbabane (Embabaan), 4500 ft., *Burtt-Davy*, 2769! Hlakikulu, *Miss Stewart*, 29! Devils Bridge, *Galpin*, 723.

70. E. Boltoni (Harv. MSS.); rhizome stout and woody, much thickened at the nodes; leaves 2–3 in a fascicle, erect, elongate-linear, acute, closely veined, $\frac{1}{3}$–1 ft. long, with 2 or 3 lanceolate exterior sheaths; scapes lateral, erect, $\frac{1}{2}$–1 ft. high, with numerous lanceolate, somewhat imbricate sheaths; racemes ovoid or oblong, dense, 1–2 in. long; bracts linear-lanceolate, acuminate, $\frac{1}{3}$–$\frac{1}{2}$ in. long; pedicels $\frac{1}{3}$ in. long; flowers medium-sized, green with dark purple lip; sepals ovate-lanceolate, acuminate, over $\frac{1}{3}$ in. long, lateral broader and rather longer than the dorsal; petals elliptic-lanceolate, acute, much shorter than the sepals; lip broadly ovate-orbicular, much narrowed at the base, somewhat 3-lobed, $\frac{1}{4}$ in. long; side lobes very short, broadly rounded; front lobe broadly ovate, apiculate, broader than long; disc with 3–5 slightly thickened nearly smooth keels; spur obsolete; column clavate, $\frac{1}{6}$ in. long, with a distinct foot.

COAST REGION: Albany Div.; in grassy spots on the flat summit of the hills, Featherstone Kloof, near Grahamstown, *MacOwan*, 681! *Bolton*! Fullers, near Grahamstown, *Reade*! *and in Herb. Bolus*, 1281! Stockenstrom Div.; Katberg, 2000 ft., *Hutton*!

The type specimen is in the Trinity College Herbarium, Dublin. The species has been more or less confused with the preceding.

71. E. tabularis (Bolus in Trans. S. Afr. Phil. Soc. v. 108, t. 1); rhizome creeping, stout, scaly, irregularly thickened between the nodes; leaves 1 or 2, radical, erect, linear-lanceolate, acute, 3–4 in. long at flowering time (older not seen); scape appearing with the young leaves and from the same sheath, erect, stout, $\frac{3}{4}$–1 ft. high, with 2 or 3 amplexicaul acute sheaths on the lower part, 3–10-flowered; bracts lanceolate, acuminate, $\frac{1}{4}$–$\frac{1}{2}$ in. long; pedicels rather shorter than the bracts; flowers dull yellow with an orange-coloured keel on the lip; sepals elliptic-oblong, obtuse or apiculate, somewhat spreading, lightly keeled behind, about $\frac{3}{4}$ in. long; petals rather narrower than the sepals and less spreading, otherwise very similar; lip 3-lobed, about $\frac{2}{3}$ in. long, not saccate at the base; side lobes involute, broadly rounded, obtuse or apiculate at the apex;

front lobe suborbicular, crenulate; disc with an obtuse fleshy
keel from its base to the middle of the front lobe, bidentate at the
apex; column clavate, ⅓ in. long, with a short foot; anther-case
obtuse.   *Bolus, Orch. Cape Penins.* 108, *t.* 1, *and Journ. Linn. Soc.*
xxv. 184; *Durand & Schinz, Conspect. Fl. Afr.* v. 26.   *Satyrium
tabulare, Linn. f. Suppl.* 402.   *Serapias tabularis, Thunb. Prodr.* 3.
Cymbidium tabulare, Sw. in Vet. Acad. Handl. Stockh. 1880, 238;
Bolus in Journ. Linn. Soc. xx. 471.

COAST REGION: Cape Div.; Table Mountain, at summit, *Thunberg*! *Harvey*!
above Klassenbosch, 2300 ft., and Muizenberg, 1400 ft., *Bolus*, 4844! Caledon
Div.; Genadendal, without collector! Swellendam Div.; summit of Craggy Peak,
near Swellendam, *Burchell*, 7358!

VI. LISSOCHILUS, R. Br.

Sepals subequal, free, spreading or reflexed, the lateral sometimes
adnate to the foot of the column.   *Petals* erect, generally much
larger and broader than the sepals and differently coloured.   *Lip*
continuous with the foot of the column, more or less distinctly
3-lobed; base variously saccate or spurred; side lobes erect or
spreading; middle one spreading or recurved; disc variously
cristate, lamellate or sometimes smooth.   *Column* erect, clavate,
more or less produced into a foot at the base; anther-bed oblique,
erect, entire.   *Anther* terminal, operculate, incumbent, semiglobose,
conical or more or less bilobed at the apex, imperfectly 2-celled;
pollinia 4, ovoid, united in pairs, affixed to a broad stipes and gland.
Capsules oblong or somewhat elongate, with prominent thickened
angles.

Terrestrial herbs; stems creeping, often thickened into rhizomes or tubers,
leafy at the base; leaves elongate, lanceolate or linear, plicate; scapes usually tall,
variously sheathed below, loosely racemose above; flowers generally medium-sized
or large, rarely small; bracts small or narrow.

DISTRIB.   Species about 100, exclusively African and mostly continental, so far
as at present known, two or three, however, occur in the Mascarene Islands.
Four of the 12 South African species occur in Tropical Africa.

Petals about ⅓ in. long:
  Spur oblong or linear-oblong:
    Spur straight, about ⅛ in. long   ...   . (1) **clitellifer.**
    Spur somewhat curved, ⅙ in. long   ...   . (2) **Rehmannii.**
    Spur shortly saccate or subobsolete   ...   . (3) **platypetalus.**
Petals about ½ in. long:
  Sepals subconnivent:
    Lip shorter than the sepals and strongly 3-lobed   ... (4) **æqualis.**
    Lip longer than the petals and obscurely 3-lobed   ... (5) **transvaalensis.**
  Sepals spreading or reflexed:
    Leaves elongate-linear, ¼–1 in. broad:
      Petals yellow, veined with brown ...   . (6) **parviflorus.**
      Petals clear yellow ...   . (7) **streptopetalus.**
    Leaves elongate-lanceolate, 1–2 in. broad   ...   ... (8) **Krebsii.**

Petals ¾–1 in. long :
   Sepals distinctly shorter than the petals :
      Lip subentire ...   . (9) **speciosus.**
      Lip with auriculate side lobes...   ...   . (10) **Wakefieldii.**
   Sepals about as long as the petals :
      Sepals spathulate-oblong, obtuse ...   . (11) **Buchanani.**
      Sepals ovate-oblong, acute or subacute :
         Spur of lip conical ; leaves and bracts broad ... (12) **Sandersoni.**
         Spur of lip saccate ; leaves narrow ; bracts narrow
           and acuminate ...   . (13) **arenarius.**

1. **L. clitellifer** (Reichb. f. in Linnæa, xx. 687) ; rhizome stout, much thickened at the nodes ; leaves in a fascicle of 4 to 6, usually small or undeveloped at flowering time, ensiform, acute, somewhat curved, more or less conduplicate, with 5 prominent veins, 2½–4 in. long ; scapes ½–1 ft. long, with several loose spathaceous sheaths below ; raceme long, somewhat lax, many-flowered ; bracts lanceolate or linear-lanceolate, acuminate, ¼–½ in. long ; pedicels slender, ½ in. or more long ; flowers yellow or brownish ; sepals ovate, acute, reflexed, about ¼ in. long ; petals broadly ovate, subacute, erect, twice as broad as the sepals ; lip strongly 3-lobed, nearly as long as the petals ; side lobes broad, obtuse, adnate to the foot of the column ; front lobe ovate-oblong, subobtuse, very convex ; disc with 5 thickened verrucose or crenulate keels ; sac short and obtuse ; column broadly oblong, about ⅛ in. long ; capsules ellipsoid-oblong. *Durand & Schinz, Conspect. Fl. Afr.* v. 28. *Eulophia platypetala, Krauss in Flora,* 1845, 305, *and in Beitr. Fl. Cap- und Natal.* 157, *not of Lindl. E. clitellifer, Bolus in Journ. Linn. Soc.* xxv. 184.

KALAHARI REGION : Transvaal ; flat above Sterk Spruit, *Sanderson,* 2 !
EASTERN REGION : Natal ; Pietermaritzburg, 2000–3000 ft., *Sutherland* ! near Durban, *Krauss,* 406 ! *McKen,* 5 ! 739 ! Attercliffe, 800 ft., *Sanderson,* 475 ! Clairmont Flat, near Durban, *Sanderson,* 476 ! Inanda, *Wood,* 170 ! Dargle Farm, *Mrs. Fannin,* 153 ! Inchanga, 2300 ft., *Engler,* 2661 ! and without precise locality, *Gueinzius* ! *Buchanan* ! *Sanderson,* 705 ! *Mrs. K. Saunders* !

2. **L. Rehmannii** (Rolfe) ; rhizome and leaves not seen ; scapes erect, somewhat slender, about 1 ft. high, with 2 or 3 loose tubular sheaths at the base ; racemes lax, about 4 in. long, many-flowered ; bracts oblong-lanceolate, acuminate, ¼–⅓ in. long ; pedicels slender, ½–⅔ in. long ; flowers small ; sepals elliptic-oblong, acute or apiculate, ¼–⅓ in. long ; petals ovate or orbicular-ovate, subobtuse or slightly apiculate, very little longer than the sepals and over twice as broad ; lip 3-lobed, rather longer than the petals ; side lobes short, broad and truncate ; front lobe reflexed, obovate, truncate, somewhat undulate, nearly as broad as long ; disc with 7 approximate elevated closely verrucose keels from the base to beyond the middle ; spur oblong or linear-oblong, ⅛ in. long ; column broadly clavate, over ⅛ in. long.

KALAHARI REGION : Transvaal ; hills above Aapies River, *Rehmann,* 4297 ! hills near Pretoria, *McLea in Herb. Bolus,* 5819 A ! kopjie at Pretoria, *Miss E. Tennant,* 4040 ! Koodoos Poort, near Pretoria, *Reck,* 1004 !

3. **L. platypetalus** (Lindl. in Hook. Comp. Bot. Mag. ii. 204);
rhizome stout, much thickened at the nodes ; leaves 3 or 4 in a fascicle, with some exterior imbricate sheaths, ensiform or broadly
ensiform, acute, rigid, usually more or less recurved, 2–5 in. long,
primary nerves scarcely more prominent than the others ; scapes
lateral, erect, rather stout, $\frac{1}{2}$–1$\frac{1}{4}$ ft. long, with several spathaceous
sheaths, the lower imbricate ; racemes usually lax and many-flowered ;
bracts ovate-lanceolate, acute or acuminate, $\frac{1}{6}$–$\frac{1}{4}$ in. long ; pedicels
$\frac{1}{3}$–$\frac{1}{2}$ in. long ; flowers small ; sepals ovate, apiculate or shortly
acuminate, reflexed, $\frac{1}{4}$ in. long ; petals broadly ovate, subobtuse,
about twice as broad as the sepals, spreading ; lip strongly 3-lobed,
shorter than the petals ; side lobes broadly rounded or nearly
truncate ; front lobe broadly oblong or orbicular-oblong, obtuse ;
disc very convex, with 3 much thickened and verrucose keels in
front and more numerous somewhat verrucose veins behind ; sac
short and obtuse ; column broadly oblong, about $\frac{1}{4}$ in. long. *Drège,
Zwei Pfl. Documente*, 130, 131, 136 ; *Durand & Schinz, Conspect. Fl.
Afr.* v. 30. *Eulophia tuberculata, Bolus in Journ. Linn. Soc.* xxv.
184, *and Ic. Orch. Austr.-Afr.* ii. *t.* 22.

COAST REGION : Uitenhage Div. ; hills near the Zwartkops River, *Drège*, 2208 !
Zeyher, 4 ! 13 ! 1120 ! between Sundays River and Addo, 1000–2000 ft., *Drège* !
and without precise locality, *Tredgold*, 35 ! Alexandria Div. ; Zuurberg Range,
2000–3000 ft., *Drège*, 2208 ! Albany Div. ; Zwarte Hoogte, *Burke* ! grassy rocks
near Grahamstown, 1800–2000 ft., *MacOwan*, 1046 ! *Burke* ! *Bolton* ! King
Williamstown Div. ; Blue Stone Quarry, near King Williamstown, *Sim*, 2.
Eastern frontier of Cape Colony, *Hutton* !
KALAHARI REGION : Transvaal ; hills near Pretoria, 4500 ft., *Miss Leendertz*,
281 ! *McLea in Herb. Bolus*, 5819 ! *Burtt-Davy*, 1049 ! Daspoort Rand, *Burtt-
Davy*, 2268 !

4. **L. æqualis** (Lindl. in Hook. Comp. Bot. Mag. ii. 204);
rhizomes stout, thickened at the nodes ; leaves in fascicles of about
6, ensiform-linear, acute, somewhat arching, with 3 prominent veins,
conduplicate, 5–8 in. long, with several spathaceous imbricate sheaths
below ; scapes lateral, erect, 1–1$\frac{1}{2}$ ft. high, rather stout, with several
spathaceous sheaths ; racemes rather short, many-flowered ; bracts
oblong-lanceolate, acuminate, $\frac{1}{2}$–$\frac{3}{4}$ in. long ; pedicels stout, about
$\frac{1}{2}$ in. long ; flowers medium-sized ; sepals somewhat spreading,
broadly oblong, somewhat apiculate, about $\frac{1}{2}$ in. long ; petals suberect, elliptic-oblong, subobtuse, rather broader than the sepals ;
lip strongly 3-lobed, about as long as the petals ; side lobes erect,
broadly oblong, obtuse, rather short ; front lobe broadly elliptic,
obtuse, convex ; disc with about 5 crenulate keels ; spur saccate,
obtuse, short ; column broadly oblong, about $\frac{1}{4}$ in. long. *Durand &
Schinz, Conspect. Fl. Afr.* v. 27. *Eulophia æqualis, Bolus in Journ.
Linn. Soc.* xxv. 184, *and Ic. Orch. Austr.-Afr.* ii. *t.* 21 ; *Schlechter
in Engl. Jahrb.* xx. 50, 8.

COAST REGION: Uitenhage Div. ; Zuurberg Range, between Enon and Drei
Fontein, *Drège*, 8273 ! Albany Div. ; Brookhuizens Poort, 2000 ft., *MacOwan*, 80 !
and in Herb. Bolus, 1284. near Grahamstown, *MacOwan, Schönland*. Komgha

Div. ; grassy hill near the mouth of the Kei River, *Flanagan in Herb. Bolus,* 1299 ! British Kaffraria, *Cooper,* 1877 ! eastern frontier of Cape Colony, *Hutton* ! *Mrs. Holland,* 13 !
KALAHARI REGION : Transvaal ; near Barberton, 3000–4000 ft., *Culver,* 1. EASTERN REGION : Transkei ; Umnyolo, near Bazeia, 3000 ft., *Baur,* 754 ! Krielis Country, *Hutton* ! Natal ; Attercliffe, 800 ft., *Sanderson,* 495 ! near Durban, 3000–4000 ft., *Sutherland* ! Inchanga, 2500 ft., *Engler,* 2643 ! near Krantz Kloof, 1500 ft., *Schlechter,* 3200 ! and without precise locality, *Mrs. Fannin,* 23 !

5. **L. transvaalensis** (Rolfe); leaves in fascicles of about 3, elongate-linear, acuminate, with 3 prominent veins, conduplicate below, $\frac{3}{4}$–1 ft. long, with several spathaceous imbricate sheaths below ; scapes lateral, erect, over 1 ft. high (base not seen), with a few spathaceous sheaths below ; racemes 4–6 in. long, somewhat lax, many-flowered ; bracts ovate to ovate-lanceolate, acuminate, $\frac{3}{4}$–1 in. long ; pedicels rather slender, 6–8 lin. long ; flowers medium-sized ; sepals subconnivent, broadly oblong, apiculate, 7–8 lin. long ; petals elliptic-obovate, abruptly acuminate, rather shorter than the sepals ; lip obscurely 3-lobed, $\frac{3}{4}$ in. long ; side lobes short, rounded at the apex ; front lobe ovate-oblong, obtuse, undulate ; disc with 3 high, very thin crenulate keels from the middle to near the apex, and numerous tubercles in front ; spur broadly conical, obtuse, 2 lin. long ; column clavate, about 5 lin. long.

KALAHARI REGION : Transvaal ; Izaneen, Zoutpansberg, 2500 ft., *Burtt-Davy,* 2900 !

6. **L. parviflorus** (Lindl. Gen. & Sp. Orch. 191) ; tubers sub-globose, large, 3–4-leaved ; leaves elongate-linear, acute, dark green, articulated at the base, $\frac{1}{2}$–1 ft. long, $\frac{1}{3}$–$\frac{1}{2}$ in. broad ; scapes lateral, erect, longer than the leaves, with a few spathaceous sheaths below ; raceme lax, several- to many-flowered ; bracts ovate-lanceo-late to elliptic-oblong, acute, $\frac{1}{3}$–$\frac{1}{2}$ in. long ; pedicels $\frac{1}{2}$–1 in. long ; flowers medium-sized ; sepals broadly elliptic-oblong, obtuse, over $\frac{1}{2}$ in. long, green striped with dull brown ; petals suborbicular, obtuse, as long as the sepals, yellow, closely veined and more or less suffused with brown ; lip 3-lobed, nearly as long as the petals ; side lobes suborbicular, whitish, closely lined with red-purple ; front lobe broadly elliptic-oblong, obtuse, bright yellow ; disc with 3 obtuse keels, somewhat elevated in the middle ; spur broadly conical, obtuse, $\frac{1}{6}$ in. long ; column broadly oblong, $\frac{1}{4}$ in. long. *Lindl. Bot. Reg.* xxiv. *Misc.* 14 ; *Maund, Botanist,* iv. *t.* 172 ; *Rolfe in Gard. Chron.* 1893, xiii. 684.

COAST REGION : Uitenhage Div. ; Stony Vale, *Gill* ! Algoa Bay, *Loddiges* !

7. **L. streptopetalus** (Lindl. Gen. & Sp. Orch. 191) ; pseudobulbs ovoid or ovoid-oblong, 1$\frac{1}{2}$–3 in. long, with several ovate strongly-veined imbricate sheaths, about 4- or 5-leaved ; leaves elongate-linear, acute, somewhat arching or suberect, strongly-veined, con-

duplicate below, articulated above the base, $\frac{1}{2}$–1 ft. long, $\frac{1}{2}$–1 in. long ; scapes erect, $1\frac{1}{2}$–$2\frac{1}{4}$ ft. high, with numerous spathaceous sheaths below ; racemes long and lax, many-flowered ; bracts oblong or oblong-lanceolate, acute, $\frac{1}{3}$–$\frac{1}{2}$ in. long; pedicels $\frac{1}{4}$–$\frac{1}{2}$ in. long; flowers medium-sized ; sepals elliptic-oblong, apiculate or subobtuse, green with a few light brown spots, $\frac{1}{3}$–$\frac{1}{2}$ in. long; petals sub-orbicular with a broadly clawed base, obtuse, about as long as the sepals, bright yellow; lip strongly 3-lobed, about as long as the petals ; side lobes erect, broadly oblong, very obtuse, green with some purple veining; front lobe ovate-orbicular, obtuse, somewhat reflexed at the sides, bright yellow; disc convex and thickened at the base ; spur broadly conical, obtuse, $\frac{1}{8}$ in. long ; column oblong, about $\frac{1}{6}$ in. long. *Drège, Zwei Pfl. Documente*, 130, 142, 156 ; *Durand & Schinz, Conspect. Fl. Afr.* v. 31, *partly.* *Eulophia streptopetala, Lindl. Bot. Reg. t.* 1002 ; *Bot. Mag. t.* 2931; *Bolus in Journ. Linn. Soc.* xxv. 183, *and .Ic. Orch. Austr.-Afr.* ii. *t.* 12 (*excl. syn. L. parviflorus*).

SOUTH AFRICA : without locality, *Zeyher,* 1592 ! 3896 ! *Mrs. Bowker* !
COAST REGION : Uitenhage Div. ; near Uitenhage, *Burchell,* 4262 ! *Bolus,* 1551. limestone hill near the mouth of the Zwartkops River, *Drège,* 4776 ! forests near the Zwartkops River and Addo, *Zeyher,* 609 ! Sand Fountain, *Burke* ! Albany Div. ; Howisons Poort, *Hutton* ! Coldstream, near Grahamstown, *Schönland,* 444. King Williams Town Div. ; King Williams Town, *Sim,* 28. Peddie Div. ; Fish River hills near Trumpeters Drift, *Drège* ! Komgha Div.; woods near Komgha, 2000 ft., *Flanagan,* 345. Eastern District of Cape Colony, *Prior* ! and without precise locality, *Miss Bowker* !
CENTRAL REGION : Somerset Div. ; Somerset, *Bowker* !
KALAHARI REGION: Transvaal ; Magalies Berg, *Burke,* 409 ! hills near Barberton, 2800–3300 ft., *Galpin,* 669.
EASTERN REGION : Transkei ; around Kentani, 1200 ft., *Miss Pegler,* 232! Natal ; Umzimkulu River, *Drège* !

Originally described from a garden specimen said to have been received from Brazil, which is clearly an error.

8. L. Krebsii (Reichb. f. in Linnæa, xx. 685) ; pseudobulbs ovoid-oblong, 2–3 in. long, about 5–7-leaved ; leaves elongate-lanceolate, acuminate, arching, $\frac{1}{2}$–$1\frac{1}{2}$ ft. long, $1\frac{1}{4}$–$2\frac{1}{2}$ in. broad, with 5–7 prominent veins, conduplicate below, articulated above the base ; scapes lateral, erect, $2\frac{1}{2}$–$3\frac{1}{2}$ ft. long, with several spathaceous sheaths ; racemes long, lax, many-flowered ; bracts ovate-lanceolate, acuminate, $\frac{3}{4}$–1 in. long; pedicels $\frac{1}{2}$–$\frac{3}{4}$ in. long; flowers medium-sized ; sepals oblong or elliptic-oblong, obtuse or apiculate, $\frac{1}{2}$–$\frac{3}{4}$ in. long, more or less reflexed, green barred with brown, sometimes purple-brown ; petals ovate-orbicular, spreading, rather longer than the sepals, bright yellow; lip strongly 3-lobed, about as long as the petals ; side lobes erect, broadly oblong or nearly quadrate, obtuse, yellow with brown stripes and apex, front lobe broadly ovate, obtuse, somewhat reflexed at the sides, bright yellow ; disc convex and thickened at the base, with 4 very obtuse keels; spur broadly conical, obtuse, about $\frac{1}{4}$ in. long; column oblong, $\frac{1}{4}$ in. long. *Bot. Mag. t.* 5861 ; *Rolfe in Dyer, Fl. Trop. Afr.* vii. 91 ; *Durand &*

Schinz, Conspect. Fl. Afr. v. 29; *Kränzl. in Ann. Naturhist.*
Hofmus. Wien, xx. 11. *L. Krebsii, var. purpurata, Ridl. in Gard.*
Chron. 1885, xxiv. 102; *Will. Orch. Alb.* vi. *t.* 259. *L. Græfei,*
Kränzl. in Gard. Chron. 1892, xi. 749, *and in Reichb. f. Xen.*
Orch. iii. 125, *t.* 272. *Eulophia Krebsii and var. purpurata, Bolus*
in Journ. Linn. Soc. xxv. 185.

KALAHARI REGION : Transvaal; Waterval River, Lydenburg District, *Wilms,*
1386 ! Ohrigstad Valley, Lydenburg, *Burtt-Davy,* 7320 ! Magaliesberg, *Burke* !
EASTERN REGION : Natal ; Attercliffe, 800 ft., *Sanderson,* 491 ! between Pietermaritzburg and Greytown, *Wilms,* 2280 ! near Tugela River, *Gerrard,* 1817 !
Ligombwe, *Penther,* 231 ! and without precise locality, *Sanderson* ! *Mrs. Fannin,*
74 ! *Mrs. K. Saunders* !

Also in Tropical Africa.

9. **L. speciosus** (R. Br. ex Lindl. Coll. Bot. t. 31); pseudobulbs
ovoid, 2 in. or more long, with a few ovate sheaths, 3–5-leaved ;
leaves elongate-linear, acute, somewhat fleshy, without prominent
veins, conduplicate below, not articulated above the base, $\frac{1}{2}$–1 ft. or
more long, $\frac{3}{4}$–1 in. or more broad ; scapes erect, stout, $1\frac{1}{2}$–$2\frac{1}{2}$ ft.
long, with several spathaceous sheaths ; racemes long, somewhat
lax, many-flowered ; flowers medium-sized ; bracts ovate-oblong to
ovate-lanceolate, acuminate, $\frac{3}{4}$–1 in. long ; pedicels $\frac{3}{4}$–1 in. long ;
sepals ovate or ovate-oblong, subacute or acute, reflexed, green,
about $\frac{1}{2}$ in. long ; petals spreading, broadly ovate or ovate-suborbicular, subobtuse, about $\frac{3}{4}$ in. long, bright yellow ; lip 3-lobed,
nearly as long as the petals ; side lobes suberect, short and transversely oblong, white with a few reddish lines ; front lobe broadly
elliptic, obtuse, reflexed at the sides, yellow with a few reddish lines
at the base ; disc convex, with 3 obtuse keels ; spur very short,
broadly conical, obtuse ; column oblong, $\frac{1}{4}$ in. long. *Lindl. Bot.*
Reg. t. 573, *and Gen. & Sp. Orch.* 191 ; *Durand & Schinz, Conspect. Fl.*
Afr. v. 30. *Satyrium giganteum, Linn. f. Suppl. Pl.* 402. *Limodorum*
giganteum, Thunb. Prodr. 4. *Cymbidium giganteum, Sw. in Schrad.*
Journ. ii. 224 ; *Thunb. Fl. Cap.* i. 129. *Cyrtopera ? gigantea, Lindl.*
Gen. & Sp. Orch. 190. *Eulophia speciosa, Bolus in Journ. Linn. Soc.*
xxv. 184, *and Ic. Orch. Austr.-Afr.* ii. *t.* 13.

VAR. β, **Culveri** (Schlechter in Engl. Jahrb. xx. 50, 10); more slender than the
type, with much smaller flowers and minute linear bracts.

SOUTH AFRICA : without locality, *Griffin* (the type specimen) ! *Mrs. Bowker* !
COAST REGION : Uitenhage Div. : sand dunes near the mouth of the Coega
River, *Ecklon & Zeyher,* 3894. Port Elizabeth Div. ; Sandy hills at Krakakamma,
Zeyher, 3895 ! Bathurst Div. ; sand hills near the mouth of the Kleinmond and
Kowie Rivers, *Atherstone,* 12 ! 28 ! 58 ! *MacOwan* ! 401 ! Albany Div. ; without
precise locality, *Miss Bowker* ! *Hutton* ! Komgha Div. ; Kei River Mouth,
Flanagan, 1025. Eastern District of Cape Colony, *Prior* !
CENTRAL REGION : Somerset Div. ; Somerset East, *Bowker* !
KALAHARI REGION : Transvaal ; Pietersburg District, hills near Potgieters Rust,
3600 ft., *Bolus,* 11165, near Barberton, 2000–2900 ft., *Galpin,* 668. Var. β,
near Barberton, 2000–3000 ft., *Culver,* 62 !
EASTERN REGION : Transkei ; rocks near Butterworth, Krielis Country,
Mrs. Barber, 9 ! coast near Kentani, *Miss Pegler,* 782. Natal ; bush near Durban,
Sanderson, 111 ! 493 ! *Plant,* 50 ! *Wood,* 707 ! Clairmont, Inanda, *Wood,*

1090! near the Lighthouse, Durban, *Wilms,* 2279; sand dunes near Durban, *Schlechter,* 3151! near Tugela River, *Gerrard,* 1815!

Also found in Rhodesia.

10. **L. Wakefieldii** (Reichb. f. & S. Moore in Journ. Bot. 1878, 136); rhizome stout, thickened at the nodes; leaves 3 or 4 in a fascicle, spreading, elongate-linear, acute or acuminate, 6–12 in. long, $\frac{1}{2}$–1 in. broad, with about 5 slightly prominent veins, protected at the base by a few spathaceous imbricate sheaths; scapes lateral, erect, $2\frac{1}{2}$–4 ft. long, with several spathaceous sheaths below; racemes elongate, lax, many-flowered; bracts lanceolate or ovate-lanceolate, acuminate, $\frac{1}{4}$–$\frac{1}{2}$ or rarely $\frac{3}{4}$ in. long; pedicels $\frac{1}{2}$–1 in. long; flowers medium-sized; sepals ovate-oblong, apiculate or shortly acuminate, reflexed, $\frac{1}{4}$–$\frac{1}{3}$ in. long, green; petals broadly ovate, subobtuse or apiculate, about $\frac{3}{4}$ in. long, bright yellow; lip pandurate-oblong, about as long as the petals; side lobes auriculate or rounded, adnate to the base of the column, bright yellow; front lobe elliptic-oblong, obtuse; disc convex, with 5–7 fleshy somewhat crenulate keels; spur broadly conical, obtuse, about $\frac{1}{6}$ in. long; column clavate, $\frac{1}{4}$ in. long, its foot very short. *Reichb. f. Otia Bot. Hamb.* ii. 76; *Rolfe in Dyer, Fl. Trop. Afr.* vii. 95. *L. dispersus, Rolfe in Gard. Chron.* 1893, xiii. 684. *Eulophia dispersa, N. E. Br. in Kew Bulletin,* 1892, 127.

TRANSVAAL REGION : Transvaal; Potgieters Rust, *Miss Leendertz,* 1929!
EASTERN REGION : Delagoa Bay, *Mrs. Monteiro*!

Also in Tropical Africa.

11. **L. Buchanani** (Reichb. f. Otia Bot. Hamb. i. 64); leaves about 3 or 4, long-petioled, somewhat arching, elongate-lanceolate, acuminate, with about 5 prominent veins, conduplicate at the base, 2–3 ft. long, 2–$2\frac{1}{2}$ in. broad; scape very stout, erect, 3–4 ft. high, with spathaceous imbricate sheaths below; racemes 6–9 in. long, rather dense, many-flowered; bracts elliptic-oblong or oblong-lanceolate, acute or subacute, $\frac{1}{3}$–$\frac{3}{4}$ in. long; pedicels slender, about $\frac{3}{4}$ in. long; flowers yellow; sepals subspathulate-oblong, obtuse, nearly 1 in. long; petals elliptic or elliptic-oblong, obtuse, twice as broad as the sepals; lip rather longer than the petals, elliptic-oblong, obtuse, obscurely 3 lobed; side lobes narrow, undulate; front lobe broadly oblong or orbicular-oblong, obtuse, very undulate; disc with 3 thickened crenulate keels behind, and 5–7 taller, more crenulate and undulate keels in front; spur broadly conical, obtuse, nearly $\frac{1}{4}$ in. long; column clavate, $\frac{1}{2}$ in. long. *Durand & Schinz, Conspect. Fl. Afr.* v. 28. *Eulophia Buchanani, Bolus in Journ. Linn. Soc.* xxv. 185, *and Ic. Orch. Austr.-Afr.* ii. *t.* 23.

EASTERN REGION : Pondoland; marshy places at the mouth of the St. John's River, *Flanagan,* 2555, *Galpin,* 3414. Natal; swamps at Inanda, about 12 miles from the sea, *Wood,* 845! near Dumisa, 2000 ft., *Rudatis,* 612! and without precise locality, *Sanderson,* 492! *Buchanan,* 1! 2! *Mrs. K. Saunders*! Zulu-land; *Haygarth in Herb. Wood,* 7547.

Also found in Rhodesia.

12. L. Sandersoni (Reichb. f. Otia Bot. Hamb. i. 62); rhizome stout, subterranean; leaves 3 or 4 in a fascicle, long-petioled, suberect or somewhat arching, elongate-lanceolate, acute, with 5–9 prominent veins, 3–4 ft. long, 2–4 in. broad; scapes lateral, erect, very stout, 4–5 ft. high, with strong spathaceous sheaths; raceme long, many-flowered; bracts ovate-oblong or elliptic-oblong, acute or acuminate, concave or convolute, $\frac{1}{2}$–1 in. long; pedicels about 1 in. long; flowers large; sepals ovate-oblong, acute, more or less reflexed, $\frac{3}{4}$–1 in. long, green with brown stripes; petals spreading, broadly ovate, subobtuse, 1–1$\frac{1}{4}$ in. long, white; lip strongly 3-lobed, longer than the petals; side lobes broadly rounded, erect, green striped with brown; front lobe broadly ovate-orbicular, obtuse, lilac-purple with numerous radiating darker stripes; disc with 3–5 erect crested keels, light green; spur broadly conical, obtuse, $\frac{1}{4}$ in. long; column clavate, narrowly winged above, $\frac{3}{4}$ in. long. *Reichb. f. in Gard. Chron.* 1885, xxiv. 17 *(Saundersonii by error)*; *Bot. Mag. t.* 6858; *Rolfe in Dyer, Fl. Trop. Afr.* vii. 85. *L. porphyroglossus, Durand & Schinz, Conspect. Fl. Afr.* v. 30, *partly, not of Reichb. f. Eulophia porphyroglossa, Bolus in Journ. Linn. Soc.* xxv. 185, *partly.*

EASTERN REGION : Natal; The Bluff, *Sanderson*, 1002! Inanda, *Wood*, 374! and without precise locality, *Lyle*! *Mrs. K. Saunders*!

Also in Tropical Africa.

13. L. arenarius (Lindl. in Journ. Linn. Soc. vi. 133); rhizomes much thickened at the nodes; leaves linear or elongate-lanceolate, acute, immature at flowering time, ultimately $\frac{3}{4}$–1 ft. or more long; scapes lateral, erect, $\frac{3}{4}$–2$\frac{1}{2}$ ft. long, with several spathaceous sheaths below; racemes lax, usually many-flowered; bracts lanceolate or linear-lanceolate, acuminate, $\frac{1}{2}$–1 in. broad; pedicels $\frac{3}{4}$–1 in. long; flowers large, purple with dusky or olivaceous sepals, and some yellow in the sac of the lip, sometimes lilac and white; sepals ovate-triangular, acuminate, reflexed, $\frac{1}{2}$–$\frac{3}{4}$ in. long; petals spreading, suborbicular or broadly elliptic, $\frac{1}{2}$–$\frac{3}{4}$ in. long; lip 3-lobed, very broad, rather longer than the petals; side lobes broadly rounded, very obtuse; front lobe broadly quadrate, obtuse or emarginate; disc with a pair of quadrate or transversely oblong erect calli in front of the sac, and 3 somewhat thickened veins extending towards the apex; spur saccate, very obtuse, about $\frac{1}{4}$ in. broad by scarcely as long; column clavate, about $\frac{1}{3}$ in. long, very broad at the apex. *Reichb. f. Otia Bot. Hamb.* i. 61, ii. 75; *N. E. Br. in Gard. Chron.* 1885, xxiv. 307; *Rolfe in Dyer, Fl. Trop. Afr.* vii. 82; *Durand & Schinz, Conspect. Fl. Afr.* v. 27. *Limodorum cucullatum, Afzel. ex Sw. in Vet. Acad. Handl. Stockh.* 1800, 243, *name only*; *Pers. Syn.* ii. 521. *Eulophia arenaria, Bolus in Journ. Linn. Soc.* xxv. 185.

EASTERN REGION : Natal; near Tongaat and the Tongaat River, 500 ft., *Sanderson*, 1017! *Mrs. K. Saunders*! *Wood in MacOwan & Bolus, Herb. Norm. Austr.-Afr.* 1368! Umvoti, *Mrs. K. Saunders*! Inanda, *Wood*, 456! 820!

Also in East and West Tropical Africa, with almost invariably larger flowers.

VII. ANSELLIA, Lindl.

Sepals and *petals* subequal, free, spreading. *Lip* articulated to the foot of the column, 3-lobed; side lobes erect, parallel; front lobe oblong or rounded; disc with 2 or 3 parallel keels. *Column* erect, equalling the side lobes of the lip, slightly curved, semiterete, with acute margins; base produced into a very short broad concave or slightly 2-lobed foot; anther-bed entire, scarcely prominent. *Anther* terminal, operculate, incumbent, convex or crowned with an obtuse conical appendage, imperfectly 2-celled; pollinia 2 or 4, confluent in pairs, waxy, ovate, rounded, attached to a short broad stipes and gland. *Capsule* oblong, without beak.

Epiphytic herbs; stems tall, thickened or somewhat fusiform, leafy; leaves distichous, elongate, plicate-veined; panicles terminal, more or less branched; flowers rather large, yellow, more or less blotched with brown, pedicelled; bracts small.

DISTRIB. An African genus of about 6 species, only one of which occurs within our limits.

1. **A. gigantea** (Reichb. f. in Linnæa, xx. 673); stems elongate, terete or somewhat sulcate, ⅓–1 ft. long, with 6 to many leaves on the upper part or near the apex and numerous imbricate membranous sheaths below; leaves distichous, linear-oblong to elliptic-lanceolate, subacute, ¼–1 ft. long, ½–1½ in. broad, with 3–5 prominent veins; panicle terminal, ½–1 ft. long, usually with several branches, rarely reduced to a simple raceme, with a few short sheaths below; bracts triangular-ovate, subacute, ⅙ in. long; pedicels slender, 1–1¼ in. long; flowers medium-sized, light yellow, more or less barred or blotched with light dusky brown; sepals and petals spreading, oblong or elliptic-oblong, obtuse, about ¾ in. long; lip 3-lobed, rather shorter than the sepals; side lobes erect, oblong, obtuse; front lobe recurved, elliptic-oblong, obtuse or emarginate; disc with 3 prominent crenulate keels; column clavate, ⅓ in. long. *Reichb. f. Xen. Orch.* ii. 18; *Bolus in Journ. Linn. Soc.* xxv. 185. *Var. citrina, Reichb. f. Xen. Orch.* ii. 18; *Saund. Ref. Bot.* ii. *t.* 136. *A. africana, var., Bolus, 1c. Orch. Austr.-Afr.* ii. *t.* 29; *Durand & Schinz, Conspect. Fl. Afr.* v. 32. *A. africana, var. natalensis, Hook. Bot. Mag. t.* 4965, *fig.* 3. *Cymbidium Sandersoni, Harv. Gen. S. Afr. Pl. ed.* 2, 360.

KALAHARI REGION: Transvaal; Waterval Boven, in Lydenburg District, *Wilms*, 1887! near Barberton, 1800 ft., *Galpin*, 1004; Avoca, 2000 ft., *Culver*, 55.
EASTERN REGION: Natal; between Durban and Attercliffe, 0–800 ft., *Sanderson*, 474! Inanda, *Wood*, 1342! Delagoa Bay, *Junod*, 281! *Stirrat*, 2988! near Lorenzo Marques, *Bolus*, 9786, and without precise locality, *Mrs. K. Saunders*! and in Herb. *Bolus*, 6060.

VIII. **POLYSTACHYA,** Hook.

Sepals connivent or somewhat spreading; dorsal free; lateral broader, sometimes very broad, adnate to the foot of the column. *Petals* usually narrower than the dorsal sepal. *Lip* superior, articulated to the foot of the column, 3-lobed or entire from a cuneate base. *Column* usually short, very broad, not winged, produced into a long foot at the base; anther-bed short, truncate. *Anther* terminal, operculate, incumbent, very convex, 1-celled or imperfectly 2-celled; pollinia 4, waxy, broadly ovate, sometimes united in pairs, affixed to a short stipes and gland. *Capsule* oblong or fusiform, sometimes elongate.

Epiphytic herbs; stems often short, sometimes thickened into pseudobulbs, mostly leafy; leaves distichous, oblong or narrow, often many-nerved, contracted into sheaths at the base; peduncle terminal, with a few sheaths below, paniculate or racemose; flowers small or medium-sized; bracts small.

DISTRIB. Species over 100, widely diffused through the tropics, though the great majority are African. The 10 South African species, so far as known, are endemic.

Stems or pseudobulbs 2–4-leaved :
  Flowers about ¼ in. long :
    Leaves linear-oblong, ¼–¾ in. broad ; inflorescence
      racemose ... ... (1) **rigidula.**
    Leaves oblong, ½ – 1¼ in. broad ; inflorescence
      paniculate :
      Side lobes of lip very short and broad ... ... (2) **tricruris.**
      Side lobes of lip oblong, longer than broad ... (3) **similis.**
  Flowers ¼–½ in. long :
    Sepals green or brownish-green :
      Pseudobulbs ovate-oblong, ¾–2 in. long ... ... (4) **Sandersoni.**
      Pseudobulbs cylindric, 3 in. or more long :
      Leaves obtuse ; limb of lip ovate, obtuse ... (5) **natalensis.**
      Leaves unequally 2-lobed ; limb of lip trans-
        versely rhomboid, acuminate ... ... (6) **transvaalensis.**
    Flowers white :
      Lip with oblong obtuse front lobe ... ... (7) **ottoniana.**
      Lip with setaceous-acuminate front lobe ... ... (8) **glaberrima.**
    Flowers deep yellow ... ... (9) **pubescens.**
  Stems slender, 1-leaved ... ... (10) **Gerrardi.**

1. **P. rigidula** (Reichb. f. in Flora, 1867, 117); stems tufted, scarcely thickened, 1–2 in. long, 2–4-leaved ; leaves narrowly oblong, obliquely and minutely bilobed, cuneate at the base, coriaceous, 3–5 in. long, ¼–¾ in. broad ; scapes 5–7 in. high, erect, slender, unbranched, with 2 or 3 narrow sheaths below, slightly pubescent above ; raceme ¾–1¼ in. long, 8–15-flowered ; bracts ovate triangular, acuminate, 1/12–1/10 in. long ; fruiting pedicels ¼ in. long ; dorsal sepal ovate, subacute, concave, 1/12 in. long ; lateral sepals triangular-ovate, acute, 1/10 in. long ; petals oblong, subobtuse, 1/12 in.

long; lip $\frac{1}{8}$ in. long, 3-lobed, cuneate at the base, front lobe broad, obtuse, crenulate; side lobes oblong, obtuse and slightly falcate; disc slightly corrugated and with a pair of erect papillæ near the base; column stout, over half as long as the dorsal sepal. *Bolus in Journ. Linn. Soc.* xxv. 186; *Durand & Schinz, Conspect. Fl. Afr.* v. 36; *Kränzl. in Reichb. f. Xen. Orch.* iii. 72, *t.* 237, *fig.* 1.

EASTERN REGION: Natal; near the Tugela River, *Gerrard*, 1812!

2. **P. tricruris** (Reichb. f. in Flora, 1867, 118); stems slightly thickened at the base, 2–5 in. long, 3–4-leaved; leaves lanceolate-oblong to cuneate-oblong, obliquely and shortly bilobed at the apex, cuneate at the base, coriaceous, 3–8 in. long, $\frac{1}{2}$–$1\frac{1}{4}$ in. broad; scapes $\frac{3}{4}$–$1\frac{1}{4}$ ft. high, erect, slender, with several short lateral branches, covered with ancipitous membranous imbricate sheaths; panicle 3–8 in. long, many-flowered, slightly pubescent, side branches $\frac{1}{2}$–$1\frac{1}{4}$ in. long; bracts triangular-ovate, long-acuminate, $\frac{1}{12}$ in. long; pedicels $\frac{1}{6}$–$\frac{1}{4}$ in. long; dorsal sepal ovate-oblong, shortly acuminate, concave, $\frac{1}{10}$ in. long; lateral sepals triangular, shortly acuminate, $\frac{1}{8}$–$\frac{1}{6}$ in. long; petals subspathulate-oblong, apiculate, $\frac{1}{10}$ in. long; lip $\frac{1}{6}$–$\frac{1}{8}$ in. long, 3-lobed, cuneate below; front lobe suborbicular, crenulate, shortly bilobed; side lobes shortly oblong, subobtuse; disc wholly pubescent, and with a prominent rounded keel extending from the middle to the base; column short, less than half as long as the dorsal sepal. *Bolus in Journ. Linn. Soc.* xxv. 186; *Durand & Schinz, Conspect. Fl. Afr.* v. 37.

EASTERN REGION: Natal; near the Tugela River, *Gerrard*, 1813!

Reichenbach described this species as from "Natal, *Sanderson*," the type being at Kew, but the only specimen answering to the description is from Gerrard, and it is believed that it is incorrectly attributed to the former collector.

3. **P. similis** (Reichb. f. Otia Bot. Hamb. ii. 112); stems tufted, narrowly fusiform, 1–$2\frac{1}{2}$ in. long, 2–5-leaved; leaves oblong or elliptic-oblong, unequally and minutely bilobed, cuneate at the base, coriaceous 2–5 in. long, $\frac{1}{2}$–$1\frac{1}{4}$ in. broad; scapes 6–10 in. high, erect, rather stout, often somewhat branched above, slightly puberulous, with 2 or 3 narrow sheaths below; racemes or panicle-branches $\frac{1}{2}$–$1\frac{1}{4}$ in. long, 10–15-flowered; bracts deltoid-triangular, acute or acuminate, $\frac{1}{12}$–$\frac{1}{10}$ in. long; pedicels $\frac{1}{6}$–$\frac{1}{4}$ in. long; dorsal sepal ovate, subobtuse, $\frac{1}{12}$ in. long; lateral sepals triangular-ovate, apiculate, $\frac{1}{10}$ in. long; petals linear-oblong, apiculate, $\frac{1}{12}$ in. long; lip $\frac{1}{8}$ in. long, 3-lobed, broadly cuneate at the base; front lobe orbicular, crenulate, obtuse; side lobes broadly oblong, obtuse, very short; disc slightly verrucose; column stout, about half as long as the dorsal sepal. *Bolus in Journ. Linn. Soc.* xxv. 186, *and Ic. Orch. Austr.-Afr.* ii. *t.* 33; *Durand & Schinz, Conspect. Fl. Afr.* v. 37.

EASTERN REGION: Pondoland; mountains near the mouth of St. Johns River, 1000 ft., *Flanagan in Herb. Bolus*, 2551, 8732! Natal; Klip Fontein, *Saunders*, 4! Riet Valley, *Mrs. Rathbone*! Tongaat, *Mrs. K. Saunders in Herb. Bolus*, 6117, and without precise locality, *Sanderson*! Delagoa Bay; at Masinga, *Schlechter*, 12138!

4. P. Sandersoni (Harv. Thes. Cap. ii. 49, t. 177); stems erect, slender above, somewhat thickened at the base, about 2 in. long, 2-leaved or rarely 3-leaved, with 3 or 4 spathaceous imbricate sheaths below; leaves linear-oblong, subobtuse, $3\frac{1}{2}$–5 in. long, about $\frac{1}{3}$ in. broad, somewhat narrowed and conduplicate towards the base; scapes erect, somewhat flattened and with 1 or 2 narrow sheaths below, about 4 in. long, sometimes branched, softly pubescent throughout; raceme or panicle loosely many-flowered; bracts broadly triangular-ovate, concave, narrowly acuminate and somewhat reflexed above, $\frac{1}{8}$–$\frac{1}{4}$ in. long; pedicels slender, about $\frac{1}{4}$ in. long; flowers medium-sized; sepals broadly triangular-ovate, shortly acuminate, about $\frac{1}{4}$ in. long, dorsal narrower and rather shorter than the lateral; petals linear-subspathulate or slightly broader above, subacute, about $\frac{1}{6}$ in. long; lip 3-lobed, about $\frac{1}{4}$ in. long; side lobes broadly rounded, narrowed behind; front lobe ovate, subobtuse; disc strongly pulverulent and somewhat verrucose; column very short and broad, its foot $\frac{1}{8}$ in. long. *Bolus in Journ. Linn. Soc.* xxv. 186, *and Ic. Orch. Austr-Afr.* ii. *t.* 31; *Durand & Schinz, Conspect. Fl. Afr.* v. 37.

COAST REGION: Komgha Div. ; near Komgha, *Flanagan.*

EASTERN REGION: Transkei; near Kentani, *Miss Pegler*, 203. Natal; Umtwalumi River, *McKen*! *Sanderson*, 895! Attercliffe, 600 ft., *Sanderson*, 134! and without precise locality, *Sanderson*, 1016! 1041! Zululand; Eshowe, *Maxwell.*

5. P. natalensis (Rolfe); stems tufted, erect, cylindric, 3–6 in. long, 2-leaved at the apex, with narrow sheaths below; leaves oblong, obtuse, coriaceous, $2\frac{1}{4}$–3 in. long, $\frac{1}{2}$–$\frac{3}{4}$ in. broad; scapes erect, 2–3 in. long, 3–5-flowered, with 2 or 3 narrow sheaths below; bracts acuminate from a broadly triangular concave base, $\frac{1}{8}$–$\frac{1}{6}$ in. long; pedicels $\frac{1}{4}$ in. long; flowers medium-sized, brownish-green or brownish-purple; dorsal sepal ovate-oblong, subobtuse, $\frac{1}{4}$–$\frac{1}{3}$ in. long; lateral sepals broadly triangular, subobtuse, $\frac{1}{2}$ in. long; petals narrowly subspathulate, apiculate, nearly $\frac{1}{4}$ in. long; lip spathulate, over $\frac{1}{3}$ in. long, with a broad claw; limb broadly ovate, subobtuse, somewhat undulate, $\frac{1}{4}$ in. broad; disc with a quadrate erect callus below the middle of the limb; column very short and broad, foot $\frac{1}{4}$ in. long; capsules elliptic-oblong, with prominent angles, 1 in. long.

EASTERN REGION: Natal; Richmond, 2500 ft., *Sanderson*, 823!

6. P. transvaalensis (Schlechter in Engl. Jahrb. xx. Beibl. 50, 28); rhizome creeping, with short internodes; pseudobulbs approximate, cylindric, narrowed towards the apex, $1\frac{1}{4}$–$4\frac{3}{4}$ in. long, with 2 acuminate sheaths below, 2–3-leaved; leaves narrowly oblong, unequally 2-lobed at the apex, 1–$3\frac{1}{2}$ in. long, $\frac{1}{4}$–$\frac{3}{4}$ in. broad; scapes

as long as or longer than the leaves ; racemes simple or very rarely
branched, few- to several-flowered ; bracts spreading, lanceolate,
aristate, about half as long as the pedicels ; flowers medium-sized ;
sepals green ; dorsal triangular-lanceolate, very acute, about ⅙ in.
long ; lateral obliquely falcate-ovate, acuminate, ½ in. long, much
broader than the dorsal ; petals white, linear at the base, dilated
into a rhomboid-ovate limb above, acuminate, ¼ in. long ; lip white,
cuneate at the base, dilated into a transversely rhomboid acuminate
limb above ; disc somewhat velvety with very short hairs, and an
obscure green callus in the centre, over ⅓ in. long by nearly as
broad ; column ⅙ in. long, its foot over twice as long ; capsule
trigonous, over ½ in. long, narrowed at the base, broad above.

KALAHARI REGION : Transvaal ; Moodies, near Barberton, 4500 ft., *Culver, 56,*
and Houtbosch Berg, 6500 ft., in primeval forest, *Schlechter.*

Only known to me from the description.

7. **P. ottoniana** (Reichb. f. in Hamb. Gartenz. 1855, 249) ;
rhizome stout and woody ; pseudobulbs approximate, ovoid, ½–¾ in.
long, covered with imbricate sheaths, 2–3-leaved at the apex ; leaves
linear or linear-oblong, subobtuse, coriaceous, 1½–6 in. long, ⅙–⅓ in.
broad ; scapes erect, 1–4 in. long, puberulous ; racemes simple,
sometimes reduced to a single flower ; bracts triangular-ovate,
acuminate, 1/12–⅓ in. long ; pedicels slender, ¼–½ in. long ; flowers
rather large, white, with a deep yellow line on the front lobe of the
lip ; sepals somewhat spreading, triangular-ovate or triangular-oblong,
subobtuse, ⅓–½ in. long ; petals oblong, obtuse, sometimes about as
long as the sepals ; lip 3-lobed, much shorter than the petals ; front
lobe broadly oblong, obtuse, revolute ; side lobes small and obtuse ;
disc pubescent, with a thickened central keel ; column very short
and stout, its foot ⅙ in. long ; mentum short and rounded ; capsules
oblong, ¾–1 in. long. *Reichb. f. in Bonpl.* 1855, 217 ; *Walp. Ann.*
vi. 638 ; *Bolus in Journ. Linn. Soc.* xxv. 186, *and in Ic. Orch.
Austr.-Afr.* ii. *t.* 32 ; *Durand & Schinz, Conspect. Fl. Afr.* v. 36 ;
Schlechter in Engl. Jahrb. xx. *Beibl.* 50, 29. *P. capensis, Sond. ex
Harv. Thes. Cap.* ii. 51, *t.* 179. *Angræcum?* sp., *Lindl. in Hook.
Comp. Bot. Mag.* ii. 205.

COAST REGION : Knysna Div. ; Knysna Forest, *Barkly*! Albany Div. ; Black
Horse Hill and near Blue Krantz, *Atherstone,* 19 ! between Grahamstown and the
sea, *MacOwan,* 1526 ! *Ecklon and Zeyher! Pym,* and without precise locality,
Mrs. Hutton! *Miss Bowker*! British Kaffraria ; Rivi Forest, *Kirk*! and without
precise locality, *Cooper,* 203. Komgha Div. ; near Komgha, *Flanagan.* Eastern
frontier of Cape Colony, *Mrs. Barber*!
KALAHARI REGION : Transvaal ; Houtbosch Berg, 5000–6000 ft., *Schlechter* ; near
Barberton, 3500–6000 ft., *Galpin,* 671 ; near Haenertsburg, *Burtt-Davy,* 3026, and
without precise locality, *Randall*!
EASTERN REGION : Transkei ; between Gekau (Geua) River and Bashee River,
Drège! 8270 ! Fort Bowker, woods on Bashee River, *Bowker,* 600 ! Bazeia, 2000–
3500 ft., *Baur,* 238 ! 740 ! near Kentani, *Miss Pegler,* 617. Natal ; near Durban,
Sanderson, 560 ! rocks at Assegai Kraal, *McKen in Herb. Sanderson,* 560 ! 896 !
Sanderson in Herb. Wood, 1041! Enon, *Mrs. Saunders in Herb. Bolus,* 5946 !
Swaziland ; in kloof at Embabaan, *Burtt-Davy,* 2877 !

8. P. glaberrima (Schlechter in Engl. Jahrb. xx. Beibl. 50, 11) ; a dwarf herb, 2¾–3½ in. high ; pseudobulbs ovoid, ⅓–⅔ in. long, 2–3-leaved ; leaves erect, linear, shortly emarginate, 1¼–2½ in. long, about ¼ in. broad ; scape terminal, erect, slender, very glabrous, shorter than the leaves, 1–2-flowered, with a small membranous sheath below the middle ; flowers shortly pedicelled, white ; dorsal sepal ovate, very acute ; lateral sepals about as long as the dorsal, obliquely ovate-falcate, attenuate and very acute at the apex, the front margin much broadened at the base, about ⅓ in. long; petals linear, dilated above the middle, acute, scarcely as long as the sepals ; lip erect, rhomboid, elongate and setaceous-acuminate at the apex, very glabrous, with a thickened middle nerve ; column slender, half as long as the lip, with an elongated foot ; mentum obtuse.

KALAHARI REGION : Transvaal ; Saddleback Mountain, near Barberton, 4000–5000 ft., *Culver*, 8.

Not seen, but said to be closely allied to the preceding.

9. P. pubescens (Reichb. f. in Walp. Ann. vi. 643); rhizome stout and woody ; pseudobulbs approximate, ovate or thickened at the base, narrowed above, 1–1½ in. long, with several broad imbricate sheaths, 2–3-leaved ; leaves elliptic-oblong to linear-oblong, subobtuse, often somewhat recurved at the apex, coriaceous, 1½–6 in. long, ⅓–¾ in. broad ; scapes erect, with a few lanceolate acuminate sheaths below, 3–8 in. high, puberulous or pubescent ; racemes usually elongated and many-flowered ; bracts triangular-ovate to triangular-lanceolate, very acuminate, ⅙–¼ in. long, concave at the base, often recurved ; pedicels about ⅓ in. long ; flowers medium sized, deep yellow with some brown stripes on the lateral sepals and lip ; sepals ovate or ovate-oblong, acute, spreading, lateral about ⅓ in. long, dorsal rather shorter and narrower ; petals spathulate-oblong, obtuse, rather shorter than the dorsal sepal ; lip 3-lobed, rather shorter than the petals ; front lobe ovate, subacute ; side lobes usually smaller and more obtuse ; disc pubescent or villous ; column very short and broad. *Bot. Mag. t.* 5586 ; *Lindenia*, iv. *t.* 170 ; *Bolus in Journ. Linn. Soc.* xxv. 186, *and Ic. Orch. Austr.-Afr.* ii. *t.* 30 ; *Durand & Schinz, Conspect. Fl. Afr.* v. 36 ; *Cogn. Dict. Ic. Orch., Polyst. t.* 1 ; *Schlechter in Engl. Jahrb.* xx. Beibl. 50, 29. *P. lindleyana, Harv. Thes. Cap.* ii. 50, *t.* 178. *Epiphora pubescens, Lindl. in Hook. Comp. Bot. Mag.* ii. 201 ; *Sond. in Linnæa,* xix. 71 ; *Drège, Zwei Pfl. Documente,* 127, 141. *Lissochilus sylvaticus, Eckl. ex Sond. in Linnæa,* xix. 71.

SOUTH AFRICA : without locality, *Prior* !
COAST REGION : Uitenhage Div. ; Van Stadens River, *Drège.* Port Elizabeth Div. ; around Krakakamma, *Burchell,* 4547 ! *Mrs. Holland,* 3 ! Bathurst Div. ; between Kowie and Kap River, *Drège,* 4578 ! Albany Div. ; wooded kloof west of Grahamstown, *Burchell,* 3590 ! near Grahamstown, *Atherstone,* 24 ! Fullers, *MacOwan,* 110 ! Stones Hill Range, on trees and rocks, 1800–2500 ft., *Galpin,* 298 ; New Years River, *Mrs. Barber,* 261 ! Kafferland, *Brownlee* ! Eastern frontier of Cape Colony, *Hutton* !
CENTRAL REGION : Somerset Div. ; Somerset, *Bowker* !
KALAHARI REGION : Transvaal ; Houtbosch Berg, 6000 ft., *Schlechter.*

10. **P. Gerrardi** (Harv. Thes. Cap. ii. 49, t. 176) ; rhizomes stout and woody ; pseudobulbs approximate, linear-oblong or somewhat elongated, thickened at the base, 3–4½ in. long, with slender basal sheaths when young which disappear early ; leaves solitary, linear-oblong or narrowly oblong, sessile, obtuse or subobtuse, 5–7 in. long, ½–¾ in. broad, coriaceous ; scapes erect, rather slender, 3–6 in. long, with 2 or 3 lanceolate basal sheaths, usually somewhat branched at the apex ; racemes or panicles usually short, somewhat dense and many-flowered ; bracts broadly triangular-ovate, acute, very concave, spreading, about $\frac{1}{12}$ in. long ; pedicels slender, ½ in. long ; flowers yellowish-green, medium-sized ; dorsal sepal elliptic-lanceolate, acute, ⅙ in. long ; lateral sepals triangular-ovate, acute, longer than and twice as broad as the dorsal ; petals oblanceolate, acute, as long as the dorsal sepal, slender at the base ; lip 3-lobed, about as long as the petals, cuneate at the base ; side lobes short and broadly rounded ; front lobe triangular, acute ; disc verrucose and puberulous ; column very short and broad, its foot about ⅛ in. long ; capsules narrowly oblong, over ¾ in. long. *Bolus in Journ. Linn. Soc.* xxv. 186 ; *Durand & Schinz, Conspect. Fl. Afr.* v. 35.

P. grandiflora, Lindl., Bot. Mag. t. 3707, a West Tropical African species, is included by Bolus (Journ. Linn. Soc. xxv. 186) as a native of Natal, but I have not seen a specimen and suspect some mistake in the record.

IX. ANGRÆCUM, Bory.

Sepals and *petals* subequal, free, spreading or connivent. *Lip* affixed to the base of the column and continuous with it, produced at the base into an elongated or saccate spur, sometimes much elongated ; side lobes small or obsolete, situated at the mouth of the spur ; limb spreading or erect, entire or 3-lobed. *Column* very short, broad, concave in front, without wings and footless ; anther-bed truncate, entire. *Anther* terminal, operculate, incumbent, convex, often produced in front, scarcely 2-celled ; pollinia 2, globose, sulcate ; stipes single, clavate or slender ; gland squamiform, simple. *Capsule* oblong or fusiform.

Epiphytic herbs ; stems leafy or rarely leafless, short or sometimes elongated, not thickened into pseudobulbs ; leaves distichous, coriaceous or fleshy, articulated to a persistent sheath, often more or less obliquely 2-lobed at the apex ; peduncles lateral, simple ; flowers usually racemose, sometimes solitary, often small or medium-sized, usually white, sometimes buff or greenish ; bracts generally small.

DISTRIB. A genus of nearly 100 species, widely diffused in Continental Africa and the Mascarene Islands, with an outlying species in China and Japan, and another in the Philippines.   Ten of the 11 South African species are endemic.
The genus is here limited to species having a single stipes to the pollinia.

Stems leafy :
  Leaves flat :
    Flowers large ; spur over 1 in. long :
      Flowers in pendulous racemes ... ... (1) **Mystacidii.**

      Flowers solitary ... ... (2) **conchiferum.**

    Flowers small ; spur under ½ in. long :
      Stems very short, with filiform erect inflores-
      cence :
        Leaves 1–2 in. long :
          Leaves lanceolate ; spur of lip saccate ... (3) **sacciferum.**

          Leaves linear ; spur of lip oblong ... ... (4) **pusillum.**

        Leaves 3–5 in. long ... ... (5) **Burchellii.**
      Stems elongated, with rather stout, often spreading
      racemes :
        Front lobe of lip as long as the side lobes ... (6) **tricuspe.**

        Front lobe of lip much shorter than the side lobes (7) **bicaudatum.**

  Leaves terete :
    Sepals broadly ovate ; lip with rounded basal angles (8) **tridentatum.**

    Sepals ovate-lanceolate; lip with subulate basal
    angles ... ... (9) **Bolusii.**

  Leaves equitant ... ... (10) **Maudæ.**

Stems leafless ... ... (11) **Chiloschistæ.**

1. **A. Mystacidii** (Reichb. f. in Linnæa, xx. 677) ; stem stout, very short, with flexuous roots ; leaves distichous, narrowly ensiform or cuneate-oblong, distinctly and unequally 2-lobed at the apex, 2–5 in. long, ⅓–¾ in. broad, with subobtuse lobes ; scapes suberect, rather slender, 4–6 in. long, with 2 or 3 ovate sheaths below ; racemes distichous, lax, many-flowered ; bracts spreading, broadly triangular-ovate, obtuse, concave, ¼ in. long ; pedicels slender, ½ in. or more long ; flowers rather large, white ; sepals and petals spreading, lanceolate-oblong, subacute, ⅓–½ in. long ; lip lanceolate-oblong, about as large as the petals but more acute ; spur slender, cylindric, curved, 2–2½ in. long ; column short and broad ; rostellum curved, with a linear claw and a small ovate-oblong limb ; pollinia globose, attached by a long curved stipes to a single ovate-oblong gland. *Bolus in Journ. Linn. Soc.* xxv. 186 ; *Durand & Schinz, Conspect. Fl. Afr.* v. 44.   *A. Saundersiæ, Bolus in Hook. Ic. Pl. t.* 1728 ; *Journ. Linn. Soc.* xxv. 186, *and Ic. Orch. Austr.-Afr.* ii. *t.* 2 ; *Durand and Schinz, Conspect. Fl. Afr.* v. 45.

COAST REGION : Komgha Div. ; near Komgha, in woods, 1800 ft., *Flanagan,* 825.
EASTERN REGION : Transkei ; Kentani, 1200 ft., *Miss Pegler,* 471 ! Natal ; ravines around Durban, and at Palmiet and Umbilo Rivers, *Sanderson,* 892 ! Bishopstowe, *Sanderson,* 1001 ! and without precise locality, *Gueinzius, Mrs. K. Saunders* !

2. **A. conchiferum** (Lindl. in Hook. Comp. Bot. Mag. ii. 205);
stems erect, about 1½ in. high, with numerous flexuous roots;
leaves distichous, linear, unequally 2-lobed, coriaceous, ¾–1¼ in.
long, with verrucose sheaths; scapes filiform, about ¾ in. long,
1-flowered; bracts ovate, cucullate, membranous; sepals and petals
reflexed, setaceous-acuminate, nearly ¾ in. long; lip with a broad
ovate-acuminate crenulate membranous limb, nearly ½ in. long,
produced below into a funnel-shaped tube, and thence into a slender
acuminate spur, about 1½ in. long; column very short and broad;
anther not seen. *Drège, Zwei Pfl. Documente,* 125; *Reichb. f. in
Linnæa,* xx. 677; *Reichb. f. in Walp. Ann.* vi. 906; *Bolus in Journ.
Linn. Soc.* xxv. 186, *and Ic. Orch. Austr.-Afr.* ii. *t.* 4; *Durand &
Schinz, Conspect. Fl. Afr.* v. 41.

SOUTH AFRICA : without locality, *Mund & Maire*!
COAST REGION : Knysna Div.; Bosch River (Outeniqualand), under 500 ft.,
Drège, 8268! near Knysna, *Beck.* King Williams Town Div.; Perie Forest, near
King Williams Town, 2000–3000 ft., *Flanagan,* 2233, *Sim,* 19, 261.
EASTERN REGION : Natal, *Mrs. K. Saunders*!

3. **A. sacciferum** (Lindl. in Hook. Comp. Bot. Mag. ii. 205); a
small epiphyte, about 2 in. high, with very short stems and
numerous slender roots; leaves distichous, oblong or lanceolate,
obtuse, obliquely emarginate, coriaceous, ¾–2 in. long; scapes
numerous, filiform, somewhat spreading, rather longer than the
leaves, with 2 or 3 small sheaths below, 2–3-flowered-near the apex;
bracts sheath-like, small, brown; pedicels curved, very short;
flowers small, often secund, greenish-yellow; sepals recurved, about
⅙ in. long, dorsal ovate, obtuse, lateral oblong, obtuse and somewhat
oblique; petals reflexed, ovate-lanceolate, acute, about as long as
the sepals; lip cucullate, ovately trowel-shaped, obtuse, fleshy, nearly
as long as the sepals; spur saccate, somewhat curved, very obtuse,
about as long as the limb; column very short and broad; anther
nearly quadrate; pollinia subglobose, attached to a short obovate
stipes; gland small. *Reichb. f. in Walp. Ann.* vi. 906; *Drège,
Zwei Pfl. Documente,* 124, 146; *Bolus in Journ. Linn. Soc.* xxv. 186,
and Ic. Orch. Austr.-Afr. i. *t.* 10; *Durand & Schinz, Conspect. Fl.
Afr.* v. 45; *Schlechter in Engl. Jahrb.* xx. *Beibl.* 50, 29.

SOUTH AFRICA : without locality, *Mund*! *Pappe,* 79!
COAST REGION : Knysna Div.; in the forest by the Quarry near Knysna,
Burchell, 5416! Karratera River, *Drège,* 8269! Outeniqualand, *Drège,* 8271a!
Albany Div.; Coldstream, near Grahamstown, *Bolus,* 6239! Berg Plaats, near
Grahamstown, 700–800 ft., *Glass*!
KALAHARI REGION : Transvaal; upper wooded ravines at Barberton, 4650 ft.,
Galpin, 1103! Houtbosch, *Rehmann,* 5859! *Schlechter*; Bawendaland; Litonandoa
River, 2000 ft., *Schlechter.*
EASTERN REGION : Transkei; between the Gekau (Geua) and Bashee Rivers,
Drège! between the Kei and Bashee Rivers, *Drège,* 8271b! Natal; Durban,
Sanderson, 827! Richmond, *Sanderson,* 1043!

4. **A. pusillum** (Lindl. in Hook. Comp. Bot. Mag. ii. 205);
stems very short, with several slender flexuous roots at the base;

leaves several, linear or lanceolate-linear, acute, narrowed at the
base, often falcately curved, 1¼–2½ in. long; scapes erect, filiform,
rather longer than the leaves, laxly many-flowered; bracts obovate,
often 2-lobed with acute somewhat toothed lobes, hyaline; flowers
very small, white; sepals spreading, about ₁₂ in. long, subobtuse,
lateral elliptic-oblong, dorsal ovate and broader; petals ovate, subacute, rather narrower than the sepals; lip suborbicular, concave,
obtuse, rather fleshy, about as long as the petals; spur oblong,
obtuse, somewhat curved, about as long as the limb, slightly inflated
at the apex; column very short and stout; pollinia obovoid, subsessile on a single broadly oblong gland; capsule broadly elliptic,
⅛ in. long. *Drège, Zwei Pfl. Documente,* 124, 125; *Drège in Linnæa,*
xx. 217; *Bolus in Journ. Linn. Soc.* xxv. 186, *and Ic. Orch. Austr.-*
Afr. i. *t.* 54; *Durand & Schinz, Conspect. Fl. Afr.* v. 45; *Schlechter*
in Engl. Jahrb. xx. *Beibl.* 50, 29.

COAST REGION: Swellendam Div.; near Grootvaders Bosch, *Zeyher,* 3891!
Knysna Div.; in the forest by the Quarry near Knysna, *Burchell,* 5414! Koratra
(Karratera) River, *Drège,* 3582a! near Bosch River, *Drège,* 3582b! Uitenhage
Div.; at Uitenhage, *West.* Port Elizabeth Div.; around Krakakamma, *Burchell,*
4548! Alexandria Div.; Olifants Hoek Forest, *Ecklon & Zeyher,* 25! Albany
Div.; Kowie River, near Grahamstown, *Miss Atherstone!* Komgha Div.; in woods
near Komgha, 2000 ft., *Flanagan,* 888.

KALAHARI REGION: Transvaal; Houtbosch Berg, 6300 ft., *Schlechter.*

EASTERN REGION: Transkei; Krielis Country, *Bowker,* 273! Zululand, on very
high trees, *Mrs. K. Saunders!*

5. **A. Burchellii** (Reichb. f. in Flora, 1867, 117); roots numerous,
rather slender; stem very short; leaves elongate-linear, subacute,
somewhat arching, 3–6 in. long, about ⅙ in. broad; scapes filiform,
3–4 in. long, with 2 or 3 tubular basal sheaths; racemes several- to
many-flowered; bracts spathaceous, apiculate, ₁₂ in. long; sepals
and petals oblong, acute; lip boat-shaped, uncinate at the apex;
spur cylindric, obtuse, shorter than the pedicel; capsule broadly
elliptic, ¼ in. long. *Bolus in Journ. Linn. Soc.* xxv. 186; *Durand &*
Schinz, Conspect. Fl. Afr. v. 40.

COAST REGION: George Div.; forest near George, *Burchell,* 5841!

The specimens at Kew are in fruit, but the flowers were described by
Reichenbach.

6. **A. tricuspe** (Bolus in Journ. Linn. Soc. xxv. 163, 164, 186,
fig. 1); a robust epiphyte, 6–8 in. high; leaves ligulate, obtuse,
rigid, many-nerved, 4–4¾ in. long, about ⅓ in. broad; racemes
ascending, straight, many-flowered, rather shorter than the leaves;
bracts broadly ovate, persistent, ₂₀ in. long; pedicels about ⅓ in.
long; sepals spreading, ⅙ in. long, lateral lanceolate, acuminate,
recurved at the apex, dorsal linear-lanceolate, acute; petals
spreading, narrowly lanceolate, acute, shorter than the sepals, lip
deflexed, as long as the petals, broadly oblong below, tricuspidate
from about the middle with narrow acute lobes, the middle one

over twice as long as the lateral; spur pendulous, filiform, some-what curved, twice as long as the lip; column short and broad; rostellum apiculate; pollinia ovoid, attached by a single stipes, which is shortly bifid at the apex, to a small oblong gland. *Durand & Schinz, Conspect. Fl. Afr.* v. 46; *Schlechter in Engl. Jahrb.* xx. *Beibl.* 50, 29.

KALAHARI REGION : Transvaal; Houtbosch Berg, 6200 ft., *Schlechter*, 4698.
EASTERN REGION : Griqualand East; Malowe Forest, 4000 ft., *Tyson*, 3081 !
Natal ; without precise locality, *Cooper*, 1398 ! *McKen*, 14 ! *Wood* !

7. **A. bicaudatum** (Lindl. in Hook. Comp. Bot. Mag. ii. 205); stems cylindrical, $\frac{1}{3}$–1 ft. high, sometimes branched; leaves dis-tichous, linear-oblong, obliquely 2-lobed, coriaceous, $2\frac{1}{2}$–$4\frac{1}{2}$ in. long, $\frac{1}{3}$–$\frac{1}{2}$ in. broad; spikes horizontal or ascending, $1\frac{1}{2}$–3 in. long, many-flowered; bracts tubular, very obtuse, membranous, very short; pedicels $\frac{1}{8}$ in. long; flowers small, light buff; sepals ovate, obtuse, spreading, about $\frac{1}{8}$ in. long; petals $\frac{1}{8}$ in. long, much narrower than the sepals; lip longer than the sepals, oblong at the base, 3-partite above; side lobes linear, twice forked at the apex; front lobe ovate, obtuse, rather shorter than the lateral; spur straight, cylindric, about $\frac{1}{3}$ in. long; column very short; rostellum beak-like, incurved; pollinia globose, attached by a clavate stipes to a single broad gland. *Sond. in Linnæa*, xix. 75; *Drège, Zwei Pfl. Documente*, 130; *Reichb. f. in Walp. Ann.* vi. 905; *Harv. Thes. Cap.* ii. 5, *t.* 108; *Bolus in Journ. Linn. Soc.* xxv. 186, *and Ic. Orch. Austr.-Afr.* ii. *t.* 3; *Durand & Schinz, Conspect. Fl. Afr.* v. 40. *Listrostachys bicaudata, Finet in Bull. Soc. Bot. Fr.* liv. *Mém.* ix. 51, *t.* 10, *fig.* 14–16. *Eulophia angustifolia, Eckl. & Zeyh. ex Finet, l.c.*

SOUTH AFRICA : without locality, *Mrs. Holland*, 10 ! *Hallack*, 3 ! *Mund* ! *Mrs. Bowker* !
COAST REGION : Knysna Div. ; Forest Hall, *Miss Newdigate* ! *Marloth in Herb. Bolus*, 6209. Uitenhage Div. ; hills near the Coega River, *Zeyher*, 1119 ! 3888 ! limestone hills near the mouth of the Zwartkops River, *Drège* ! ravines near Uitenhage, *Prior* ! Port Elizabeth Div.; around Krakakamma, *Burchell*, 4553 ! *Pappe*, 80 ! Albany Div. ; near Grahamstown, *MacOwan*, 1459 ! Amos Kloof, 2000 ft., *Galpin*, 297. Komgha Div. ; woods near Komgha, 2000 ft., *Flanagan*, 636, and without precise locality, *Mrs. Barber*, 509 !
CENTRAL REGION : Somerset Div. ; Somerset East, *Bowker* !
EASTERN REGION : Natal; Bishopstowe and Richmond, *Sanderson*, 832 ! and without precise locality, *Mrs. K. Saunders* ! Zululand ; Ingoma, *Gerrard*, 1558 ! Eshowe, *Mrs. K. Saunders.*

8. **A. tridentatum** (Harv. Thes. Cap. ii. 6, in note); stems rather stout, with flexuous branches, 2–6 in. long; leaves rather crowded, rather stout, subterete, often curved, $1\frac{1}{2}$–3 in. long, with verrucose sheaths; flowers borne in axillary fascicles, pale waxy yellow, small; sepals broadly ovate, subacute, about $\frac{1}{12}$ in. long; petals ovate-lanceolate, subobtuse or sometimes with a small tooth near the apex, much narrower than the petals; lip as long as the sepals, broadly oblong, tricuspidate at the apex, with acute lobes and

rounded basal angles; spur straight, cylindric, $\frac{1}{4}$ in. long; column short and stout; rostellum subacute; pollinia globose, attached by a short clavate stipes to a single broad gland. *Bolus in Journ. Linn. Soc.* xxv. 186; *Durand & Schinz, Conspect. Fl. Afr.* v. 47.

EASTERN REGION: Natal; jutting rocks at Holderness, Krans Kloof, 1000 ft., *Sanderson,* 562!

9. **A. Bolusii** (Rolfe); stems rather slender, with flexuous branches, 3–5 in. long; leaves somewhat lax, slender, semi-cylindrical, depressed and traversed by a narrow channel on the upper surface, somewhat curved, 2½–4 in. long, with striate sheaths; flowers cream-coloured or pale waxy yellow, small, borne in axillary fascicles or 2–3 on very short peduncles; lateral sepals spreading, ovate-lanceolate, acute, subcordate at the base, reflexed at the apex, about $\frac{1}{8}$ in. long; dorsal ovate, acute, as long as the lateral, recurved; petals triangular-lanceolate, acute, recurved, rather shorter and narrower than the sepals; lip spreading, rather longer than the sepals, oblong-lanceolate, with a pair of oblong spreading lobes at the base, 3-lobed near the apex, with narrow spreading acute side lobes and a broader acuminate recurved front lobe; spur cylindric, straight, pendulous, about $\frac{1}{4}$ in. long; column short, broad; rostellum broadly ovate at the base, acuminate above; pollinia spherical, attached to a single rhomboid-cuneate stipes, with a single ovate-elliptic gland. *A. tridentatum, Bolus, Ic. Orch. Austr.-Afr.* i. *t.* 53, *partly, not of Harv.*

EASTERN REGION: Zululand; near Eshowe, *Maxwell in Herb. Bolus,* 6319!

Much more slender than *A. tridentatum,* Harv., with which it is united by Dr. Bolus, and with much broader sepals and a differently shaped lip.

10. **A. Maudæ** (Bolus, Ic. Orch. Austr.-Afr. i. t. 9); a dwarf epiphyte, about 2 in. high; roots numerous, filiform; stem very short, erect, scarcely $\frac{1}{2}$ in. high; leaves about 4, distichous, equitant, very fleshy, oblong-lanceolate, subobtuse, rarely somewhat falcate, $\frac{3}{4}$–1 in. long; spikes erect from the axils of the fallen leaves, 1½–2 in. long, rather slender, clothed with numerous imbricate sheaths below; spikes rather densely many-flowered; bracts ovate, acute, enfolding the pedicel and base of the flower, about $\frac{1}{4}$ in. long; pedicels short and stout; flowers very small, white; sepals somewhat spreading, about $\frac{1}{6}$ in. long, lateral narrowly lanceolate, acuminate, dorsal broader, acute; petals somewhat spreading, lanceolate, acute; lip about as long as the petals, entire, broadly ovate at the base, very acuminate above; spur cylindric, obtuse, slightly curved, rather shorter than the limb; column oblong, narrowed at the apex, about $\frac{1}{4}$ as long as the sepals; rostellum subulate, very acuminate, sharply upcurved at the apex; pollinia subglobose, attached by a clavate stipes to a single gland.

EASTERN REGION: Zululand; at Eshowe, *Mrs. K. Saunders!* and *in Herb. Bolus,* 6270!

11. A. Chiloschistæ (Reichb. f. in Linnæa, xx. 678); a very
slender leafless epiphyte ; roots numerous, slender ; stem very short ;
racemes spreading, very slender, 3–12, from the summit of the stem,
2–4 (or occasionally 8) in. long, lax, many-flowered ; bracts lanceolate, very minute ; pedicels very short ; flowers minute, white ;
sepals broadly oblong or obovate-oblong, very obtuse, under $\frac{1}{12}$ in.
long ; petals similar but a little smaller; lip transversely oblong,
very obtuse, about as long as the petals ; sac subglobose, rather
longer than the limb, naked within ; column short and stout,
somewhat hooded through the infolding of the top and sides, the
latter being winged and having a free basal angle ; anther large,
nearly square ; pollinia 2, subglobose, attached to a single clavate
stipes ; gland ovate, small. *Bolus in Journ. Linn. Soc.* xxv. 186,
and Ic. Orch. Austr.-Afr. i. *t.* 6 ; *Rolfe in Dyer, Fl. Trop. Afr.* vii.
149 ; *Durand & Schinz, Conspect. Fl. Afr.* v. 40.

EASTERN REGION : Natal ; near Tongaat, *Mrs. K. Saunders*! *and in Herb.
Bolus,* 6219 ! and without precise locality, *Gueinzius*! *Wood*!

Also in Tropical Africa and Madagascar.

X. LISTROSTACHYS, Reichb. f.

Sepals and *petals* subequal, free, spreading or subconnivent. *Lip*
affixed to the base of the column and continuous with it, produced
at the base into a cylindrical, often long and slender spur ; side
lobes small or obsolete, situated at the mouth of the spur ; limb
spreading or erect, entire or obscurely 3-lobed. *Column* very short,
broad, concave in front, without wings, footless ; anther-bed truncate, entire. *Anther* terminal, operculate, incumbent, convex, often
produced in front, scarcely 2-celled ; pollinia 2, globose, sulcate,
situated upon a pair of usually slender stipites, distinct or only
united at the base, where they are attached to a single variously-
shaped gland. *Capsule* oblong or fusiform.

Epiphytic herbs ; stems leafy, usually short, not thickened into pseudobulbs ;
leaves distichous, coriaceous or fleshy, articulated to a persistent sheath; peduncles
lateral, simple ; flowers racemose or capitate, very various in size ; bracts small or
medium-sized.

DISTRIB. Species about 70, exclusively African or Mascarene, the majority
being Tropical African. The single South African species is endemic.

1. L. arcuata (Reichb. f. in Walp. Ann. vi. 907); stems short
and stout ; leaves distichous, linear-oblong, emarginate, 2–6 in.
long, $\frac{1}{2}$–1 in. broad, very coriaceous ; scapes ascending, stout, $2\frac{1}{2}$–5
in. long ; racemes distichous, dense, many-flowered ; bracts ovate
or roundish-ovate, concave, membranous, $\frac{1}{4}$–$\frac{1}{2}$ in. long ; pedicels
stout, about $\frac{3}{4}$ in. long; flowers white ; sepals recurved, linear-
lanceolate, very acuminate, fleshy, $\frac{1}{2}$–$\frac{3}{4}$ in. long ; petals rather
smaller than the sepals, otherwise similar ; lip similar to the sepals

but with a broader concave base, revolute; spur curved, tapering
upwards from a stout base, 1–1¼ in. long; column short and stout;
rostellum elongate, beaked; pollinia spherical, attached by a pair
of curved linear stipites to a single oblong gland; capsule oblong,
1–1¼ in. long. *Durand & Schinz, Conspect. Fl. Afr.* v. 48. *Angræcum
arcuatum, Lindl. in Hook. Comp. Bot. Mag.* ii. 204; *Drège, Zwei Pfl.
Documente,* 130, 132, 136, 141; *Paxt. Fl. Gard.* ii. 120, *fig.* 199;
Harv. Thes. Cap. ii. 5, *t.* 107; *Bolus in Journ. Linn. Soc.* xxv. 186,
and *Ic. Orch. Austr.-Afr.* ii. *t.* 1; *Schlechter in Engl. Jahrb.* xx.
Beibl. 50, 29.

SOUTH AFRICA : without locality, *Prior*! *Mrs. Bowker*!
COAST REGION : Knysna Div. ; Forest Hall, *Miss Newdigate*! Uitenhage Div. ;
Addo, 1000–2000 ft., *Drège*! limestone hills near the mouth of the Zwartkops
River, *Drège,* 4580! *Prior*! hills near Koega River, *Zeyher,* 1117! Uitenhage,
Zeyher, 3890! *Tredgold,* 35! Port Elizabeth Div. ; Port Elizabeth, *Mrs. Holland,*
8! Alexandria Div. ; Olifants Hoek, *Atherstone,* 20! Zuurberg Range, *Drège.*
Bathurst Div. ; Glenfilling, *Drège*! Albany Div. ; wooded cliffs west of Grahams-
town, *Burchell,* 3591!
CENTRAL REGION : Somerset Div. ; Somerset East, *Bowker*!
KALAHARI REGION : Transvaal ; Golden Valley, Sheba Hill, near Barberton,
3500 ft., *Galpin,* 850, *Culver,* 73. Bawendaland ; Zindi River, 2500 ft.,
Schlechter.
EASTERN REGION : Pondoland, *Tillet*! Natal ; Inanda, *Wood,* 927! Durban,
Sanderson! 893! and without precise locality, *Gueinzius*! *Peddie*! *Mrs. K.
Saunders*!

XI. MYSTACIDIUM, Lindl.

Sepals and *petals* subequal, free, spreading or subconnivent. *Lip*
affixed to the base of the column and continuous with it, produced
at the base into a short or long spur; side lobes small or obsolete,
situated at the sides of the spur; limb spreading or erect, generally
entire. *Column* very short, broad, concave in front, without wings
and footless; anther-bed truncate, entire. *Anther* terminal,
operculate, incumbent, convex, often produced in front, scarcely
2-celled; pollinia 2, globose, sulcate, situated upon a pair of slender
stipites, which are attached to separate oblong or squamiform glands.
Capsule oblong or fusiform.

Epiphytic herbs ; stems leafy, short or long, not thickened into pseudobulbs ;
leaves distichous, coriaceous or fleshy, articulated to a persistent sheath ;
peduncles lateral, simple ; flowers generally small or medium-sized, racemose or
solitary ; bracts usually small.

DISTRIB. Species about 50, African or Mascarene, with a single outlying representative in Ceylon. The nine South African species are endemic.

Stems leafy :
  Rostellum without bearded appendages :
    Plants with distinct more or less elongated stems :
      Lip trowel-shaped ; spur falcately incurved ... (1) **Gerrardi.**

      Lip fan-shaped ; spur slightly incurved ... ... (2) **Pegleræ.**

    Plants with very short or nearly obsolete stems :
      Leaves 3½–5 in. long ... ... (3) **Millari.**

Leaves 1–2½ in. long:
    Sepals and petals obtuse ; spur of lip subclavate
        and nearly straight ...   . (4) **caffrum.**

    Sepals and petals acute ; spur of lip slender and
        curved   ...   . (5) **Flanaganii.**

Rostellum with 2 more or less barbate appendages :
    Sepals and petals broad and subobtuse   ...   ... (6) **pusillum.**

Sepals and petals narrow and acute or acuminate :
    Leaves ¾–1¼ in. long; spur of lip about 1¼ in.
        long   ...   . (7) **venosum.**

    Leaves 1¼–4 in. long ; spur of lip 1½–2¼ in. long ... (8) **filicorne.**
Stems leafless   ...   . (9) **gracile.**

1. **M. Gerrardi** (Bolus in Journ. Linn. Soc. xxv. 187) ; stems
climbing, emitting roots at intervals, slender, 2–6 in. long ; leaves
distichous, strap-shaped, obliquely emarginate or shortly 2-lobed,
coriaceous, 2–4 in. long, ⅙–¼ in. broad ; racemes flexuous, rather
shorter than the leaves, lax, 5–8-flowered ; bracts ovate-oblong,
acute, minute ; pedicels rather longer than the bracts ; flowers
white or cream-coloured ; sepals somewhat spreading, ⅙–¼ in. long,
lateral lanceolate-oblong, subobtuse, somewhat concave, dorsal
ovate, very obtuse ; petals broadly ovate or suborbicular, obtuse,
rather shorter than the sepals ; lip transversely oblong, obscurely
3-lobed, rather longer than the sepals ; side lobes rounded ; front lobe
obscurely apiculate ; spur cylindric with a wide mouth, obtuse,
incurved or hooked, about twice as long as the limb ; column broad
and short ; anther very obtuse ; pollinia 2, distant, nearly hemi-
spherical, each attached to a linear flat incurved stipes lying on
either margin of the rostellum ; glands 2, ovate, distant ; rostellum
beak-like, cuneate-oblong, the apex bent down into the mouth of
the spur. *Bolus in Trans. S. Afr. Phil. Soc.* xvi. 145 ; *Durand &
Schinz, Conspect. Fl. Afr.* v. 52. *Aëranthus Gerrardi, Reichb. f. in
Flora,* 1867, 117. *Angræcum Gerrardi, Bolus, Ic. Orch. Austr.-Afr.*
i. *t.* 7.

COAST REGION : Komgha Div. ; near Komgha, *Flanagan,* 1698 !
EASTERN REGION : Natal ; Krans Kop, *McKen,* 19 ! Deepdene, Richmond,
Sanderson, 831 ! and without precise locality, *Mrs. K. Saunders* ! Zululand ;
woods near Eshowe, *Mrs. Charles Saunders,* 6218 (ex *Bolus*), and without precise
locality, *Gerrard,* 1819 !

2. **M. Pegleræ** (Bolus in Trans. S. Afr. Phil. Soc. xvi. 146) ; a
glabrous epiphyte with slender aërial roots ; stems strong, leafy,
clothed with the sheaths of fallen leaves below, 2–2¼ in. long ;
leaves distichous, ligulate, unequally 2-lobed, with obtuse lobes,
narrowed below, semi-amplexicaul, coriaceous, 1¼–2 in. long, ¼–⅓ in.
broad ; racemes somewhat spreading ; rhachis zigzag or flexuous,
1¾–2 in. long, 5–6-flowered ; flowers erect, white ; sepals spreading,
about ¹⁄₁₀ in. long ; lateral lanceolate, acute ; dorsal oblong or ovate,
obtuse, rather shorter than the lateral ; petals somewhat spreading,
rhomboid, with rounded angles, shorter than the sepals, nearly as

broad as long; lip arcuately deflexed, subflabellate, with rounded angles, about ¼ in. long, front margin obscurely 3-lobed, side lobes somewhat undulate ; spur cylindric, somewhat curved, about ¼ in. long ; column decurved, oblong, emarginate ; rostellum deflexed, lanceolate, acute ; pollinia globose, affixed to a pair of slender stipites ; glands ovate, minute. *Bolus, Ic. Orch. Austr.-Afr.* ii. *t.* 6.

EASTERN REGION: Transkei ; on trees in forest near Kentani, 1000–1200 ft., *Miss Pegler,* 993.

3. M. Millari (Bolus in Trans. S. Afr. Phil. Soc. xvi. 147);

a glabrous caulescent epiphyte with rather stout aërial roots ; stems stout, clothed with the sheaths of the fallen leaves below, 1¾–2 in. long; leaves 2 or 3, usually at the summit of the stem, erect or spreading, ligulate, subundulate, equally 2-lobed, with narrow sub-obtuse lobes, coriaceous, depressed along the centre, 3¼–5½ in. long, over ½ in. broad ; racemes suberect, ½–1¼ in. long, rigid, 7–10-flowered ; flowers mostly erect, white; pedicels about ¼ in. long ; sepals spreading, about ¼ in. long, lateral oblong-lanceolate, sub-acute, dorsal oblong, very obtuse, somewhat concave, with a small marginal tooth at each side near the base, about twice as broad as the lateral ; petals somewhat spreading, obliquely ovate, sub-acute, rather shorter and narrower than the dorsal sepal ; lip ovate, obtuse, concave, decurved at the apex, as long as the petals ; spur inflated at the base, then cylindric, incurved, ¾ in. long ; column decurved, oblong or subrhomboid, dilated below the middle ; pollinia lenticular, affixed to a pair of slender stipes ; glands ovate ; rostellum without appendages. *Bolus, Ic. Orch. Austr.-Afr.* ii. *t.* 5.

EASTERN REGION: Natal; on trees in ravines near Durban, 500 ft., *Millar in Herb. Wood,* 8437 ! Palmiet and Umbilo Rivers, *Sanderson,* 892 A !

4. M. caffrum (Bolus in Trans. S. Afr. Phil. Soc. xvi. 145); a

dwarf epiphyte ; stems short and stout, climbing, about ½ in. long, with numerous slender roots ; leaves distichous, 3–4, linear-oblong or strap-shaped, obliquely 2-lobed, with obtuse lobes, somewhat narrowed at the base, coriaceous, 2–2½ in. long, ¼–⅓ in. broad ; scapes spreading or arcuate, 1¾–2 in. long, flexuous, rather dense and many-flowered ; bracts cucullate, minute ; pedicels slender, ⅓ in. long ; flowers medium-sized, white ; sepals spreading, elliptic-oblong, obtuse, about ¼ in. long, the dorsal rather broader than the lateral ; petals broadly elliptic, obtuse, as long as the petals and rather broader ; lip ovate-orbicular, obtuse or emarginate, rather smaller than the petals ; spur funnel-shaped at the base, then narrowed, slightly thickened at the apex, obtuse, rather longer than the pedicels ; column short, abruptly deflexed at the apex ; rostellum somewhat elongate, beaked ; pollinia spherical-compressed, attached to two distinct filiform caudicles ; glands flattened at the base. *Angræcum caffrum, Bolus, Ic. Orch. Austr.-Afr.* i. *t.* 8.

EASTERN REGION : Pondoland ; at Emagushen, *Tyson*, 2841.   Gr.qualand East ;
near Fort Donald, 4340 ft., *Tyson*, 1607. Natal; without precise locality,
Mrs. K. Saunders ! *Gerrard*, 2184 !

5. M. **Flanaganii** (Bolus in Trans. S. Afr. Phil. Soc. xvi. 145);
stem very short, with several very flexuous fleshy roots ; leaves 3–5,
oblong or lanceolate-oblong, subobtuse or obscurely 2-lobed,
leathery, 1–1½ in. long, ¼–⅓ in. broad ; racemes several, slender,
arching, sometimes flexuous, ½–2½ in. long, 5–10-flowered ; bracts
minute, cup-shaped, caducous ; pedicels slender, ⅓ in. long ; flowers
small, pale greenish-yellow ; lateral sepals decurved and spreading,
narrowly rhomboid, acuminate, narrowed at the base, about ¼ in.
long ; dorsal sepal ovate-lanceolate, acute, recurved at the apex,
shorter than the lateral ; petals spreading, ovate-lanceolate, acute,
somewhat undulate, recurved at the apex, as long as the dorsal
sepal ; lip ovate-lanceolate, subacute, about as long as the petals,
recurved ; spur filiform, elongated, curved, nearly 1 in. long; column
short, broad ; anther large ; rostellum projecting forwards, curved,
oblong, obtuse, with two small lobes on each side ; pollinia ovoid,
with distinct filiform stipes and separate ovate glands ; capsule
elliptic, subcompressed, ribbed, ¼ in. long.   *Angræcum Flanagani*,
Bolus, Ic. Orch. Austr.-Afr. i. *t.* 52.

COAST REGION : Komgha Div. ; in woods near Komgha, 2000 ft., *Flanagan*,
1027 !

6. M. **pusillum** (Harv. Thes. Cap. ii. 47, t. 173); stems very
short, with several slender flexuous roots ; leaves 2–3, oblong or
elliptic-oblong, obtuse or unequally emarginate, coriaceous, ½–1 in.
long ; racemes 2–4, slender, spreading, longer than the leaves,
loosely 3–5-flowered ; bracts hooded, papery, small ; flowers pale
greenish-yellow, small ; sepals spreading, about ₁₂ in. long,
lateral elliptic-lanceolate, subobtuse, dorsal broadly elliptic, obtuse ;
petals oblong, obtuse, rather shorter than the sepals ; lip ovate-
oblong, obtuse, somewhat undulate, as long as the petals ; spur
elongated, curved, funnel-shaped at the base, then tapering upwards
and filiform, about ½ in. long ; column short and broad ; rostellum
projecting forwards, 3-partite, with slender slightly bearded side
lobes and lanceolate acute smooth front lobe ; pollinia subglobose,
attached by filiform stipites to separate ovate-lanceolate glands.
Bolus in Journ. Linn. Soc. xxv. 187, *and Ic. Orch. Austr.-Afr.* i.
t. 57 ; *Durand & Schinz, Conspect. Fl. Afr.* v. 54.   *Aëranthus
pusillus, Reichb. f. in Flora*, 1867, 117.

COAST REGION : Komgha Div. ; on trees near the mouth of the Kei River,
100 ft., *Flanagan*, 1808.
EASTERN REGION : Natal ; Tongaat, *Mrs. K. Saunders* ! Tugela, *Mrs. K.
Saunders* ! and without precise locality, *Gerrard* !

Sanderson, 827, in the Trinity College Herbarium, Dublin, may be a reduced
form of this. It is labelled " Kranskloof ? or Maritzburg ? flowered at Durban,
Jan. 1865 ; flower pale green,"

7. M. venosum (Harv. MSS.) ; stem very short, with numerous
slender flexuous roots ; leaves 2–4, oblong, unequally 2-lobed at the
apex, coriaceous, suberect or spreading, 1–1¾ in. long; racemes
slender, arching, 1¾–2¼ in. long, loosely 5–8-flowered ; bracts
ochreate, minute, deciduous ; flowers small, white ; sepals spreading
or recurved, oblong-lanceolate, acuminate, ¼–⅓ in. long, lateral
broader than the dorsal at the base ; petals linear-lanceolate, acute,
as long as the sepals ; lip orbicular at the base round the wide
orifice of the spur, with one or two small lobules, then suddenly
contracted into a narrow lanceolate acute limb, about as long as
the sepals; spur elongated, curved, gradually narrowed from the
orifice, 1½–2 in. long ; column short and stout ; rostellum beaked,
projecting forwards, 3-fid, with linear shortly bearded side lobes,
and an acuminate smooth front lobe; pollinia broadly ellipsoid,
attached by filiform stipites to separate oblong glands. *M. gracile*,
Bolus, Ic. Orch. Austr.-Afr. i. *t.* 56, *not of Harv.*

CoAST REGION: Komgha Div. ; near Komgha, *Flanagan in MacOwan & Bolus,
Herb. Norm. Austr.-Afr.* 1369 !
EASTERN REGION: Natal ; without precise locality, *Mrs. K. Saunders*! *Fannin,*
86 ! Zululand, *Mrs. K. Saunders*!

8. M. filicorne (Lindl. in Hook. Comp. Bot. Mag. ii. 206) ; stems
short and stout, with several rather stout flexuous roots below ;
leaves 3–5, distichous, oblong or obovate-oblong, emarginate or
obtusely 2-lobed at the apex, coriaceous, 2–5 in. long, ½–1 in. broad ;
racemes several, arching or pendulous, sometimes suberect, flexuous,
3–8 in. long, loosely many-flowered ; bracts ochreate, ⅛ in. long ;
pedicels slender, over ½ in. long; flowers rather large, often secund,
creamy-white ; sepals spreading or recurved, linear-lanceolate, acute,
⅓–½ in. long, dorsal rather broader than the lateral ; petals similar
to the lateral sepals but rather smaller, spreading or recurved ; lip
lanceolate or ovate-lanceolate, acuminate, about as long as the
petals, with a pair of minute basal teeth ; spur slender, cylindric,
curved, 1¼–1¾ in. long, narrowly funnel-shaped at the mouth ;
column short and stout, somewhat constricted below the anther;
rostellum beaked, projecting forwards, 3-partite, with the lateral
lobes papillose and the middle lobe lanceolate and smooth ; pollinia
broadly ellipsoid, attached by curved slender stipites to separate
lanceolate-oblong glands. *Drège, Zwei Pfl. Documente*, 124, 134,
142, 151 ; *Harv. Thes. Cap.* ii. 48, *t.* 175 ; *Bolus in Journ. Linn.
Soc.* xxv. 187, *and Ic. Orch. Austr.-Afr.* i. *t.* 55. *M. longicornu,
Durand. & Schinz, Conspect. Fl. Afr.* v. 53. *Epidendrum capense,
Linn. f. Suppl.* 407. *Limodorum longicorne, Thunb. Prodr. Pl. Cap.*
3 (*excl. syn.*), *and Fl. Cap. ed. Schult.* 28. *L. longicornu, Sw. in
Schrad. Journ. Bot.* ii. 230. *Eulophia longicornis, Spreng. Syst. Veg.*
iii. 720. *Angræcum capense, Lindl. Gen. & Sp. Orch.* 248. *Aëranthus
filicornis, Reichb. f. in Walp. Ann.* vi. 900.

SOUTH AFRICA: without locality, *Thunberg* ! *Mrs. Holland* ! *Mrs. Bowker* !
COAST REGION : Knysna Div. ; Koratra (Karratera) River, *Drège*! Knysna
Forest, *Barkly* ! Uitenhage Div. ; near Uitenhage, *Burchell*, 4247 ! dense forests

of Addo, *Zeyher*, 1118! Zwartkops River, *Zeyher*, 3889! Enon, at Olyfhout
Kloof and Olifants Kloof, *Drège*! Coega River, *Zeyher*! Port Elizabeth Div.;
Port Elizabeth, *Walsh*! Bathurst Div.; Lushington Valley, Jagers Drift, and
woods near Riet River, *Atherstone*, 18! near Kowie River, *Ecklon & Zeyher*, 16!
MacOwan, 312! Albany Div.; near Grahamstown, *Bolton*! Howisons Poort,
Hutton! and without precise locality, *Atherstone*, 54! *Bowie*! King Williamstown
Div.; King Williamstown, *Brownlee*! Peddie Div.; Fish River Hills, near
Trumpeters Drift, *Drège*. British Kaffraria, *Cooper*, 3617!
CENTRAL REGION: Somerset Div.?; Somerset East? *Bowker*!
EASTERN REGION: Pondoland; St. Johns River, *Drège*. Natal; Tugela River,
Gerrard, 1816! *Mudd*! Source of Nototi River, *Gerrard*! Bishopstowe, *Sanderson*,
1001! and without precise locality, *Sanderson*, 504! *Trimen*! *Buchanan*!

9. M. gracile (Harv. Thes. Cap. ii. 48, t. 174); a leafless epiphyte,
with very short stems and numerous slender flexuous roots; scapes
somewhat arching, very slender, 1–1$\frac{1}{4}$ in. long, racemes one-sided,
lax, 4–8-flowered; bracts obliquely cup-shaped, about $\frac{1}{12}$ in. long;
pedicels slender, $\frac{1}{4}$ in. long; sepals and petals subconnivent, ovate-
lanceolate, subobtuse, about $\frac{1}{8}$ in. long; lip ovate-oblong, subobtuse,
about as long as the petals, with very short side lobes at the mouth
of the spur; spur curved, $\frac{3}{4}$ in. long, funnel-shaped at the mouth,
narrowed and very slender above; column short and stout; rostellum 3-partite, front lobe triangular, acute, side lobes clavate,
with papillose apex; pollinia subglobose, attached by slender curved
stipites to separate oblong glands. *Bolus in Journ. Linn. Soc.* xxv.
187; *Durand & Schinz, Conspect. Fl. Afr.* v. 53. *Aëranthus gracilis*,
Reichb. f. in Flora, 1867, 117.

COAST REGION: Bedford Div.; Kaga Berg Range, in primæval forest, 4500 ft.,
Weale in Herb. Hance, 16338! Stutterheim Div.; Klaklazele Berg, *Cooper*, 271!
EASTERN REGION: Natal; Dargle Farm, *Mrs. Fannin*, 95! and without precise
locality, *Sanderson*!

Bolus (Journ. Linn. Soc. xxv. 187) includes *Mystacidium Meirax*, Bolus
(*Aëranthus Meirax*, Reichb. f. in Flora, 1885, 540) as doubtfully South African,
but it was described in a paper on Comoro Island Orchids, and probably came
from there, though no locality is given.

XII. CORYMBIS, Thouars.

Sepals and *petals* narrow and approximate below, somewhat
spreading above. *Lip* erect from the base, linear, channelled,
dilated at the apex into a short recurved limb. *Column* elongated,
erect, terete, clavate at the apex, terminating in two erect lobes or
auricles; anther-bed short. *Anther* erect, narrow, acuminate,
about as long as the column; cells contiguous; pollinia granular,
affixed to a subulate stipes, with a peltate gland, descending behind
the rostellum. *Stigma* broad, often thickened at the lower margin;
rostellum erect, acuminate, bifid after the removal of the pollinia.
Capsule linear, subterete, crowned with the column and the remains
of the persistent perianth.

Tall, erect, terrestrial herbs, with leafy, sometimes branched stems, and numerous fibrous roots ; leaves large, elliptic-lanceolate, acute, plicate-veined, sessile or petiolate, enlarged at the base into an amplexicaul sheath ; flowers large or medium-sized, loosely arranged in short subcorymbose axillary or terminal panicles or racemes; bracts small, ovate.

DISTRIB. A genus of about 15 species, widely dispersed through the tropics.

1. **C. Welwitschii** (Reichb. f. in Flora, 1865, 183) ; stems erect, 3–6 ft. high ; leaves elliptic-lanceolate or broadly elliptic, acuminate, 7–12 in. long, $1\frac{1}{2}$–$2\frac{1}{2}$ in. broad ; panicles axillary and terminal, shortly pedunculate ; bracts ovate-lanceolate, acuminate, $\frac{1}{2}$–$\frac{3}{4}$ in. long ; flowers greenish-white ; sepals and petals narrowly elongate-linear, narrowly lanceolate near the apex, $2\frac{1}{2}$ in. long ; lip similar to the sepals, but the apical part of the limb is ovate and about four times as broad ; column $1\frac{3}{4}$ in. long ; capsule linear-oblong, 1 in. long. *Rolfe in Bolet. Soc. Brot.* ix. 141, *and in Dyer, Fl. Trop. Afr.* vii. 180 ; *Durand & Schinz, Conspect. Fl. Afr.* v. 56 ; *Bolus, Ic. Orch. Austr.-Afr.* ii. *t.* 98. *Corymbis disticha, Lindl. Fol. Orch. Corymbis,* 1, *partly; Journ. Linn. Soc.* vi. 138, *partly. Corymbis corymbosa, Durand & Schinz, Conspect. Fl. Afr.* v. 56, *partly. Corymborchis Welwitschii, O. Kuntze, Rev. Gen. Pl.* ii. 658.

EASTERN REGION : Pondoland ; mouth of St. Johns River, 300 ft., *Flanagan in Herb. Bolus,* 8703 !

Also in Tropical Africa.

XIII. ZEUXINE, Lindl.

Sepals subequal ; dorsal sepal erect, concave ; lateral spreading, free. *Petals* narrow, often cohering with the dorsal sepal into a galea. *Lip* very shortly adnate to the base of the column, erect, concave or subsaccate at the base, with or without a pair of calli inside, more or less contracted in the middle, dilated at the apex into a small entire or larger 2-lobed spreading limb. *Column* very short ; anther-bed short, continuous with the margin of the rostellum. *Anther* erect or inclined in front, oblong, shortly apiculate ; cells contiguous with the outer valves, broad ; pollinia granular, 2, affixed to a linear or cuneate stipes, with a broad rounded gland, descending behind the rostellum. *Stigmas* 2, lateral ; rostellum erect, short and broad or minute. *Capsule* small, erect, ovoid or subglobose.

Slender or dwarf terrestrial herbs, with short creeping rhizomes, and ascending or erect simple stems ; leaves sessile on the broad sheath and linear, or petiolate with an ovate or lanceolate limb, often membranous ; flowers small, numerous, arranged in dense or slender spikes ; bracts membranous, mostly shorter than the flowers.

DISTRIB. Species about 30, mostly Indian and Malayan, with a few Tropical African representatives and one endemic South African species.

1. **Z. cochlearis** (Schlechter in Engl. Jahrb. xx. Beibl. 50, 11);
a terrestrial herb, with a tuft of thick brown hairy roots; stems
erect, terete, leafy, 2–12 in. high, with a few reduced sheaths at the
base; leaves numerous, suberect, linear-lanceolate, acute, amplexicaul
at the base, $\frac{3}{4}$–2 in. long; spikes oblong or narrowly ovate, 1–2$\frac{1}{4}$ in.
long, dense-flowered; bracts lanceolate or linear-lanceolate, acumi-
nate, $\frac{1}{4}$–$\frac{1}{3}$ in. long; flowers shortly pedicelled, about $\frac{1}{8}$ in. long, white,
with a yellow lip; sepals connivent, ovate-lanceolate, subobtuse, con-
cave, the dorsal somewhat gibbous at the base; petals adherent to the
margin of the dorsal sepal, ovate-oblong, obtuse, slightly shorter than
the sepals, the outer margin undulate; lip somewhat fleshy, pandurate,
obtuse, about as long as the sepals, concave at the base, and with a
pair of oblong calli, margins incurved and irregularly crenulate;
column short and broad; rostellum 2-partite with subulate acuminate lobes; anther rounded; pollinia pyriform, attached to a broad
flattened stipes, narrowing to a small gland at the base. *Bolus, Ic.
Orch. Austr.-Afr.* i. *t.* 58.

EASTERN REGION : Natal ; marshy sandy places near the seashore at the mouth
of the Umgeni River, near Durban, *Schlechter*, 3001, 3002 ! *Wood*, 5321.

XIV. PLATYLEPIS, A. Rich.

Sepals subequal, narrow, free, connivent round the column;
lateral united at the base into a very short chin. *Petals* narrow,
slightly cohering with the dorsal sepal into a hood. *Lip* sessile at
the base of the column, erect, channelled, broadly ventricose at the
base, cohering with the margins of the column; limb very little
dilated, shorter than the sepals. *Column* elongated, subterete;
anther-bed oblong, erect behind the rostellum. *Anther* erect,
acuminate, as long as the rostellum ; cells distinct; pollinia 2,
sectile or granular, adhering to the lobes of the rostellum. *Stigma*
broad, papillose ; rostellum erect, 2-lobed, with lanceolate acuminate
lobes. *Capsules* oblong, shortly contracted at the apex.

Terrestrial herbs, with creeping rhizomes and ascending leafy stems ; leaves
petiolate, ovate or ovate-lanceolate, membranous ; flowers narrow, shortly
pedicelled, arranged in dense spikes ; bracts ovate.

DISTRIB. Species 6, four being natives of the Mascarene Islands, one Tropical
African, and one South African.

1. **P. australis** (Rolfe in Kew Bulletin, 1906, 378); rhizome
creeping; stem ascending, leafy; leaves petiolate, ovate, subacuminate, membranous, 15–20-nerved, $\frac{3}{4}$–1$\frac{1}{2}$ in. long, 1–2 in. broad;
petiole 1$\frac{1}{4}$–2 in. long, dilated at the base into a tubular membranous
sheath ; scapes erect, 6–12 in. high, with several distant spathaceous
sheaths below ; racemes oblong or elongate, 2–4$\frac{1}{2}$ in. long, many-

flowered; bracts ovate, acute, glandular-pubescent, $\frac{1}{4}$–$\frac{1}{2}$ in. long; pedicels $\frac{1}{4}$ in. long, sparsely glandular-pubescent; flowers green, with a white upper half to the lip; dorsal sepal erect, oblong, sub-obtuse, $\frac{1}{4}$ in. long, glandular-pubescent outside; lateral subconnivent, suddenly reflexed about the middle, oblong, subobtuse; petals spathulate-linear, obtuse, cohering with the dorsal sepal into a narrow hood; lip $\frac{1}{4}$ in. long, erect, ventricose and 2-gibbous at the base, lightly adhering to the margins of the column; limb elliptic-oblong, constricted at the base, recurved at the apex; column nearly $\frac{1}{4}$ in. long, clavate. *P. glandulosus, Bolus in Journ. Linn. Soc.* xxv. 187, *and Ic. Orch. Austr.-Afr.* i. *t.* 11, *not of Reichb. f.; Durand & Schinz, Conspect. Fl. Afr.* v. 58.

EASTERN REGION: Natal; in the bush swamp at the head of Durban Bay, *Sanderson*, 1048! *Wood*, 812! 4122! and in *MacOwan & Bolus, Herb. Norm. Austr.-Afr.*, 1008! and without precise locality, *Mrs. K. Saunders*!

XV. POGONIA, Juss.

Sepals subequal, free, erect or rarely spreading. *Petals* similar to the sepals, or often broader or shorter, erect or declinate. *Lip* erect from the base of the column, free, without a spur, sessile or unguiculate, entire or 3-lobed; side lobes enfolding the column; disc variously crested or lamellate. *Column* elongated, slightly clavate at the apex, without wings; anther-bed more or less elevated, entire or denticulate. *Anther* somewhat stipitate on the margin of the anther-bed, more or less incumbent, obtusely conical or subglobose, imperfectly 2-celled; pollinia granular, 2, sometimes confluent into 1, not tailed, free or sessile on the rostellum. *Stigma* oblong or broad; rostellum short.

Terrestrial herbs; rhizomes tuberiferous; leaves 1 to few, very various in shape, often not appearing until after the flowers; flowers medium-sized or large, solitary or in erect or pendulous racemes; bracts often small.

DISTRIB. Species about 80, widely dispersed through the tropics, with a few temperate representatives in North and South America and in Japan. The single South African species belongs to the section *Nervilia*.

1. **P. purpurata** (Reichb. f. & Sond. in Flora, 1865, 184); a dwarf deciduous herb, with spherical or ovoid tubers $\frac{1}{2}$–$\frac{3}{4}$ in. in diam.; leaf solitary, radical, petiolate, elliptic-ovate, acute, plicate; blade about $2\frac{1}{2}$ in. long, 1 in. broad; petiole 1 in. long, with a lanceolate acute sheath; scape appearing before the leaf, erect, 6–9 in. high, with 2 or 3 distant sheaths; raceme loosely 3–5-flowered; bracts narrowly linear, membranous, $\frac{1}{3}$–$\frac{1}{2}$ in. long; pedicels slender, $\frac{1}{4}$ in. long; flowers spreading, medium-sized, pale-green, with some narrow radiating veins on the lip; sepals spreading, narrowly

lanceolate, acuminate, about $\frac{3}{4}$ in. long; petals rather broader than the sepals, otherwise similar; lip strongly 3-lobed, nearly as long as the sepals, cuneate at the base; side lobes broadly oblong, generally obtuse or rounded at the apex, erect; front lobe subulate-oblong, acute, longer than the side lobes, somewhat recurved; column clavate, very slender at the base, about $\frac{1}{4}$ in. long, suddenly dilated at the apex; anther incumbent, hinged at the apex of the column; pollen granular; capsule ovoid or ellipsoid, shortly pedicelled, about $\frac{1}{2}$ in. long. *Bolus in Journ. Linn. Soc.* xxv. 187, *and Ic. Orch. Austr.-Afr.* i. *t.* 12; *Durand & Schinz, Conspect. Fl. Afr.* v. 59; *Schlechter in Engl. Jahrb.* xx. *Beibl.* 50, 29.

KALAHARI REGION: Transvaal; hill tops near Orotava Mine, Barberton, 4000 ft., *Culver*, .57! Mundt's Farm, near Pretoria, 4700 ft., *Schlechter*, 3704! Magaliesberg, *Zeyher*, 1584! Aapies River, foot of Magaliesberg Range, 4800 ft., *Schlechter*, 3663.

XVI. BRACHYCORYTHIS, Lindl.

Sepals free, connivent or the lateral ultimately spreading; lateral oblique, often broader than the dorsal, sometimes falcate and ascending. *Petals* usually oblique and similar to the dorsal sepal or narrower, incurved over the column. *Lip* continuous with the base of the column, free, spreading or incurved; claw broad, somewhat fleshy, concave or gibbous at the base, but not spurred; limb dilated, tridentate or trilobed. *Column* short and broad, anther-bed erect. *Anther* erect or somewhat reclinate, broad; cells parallel; apex inferior, adnate to the short side lobes of the rostellum; pollinia granular; caudicles short; glands large, contiguous; staminodes lateral, small, rounded or auriculate. *Stigma* pulvinate, fleshy or concave, often large; rostellum short, trilobed; middle lobe erect between the anther-cells, somewhat plicate; side lobes short, suberect. *Capsule* narrowly oblong.

Terrestrial, usually very leafy herbs, with undivided fusiform or ovoid tubers; leaves sessile, generally numerous and imbricate, gradually decreasing upwards into the bracts; spikes or racemes usually dense and many-flowered, rarely somewhat lax; bracts ovate or ovate-lanceolate, often somewhat leafy.

DISTRIB. Species about 22, exclusively Continental African, 4 of which occur within our limits, 3 of them being endemic.

Leaves cauline, very numerous:
  Whole plant glabrous or nearly so:
    Leaves $1\frac{1}{2}$–2 in. long; bracts $\frac{1}{2}$–$1\frac{1}{4}$ in. long; lateral
     sepals $\frac{1}{4}$ in. long ...   . (1) **ovata.**

    Leaves $\frac{3}{4}$–$1\frac{1}{4}$ in. long; bracts $\frac{1}{2}$–$\frac{3}{4}$ in. long; lateral
     sepals $\frac{1}{8}$ in. long ...   . (2) **Allisoni.**

  Whole plant very pubescent   ...   . (3) **pubescens.**

Leaves radical or subradical, few:
  Leaves 2 or 3, oblong or lanceolate, suberect ...   ... (4) **Tysoni.**

  Lower leaf ovate, spreading or recurved   ...   ... (5) **virginea.**

1. B. ovata (Lindl. Gen. & Sp. Orch. 363); a robust glabrous herb, $\frac{3}{4}$–$1\frac{1}{2}$ ft. high, with densely leafy stems, and 3 or 4 much thickened fusiform roots; leaves very numerous, sessile, somewhat spreading, ovate or lanceolate-ovate, acute or acuminate, lower $1\frac{1}{2}$–2 in. long, $\frac{1}{2}$–$1\frac{1}{4}$ in. broad, upper gradually decreasing into the leafy bracts; racemes dense, leafy, 4–9 in. long, with very numerous flowers; bracts sessile, spreading, ovate-lanceolate or broadly lanceolate, acuminate, $\frac{1}{2}$–$1\frac{1}{4}$ in. long; pedicels stout, $\frac{1}{2}$–$\frac{3}{4}$ in. long; flowers medium-sized, purple or lilac-purple, with a yellow keel and a few yellow spots on the lip; dorsal sepal broadly elliptic-oblong, subobtuse, somewhat cucullate, $\frac{1}{4}$ in. long; lateral sepals obliquely falcate-ovate, ascending, subobtuse, slightly longer than the dorsal; petals obliquely ovate, subobtuse, as long as the dorsal sepal, curved over the column at the apex; lip suberect from the curved base, obovate-oblong, apiculate or shortly 3-dentate at the apex, recurved at the sides, nearly $\frac{1}{2}$ in. long, base gibbous and concave; disc with a narrow keel extending to the apex; column broad, $\frac{1}{6}$ in. long. *Drège, Zwei Pfl. Documente,* 149, 151; *Harv. Thes. Cap.* i. 34, *t.* 53; *Bolus in Journ. Linn. Soc.* xxv. 205, *and Ic. Orch. Austr.-Afr.* i. *t.* 62; *Kränzl. Orch. Gen. et Sp.* i. 541, *and in Ann. Naturhist. Hofmus. Wien,* xx. 3; *Durand & Schinz, Conspect. Fl. Afr.* v. 115. *Platanthera ovata, Schlechter in Engl. Jahrb.* xx. *Beibl.* 50, 12, 30.

COAST REGION: Stockenstrom Div.; Kat Berg, 2000 ft., *Hutton*! Stutterheim Div.; Dohne Mountain, 3600 ft., *Bolus.* King Williamstown Div.; Perie Forest, *Sim*, 26. Komgha Div.; near mouth of Kei River, *Flanagan*, 1300.

KALAHARI REGION: Transvaal; Mauch Berg, near Lydenberg, 5000–5850 ft., *Atherstone*! near Lydenburg, 5000 ft., *Schlechter*, 3922; Mac Mac, *Mudd*!

EASTERN REGION: Transkei; Fort Bowker, *Bowker*, 28! 621! near Butterworth, *Mrs. Barber*, 16! 28! Tembuland; in grassy places, *Bolus.* Griqualand East; near Newmarket, *Krook, Penther*, 115. Pondoland; between Umtata River and Umsikaba River, 1000–2000 ft., *Drège*, 4569! near Umkwani River, 200 ft., *Tyson*, 2670. Natal; Bellair and Fields Hill, 800 ft., *Sanderson*, 211! Attercliffe, 800 ft., *Sanderson*, 482! near Durban, *Gerrard*, 735! Highlands, *Gerrard*, 1540! Inanda, *Wood*, 424! 1169! and without precise locality, *Buchanan*! *Gerrard*, 606! Swaziland; mountain sides at 4000 ft., *Galpin*, 725! near Pinetown, 800 ft., *Wood*, 540! Zululand, *Gerrard*, 606!

2. B. Allisoni (Rolfe); a glabrous herb, about 9 in. high, with densely leafy stems; leaves very numerous, sessile, somewhat spreading, ovate or lanceolate-ovate, acute or acuminate, lower $\frac{3}{4}$–$1\frac{1}{4}$ in. long, $\frac{1}{2}$–$\frac{3}{4}$ in. broad, upper decreasing into the somewhat leafy bracts; raceme dense, 4 in. long, many-flowered; bracts sessile, spreading, ovate or lanceolate-ovate, acute or acuminate, $\frac{1}{2}$–$\frac{3}{4}$ in. long; pedicels stout, $\frac{1}{4}$–$\frac{1}{3}$ in. long; flowers dull bluish-purple (*Allison*); dorsal sepal broadly elliptic-oblong, subobtuse, somewhat concave, over $\frac{1}{6}$ in. long; lateral sepals obliquely ovate, ascending and slightly falcate, subobtuse, longer than the dorsal sepal; petals obliquely ovate, subobtuse, rather longer than the dorsal sepal, somewhat incurved at the apex; lip incurved from the base, suberect at the apex, obovate, obtuse and

obscurely tridentate, recurved at the sides, about $\frac{1}{3}$ in. long, gibbous
at the base and broadly concave owing to the somewhat dilated
sides ; disc with an obtuse keel extending to the apex ; column
broad, $\frac{1}{8}$ in. long.

EASTERN REGION : Natal ; in damp places at Oliviers Hoek, sources of Tugela
River, 4000 ft., *Allison* !

3. **B. pubescens** (Harv. Thes. Cap. i. 35, t. 54) ; a stout erect
herb, 1–2 ft. high, densely leafy, and pubescent all over; leaves
sessile, ovate or ovate-lanceolate, acute or acuminate, $\frac{3}{4}$–1$\frac{1}{4}$ in. long,
decreasing into short sheaths below and into the bracts above ;
racemes 3–8 in. long, dense, many-flowered ; bracts lanceolate,
acuminate, $\frac{1}{2}$–1 in. long ; flowers medium-sized, white and flesh-
coloured ; dorsal sepal elliptic-oblong, obtuse, concave, $\frac{1}{4}$ in. long ;
lateral sepals obliquely semi-ovate, obtuse, falcately ascending, rather
longer than the dorsal ; petals obliquely semiovate-oblong, subobtuse,
rather shorter than the dorsal sepal, incurved over the column at
the apex ; lip suberect from the curved base, $\frac{1}{3}$ in. long ; cuneately
dilated and 3-lobed at the apex, with a broadly triangular front
lobe and rounded side lobes, base gibbous and subsaccate with a
pair of rounded auricles ; disc obtusely carinate, with a pair of
rounded crests near the base ; column about $\frac{1}{6}$ in. long. *Bolus in
Journ. Linn. Soc.* xxv. 205 ; *Ridl. in Journ. Bot.* 1895, 295 ; *Rolfe
in Dyer, Fl. Trop. Afr.* vii. 201 ; *Kränzl. Orch. Gen. et Sp.* i. 542 ;
Durand & Schinz, Conspect. Fl. Afr. v. 115. *Peristylus hispidulus
and var. minor, Rendle in Journ. Linn. Soc.* xxx. 398. *Platanthera
Brachycorythis, Schlechter in Engl. Jahrb.* xx. *Beibl.* 50, 12, 30.

KALAHARI REGION : Transvaal ; Diamond Fields, *Tuck*, 1 ! Pilgrims Rest,
Greenstock ! near Lydenburg, 5000 ft., *Schlechter*, 3926 !
EASTERN REGION : Natal ; near Durban, *Sanderson*, 178 ! *Gerrard*, 719 ! Dargle
Farm, *Mrs. Fannin*, 80 ! Inanda, *Wood*, 1659 ! and without precise locality,
Cooper ! *Buchanan* ! Zululand, *Gerrard*, 630 !

Also in Tropical Africa.

4. **B. Tysoni** (Bolus in Journ. Linn. Soc. xx. 485) ; a slender
glabrous sparsely leafy herb, $\frac{3}{4}$–1$\frac{1}{4}$ ft. high, and a pair of ovoid
tubers ; leaves 2 or rarely 3, near the base of the stem, lanceolate-
oblong, acute or subacute, attenuate at the base, suberect or rarely
spreading, 1$\frac{1}{2}$–5 in. long, $\frac{1}{4}$–$\frac{1}{2}$ in. broad ; scapes slender, erect, $\frac{3}{4}$–1$\frac{1}{4}$
ft. high, with several lanceolate acuminate sheaths, rarely leaf-like
below ; racemes 2–6 in. long, lax, few- to many-flowered ; bracts
sessile, ovate-lanceolate, acuminate, $\frac{1}{3}$–$\frac{1}{2}$ in. long ; pedicels rather
stout, $\frac{1}{4}$–$\frac{1}{3}$ in. long ; flowers small, light green, with some brown on
the sepals, and the lip white with a lilac-coloured band at the base ;
dorsal sepal ovate or lanceolate-ovate, acute or acuminate, $\frac{1}{4}$ in.
long, somewhat cucullate ; lateral sepals rather narrower than the
dorsal, oblique and somewhat spreading ; petals obliquely ovate,
acute, shorter than the dorsal sepal, incurved over the column at

the apex; lip suberect from the curved base, $\frac{1}{3}$ in. long; limb broadly dilated, suborbicular, apiculate or cuspidate, minutely crenulate, somewhat reflexed at the sides; base slightly gibbous and concave; disc with a slender keel extending to the apex; column broad, $\frac{1}{6}$ in. long. *Bolus in Journ. Linn. Soc.* xxv. 205, *and Ic. Orch. Austr.-Afr.* i. *t.* 63; *Durand & Schinz, Conspect. Fl. Afr.* v. 115. *Neobolusia Tysoni, Schlechter in Engl. Jahrb.* xx. *Beibl.* 50, 6, 30; *Kränzl. Orch. Gen. et Sp.* i. 550, *and in Ann. Naturhist. Hofmus. Wien,* xx. 3.

COAST REGION : Bedford Div. ; summit of Kaga Berg (not Kat Berg), 3200 ft., *MacOwan*, 1109! at 3000 ft., *Hutton*! British Kaffraria ; mountain sides, *Mrs. Barber*, 40 !

KALAHARI REGION : Orange River Colony ; slopes at foot and near summit of Quaqua Mountain, Witzies Hoek, *Thode*, 51! Transvaal ; Botsabelo, 5500 ft., *Schlechter*, 4062! in marshes, Houtbosch (Woodbush) Mountains, 8800 ft., *Schlechter*, 4447 ! *Rehmann*, 5848! Belfast, *Burtt-Davy*, 1302 !

EASTERN REGION : Tembuland ; Xalanga, 4800 ft., *Bolus* ; near Maclear, *Bolus* ; near Umtata, *Schlechter*, 6343. Griqualand East ; near Kokstad, 5000 ft., *Tyson*, 1083 ! *and in MacOwan & Bolus, Herb. Norm. Austr.-Afr.*, 479 ! *Krook, Penther*, 218 ; Newmarket, *Penther*, 114 ; Nalogha, *Penther*, 176. Natal ; sources of Tugela River at Oliviers Hoek, 4000 ft., *Allison* !

5. **B. virginea** (Rolfe) ; a slender suberect herb, about $\frac{1}{2}$ ft. high, with a single radical leaf and an ovoid tuber at the base ; leaf shortly petiolate, broadly ovate, acute or shortly acuminate, submembranous, $\frac{3}{4}$–1 in. long, about $\frac{1}{2}$ in. broad ; scape slender, somewhat flexuous, glabrous, with one lanceolate acuminate sheath about the middle ; raceme short, about 3–5-flowered ; bracts ovate, acuminate, about $\frac{1}{4}$ in. long ; flowers with white sepals, and pink petals and lip, the latter with a green central nerve and base ; dorsal sepal ovate-oblong, subacute, over $\frac{1}{3}$ in. long, 3-nerved ; lateral sepals somewhat oblique and spreading, otherwise similar to the dorsal ; petals broadly ovate, subobtuse, about $\frac{1}{3}$ as long as the sepals, somewhat oblique at the base, 2-nerved ; lip suberect from the curved base, two-thirds as long as the sepals ; limb broadly ovate, subentire, subobtuse, $\frac{1}{6}$ in. broad, base somewhat gibbous and concave, somewhat dilated at the sides ; disc with an obtuse keel at the base ; column short and broad. *Platanthera virginea, Bolus, Ic. Orch. Austr.-Afr.* i. *t.* 60.

KALAHARI REGION: Orange River Colony ; sides and summit of the Mont-aux-Sources, 8000–9000 ft., *Flanagan*, 1982 !

EASTERN REGION: Natal ; Van Reenans Pass, 6000–7000 ft., *Schlechter*.

An anomalous species, which Dr. Bolus referred somewhat doubtfully to *Platanthera*. It, however, agrees better with *Brachycorythis* in floral structure.

XVII. PLATANTHERA, L. C. Rich.

Sepals unequal, free; lateral more or less spreading or reflexed. *Petals* simple, usually narrower than the dorsal sepal and adpressed to it, forming a galea. *Lip* continuous with the column, sometimes shortly adnate to it, produced at the base into a short or much elongated spur; limb spreading or pendulous, narrow or broad, entire or 3-lobed; side lobes sometimes fimbriate or pectinate. *Column* short, footless; anther-bed erect, short or scarcely as long as the anther. *Anther-cells* parallel or diverging; apex inferior, short and adnate to the side lobes of the rostellum; pollinia granular, with short caudicles and exserted naked glands; staminodes lateral, small, rounded or auriculate. *Stigmas* sessile or subsessile, more or less confluent, often pulvinate; rostellum 3-lobed; middle lobe subulate or tooth-like, situated between the anther-cells; side lobes short. *Capsule* elliptic or oblong.

Erect terrestrial herbs, with ovoid-globose or rarely somewhat lobed tubers; leaves radical or cauline, spreading or suberect; flowers small or medium-sized, numerous, borne in dense or lax spikes or racemes; bracts narrow or rarely somewhat leafy.

DISTRIB. Species about 60, mostly in temperate and subtropical regions of the northern hemisphere, with a few somewhat anomalous species in Africa, two of which occur within our limits, one being endemic.

Sepals ¼ in. long; lip ovate-oblong, subentire   ...   ...   (1) **tenuior.**

Sepals ⅛ in. long; lip 3-lobed from a narrow base   ...   (2) **macowaniana.**

1. **P. tenuior** (Schlechter in Engl. Jahrb. xx. Beibl. 50, 12, 30); an erect, somewhat slender herb, ½–1¼ ft. high, with leafy stems; leaves sessile, suberect, ovate or ovate-lanceolate, acute or acuminate, ¾–1½ in. long, decreasing into sheaths below and into the bracts above; racemes 1–4½ in. long, somewhat dense, many-flowered; bracts suberect, lanceolate or ovate-lanceolate, acute or acuminate, ⅓–1 in. long; pedicels ¼–⅓ in. long; flowers medium-sized, purple; dorsal sepal elliptic-oblong, obtuse, concave, ¼ in. long; lateral sepals obliquely semiovate-oblong, subobtuse, somewhat spreading, rather longer than the dorsal; petals obliquely oblong, obtuse, ¼ in. long; lip rather longer than the dorsal sepal, suberect, entire, oblong or ovate-oblong, obtuse; minutely crenulate; disc with a fleshy central nerve and a pair of narrow erect lamellæ at the base; spur conic-oblong, obtuse, somewhat curved, as long as the limb; column stout, ⅙ in. long. *Rolfe in Dyer, Fl. Trop. Afr.* vii. 205; *Bolus, Ic. Orch. Austr.-Afr.* i. t. 61. *Brachycorythis tenuior, Reichb. f. in Flora,* 1865, 183, *and in Otia Bot. Hamb.* ii. 104; *Kränzl. Orch. Gen. et Sp.* i. 543, *and in Ann. Naturhist. Hofmus. Wien,* xx. 3; *Durand & Schinz, Conspect. Fl. Afr.* v. 115. *Habenaria tenuior, N.E. Br. in Gard. Chron.* 1885, xxiv. 307; *Bolus in Journ. Linn. Soc.* xxv. 192.

KALAHARI REGION : Transvaal ; Stryd Poort, Makapans Berg, *Rehmann*, 5390 !
near Marabastad, 4700 ft., *Schlechter*, 4349 ; Middelburg, *Hewitt*, 8042 !
EASTERN REGION : Natal ; between Maritzburg and Bishopstowe, 2000 ft.,
Sanderson, 1046 ! Inanda, *Wood*, 714 ! near Umkomanzi River, *Krook, Penther*,
110 ; near Estcourt, *Krook, Penther*, 273, and without precise locality, *Mrs.
K. Saunders*!

Also in Tropical Africa, in the Mozambique District.

2. **P. macowaniana** (Schlechter in Engl. Jahrb. xx. Beibl. 50, 12) ;
an erect stout glabrous herb, 4–8 in. high, with very leafy stems
and several clavate or much thickened roots; leaves cauline,
numerous, sessile, suberect, ovate-lanceolate, acute or acuminate,
1–2 in. long, $\frac{1}{4}$–$\frac{1}{2}$ in. broad, gradually decreasing upwards into the
leafy bracts ; racemes dense, leafy, 2–4 in. long, with very numerous
flowers ; bracts sessile, spreading, ovate-lanceolate or narrow from a
broad base, acuminate, $\frac{1}{4}$–$\frac{1}{2}$ in. long ; pedicels rather stout, $\frac{1}{6}$–$\frac{1}{4}$ in.
long ; flowers small, pale green, with brown sepals and a light
purple anther ; sepals subconnivent, ovate, obtuse, $\frac{1}{6}$ in. long, the
dorsal somewhat cucullate ; petals ovate, obtuse, about as long as
the sepals ; lip about as long as the sepals, recurved, 3-lobed from a
narrow base, somewhat fleshy ; side lobes triangular-oblong, obtuse,
shorter than the oblong obtuse front lobe ; spur ovoid-oblong,
obtuse, rather shorter than the limb ; column very short and broad.
Bolus, Ic. Orch. Austr.-Afr. i. t. 59. *Brachycorythis macowaniana,
Reichb. f. Otia Bot. Hamb.* ii. 104 ; *Kränzl. Orch. Gen. et Sp.* i. 545,
and in Ann. Naturhist. Hofmus. Wien, xx. 3 ; *Durand & Schinz,
Conspect. Fl. Afr.* v. 115. *Habenaria macowaniana, N. E. Br.
in Gard. Chron.* 1889, v. 168; *Bolus in Journ. Linn. Soc.* xxv. 192.
*Gymnadenia macowaniana, Schlechter in Verhandl. Bot. Ver.
Brandenb.* xxxv. 46.

COAST REGION : Riversdale Div. ; slopes of the Langeberg Range near Riversdale,
1000 ft., *Schlechter*, 1900 ! Knysna Div. ; grassy mountain near Knysna, 1000 ft.,
Newdigate in MacOwan Herb. Norm. Austr.-Afr., 1733 ! Knysna, *Schlechter,
Penther*, 283 ; Silver River, *Penther*, 55. Albany Div. ; Featherstone Kloof, near
Grahamstown, 2200 ft., *MacOwan*, 2627 ! Grahamstown, *Tuck* ! *Glass*.

XVIII. SCHIZOCHILUS, Sond.

Sepals free, connivent or ultimately spreading, subequal. *Petals*
oblique, smaller than the sepals. *Lip* united to the base of the
column, spreading; claw broadly concave, produced at the base
into a spur, contracted at the mouth ; limb entire or trifid.
Column short; rostellum broad, entire, elevated and subgaleate ;
stigma near the base of the column, elevated, fleshy and concave.
Anther erect or slightly reclinate, broad; cells elevated, parallel,
adnate at the apex to the short side lobes of the rostellum ; pollinia
in separate cells, coarsely granular ; caudicles short ; glands distinct.
Capsule ovoid or oblong, short, straight.

Glabrous terrestrial herbs, with ovoid-oblong tubers ; leaves radical or subradical, small or narrow, the upper reduced to short sheaths ; flowers small,

densely spicate, subsecund, yellow or white ; bracts narrow, shorter than the
flowers.

DISTRIB.   Species 11, all but the Rhodesian *S. Cecili*, Rolfe, limited to extra-
tropical South Africa.

Sepals 3–3½ lin. long :
  Spikes more or less elongated or narrow :
    Lip markedly 3-lobed :
      Leaves 2–4 lin. broad :
        Spur half as long as the limb of the lip ;   disc
          with a small erect basal tubercle ...   ... (1) **Zeyheri.**

        Spur nearly as long as the limb of the lip ;   disc
          with a short keel and two minute basal
          tubercles   ...   . (2) **Sandersoni.**

      Leaves 1½–2 lin. broad   ...   . (3) **strictus.**

    Lip shortly 3-lobed ; disc with three narrow basal
      keels   ...   . (4) **trilobus.**

  Spikes subcapitate, ¾ in. or more broad   ...   ... (5) **flexuosus.**

Sepals 2–2½ lin. long :
  Scapes 1–1½ ft. high ; leaves few :
    Raceme rather lax ; flowers orange ; spur rather
      longer than the limb of the lip ...   . (6) **Rehmanni.**

    Raceme rather dense ; flowers white ; spur scarcely
      one-eighth as long as the limb of the lip   ... (7) **transvaalensis.**

  Scape 6–10 in. high ; leaves several :
    Leaves 2–6 lin. broad ...   . (8) **Bulbinella.**

    Leaves 1½–2 lin. broad :
      Spikes short and broad ; spur one-quarter as
        long as the limb of the lip   ...   . (9) **angustifolius.**

      Spikes more elongated or narrow ; spur half as
        long as the limb of the lip   ...   . (10) **Gerrardi.**

1. **S. Zeyheri** (Sond. in Linnæa, xix. 78) ; plant ½–1¼ ft. high,
with 2 ovoid tubers ; leaves 3–8, radical or subradical, linear or
linear-oblong, obtuse or subacute, 1–4½ in. long, 2–6 lin. broad ;
scapes ½–1¼ ft. high, with 3 to 6 lanceolate sheaths ; racemes oblong,
somewhat broad, ¾–2 in. long, rather dense ; bracts ovate or ovate-
lanceolate, acuminate, ¼–⅓ in. long ; flowers yellow or rarely
porcelain-white with a yellow lip ; sepals ovate-oblong, subobtuse
or apiculate, over ¼ in. long, prominently 3-nerved ; petals narrowly
ovate, acute, more than half as long as the sepals, 1-nerved ; lip
about as long as the sepals, pubescent, 3-lobed ; side lobes falcate-
oblong, semiovate, subobtuse ; front lobe linear-oblong, obtuse or
subacute, about twice as long as the side lobes ; disc with a small
erect tubercle at the base ; spur cylindrical or subcompressed,
oblong, obtuse, about half as long as the limb ; column very short.
Bolus in Journ. Linn. Soc. xxv. 205, *and Ic. Orch. Austr.-Afr.* i. *t.* 18
(*excl. Sanderson and Wood,* 478) ; *Durand & Schinz, Conspect. Fl.
Afr.* v. 116.   *Brachycorythis Zeyheri, Reichb. f. in Flora,* 1867, 117.
Platanthera Zeyheri, Schlechter in Engl. Jahrb. xx. *Beibl.* 50, 12.

COAST REGION : Fort Beaufort Div. ; Winterberg Range, *Zeyher*! *Mrs. Barber*,
31 ! 240 ! 518 ! Stockenstrom Div. ; in marshes at the summit of Elands Berg,

6000 ft., *Scully*, 405 ! *and in MacOwan & Bolus, Herb. Norm. Austr.-Afr.*, 1378 ! Kat Berg, 4000–5000 ft., *Zeyher*, 56 ! King Williamstown Div. ; Perie Forest, *Sim*.

KALAHARI REGION : Orange River Colony ; Oliviers Hoek Pass, summit of the Drakensberg Range, *Wood*, 3425 !

EASTERN REGION : Transkei ; Fakus Territory, *Sutherland* ! top of Bazeia Mountain, 4000 ft., *Baur*, 630 ! 631 ! Griqualand East ; Klein Pot River, Maclear District, 4550 ft., *Galpin*, 6841 ! near Kokstad, *Tyson*, 1600. Natal ; above Karkloof, *Buchanan in Herb. Sanderson*, 1080 ! and without precise locality, *McKen*, 3 !

2. **S. Sandersoni** (Harv. MSS.) ; plant 1–1¾ ft. high ; leaves 2–4, radical or subradical, linear or linear-oblong, acute or subacute, 1⅓–4 in. long, 2–4 lin. broad ; scapes 1–1¾ ft. high, with 6–8 narrow acuminate sheaths ; racemes oblong or somewhat elongate, 1–4 in. long, dense or somewhat lax ; bracts ovate or ovate-lanceolate, acuminate, ¼–⅓ in. long ; flowers yellow (*Sanderson*) ; sepals oblong or ovate-oblong, subobtuse, ¼–⅓ in. long, 3-nerved ; petals narrowly ovate, acute, more than half as long as the sepals, 1-nerved ; lip rather shorter than the sepals, 3-lobed, pubescent ; side lobes falcate-oblong, subobtuse ; front lobe linear-oblong, subobtuse, nearly twice as long as the side lobes ; disc with a short keel and two minute tubercles at the base ; spur cylindrical, obtuse, nearly as long as the limb ; column short.

EASTERN REGION : Natal ; near stream on the west side of Fields Hill, 1000 ft., *Sanderson*, 564 ! near Durban, *Gerrard*, 2176 ! Inanda, *Wood*, 478 !

3. **S. strictus** (Rolfe) ; plant straight, ¾–1¼ ft. high ; leaves 5–8, subradical, linear, acute, 1¾–4 in. long, ⅛–⅙ in. broad ; scapes ¾–1¼ ft. high, with 6 to 8 narrow acuminate sheaths ; racemes oblong, ¾–1½ in. long (mature not seen), dense ; bracts ovate, acuminate, ¼–⅓ in. long ; flowers apparently yellow ; sepals ovate-oblong or ovate-lanceolate, subacute, over ¼ in. long, 3-nerved ; petals ovate-oblong, acute, more than half as long as the sepals, 1-nerved ; lip about as long as the sepals, 3-lobed, pubescent ; side lobes falcate-oblong, subacute ; front lobe linear-oblong, subobtuse, much longer than the side lobes ; disc with a narrowly oblong tubercle at the base ; spur cylindrical, oblong, obtuse, over half as long as the limb ; column very short. *Platanthera Zeyheri*, *Schlechter in Engl. Jahrb.* xx. *Beibl.* 50, 30 (*not elsewhere*).

KALAHARI REGION : Transvaal ; Klein Olifants River, 5500–6000 ft., *Schlechter*, 4028 ! O'Neills Farm, Lydenburg District, *Wilms*, 1397 !

Very similar to *S. Zeyheri* in floral structure, but markedly different in its straight habit and very narrow leaves.

4. **S. trilobus** (Rolfe) ; plant about 1 ft. high ; leaves 3–4, radical, linear-oblong, subacute, 2–3 in. long, ⅙–¼ in. broad ; scapes about 1 ft. high, with about 6 narrow acuminate sheaths ; racemes oblong, recurved (in specimens seen), 1–1½ in. long, rather dense ; bracts ovate, shortly acuminate, ¼–⅓ in. long ; flowers apparently yellow ; sepals ovate or ovate-oblong, subobtuse, ¼ in. long,

3-nerved ; petals narrowly ovate, acute, two-thirds as long as the
sepals, 1-nerved ; lip rather shorter than the sepals, shortly 3-lobed,
slightly pubescent ; side lobes broadly oblong, obtuse ; front lobe
broadly oblong, obtuse, rather longer than the side lobes ; disc
with 3 narrow keels at the base, the median longer and rather
slender ; spur cylindrical, subobtuse, about as long as the limb ;
column short and broad.

EASTERN REGION : Natal ; Dargle Farm, *Mrs. Fannin*, 8 !

5. S. flexuosus (Harv. MSS.); plant $\frac{3}{4}$–1 ft. high ; leaves 3–5,
radical or subradical, linear or linear-oblong, subacute, 1$\frac{1}{2}$–3 in.
long, $\frac{1}{6}$–$\frac{1}{3}$ in. broad ; scapes $\frac{3}{4}$–1 ft. high, somewhat flexuous,
with 4–6 lanceolate sheaths ; racemes oblong or subcapitate,
$\frac{1}{2}$–1$\frac{1}{2}$ in. long, somewhat recurved, dense ; bracts ovate, acuminate, $\frac{1}{4}$–$\frac{1}{3}$ in. long ; flowers white (*Wood*); sepals ovate or
ovate-oblong, subobtuse, $\frac{1}{4}$ in. long, 3 - nerved ; petals broadly
subacute, half as long as the sepals, 1-nerved ; lip nearly as
long as the sepals, 3-lobed, slightly pubescent ; side lobes falcate-
ovate, subobtuse ; front lobe linear-oblong, subobtuse, twice as long
as the side lobes ; disc with 3 minute tubercles at the base ; spur
cylindrical, obtuse, nearly as long as the limb ; column short and
broad.

EASTERN REGION : Natal ; Dargle Farm, *Mrs. Fannin*, 56 ! amongst grass on a
dry hill at Liddesdale, 4000 ft., *Wood*, 3934 !

Based on a specimen collected by Mrs. Fannin, in which the colour of the flower
is not noted, but the one collected by Wood apparently represents the same
species.

6. S. Rehmanni (Rolfe); plant 1–1$\frac{1}{2}$ ft. high ; leaves 2–3,
radical or subradical, linear or linear-oblong, subacute, 1$\frac{1}{2}$–5 in.
long, $\frac{1}{6}$–$\frac{1}{4}$ in. broad ; scapes 1–1$\frac{1}{2}$ ft. high, with 3–5 narrow
acuminate sheaths ; racemes oblong, 1–3 in. long, rather lax ;
bracts ovate or ovate-lanceolate, acuminate, $\frac{1}{4}$–$\frac{1}{3}$ in. long ; flowers
orange (*Mudd*); sepals oblong, subobtuse, about 2$\frac{1}{2}$ lin. long,
3-nerved ; petals narrowly ovate-oblong, acute, two-thirds as long
as the sepals, 1-nerved ; lip as long as the sepals, 3-lobed,
pubescent ; side lobes falcate-ovate, subobtuse ; front lobe oblong,
subobtuse, nearly twice as long as the side lobes ; disc with 3
prominent tubercles at the base ; spur cylindrical, obtuse, rather
longer than the limb ; column short and stout.

KALAHARI REGION : Transvaal ; Houtbosch, *Rehmann*, 5849 ! near Lydenburg,
Atherstone ! Mac Mac, *Mudd* !

7. S. transvaalensis (Rolfe); plant $\frac{3}{4}$–1$\frac{1}{2}$ ft. high ; leaves
linear or linear-oblong, acute or subacute, 1$\frac{1}{4}$–3 in. long, $\frac{1}{6}$–$\frac{1}{3}$ in.
broad ; scapes $\frac{3}{4}$–1$\frac{1}{2}$ ft. high, somewhat flexuous, with 6–8 very
narrow acuminate sheaths ; racemes elongate, somewhat narrow,
1$\frac{1}{2}$–3$\frac{1}{2}$ in. long, rather dense ; bracts ovate, acute, $\frac{1}{6}$–$\frac{1}{4}$ in.

long; flowers white (*Mudd*); sepals ovate or ovate-oblong, sub-obtuse, $\frac{1}{6}$ in. long, 3-nerved; petals broadly ovate, acute or sub-acute, two-thirds as long as the sepals, 1-nerved; lip as long as the sepals, shortly 3-lobed, somewhat fleshy and pubescent; side lobes semiovate or broadly rounded, obtuse; front lobe ovate or tri-angular-ovate, obtuse, 3 times as long as the side lobes; disc with 3 fleshy tubercles at the base; spur conical, very minute, scarcely one-eighth as long as the lip; column short and very broad.

KALAHARI REGION : Transvaal ; near Lydenburg, *Atherstone* ! Mac Mac, *Mudd* ! Graskop, near Lydenburg, *Burtt-Davy*, 1464 !

8. S. Bulbinella (Bolus in Journ. Linn. Soc. xxv. 205); plant $\frac{1}{2}$–1 ft. high; leaves 4–7, radical or subradical, linear or linear-oblong, subacute, 2–3$\frac{1}{2}$ in. long, $\frac{1}{6}$–$\frac{1}{2}$ in. broad; scapes $\frac{1}{2}$–1 ft. high, with 4–6 lanceolate acuminate sheaths; racemes oblong, rather broad, 1–1$\frac{3}{4}$ in. long; bracts ovate, acute, $\frac{1}{6}$–$\frac{1}{4}$ in. long; flowers bright golden-yellow (*Galpin*); sepals broadly ovate, subobtuse, under $\frac{1}{6}$ in. long, 3-nerved; petals broadly ovate, subobtuse, more than half as long as the sepals, 1-nerved; lip as long as the sepals, obscurely or shortly 3-lobed, slightly pubescent; side lobes rounded or very obtuse, very short, sometimes subobsolete; front lobe oblong or ovate-oblong, subobtuse; disc with 3 nerves or obscure keels at the base; spur saccate, very minute; column short and very broad. *Durand & Schinz, Conspect. Fl. Afr.* v. 116. *S. Burchellii, Ind. Kew. Suppl.* i. 384 (*by error*). *Brachycorythis Bulbinella, Reichb. f. in Flora*, 1867, 116. *Platanthera Bulbinella, Schlechter in Engl. Jahrb.* xx. *Beibl.* 50, 12.

KALAHARI REGION : Transvaal ; among stones on shaded bend of hill at Bosch's Farm, near Barbérton, 4000 ft., *Galpin*, 713 !
EASTERN REGION : Transkei ; Fakus Territory, *Sutherland* ! Griqualand East ; Mount Currie, near Kokstad, 5700–6500 ft., *Tyson*, 1072 ! *and in MacOwan & Bolus, Herb. Norm. Austr.-Afr.* 478 !

9. S. angustifolius (Rolfe); plant about $\frac{1}{2}$ ft. high; leaves 5–9, radical, narrowly linear, acute, 1$\frac{1}{2}$–3 in. long, $\frac{1}{8}$–$\frac{1}{6}$ in. broad; scapes about $\frac{1}{2}$ ft. high, with 3–5 narrow acuminate sheaths; racemes oblong or subcapitate, broad, $\frac{1}{2}$–1 in. long, dense; bracts ovate, acute, $\frac{1}{6}$–$\frac{1}{4}$ in. long; flowers white (*Sankey, Wood*), white with yellow lip (*Allison*); sepals ovate or ovate-oblong, subobtuse, over $\frac{1}{8}$ in. long, 3-nerved; petals broadly ovate, subacute, not half as long as the sepals, 1-nerved; lip rather shorter than the sepals, 3-lobed, somewhat pubescent; side lobes broadly rounded or somewhat ovate, obtuse; front lobe broadly oblong, obtuse, twice as long as the side lobes; disc with 3 nerves or obscure keels at the base; spur broadly oblong, obtuse, one-quarter as long as the limb; column short and very broad.

KALAHARI REGION : Orange River Colony; marsh near Harrismith, *Sankey*, 256 !

EASTERN REGION: Griqualand East; Insiswa Mountains, *Schlechter*! Natal;
stony slopes of the Drakensberg near Tugela Falls, *Wood*, 3444! marshy places at
Oliviers Hoek, sources of Tugela River, 5000 ft., *Allison*, 6!

10. S. Gerrardi (Bolus in Journ. Linn. Soc. xxv. 205); plant
⅓–¾ ft. high; leaves 4–8, narrowly linear, acute, ¾–2½ in. long, ⅛–⅙ in.
broad; scapes ⅓–¾ ft. high, with 4–6 narrow acuminate sheaths;
racemes elongate, narrow, 1–3 in. long, dense; bracts lanceolate or
ovate-lanceolate, very acuminate, ⅙–⅓ in. long; flowers apparently
white; sepals ovate, subobtuse, somewhat concave, ⅛ in. long,
3-nerved; petals suborbicular or very broadly ovate, obtuse or
minutely apiculate, one-third as long as the sepals, 1-nerved; lip
two-thirds as long as the sepals, shortly 3-lobed, slightly pubescent,
rather fleshy; side lobes semiovate or broadly rounded, obtuse;
front lobe ovate or ovate-oblong, obtuse, twice as long as the side
lobes; disc with 3 short fleshy tubercles below the middle; spur
oblong, obtuse, about half as long as the limb; column short and
very broad; capsule oblong, nearly ¼ in. long. *Durand & Schinz,
Conspect. Fl. Afr.* v. 116. *Brachycorythis Gerrardi, Reichb. f. in
Flora,* 1867, 116. *Platanthera Gerrardi, Schlechter in Engl. Jahrb.* xx.
Beibl. 50, 12.

KALAHARI REGION: Transvaal; Devils Knuckles, near Spitz Kop, Lydenburg
District, *Wilms*, 1385!
EASTERN REGION: Griqualand East; Insiswa Mountains, *Schlechter*! Insiswa
Mountains to Umzinklowa River, *Krook in Herb. Penther*, 660! Natal; Ingoma,
Gerrard, 1542!

XIX. BARTHOLINA, R. Br.

Sepals subequal, free, erect, narrow, somewhat herbaceous. *Petals*
coloured, about as long as the sepals. *Lip* adnate to the column at
the extreme base, produced into an acute spur; limb broad,
spreading, deeply fimbriate-multifid. *Column* very short below the
anther; middle lobe of rostellum short, reflexed; stigma not produced; anther-bed erect, very long and narrow, concave, continuous at the base with the side lobes of the rostellum. *Anther*
connective not distinct from the anther-bed; cells subparallel,
adnate, apex elongate, adnate to the lobes of the rostellum; pollinia
solitary in the cells, coarsely granular; caudicles very long, dilated
into glands at the apex, included within the lobes of the rostellum.

Dwarf terrestrial herbs, with ovoid-oblong tubers; leaves solitary, basal,
spreading, reniform-orbicular and amplexicaul; flowers solitary at the apex of a
short hairy scape.

DISTRIB. Species 3, limited to extra-tropical South Africa, and chiefly
western.

Segments of the lip filiform throughout:
Sepals ⅓–½ in. long; lip ¾–1¼ in. long ...   . (1) **pectinata**.
Sepals ¼–⅓ in. long; lip ½–¾ in. long   ...   . (2) **lindleyana**.
Segments of lip shortly clavate at the apex   ...   ... (3) **Ethelæ**.

1. B. pectinata (R. Br. in Ait. Hort. Kew, ed. 2, v. 194); plant 2–9 in. high, with two ovoid tubers; leaf solitary, horizontal, sessile, broadly cordate or orbicular with a cordate base, flat or slightly convex, $\frac{1}{2}$ to more than 1 in. long by about as broad; scapes erect, 2–9 in. high, copiously pilose with long spreading hairs, 1-flowered; bracts spathaceous, oblong-lanceolate, subacute, pubescent, $\frac{1}{3}$–$\frac{1}{2}$ in. long; pedicels $\frac{1}{2}$–$\frac{3}{4}$ in. long, pubescent; flowers very large, with light green sepals and lilac or light purple petals and lip; dorsal sepal cucullate, lanceolate, subacute, pubescent, $\frac{1}{3}$–$\frac{1}{2}$ in. long; lateral sepals linear-oblong, subacute, pubescent, about as long as the dorsal; petals lanceolate-linear, long-acuminate, one-third longer than the sepals; lip spreading, fan-shaped, $\frac{3}{4}$–$1\frac{1}{4}$ in. long, divided to below the middle into numerous filiform or narrowly linear acuminate segments; disc smooth; spur curved, narrowly conical, acute or subacute, $\frac{1}{4}$–$\frac{1}{3}$ in. long; column clavate, $\frac{1}{4}$ to more than $\frac{1}{3}$ in. long. *Lindl. Gen. & Sp. Orch.* 334; *Endl. Ic. Gen. Pl. t.* 1534; *Bot. Mag. t.* 7450; *Bolus in Journ. Linn. Soc.* xxv. 188, *in Trans. S. Afr. Phil. Soc.* v. 111 (*excl. Bot. Reg. fig. & eastern habitat*); *and Orch. Cap. Penins.* 111; *Kränzl. Orch. Gen. et Sp.* i. 593. *B. burmanniana, Ker in Journ. Sci. & Arts*, iv. 204, *t.* 6, *fig.* 2; *Drège, Zwei Pfl. Documente*, 81; *Sond. in Linnæa*, xix. 83; *Durand & Schinz, Conspect. Fl. Afr.* v. 68. *Arethusa ciliaris, Linn. f. Suppl.* 405. *Orchis burmanniana, Linn. Pl. Afr. Rar.* 26, *Amœn. Acad.* vi. 108, *and Sp. Pl. ed.* ii. 1334; *Sw. in Web. & Mohr, Archiv.* i. 55, *t.* 3. *O. pectinata, Thunb. Prodr.* 4, *and Fl. Cap. ed. Schult.* 5; *Willd. Sp. Pl.* iv. 11.

South Africa: without locality, *Oldenburg*, 607! *Masson*!

Coast Region: Tulbagh Div.; plain near Tulbagh, 430 ft., *Schlechter*, 1412! Worcester Div.; Dutoits Kloof, 3000–4000 ft., *Drège*, 1235a! Paarl Div.; Drakenstein Range, *Rehmann*, 2232! Cape Div.; Cape Flats, *Pappe*, 21! Wynberg, *Pappe*! *Harvey*! near Van Kamps Bay, 50 ft., *MacOwan & Bolus, Herb. Norm. Austr.-Afr.*, 154! Muizen Berg, *Wolley-Dod*, 1749! Simons Bay, *Bulger*! near Cape Town, *Harvey*, 124! *Bauer*! Table Mountain, 300 ft., *Dümmer*, 395! and without precise locality, *Rogers*! *Pappe*! *Hooker*, 370! *Bolus*! Caledon Div.; Caledon, *Prior*!

2. B. lindleyana (Reichb. f. Otia Bot. Hamb. ii. 119, name only); plant 3–6 in. high, with two ovoid tubers; leaf solitary, horizontal, sessile, orbicular with a cordate base, flat, $\frac{1}{2}$–$\frac{3}{4}$ in. long by about as broad; scapes erect, 3–6 in. high, copiously pilose with long spreading hairs, 1-flowered; bracts spathaceous, oblong or oblong-lanceolate, subacute, pubescent, $\frac{1}{6}$–$\frac{1}{4}$ in. long; pedicels $\frac{1}{3}$–$\frac{1}{2}$ in. long, pubescent; flowers large, with green sepals and bright purple petals and lip; dorsal sepal cucullate, lanceolate, subacute, pubescent, $\frac{1}{4}$–$\frac{1}{3}$ in. long; lateral sepals linear-oblong, acute or subacute, pubescent, about as long as the dorsal; petals lanceolate-linear, long-acuminate, one-third longer than the sepals; lip spreading, fan-shaped, $\frac{1}{2}$–$\frac{3}{4}$ in. long, and divided to below the middle into numerous filiform or narrowly linear acuminate segments; disc smooth; spur curved, narrowly conical, acute or subacute, $\frac{1}{4}$–$\frac{1}{3}$ in. long; column clavate, about $\frac{1}{4}$ in. long. *Durand & Schinz, Conspect. Fl. Afr.* v.

69 (*excl. syn. Endl.*).   *B. pectinata, Lindl. in Bot. Reg. t.* 1653 (*excl. all syn.*) *and in Hook. Comp. Bot. Mag.* ii. 210; *Drège, Zwei Pfl. Documente,* 121, *not of R. Br.*

COAST REGION : Uniondale Div. ; in Langkloof, between Keurbooms River and Kromme River, 2000–3000 ft., *Drège,* 1235b ! Uitenhage Div. ; Uitenhage, collector not stated ! Port Elizabeth Div. ; sandy places at Port Elizabeth, *Kemsley* ! *Hallack* ! *Mrs. Holland,* 5 ! Albany Div. ; Mill River, 12 miles from Grahamstown, *Atherstone,* 23 ! near Grahamstown, 2200 ft., *MacOwan,* 1528 ! *Bolton* ! Kowie River, *Harvey* ! and without precise locality, *Hutton* !

Nearly allied to *B. pectinata,* R. Br., from which it was separated by Reichenbach, and chiefly differing from the western plant by its smaller flowers.

3. **B. Ethelæ** (Bolus in Journ. Linn. Soc. xx. 472) ; plant ½–1 ft. high, with 2 ovoid tubers ; leaf solitary, horizontal, sessile, orbicular, with a broadly cordate base, flat or slightly convex, ½–1¼ in. long by about as broad ; scapes erect, ½–1 ft. high, copiously pilose with long spreading hairs, 1-flowered ; bracts spathaceous, oblong, subacute, pubescent, about ¼ in. long ; pedicels ½–¾ in. long, puberulous ; flowers very large, with green sepals and pale lilac-blue petals and lip ; dorsal sepal cucullate, lanceolate, acuminate, puberulous, about ½ in. long ; lateral sepals linear-lanceolate, acuminate, puberulous, as long as the dorsal ; petals linear-lanceolate, narrowed and obtuse at the apex, rather longer than the sepals ; lip spreading, fan-shaped, ¾–1¼ in. long, and divided to below the middle into numerous filiform capitate segments ; disc smooth ; spur curved, narrowly conical, acuminate, ⅓ to nearly ½ in. long ; column clavate, apiculate, about ⅓ in. long. *Bolus in Journ. Linn. Soc.* xxv. 188, *in Trans. S. Afr. Phil. Soc.* v. 112, *t.* 3, *and Orch. Cap. Penins.* 112, *t.* 3 ; *Durand & Schinz, Conspect. Fl. Afr.* v. 68 ; *Kränzl. Orch. Gen. et Sp.* i. 594, *and in Ann. Naturhist. Hofmus. Wien,* xx. 2.

COAST REGION : Tulbagh Div. ; Saron, 800 ft., *Schlechter* ! Cape Div. ; under shrubs at the foot of a dry hill overlooking the sea between Kalk Bay and Fish Hoek, 150 ft., *Bolus,* 4850 ! *and in MacOwan & Bolus, Herb. Norm. Austr.-Afr.,* 500 ! sandy places in a valley on Muizen Berg, 900–1000 ft., *Bolus* ; Simons Berg, *Wolley-Dod,* 846 ! Simons Bay, *Prior* ! and without precise locality, *Rogers* ! *Grey* ! Caledon Div. ; Leos Kraal, *Penther,* 280 ! Knysna Div. ; Knysna, *Trimen* !

XX. HOLOTHRIX, L. C. Rich.

Sepals subequal, connivent, herbaceous, sometimes hairy. *Petals* longer than the sepals, narrow, entire or variously divided at the apex. *Lip* adnate to the base of the column, erect or spreading, concave or involute at the sides, divided into from three to many segments at the apex, produced at the base into a straight or curved spur. *Column* very short, usually auricled at the sides of the stigma ; anther-bed erect, broad, concave or almost cucullate ; connective of the anther not distinct from the anther-bed. *Anther-cells* ovoid, adnate, distinct, included ; pollinia coarsely granular, with very short caudicles, terminating in a small naked gland. *Stigma* bipartite. *Capsule* ovoid or oblong.

Terrestrial herbs, with one or two sessile ovate or orbicular-reniform radical leaves ; scapes slender, usually hairy and without sheaths ; flowers small, in slender, usually second spikes.

DISTRIB. Species about 40, exclusively African, and including 14 Tropical African representatives, 2 from Madagascar, and 1 from the Comoro Islands.

*Petals undivided at the apex :
　Petals and lip somewhat fleshy, green or greenish :
　　Lip undivided ...　...　...　...　...　...(1) **exilis.**

　　Lip 3-lobed :
　　　Lobes equal or subequal :
　　　　Flowers scarcely 2 lin. long　...　...　...(2) **villosa.**

　　　Flowers 3 lin. or more long :
　　　　Sepals about 1½ lin. long　...　...　...(3) **condensata.**

　　　　Sepals about 2½ lin. long　...　...　...(4) **lithophila.**

　　　Side lobes not half as long as the front lobe　...(5) **Thodei.**

　　Lip 5-lobed :
　　　Lobes of lip about as long as the limb, linear,
　　　　subequal :
　　　　Scape pilose :
　　　　　Bracts and sepals ciliate, with long hairs :
　　　　　　Flowers about 3 lin. long　...　...　...(6) **incurva.**

　　　　　　Flowers about 2 lin. long　..　...　...(7) **micrantha.**

　　　　　Bracts and sepals glabrous　...　...　...(8) **secunda.**

　　　　Scape very densely villous　...　...　...(9) **squamulosa.**

　　　Lobes of lip shorter than the limb, usually oblong
　　　　and unequal :
　　　　Petals 1½–2 lin. long　...　...　...　...(10) **parvifolia.**

　　　　Petals 3½–4 lin. long :
　　　　　Scape under ⅓ ft. high ; raceme second　...(11) **rupicola.**

　　　　　Scape about 1½ ft. high ; raceme not second　(12) **pilosa.**

　Petals and lip membranous, white or pink :
　　Petals entire ; lip not papillose :
　　　Lip 3-lobed ...　...　...　...　...　...(13) **Culveri.**

　　　Lips 5–7-lobed or toothed :
　　　　Raceme short, subcorymbose, few-flowered　...(14) **Mundii.**

　　　　Racemes elongate, many-flowered :
　　　　　Spur much curved or circinate, rather broad
　　　　　　and subobtuse :
　　　　　　Petals ovate-oblong ; lip with 5 to 7 unequal
　　　　　　　linear lobes ...　...　...　...　...(15) **confusa.**

　　　　　　Petals linear-oblong ; lip with 5 nearly
　　　　　　　regular oblong lobes　...　...　...(16) **lindleyana.**

　　　　　Spur nearly straight, narrow and acute :
　　　　　　Scape 4 to 9 in. high ; lip 4 to 5 lin. long,
　　　　　　　with 5 to 7 distinct teeth　...　...(17) **orthoceras.**

　　　　　　Scape 2 to 3 in. high ; lip 3 to 4 lin. long,
　　　　　　　with 7 to 9 very short teeth ...　...(18) **macowaniana.**

　　　　　Petals with a broad basal tooth in front ; lip
　　　　　　papillose ...　...　...　...　...　...(19) **aspera.**

****Petals 5–7-lobed at the apex ;
　Flowers not dimorphic :
　　Scapes without sheaths :
　　　Flowers 2–3 lin. long ; lip 5–7-lobed　...　...　(20) **multisecta.**

　　　Flowers 3–4 lin. long ; lip 8–9-lobed　...　...　(21) **Scopularia.**

　　Scapes bearing several narrow sheaths :
　　　Flowers about ⅓ in. long　...　...　...　...　(22) **schlechteriana.**

　　　Flowers about ½ in. long　...　...　...　...　(23) **grandiflora.**

　　Flowers dimorphic ; petals and lip of the upper flowers
　　　divided into very long filiform segments ...　...　(24) **Burchellii.**

1. **H. exilis** (Lindl. Gen. & Sp. Orch. 283) ; plant ¼–¾ ft. high,
with 2 small ovoid tubers ; leaves 1 or 2, orbicular or ovate,
obtuse, flat, somewhat fleshy, pilose with long hairs, ¼–½ in. long ;
scapes ¼–¾ ft. high, slender, pilose with long spreading hairs, without sheaths ; racemes ¾–2 in. long, usually lax, secund ; bracts
ovate, acuminate, about 1 lin. long ; pedicels longer than the bracts ;
flowers very small, light green, with ochreous-yellow petals and lip ;
sepals ovate, obtuse, ¾ lin. long, 1-nerved, glabrous ; petals linear-
lanceolate, obtuse, somewhat fleshy above, about 1 lin. long ; lip
about as long as the petals, 3-lobed, with very short membranous
obtuse side lobes and a linear obtuse front lobe ; spur narrowly
conical, obtuse, curved, one-third as long as the lip ; column broad,
half as long as the sepals. *Lindl. in Hook. Comp. Bot. Mag.* ii. 207 ;
Bolus in Journ. Linn. Soc. xxv. 189 ; *Durand & Schinz, Conspect. Fl.
Afr.* v. 70 ; *Schlechter in Oest. Bot. Zeitschr.* 1898, 443 ; *Kränzl.
Orch. Gen. et Sp.* i. 574, *and in Ann. Naturhist. Hofmus. Wien,* xx. 1.
H. exilis, var. typica, Schlechter in Oest. Bot. Zeitschr. 1898, 444.

VAR. β, **brachylabris** (Bolus, Ic. Orch. Austr.-Afr. i. t. 14, fig. A) ; lip entire,
ovate, with a linear obtuse apex. *Schlechter in Oest. Bot. Zeitschr.* 1898, 444 ;
Kränzl. Orch. Gen. et Sp. i. 574. *H. brachylabris, Sond. in Linnæa,* xix. 78 ;
Bolus in Journ. Linn. Soc. xxv. 189 ; *Durand & Schinz, Conspect. Fl. Afr.* v. 69.

COAST REGION : Cape Div. ; Table Mountain, 2500–3000 ft., *Schlechter.*
Riversdale Div. ; near Zoetemelks River, *Burchell,* 6738/1 ! *Schlechter,* 2461.
George Div. ; hills near George, 750 ft., *Schlechter,* 2243. Albany Div. ; Cold-
stream, near Grahamstown, 2200 ft., *Glass in Herb. Bolus,* 6237 ! and in
MacOwan & Bolus, Herb. Norm. Austr.-Afr., 1370, partly ! Var. β : Cape Div. ;
Table Mountain, 2500–3000 ft., *Schlechter,* 81 ; near Muizenberg Vley, *Wolley-Dod,*
3651 ! Riversdale Div. ; near Riversdale, *Schlechter.* Knysna Div. ; Karratera
River, 300 ft., *Schlechter,* 5885 ! Silver River, *Penther,* 290. Uitenhage Div. ;
Uitenhage, *Zeyher* ! Albany Div. ; Coldstream, near Grahamstown, 2200 ft.,
Glass in MacOwan & Bolus, Herb. Norm. Austr.-Afr., 1370, partly !

H. brachylabris, Sond., is clearly only a variety, for it has been found with the
type in several localities, and forms intermediate in the structure of the lip also
occur.

2. **H. villosa** (Lindl. in Hook. Comp. Bot. Mag. ii. 207) ; plant
⅓–1 ft. high, with two ovoid tubers ; leaves 2, unequal, broadly
ovate or ovate-orbicular, obtuse or subacute, sometimes cordate at
the base, ¾–1½ in. long, villous with very long hairs ; scapes ⅓–1 ft.
high, villous with very long spreading hairs, without sheaths ;

racemes $\frac{3}{4}$–$2\frac{1}{2}$ (rarely 4) in. long, dense or rarely lax; bracts ovate, subacute, villous, with very long hairs at the apex; pedicels rather longer than the bracts; flowers ochreous-yellow (*Bolus*); sepals ovate, subobtuse, $\frac{3}{4}$ lin. long, 1-nerved, glabrous; petals ovate-lanceolate, narrow at the apex and subobtuse, not twice as long as the sepals; lip about as long as the petals, 3-lobed to about the middle, with linear obtuse lobes; spur conical, somewhat curved, obtuse, about one-third as long as the limb; column broad, nearly half as long as the sepals; capsule oblong, $2\frac{1}{2}$ lin. long. *Drège, Zwei Pfl. Documente*, 101, *and in Linnæa*, xx. 217; *Sond. in Linnæa*, xix. 76; *Bolus in Journ. Linn. Soc.* xxv. 189, *Trans. S. Afr. Phil. Soc.* v. 117, *Orch. Cap. Penins.* 117, *and Ic. Orch. Austr.-Afr.* i. *t.* 14, *fig. B.*; *Durand & Schinz, Conspect. Fl. Afr.* v. 72; *Schlechter in Oest. Bot. Zeitschr.* 1898, 445; *Kränzl. Orch. Gen. et Sp.* i. 578. *H. parvifolia, Lindl. Gen. & Sp. Orch.* 283, *partly. H. gracilis, Lindl. in Hook. Comp. Bot. Mag.* ii. 207; *Drège, Zwei Pfl. Documente*, 89; *Bolus in Journ. Linn. Soc.* xxv. 189, *Trans. S. Afr. Phil. Soc.* v. 116, *and Orch. Cap. Penins.* 116; *Durand & Schinz, Conspect. Fl. Afr.* v. 70; *Kränzl. Orch. Gen. et Sp.* i. 579, *and in Ann. Naturhist. Hofmus. Wien*, xx. 1. *Orchis hispida, Thunb. Fl. Cap. ed. Schult.* 6, *partly. Habenaria hispida, Spreng. Tent. Suppl.* 27, *partly.*

SOUTH AFRICA: without locality, *Grey*! *Wright*, 142! *Trimen*!
COAST REGION: Malmesbury Div.; Hopefield, *Penther*, 127. Tulbagh Div.; Witzenberg Range, *Zeyher*, 3901! mountains above Mitchells Pass, 1500 ft., *Bolus*. Worcester Div.; Bains Kloof, *Hutton*! Paarl Div.; Great Drakenstein Mountains and at the foot of Paarl Mountains, under 1000 ft., *Drège*, 1253a! 1253b! 1253c! Dwars River, near French Hoek, 1000 ft., *Schlechter*. Cape Div.; Cape Flats, *Zeyher*, 4678! Devils Mountain, *Pappe*, 14! *Wolley-Dod*, 387! near Cape Town, *Harvey*! *Wilms*, 3663! *Bolus, Kässner, Schlechter*! rocky clefts on Table Mountain, 2400 ft., *Bolus*, 4655! *Rehmann*, 573! *Schlechter*, 103; Simons Bay, *Brown*! Caledon Div.; near Genadendal, *Prior*! Leos Kraal and Zonder Einde River, *Penther*, 316. George Div.; Montague Pass, 2650 ft., *Schlechter*, 5585! near George, *Penther*, 163. Knysna Div.; hills near Knysna, *Forcade*.
CENTRAL REGION: Graaff Reinet Div.; summit of Cave Mountain, 4300 ft., *Bolus*, 787! Sneeuwberg Range, *Bolus*.

3. **H. condensata** (Sond. in Linnæa, xix. 76); plant $\frac{1}{3}$–$\frac{1}{2}$ ft. high, with two ovoid tubers; leaves 2, unequal, broadly ovate or orbicular, obtuse or subacute, villous, 1–2 in. long, somewhat fleshy; scapes $\frac{1}{3}$–$\frac{1}{2}$ ft. high, rather stout, copiously villous with long spreading hairs, without sheaths; racemes 1–$2\frac{1}{2}$ in. long, dense or somewhat lax, secund; bracts ovate, acuminate, $1\frac{1}{2}$–2 lin. long, villous with long hairs; pedicels rather longer than the bracts; flowers more than $\frac{1}{4}$ in. long, greenish-yellow; sepals ovate, obtuse, glabrous, $1\frac{1}{2}$ lin. long, 1-nerved; petals linear-oblong, obtuse, $2\frac{1}{2}$ lin. long, 1-nerved; lip about as long as the petals, 3-lobed nearly to the middle; lobes oblong, obtuse, subequal; spur conical, straight, subobtuse, about half as long as the limb; column broad, nearly half as long as the sepals; capsule oblong, 3–4 lin. long. *Bolus in Journ. Linn. Soc.* xxv. 189, *in Trans. S. Afr. Phil. Soc.* v. 115, *t.* 22,

fig. 8–11, *Orch. Cap. Penins.* 115, *t.* 22, *fig.* 8–11, *and Ic. Orch.
Austr.-Afr.* ii. *t.* 36; *Durand & Schinz, Conspect. Fl. Afr.* v. 70;
Schlechter in Oest. Bot. Zeitschr. 1898, 445; *Kränzl. Orch. Gen. et
Sp.* i. 577. *H. parvifolia, Hook. Ic. Pl. t.* 103, *fig. B.*

SOUTH AFRICA : without locality, *Prior* !
COAST REGION : Cape Div. ; near Cape Town, *Harvey* ! Table Mountain, in
clefts of rock with a southern aspect, 2800 ft., *Bolus*, 4905 ! *Kässner, Schlechter,*
89, at summit, *Wolley-Dod,* 863 ! Waai Vley, *Wolley-Dod,* 2306 ! Constantia Berg,
Wolley-Dod, 2139. Caledon Div. ; Zwarteberg Range, on wet rocks at 1500 ft.,
Schlechter, 559. Swellendam Div. ; near Swellendam, *Mund* !

H. parvifolia, Hook., is quite distinct from both the plants included by Lindley
under his *H. parvifolia.*

4. H. lithophila (Schlechter in Oest. Bot. Zeitschr. 1898, 446);
plant $\frac{1}{3}$–$\frac{1}{2}$ ft. high, with ovoid tubers ; leaves 2, unequal, broadly
ovate or ovate-orbicular, obtuse or subacute, villous, 1–1$\frac{3}{4}$ in. long,
somewhat fleshy ; scapes $\frac{1}{3}$–$\frac{1}{2}$ ft. high, rather stout, villous with
spreading hairs, without sheaths ; racemes 1$\frac{1}{2}$–2$\frac{1}{2}$ in. long, rather
dense, somewhat secund ; bracts ovate or ovate-lanceolate, acuminate,
2–3 lin. long, villous with long hairs ; pedicels rather longer
than the bracts ; sepals ovate or ovate-oblong, obtuse, ciliate,
about 1$\frac{1}{2}$ lin. long, 1-nerved ; petals linear-oblong, obtuse, 2–3 lin.
long, 1-nerved ; lip scarcely as long as the petals, 3-lobed ; lobes
shorter than the limb, oblong, obtuse, subequal ; spur narrowly
conical, straight, obtuse, rather shorter than the limb ; column
broad, about one-third as long as the sepals ; capsule oblong, about
$\frac{1}{3}$ in. long. *Schlechter in Engl. Jahrb.* xxvi. 331.

COAST REGION : Caledon Div. ; above Vogelgat Lagoon, near the mouth of the
Klein River ; 3500 ft., *Schlechter,* 9556 !

5. H. Thodei (Rolfe); leaves and bulbs not seen; scapes
about 3 in. high, rather stout, densely pilose with rather stout
retrorse hairs ; racemes 1$\frac{1}{2}$ in. long, somewhat lax, many-
flowered, not secund ; bracts ovate, subacute, villous, about 1 lin.
long ; pedicels rather shorter than the bracts ; flowers very small,
greenish ; sepals ovate-oblong, subobtuse, somewhat ciliate, 1 lin.
long, 1-nerved ; petals ovate-lanceolate, narrowed upwards, subobtuse, fleshy, about 2 lin. long, 1-nerved ; lip not much longer than
the sepals, very broad, shortly 3-lobed, with three broad obtuse
fleshy lobes, the front lobe about twice as long as the lateral pair ;
spur conical, subobtuse, straight, about one-third as long as the
limb ; column broad, about one-third as long as the sepals.

KALAHARI REGION : Orange River Colony ; summit of Quaqua Mountains,
Witzies Hoek, 7500 ft., in stony and grassy places, *Thode,* 48 !

6. H. incurva (Lindl. in Hook. Comp. Bot. Mag. ii. 207) ; plant
about 5 in. high, with ovoid tubers ; leaves not seen ; scapes about
5 in. high, slender, pilose with long somewhat retrorse hairs ;
racemes about 1$\frac{1}{2}$ in. long, lax, secund ; bracts ovate, acuminate,

ciliate with long hairs, 1½ lin. long; pedicels rather shorter than the bracts; flowers very small; sepals ovate, subobtuse, over 1 lin. long, 1-nerved, with a few long apical hairs; petals narrowly linear, obtuse, nearly ¼ in. long, somewhat broader at the base; lip shorter than the petals, 5-lobed to about the middle; lobes linear or filiform, obtuse; spur broadly conical, very obtuse, curved, about one-third as long as the limb; column broad, about half as long as the sepals. *Bolus in Journ. Linn. Soc.* xxv. 189; *Durand & Schinz, Conspect. Fl. Afr.* v. 70; *Schlechter in Oest. Bot. Zeitschr.* 1898, 443; *Kränzl. Orch. Gen. et Sp.* i. 586.

COAST REGION: Stockenstrom Div.; Kat Berg, 3000–4000 ft., *Drège*, 8275!
KALAHARI REGION: Orange River Colony; grassy plains at the summit of Quaqua Mountain, near Witzies Hoek, 7500 ft., *Thode (ex Schlechter)*.
EASTERN REGION: Natal; Drakensberg Range, near Van Reenen, 5000–6000 ft., *Haygarth in Herb. Wood*, 5574 (*ex Schlechter*).

A high mountain plant. I have only seen Lindley's original specimen, and suspect that some of the localities require verification.

7. **H. micrantha** (Schlechter in Engl. Jahrb. xx. Beibl. 50, 31); plant about 8 in. high, with two large ovoid tubers; leaves 2, unequal, suborbicular, subacute, rather fleshy, somewhat concave, pilose or almost hispid, ½–¾ in. long; scapes about 8 in. long, rather stout, pilose with spreading hairs, without sheaths; racemes 4 in. long, dense, somewhat secund; bracts ovate or triangular-ovate, acute, ciliate, about 1 lin. long; pedicels shorter than the bracts; flowers very small; sepals broadly ovate, subobtuse or apiculate, ciliate at the apex, ½ lin. long, 1-nerved; petals lanceolate-linear, obtuse, rather fleshy, about 2 lin. long, 1-nerved; lip about as long as the petals, 5-lobed to the middle, lobes filiform, obtuse, subequal; spur broadly conical, obtuse, about as long as the sepals; column very broad, half as long as the sepals; capsule oblong, about 2 lin. long. *Kränzl. Orch. Gen. et Sp.* i. 585.

KALAHARI REGION: Transvaal; mountains above Heidelberg, 5400 ft., on grassy cliffs, *Schlechter*, 3522!

8. **H. secunda** (Reichb. f. Otia Bot. Hamb. ii. 119); ½–¾ ft. high, with large ovoid tubers; leaves 2, orbicular or broadly ovate-orbicular, obtuse, 1–1½ in. long; scapes erect, ½–¾ ft. high, villous at the base, without sheaths; racemes secund, 2½–4 in. long, many-flowered; bracts ovate, acuminate, ⅛–¼ in. long; pedicels ⅛–¼ in. long; sepals ovate, subacute, 1-nerved, 1/12 in. long; petals broadly lanceolate, long-acuminate, somewhat falcate, ¼ in. long, 2-nerved to the middle, the two nerves confluent into one above, the basal third adnate to the claw of the lip; lip nearly as long as the petals, 5-lobed to below the middle; lobes filiform, suborbute, the lateral somewhat diverging; spur curved, cylindrical or narrowly conical, subobtuse, ⅛ in. long; column stout, half as long as the sepals; capsule oblong, ¼ in. long. *Bolus in Journ. Linn. Soc.* xxv. 189, *and Ic. Orch. Austr.-Afr.* ii. t. 37; *Durand & Schinz, Conspect. Fl. Afr.* v. 72; *Schlechter in Oest.*

Bot. Zeitschr. 1898, 441 ; *Kränzl. Orch. Gen. et Sp.* i. 586. *Orchis secunda, Thunb. Prodr.* 4 ; *Fl. Cap. ed. Schult.* 6. *Tryphia major, Sond. in Linnæa,* xix. 82, xx. 218.

COAST REGION: Clanwilliam Div. ; Koude Berg, near Wupperthal, *Bolus,* 9094 ! near Clanwilliam, 350 ft., *Schlechter,* 8599, *Leipoldt* ; Brack Fontein, *Zeyher* ! Olifants River and near Brackfontein, *Ecklon & Zeyher,* 19 ! Cape Div. ; Table Mountain, *Thunberg* ! Swellendam Div. ; near Vormansbosch, 2000– 4000 ft., *Zeyher,* 3902. Robertson Div. ; near Ashton, 660 ft., *Marloth,* 3234. WESTERN REGION : Little Namaqualand ; Nababeep, near Ookiep ? 3100 ft., *Scully,* 149 ! *Warden in Herb. Bolus,* 6570 ; Modderfontein, *Whitehead* !

9. H. squamulosa (Lindl. in Hook. Comp. Bot. Mag. ii. 206); plant $\frac{1}{3}$–$\frac{3}{4}$ ft. high, with two large ovoid tubers ; leaves 1 or 2, orbicular or broadly cordate-orbicular, obtuse or rarely subacute, rather fleshy, flat or somewhat concave, $\frac{1}{2}$–$1\frac{1}{2}$ in. long, upper surface covered with curved subulate or broad-based chaffy scales ; scapes rather stout, $\frac{1}{3}$–$\frac{3}{4}$ ft. high, very densely villous with spreading or retrorse hairs, without sheaths ; racemes 1–4 in. long, secund, usually dense ; bracts triangular-ovate, acuminate, about 2 lin. long, very villous ; pedicels about as long as the bracts ; flowers small, light green ; sepals ovate-oblong, suboblong, villous, about $1\frac{1}{2}$ lin. long, 1-nerved ; petals linear, obtuse, twice as long as the sepals, 1-nerved ; lip about as long as the petals, 5-lobed to about the middle ; lobes linear, obtuse, subequal or the outer pair shorter ; spur conical, subobtuse, curved, about half as long as the limb ; column broad, half as long as the sepals ; capsule oblong, $\frac{1}{4}$–$\frac{1}{3}$ in. long. *Sond. in Linnæa,* xix. 76 ; *Drège, Zwei Pfl. Documente,* 114, 125, *and in Linnæa,* xx. 217 ; *Bolus in Journ. Linn. Soc.* xxv. 189, *in Trans. S. Afr. Phil. Soc.* v. 114, *and Orch. Cap. Penins.* 114 ; *Durand & Schinz, Conspect. Fl. Afr.* v. 72 ; *Schlechter in Oest. Bot. Zeitschr.* 1898, 442 ; *Kränzl. Orch. Gen. et Sp.* i. 584, *and in Ann. Naturhist. Hofmus. Wien,* xx. 2. *H. parvifolia, Lindl. Gen. & Sp. Orch.* 283, *partly. H. squamulosa, var. scabra, Bolus in Trans. S. Afr. Phil. Soc.* v. 114, *t.* 23, *fig. A, and Orch. Cap. Penins.* 114, *t.* 23, *fig. A* ; *Durand & Schinz, Conspect. Fl. Afr.* v. 72 ; *Kränzl. Orch. Gen. et Sp.* i. 585. *H. squamulosa, var. typica, Schlechter in Oest. Bot. Zeitschr.* 1898, 442. *Orchidea hispida, Burch. ex Lindl. Gen. & Sp. Orch.* 303.

VAR. β, hirsuta (Bolus in Trans. S. Afr. Phil. Soc. v. 114, t. 23, fig. B) ; surface of leaves densely villous, hairs narrow and not scale-like. *Bolus, Orch. Cap. Penins.* 114, *t.* 23 ; *Durand & Schinz, Conspect. Fl. Afr.* v. 72 ; *Kränzl. Orch. Gen. et Sp.* i. 585. *H. squamulosa, var. harveyana, Schlechter in Oest. Bot. Zeitschr.* 1898, 442. *H. harveiana, Lindl. in Hook. Comp. Bot. Mag.* ii. 206 ; *Sond. in Linnæa,* xix. 76 ; *Hook. Ic. Pl. t.* 103, *fig. A. Orchis hispidula, Linn. f. Suppl.* 401.

VAR. γ, glabrata (Schlechter in Oest. Bot. Zeitschr. 1898, 442) ; upper surface of leaves glabrous except at the villous margin. *H. Monotris, Reichb. f. Otia Bot. Hamb.* ii. 119 ; *Durand & Schinz, Conspect. Fl. Afr.* v. 71 ; *Kränzl. Orch. Gen. et Sp.* i. 581. *Monotris secunda, Lindl. in Bot. Reg. sub t.* 1701, *and Gen. & Sp. Orch.* 303.

SOUTH AFRICA : without locality, *Pappe* ! *Rogers* ! Var. β : *Wright,* 137 !

COAST REGION : Worcester Div. ; mountains near De Liefde, 2000–3000 ft., *Drège*, 1253c ! Cape Div. ; Cape Flats, *Bolus*, 3929, partly ! sandy places near Rondebosch, under 100 ft., *Bolus*, 7022, partly ! *and in MacOwan & Bolus, Herb. Norm. Austr.-Afr.*, 410 ! *Wolley-Dod*, 669 ! Devils Mountain, *Pappe*, 13 ! Wynberg and Kirstenbosch, *Zeyher*, 4677 ! *Schlechter*, 1689 ! Constantia Berg, *Schlechter*, Muizenberg Mountains, *Wolley-Dod*, 784 ! Simonstown, *Wolley-Dod*, 1814 ! Stellenbosch Div. ; near Somerset West, *Ecklon & Zeyher*, 17 ! Hottentots Holland, *Prior* ! Lowrys Pass, *Penther*, 108 ! Caledon Div. ; Palmiet River, *Penther*, 267, 270 ! near Genadendal, *Prior* ! Villiers Dorp, *Grey* ! Swellendam Div. ; Grootvaders Bosch, *Zeyher*, 1581, partly ! Riversdale Div. ; Gauritz River, *Penther*, 312. Knysna Div. ; Karratera River, under 1000 ft., *Drège*, 1253d ! Port Elizabeth Div. ; Port Elizabeth, *Mrs. Holland*, 40 ! Var. *β* : Cape Div. ; Simons Bay, *Wright* ! Cape Flats, *Harvey* ! *Bolus*, 3929, partly ! Swellendam Div. ; Grootvaders Bosch, *Zeyher*, 1581, partly ! sandy places near Rondebosch, under 100 ft., *Bolus*, 7022, partly ! Var. *γ* : Mossel Bay Div. ; between Zoute River and Duyker River, *Burchell*, 6369 !

The plants included here were referred by Lindley to three distinct species. Dr. Bolus, however, remarks that var. *β* (*H. harveiana*, Lindl.) grows with the type, and he is satisfied that there is no specific difference between them. The type has large coarse hyaline scales on the leaves, while var. *β* is pubescent, but there is an intermediate form. The two are also mixed in several of the gatherings. Var. *γ* (*Monotris secunda*, Lindl.) rests upon the original specimen collected by Burchell in a somewhat different locality, and should be re-collected. The specimen is somewhat imperfect, and I am not satisfied about the difference relied upon for its separation.

10. **H. parvifolia** (Lindl. Gen. & Sp. Orch. 283, partly) ; plant ⅓–½ ft. high, with two ovoid tubers ; leaves 1 or 2, the upper if present smaller, ovate or orbicular, obtuse, thick and fleshy, coarsely hirsute, flat or somewhat concave, 4–5 lin. long ; scapes rather stout, ⅓–½ in. long, closely villous with retrorse hairs, without sheaths ; racemes 1–3 in. long, somewhat lax, many-flowered, not secund ; bracts narrowly ovate, acute, villous with long hairs at the apex, 1–1½ in. long ; pedicels about as long as the bracts ; flowers very small, dull ochreous-yellow ; sepals ovate, obtuse, villous, ⅔ lin. long, 1-nerved ; petals falcate-linear, obtuse, about 1½ lin. long, somewhat fleshy ; lip about as long as the petals, rather fleshy, shortly 5-lobed ; lobes oblong, obtuse, the outer pair very short ; spur curved, stoutly cylindrical, obtuse, about half as long as the limb ; column stout, about half as long as the sepals. *Bolus in Journ. Linn. Soc.* xxv. 188, *in Trans. S. Afr. Phil. Soc.* v. 115, *t.* 24, *excl. syn. Linn. f.* (*not of Hook.*), *and Orch. Cap. Penins.* 115, *t.* 24, *excl. syn. Linn. f.* (*not of Hook.*). *H. hispidula, Durand & Schinz, Conspect. Fl. Afr.* v. 70, *partly* ; *Schlechter in Oestr. Bot. Zeitschr.* 1898, 443, *partly* ; *Kränzl. Orch. Gen. et Sp.* i. 579, *partly, and in Ann. Nat. Hofmus. Wien*, xx. 1. *O. hispida, Thunb. Prodr.* 4, *excl. syn. Linn. f.* ; *Fl. Cap. ed. Schult.* 6, *excl. syn. Linn. f.* *Habenaria hispida, Spreng. Tent. Suppl.* 27, *partly.*

SOUTH AFRICA: without locality, specimen (the type) from Herb. Caley in *Herb. Bentham* ! *and Herb. Lindley* !

COAST REGION : Cape Div. ; summit of Table Mountain and on rocks on the western side, 2400–3500 ft., *Thunberg* ! lower plateau of Table Mountain, above Klassenbosch, 2400 ft., *Bolus*, 7034 ! *Kässner, Schlechter*, 482 ; Waai Vley;

Wolley-Dod, 2339 ! Swellendam Div. ; Langeberg Range, near Zuurbraak, 3500 ft., *Schlechter.* Knysna Div. ; mountains near Knysna, *Forcade.*

EASTERN REGION : Natal ; Van Reenen, *Krook, Penther,* 871.

I have not seen the eastern plant, which requires verification. The species has been more or less confused with *H. villosa.*

11. **H. rupicola** (Schlechter in Engl. Jahrb. xxiv. 419) ; plant 3–5 in. high ; leaves 2, unequal, suborbicular or reniformly orbicular, subobtuse, somewhat fleshy, slightly ciliate at the margin, about 1 in. long ; scapes 3–5 in. long, rather stout, villous with spreading hairs, without sheaths ; racemes 2–2½ in. long, rather lax, secund ; bracts ovate, acute, villous (especially at the margin) with long hairs, about 2 lin. long ; pedicels rather shorter than the bracts ; flowers greenish (*Thode*) ; sepals ovate, acute, ciliate, about 1½ lin. long, 1-nerved ; petals linear-lanceolate, acute, ¼–⅓ in. long, 1-nerved ; lip rather shorter than the petals, with a broad limb, 5-lobed to the middle, with narrowly linear somewhat diverging equal lobes ; spur broadly conical, much curved, very obtuse, over half as long as the sepals ; column broad, about one-third as long as the sepals. *Schlechter in Oest. Bot. Zeitschr.* 1898, 446 ; *Kränzl. Orch. Gen. et Sp.* i. 583, *and in Ann. Naturhist. Hofmus. Wien,* xx. 1 ; *Bolus, Ic. Orch. Austr.-Afr.* ii. *t.* 40, *fig. B.*

COAST REGION: Clanwilliam Div. ; near Oliphants River, *Penther,* 198.

KALAHARI REGION : Orange River Colony ; Mont aux Sources, 8000–9000 ft., in fissures of rocks, *Thode,* 6 ! *Flanagan,* 1981.

12. **H. pilosa** (Reichb. f. Otia Bot. Hamb. ii. 119) ; leaves unknown ; scapes stout, erect, 1½ ft. high, densely pilose with long spreading or reflexed hairs, without sheaths ; racemes 8 in. long, very dense, not secund, many-flowered ; bracts triangular-lanceolate, acuminate, ¼ in. long, pilose at the margin ; sepals ovate, acuminate, saccate at the base, 1-nerved, ¼ in. long ; petals linear, obtuse, with a narrow or filiform base, over ⅓ in. long, somewhat thickened at the apex, free ; lip elliptic-oblong, over ⅓ in. long, shortly 5-lobed at the apex, with oblong obtuse lobes ; spur saccate, obtuse 1 lin. long ; column broad, one-third as long as the sepals ; capsule elliptic-oblong, nearly ⅓ in. long. *Bolus in Journ. Linn. Soc.* xxv. 189 ; *Durand & Schinz, Conspect. Fl. Afr.* v. 71 ; *Schlechter in Oest. Bot. Zeitschr.* 1898, 442 ; *Kränzl. Orch. Gen. et Sp.* i. 576. *Saccidium pilosum, Lindl. Gen. & Sp. Orch.* 301. *Orchidea pilosa, Burch. ex Lindl. Gen. & Sp. Orch.* 301.

COAST REGION: Swellendam Div. ; on a dry hill near Breede River, *Burchell,* 7483 !

Described from an imperfect specimen collected by Burchell and now preserved at Kew, which is all that is known of the species. Lindley's Herbarium only contains a sketch and a scrap of the original specimen. Lindley referred the plant to a distinct genus, *Saccidium,* on account of the sepals being saccate at the base.

13. **H. Culveri** (Bolus in Trans. S. Afr. Phil. Soc. xvi. 147) ; plant about 5 in. high ; leaves 2 (?), ovate, withering early ; scape

slender, pilose with retrorse hairs, glabrescent upwards, with remote lanceolate acuminate or aristate sheaths ; racemes nearly 2 in. long, dense, more or less secund ; bracts lanceolate, acuminate, the lower longer than the pedicel, upper rather shorter; flowers white ; sepals more than 1 lin. long, the dorsal oblong ; the lateral oblanceolate ; petals spreading, oblong, acute, 1½ lin. long, twice as broad as the sepals, minutely scabrous ; lip rather longer than the petals, sub-quadrate, 3-lobed, minutely scabrous ; side lobes spreading, tooth-like ; front lobe much larger, subtrulliform, acute ; spur nearly straight, acute, nearly 1 lin. long ; column oval.   *Bolus, Ic. Orch. Austr.-Afr.* ii. *t.* 40, *fig. A,* 1, 2, 3, 6, 8, 9.   *Deræmeria Culveri, Schlechter in Engl. Jahrb.* xxxviii. 144.

VAR. β, **integra** (Bolus, l.c. 148) ; sepals spurred at the base ; lip entire. *Bolus, Ic. Orch. Austr.-Afr.* ii. *t.* 40, *fig. A,* 4, 5, 7.

KALAHARI REGION : Transvaal ; Fig-tree Creek, near Barberton, on rocky slopes, 2000 ft., *Culver,* 84.   Var. β, growing with and close to the typical form, *Culver,* 84a.

14. H. Mundii (Sond. in Linnæa, xix. 77) ; plant 2½–5 in. high, with 2 ovoid tubers ; leaves 2, unequal, ovate or orbicular, obtuse or subacute, submembranous, ciliate, ¼–¾ in. long ; scapes 2½–5 in. high, somewhat slender, villous with spreading hairs, without sheaths ; racemes very short, subcapitate, 4–9-flowered, not secund ; bracts broadly ovate, cuspidate, very hirsute or hispid, ¾–1 lin. long ; pedicels about 1½ lin. long ; flowers very small, white ; sepals broadly ovate, apiculate or obtuse, somewhat concave, glabrous, over ½ lin. long, 1-nerved ; petals ovate-oblong, long-acuminate, rather longer than the sepals, 1-nerved ; lip broadly flabellate, about 1 lin. long, divided to about the middle into 5 oblong obtuse lobes ; spur conical, obtuse, two-thirds as long as the sepals ; column· broad, about half as long as the sepals ; capsule oblong, about ¼ in. long.   *Reichb. f. in Walp. Ann.* i. 796.   *H. Mundtii, Bolus in Journ. Linn. Soc.* xxv. 189, *in Trans. S. Afr. Phil. Soc.* v. 113, *Orch. Cap. Penins.* 113, *and Ic. Orch. Austr.-Afr.* i. *t.* 13 ; *Durand & Schinz, Conspect. Fl. Afr.* v. 71 ; *Schlechter in Oest. Bot. Zeitschr.* 1899, 18 ; *Kränzl. Orch. Gen. et Sp.* i. 586.

COAST REGION : Tulbagh Div. ; Winterhoek Mountains, on cliffs, 2500–3000 ft., *Zeyher.* Cape Div. ; Cape Flats, near Rondebosch, 100 ft., *Bolus,* 4971! *Miss Hoskins-Abrahall* ; Lion Mountain, 250 ft., *Schlechter,* 1386. Swellendam Div. ; near Swellendam, *Mund* ! *Pappe* !

15. H. confusa (Rolfe) ; plant ¼–½ ft. high, with 2 large ovoid tubers ; leaves 2, unequal, broadly ovate, obtuse or subacute, submembranous, glabrous or nearly so, ½–1 in. long ; scapes ¼–½ in. high, moderately stout, pilose or sparsely villous, without sheaths ; racemes 1–2½ in. long, somewhat lax, secund ; bracts ovate, acuminate, shortly pubescent, about 1 lin. long ; pedicels about 2 lin. long ; flowers ¼ in. long, apparently white ; sepals broadly ovate, obtuse, somewhat concave, glabrous, about 1 lin. long, 1-nerved ; petals ovate-oblong, obtuse, nearly 2 lin. long, 1-nerved ; lip broadly flabellate, with a narrow base, $\frac{1}{4}$ in. long, unequally 7-lobed to about the middle ; lobes oblong or linear-oblong, obtuse ; disc smooth ; spur conical, much curved, obtuse, one-third as long as the limb ; column broad, about half as long as the sepals. *H. aspera, Schlechter in Oest. Bot. Zeitschr.* 1899, 19, *partly.*

COAST REGION: Clanwilliam Div.; mountain sides about Clanwilliam, 250 ft., *Leipoldt in MacOwan & Bolus, Herb. Norm. Austr -Afr.*, 1757 ! stony places on Blauw Berg, 1500 ft., *Schlechter,* 8465 ! near Olifants River, 350 ft., *Schlechter,* 5036, 5077. Worcester Div. ; Hex River Valley, *Wolley-Dod,* 4054 !

This has been confused with *H. aspera,* Reichb. f., but is quite distinct in the shape of the petals, as well as in the shape and smooth disc of the lip. *Schlechter,* 8465, has much smaller flowers, which are not normally developed.

16. **H. lindleyana** (Reichb. f. Otia Bot. Hamb. ii. 119) ; plant $\frac{1}{3}$–$\frac{3}{4}$ ft. high, with 2 large ovoid tubers ; leaves 2, subequal, ovate, acute or subacute, membranous, glabrous, $\frac{3}{4}$–2 in. long ; scapes $\frac{1}{3}$–$\frac{3}{4}$ in. high, usually slender and somewhat flexuous, sparsely pubescent or glabrous, without sheaths ; racemes 1–4 in. long, somewhat dense or sometimes lax, secund ; bracts ovate or ovate-lanceolate, acute or acuminate, glabrous, 1–1$\frac{1}{4}$ lin. long ; pedicels usually longer than the bracts ; flowers $\frac{1}{4}$–$\frac{1}{3}$ in. long, white ; sepals ovate, acute or subacute, glabrous, 1$\frac{1}{2}$ lin. long, 1-nerved ; petals linear - oblong, subobtuse, about twice as long as the sepals, 1-nerved ; lip 3–4 lin. long, flabellate, unequally 5-lobed, or 3-lobed with the front lobe 3-fid to the middle ; lobes oblong, obtuse ; spur conical, much curved or circinate, one-quarter as long as the limb ; column stout, about half as long as the sepals ; capsule oblong, over $\frac{1}{3}$ in. long. *Gard. Chron.* 1888, iii. 364, 365, *fig.* 55, 56 ; *Bolus in Journ. Linn. Soc.* xxv. 190, *and Ic. Orch. Austr.-Afr.* ii. *t.* 35 ; *Durand & Schinz, Conspect. Fl. Afr.* 70 ; *Schlechter in Oest. Bot. Zeitschr.* 1899, 20 ; *Kränzl. Orch. Gen. et Sp.* i. 580. *H. secunda, Reichb. f. Otia Bot. Hamb.* ii. 119 ; *Bolus in Journ. Linn. Soc.* xxv. 189 ; *Durand & Schinz, Conspect. Fl. Afr.* v. 72 ; *Kränzl, Orch. Gen. et Sp.* i. 586. *Tryphia secunda, Lindl. in Hook. Comp. Bot. Mag.* ii. 209 ; *Drège, Zwei Pfl. Documente,* 124 ; *Sond. in Linnæa,* xix. 82 ; *Harv. Thes. Cap.* ii. 4, *t.* 105, *fig. B. T. nivea, Zeyh. ex Harv. Gen. S. Afr. Pl. ed.* i. 323. *T. sp., Drège, Zwei Pfl. Documente,* 116.

VAR. β, **parviflora** (Rolfe) ; flowers about half as large as in the type. *H. parviflora, Reichb. f. Otia Bot. Hamb.* ii. 119 ; *Bolus in Journ. Linn. Soc.* xxv. 190 ; *Durand & Schinz, Conspect. Fl. Afr.* v. 71 ; *Kränzl. Orch. Gen. et Sp.* i. 580. *Tryphia parviflora, Lindl. in Hook. Comp. Bot. Mag.* ii. 209 ; *Drège, Zwei Pfl. Documente,* 66.

SOUTH AFRICA: without locality, *Mund* ! *Mrs. Holland,* 4 ! *Mrs. Bowker* ! *Zeyher,* 3907 !
COAST REGION : Caledon Div. ; Genadendal, 2000–3000 ft., *Drège.* Knysna Div. ; Karratera River, under 1000 ft., *Drège,* 3580a ! 3580b ! Plettenbergs Bay,

forest land, *Bowie* ! Uitenhage Div. ; in forest near Zwartkops River, *Zeyher*, 137 ! *Pappe*, 16 ! near Uitenhage, *Tredgold* ! Algoa Bay, *Forbes*, 80 ! Port Elizabeth Div. ; Port Elizabeth, *Mrs. Hewitson* ! *Miss West*, 216, *Kemsley* ! Albany Div. ; Zwartwater Poort, on rocks, *Burchell*, 3415 ! Bothas Hill, *MacOwan*, 627 ! near Grahamstown, *Bolton* ! Signal Hill, 2400 ft., *Galpin*, 162, and without precise locality, *Hutton* ! *Williamson* ! *Mrs. Barber*, 256 ! *Glass*. Fort Beaufort Div. ; Winterberg Range, *Mrs. Barber*, 256 ! British Kaffraria ; *Cooper*, 1804 ! Hangmans Bush, *Cooper*, 3616 ! Eastern frontier of Cape Colony, *Prior* ! *Hutton* ! *MacOwan*, 627 !

CENTRAL REGION : Somerset Div. ; Somerset, *Bowker* ! *Mrs. Barber*, 256 ! Var. β : Willowmore Div. ; Zwanepoels Poort Mountains, 3000–4000 ft., *Drège*, 8276a ! *Marloth*, 4130 ; Witte Berg, 2000–3000 ft., *Drège*.

Var. β was originally described as a species by Lindley, but chiefly differs in its far smaller flowers.

17. H. orthoceras (Reichb. f. Otia Bot. Hamb. ii. 119) ; plant $\frac{1}{3}$–$\frac{3}{4}$ ft. high, with 2 ovoid tubers ; leaves 2, unequal, broadly ovate or elliptic-ovate, subacute, submembranous, with white reticulated veins, glabrous, 1–2 in. long ; scapes $\frac{1}{2}$–$\frac{3}{4}$ ft. high, rather slender, pubescent, without sheaths ; racemes 1–3 in. long, dense or sometimes lax, secund ; bracts ovate-lanceolate, acute, pubescent, about 1 lin. long ; pedicels about 2 lin. long ; flowers 4–5 lin. long, white, membranous ; sepals ovate, acute, glabrous, 1 lin. long, 1-nerved ; petals oblong-lanceolate, acuminate, 3 lin. long, 1-nerved ; lip 4–5 lin. long, cuneate, narrow at the base, unequally lobed ; sides lobes linear, subacute, the middle part divided into 3 or rarely 5 broadly triangular teeth ; spur narrowly conical, nearly straight, about one-third as long as the limb ; column stout, about half as long as the petals. *Bolus in Journ. Linn. Soc. xxv. 190, and Ic. Orch. Austr.-Afr. ii. t. 34 ; Durand & Schinz, Conspect. Fl. Afr. v. 71 ; Bot. Mag. t. 7523 ; Schlechter in Oest. Bot. Zeitschr. 1899, 17 ; Kränzl. Orch. Gen. et Sp. i. 582. Tryphia orthoceras, Harv. Thes. Cap. ii. 4, t. 105, fig. A.*

COAST REGION : Albany Div. ; near Grahamstown, 2000–3000 ft., *Atherstone*, 27 ! *MacOwan*, 1288 ! *Schönland, Glass, Schlechter*, 2571, *South, Hutton* ! Stockenstrom Div. ; Kat Berg, *Hutton* ! King Williamstown Div. ; Perie, 4000 ft., *Sim*, 947. British Kaffraria ; *Mrs. Barber* ! Eastern frontier of Cape Colony, *Hutton* ! KALAHARI REGION : Transvaal ; near Barberton, 3600 ft., *Culver*, 50, *Thorncroft*, 391, *Mrs. Deglon* ! Houtbosch, 5500 ft., *Schlechter*, 4738 ! (4748 *ex Schlechter*), near Mailas Kop, *Schlechter*. EASTERN REGION : Natal ; Drakensberg Range, *Fannin in Herb. Sanderson*, 706 ! Krans Kop, *McKen*, 28 ! Greytown, *McKen in Herb. Sanderson*, 1009 ! Mid Illovo, 2000 ft., *Wood*, 1869 ! Weenen, 3000–4000 ft., *Wylie in Herb. Wood*, 6764 ; sources of Polela River, 6000–7000 ft., *Evans*, 617 ; Umkomaas, in woods, 4000–6000 ft., *Wood* !

18. H. macowaniana (Reichb. f. Otia Bot. Hamb. ii. 108) ; plant 2–3 in. high, with two ovoid tubers ; leaves 2, unequal, ovate, subacute, membranous, glabrous, $\frac{1}{3}$–$\frac{1}{2}$ in. long ; scapes 2–3 in. long, rather stout, shortly pubescent, without sheaths ; racemes $\frac{1}{2}$–1 in. long, somewhat lax, secund ; bracts ovate-lanceolate, acuminate, 1$\frac{1}{2}$–2 lin. long, puberulous ; pedicels rather longer than the bracts ;

flowers 3–4 lin. long ; sepals ovate, very acuminate, glabrous, 1¼ lin.
long ; 1-nerved ; petals ovate-lanceolate, very acuminate, about 2
lin. long, 1-nerved ; lip narrowly flabellate, with a cuneate base,
3–4 lin. long, membranous, with about 7–9 short broad apical lobes
or teeth ; spur linear or very narrowly conical, obtuse, about half
as long as the limb ; column broad, half as long as the sepals. *Bolus
in Journ. Linn. Soc.* xxv. 190 ; *Durand & Schinz, Conspect. Fl. Afr.*
v. 71 ; *Schlechter in Oest. Bot. Zeitschr.* 1899, 18 ; *Kränzl. Orch.
Gen. et Sp.* i. 573.

COAST REGION : Albany Div. ; mountain rocks near Howisons Poort, 2000 ft.,
Glass, Schönland. Bedford Div. ; Kaga Berg, *Weale in Herb. MacOwan* ! Stockenstrom Div. ; Kat Berg, near Seymour, *Scully in Herb. Bolus,* 6204 ! Queenstown
Div. ; Queenstown, *Wolley-Dod* !
EASTERN REGION : Natal ; Lion River, 3000 ft., *Mrs. Fannin,* 160 !

19. **H. aspera** (Reichb. f. Otia Bot. Hamb. ii. 119) ; plant about
4 in. high, with 2 large ovoid tubers ; leaves 2, unequal, orbicular,
obtuse or subacute, somewhat fleshy, glabrous or slightly pilose,
½–¾ in. long ; scapes 4 in. high, somewhat stout, pilose or villous
with retrorse hairs, without sheaths ; racemes about 1¼ in. long, lax,
about 5-flowered ; bracts ovate, apiculate, puberulous, 1–1½ lin.
long ; pedicels longer than the bracts ; flowers somewhat fleshy,
about 2½ lin. long ; sepals ovate, obtuse, somewhat concave, glabrous,
about 1 lin. long, 1-nerved ; petals broadly triangular-ovate, with a
broad basal tooth in front, subacute, 2 lin. long, obscurely 1-nerved ;
lip about 2½ lin. long, unequally 5-lobed to about the middle ; the
central lobe linear, obtuse, the 2 lateral pairs oblong, obtuse ; disc
covered with crystalline papillæ ; spur broadly oblong, obtuse,
curved, one-third as long as the limb ; column broad, half as long as
the sepals. *Bolus in Journ. Linn. Soc.* xxv. 190 ; *Durand & Schinz,
Conspect. Fl. Afr.* v. 69 ; *Schlechter in Oest. Bot. Zeitschr.* 1899, 19,
partly ; *Kränzl. Orch. Gen. et Sp.* i. 587, *and in Ann. Naturhist. Hofmus. Wien,* xx. 2. *Bucculina aspera, Lindl. in Hook. Comp. Bot.
Mag.* ii. 209 ; *Drège, Zwei Pfl. Documente,* 95.

COAST REGION : Clanwilliam Div. ; near Oliphants River, *Penther,* 252.
WESTERN REGION : Little Namaqualand ; between Mieren Kasteel and Zwart
Doorn River, under 1000 ft., *Drège,* 8276b ! Karree Bergen, 1000 ft., among
stones, *Schlechter (ex Schlechter).*

I have only seen Drège's original specimens, some of those cited as belonging
here by Schlechter being now referred to *H. confusa,* Rolfe.

20. **H. multisecta** (Bolus in Journ. Linn. Soc. xxv. 170, 190, fig.
7) ; plant ¾–1 ft. high, with large ovoid tubers ; leaves 2, unequal,
orbicular, obtuse, rather fleshy, pilose and ciliate, ½–¾ in. long ; scapes
¾–1 ft. high, rather stout, very villous with spreading or retrorse
hairs ; racemes 2–3½ in. long, dense, secund ; bracts ovate, acute,
strongly hirsute or hispid with long hairs, about 2 lin. long ; pedicels
longer than the bracts ; flowers 2–3 lin. long ; sepals ovate, subacute, ciliate, 1–1¼ lin. long, 1-nerved ; petals 2–3 lin. long, 3-fid to

about the middle, linear-oblong below, with linear, somewhat diverging lobes; lip about as long as the petals, flabellate, with a narrow base, divided to the middle into 5–7 linear diverging lobes; spur narrowly conical, curved, obtuse, about as long as the sepals; column broad, scarcely half as long as the sepals. *Kränzl. Orch. Gen. et Sp.* i. 589. *H. multiseta, Durand & Schinz, Conspect. Fl. Afr.* v. 71. *H. Scopularia, Schlechter in Oest. Bot. Zeitschr.* 1899, 21, *partly.*

COAST REGION: Stockenstrom Div.; summit of Elandsberg, 6000 ft., *Scully*, 391! *and in MacOwan & Bolus, Herb. Norm. Austr.-Afr.*, 1371! British Kaffraria; *Mrs. Hutton*! Fort Bowker, among rocks, *Bowker*, 769, partly!
EASTERN REGION: Griqualand East; Mount Currie, 5300 ft., *Tyson*, 1542. Natal; Dargle Farm, *Mrs. Fannin*, 82! *Hallack*!

Differs from *H. Scopularia*, Reichb. f., with which Schlechter has included it, in its smaller flowers and in the fewer lobes of the lip. *Baur*, 737, which was included by Bolus, has simple petals, and does not belong here.

21. **H. Scopularia** (Reichb. f. Otia Bot. Hamb. ii. 119); plant about ¾ ft. high, with large ovoid tubers; leaves 2, unequal, orbicular, obtuse, rather fleshy, pilose, ciliate, ¾–1¼ in. long; scapes ¾ ft. high, rather stout, very villous, with spreading or retrorse hairs; racemes 2–3½ in. long, dense, secund; bracts ovate, acute, strongly hirsute or hispid with long hairs, about 2–3 lin. long; pedicels longer than the bracts; flowers 3–4 lin. long; sepals ovate or ovate-oblong, subacute, ciliate, 1¼ lin. long, 1-nerved; petals 3–4 lin. long, 3-fid to about the middle, linear-oblong below, 1-nerved, with linear, somewhat diverging lobes; lip about as long as the petals, flabellate, with a narrow base, divided to the middle into 8 to 9 linear diverging lobes; spur narrowly conical, curved, obtuse, about as long as the sepals; column broad, scarcely half as long as the sepals. *Bolus in Journ. Linn. Soc.* xxv. 190, *and Ic. Orch. Austr.-Afr.* ii. *t.* 38, *excl.* 3 *last syn.* ; *Durand & Schinz, Conspect. Fl. Afr.* v. 71; *Schlechter in Oest. Bot. Zeitschr.* 1899, 21, *partly. H. Burchellii, Kränzl. Orch. Gen. et Sp.* i. 589, *partly, not of Reichb. f. Scopularia secunda, Lindl. in Hook. Comp. Bot. Mag.* ii. 207; *Drège, Zwei Pfl. Documente*, 53.

CENTRAL REGION: Aliwal North Div.; Witte Bergen, 6000–8000 ft., *Drège*!
KALAHARI REGION: Orange River Colony; Mont aux Sources, 8000–9000 ft., *Flanagan*, 1980! Transvaal; mountains near Barberton, 4000–5000 ft., *Culver*, 72, *Galpin*, 585.

Schlechter includes the preceding and two allied Tropical African species with this, in which view I cannot concur. Kränzlin wrongly includes *H. Scopularia*, Reichb. f., as a synonym of *H. Burchellii*.

22. **H. schlechteriana** (Kränzl. ex Schlechter in Oest. Bot. Zeitschr. 1899, 21, without description); plant ½–1¼ ft. high, stout, with 2 ovoid tubers; leaves 2, unequal, orbicular, obtuse or shortly apiculate, membranous, glabrous, 1–2½ in. long; scapes ½–1¼ ft. high, pilose at the base, very slightly so above, with several ovate very acuminate sheaths; racemes 3–6 in. long, dense, somewhat

secund; bracts narrowly ovate-lanceolate, long-acuminate, 2–3 lin. long; pedicels shorter than the bracts; sepals ovate, acute, glabrous, 1¼–1½ lin. long, 1-nerved; petals 2½–3 lin. long, 5-fid to about the middle, oblong and somewhat dilated below, with filiform diverging somewhat unequal lobes; lip as long as the petals, flabellate, with a narrow base, divided to the middle into 7 filiform diverging lobes; spur narrowly conical, curved at the apex, longer than the sepals; column stout, about half as long as the sepals; capsule oblong, 4–5 lin. long. *Kränzl. Orch. Gen. et Sp.* i. 588, *and in Ann. Naturhist. Hofmus. Wien*, xx. 2.

COAST REGION: Humansdorp Div.; near Clarkson, 500 ft., among stones, *Schlechter*, 6015! *Penther*, 44. Queenstown Div.; near Queenstown, *Mrs. Barber.* Albany Div.; near Howisons Poort, 2000–3000 ft., *South*, 423!

23. **H. grandiflora** (Reichb. f. Otia Bot. Hamb. ii. 119); plant 1 ft. high, stout; leaves solitary, orbicular; scapes 1 ft. high, stout, pilose, with a few ovate acuminate sheaths; racemes 4 in. long; bracts acuminate, glabrous, as long as the pedicels; flowers ½ in. long; sepals ovate-oblong, subobtuse, glabrous, 1-nerved, dorsal 3½ lin. long, lateral 2½ lin. long; petals ½ in. long, linear-oblong below, then somewhat dilated, divided nearly to the middle into 6 linear obtuse segments; lip as long as the petals, flabellate, with a narrow base, the upper third divided into about 18 linear obtuse lobes; spur conical, obtuse, curved, 2½ lin. long; column broad, rather over 1 lin. long. *Bolus in Journ. Linn. Soc.* xxv. 190; *Durand & Schinz, Conspect. Fl. Afr.* v. 70; *Schlechter in Oest. Bot. Zeitschr.* 1899, 22; *Kränzl. Orch. Gen. et Sp.* i. 591. *Scopularia grandiflora, Sond. in Linnæa*, xix. 79.

COAST REGION: Uitenhage Div.; Uitenhage, *Wiedemann*! near Olifants River, among stones, *Zeyher.*

I have only seen a single flower of this remarkable species, which was sent by Sonder to Dr. Lindley, and is preserved in the Herbarium of the latter, now at Kew.

24. **H. Burchellii** (Reichb. f. Otia Bot. Hamb. ii. 119); plant ½–1½ ft. high, with large ovoid tubers; leaves 2, unequal, broadly ovate, orbicular or reniform, obtuse or subacute, glabrous, somewhat fleshy, ¾–3 in. broad; scape ½–1½ ft. high, pilose, with spreading or retrorse hairs, with several ovate-lanceolate acuminate sheaths; racemes 3–6 in. long, dense, somewhat secund; bracts ovate or ovate-lanceolate, very acuminate, 2–3 lin. long; pedicels longer than the bracts; flowers dimorphic, yellowish-green, the lower 3 lin. long, the upper more than twice as long owing to the much elongated lobes of the petals and lip; sepals ovate or ovate-oblong, subobtuse, glabrous, 1½–2 lin. long, 1-nerved; petals of the lower flowers 2–3 lin. long, oblong, divided to about the middle into 7 to 9 filiform more or less curved lobes, of the upper flowers ½ in. or more long, divided into 7 to 9 much elongated filiform lobes, about 4 times as

long as the limb; lip about as long as the petals, of the lower
flowers oblong, divided into about 11 to 13 short filiform, more or
less curved lobes, of the upper flowers broader, divided into 11 to
13 much elongated filiform lobes, about 4 times as long as the limb ;
spur narrowly conical, subobtuse, of the lower flowers much curved
or circinate, 2 to 3 lin. long, of the upper flowers more slender
and rather shorter; column broad, scarcely half as long as the
sepals ; capsule oblong, 3–4 lin. long. *Bolus in Journ. Linn. Soc.*
xxv. 189, *and Ic. Orch. Austr.-Afr.* ii. *t.* 39; *Durand & Schinz,
Conspect. Fl. Afr.* v. 69 ; *Schlechter in Oest. Bot. Zeitschr.* 1899, 22 ;
Kränzl. Orch. Gen. et Sp. i. 589, *partly, and in Ann. Naturhist.
Hofmus. Wien,* xx. 2. *Scopularia Burchellii, Lindl. in Bot. Reg.
sub t.* 1701, *and Gen. & Sp. Orch.* 304; *Sond. in Linnæa,* xix. 79 ;
Drège, Zwei Pfl. Documente, 130, 137, 138 (*Scopolia*); *Drège in
Linnæa,* xx. 217. *Orchidea pectinata, Burchell ex Lindl. Gen. &
Sp. Orch.* 304.

SOUTH AFRICA: without locality, *Masson* ! *Mrs. Barber,* 75 ! *Mrs. Bowker* !
COAST REGION : Swellendam Div. ; Riet Kuil, near Buffeljagts River, *Zeyher,*
3904 ! 3905 ! Riversdale Div. ; near Zoetemelks River, *Burchell,* 6709 ! *Bolus,*
11382. Oudtshoorn Div. ; near Oudtshoorn, 1100 ft., *Miss Hope.* Mossel Bay
Div. ; between the landing place at Mossel Bay and Cape St. Blaize, *Burchell,*
6274 ! near Groot Brack River, *Schlechter,* 5574. Uitenhage Div. ; Zuurberg
Range, 2500–3500 ft., *Drège,* 2209a ! Klein Bruitjes Hoek, 3000–4000 ft., *Drège* ;
near mouth of the Zwartkops River, *Drège,* 2209 ! *Zeyher,* 1029 ! near Uitenhage,
Pappe, 12 ! *Tredgold,* 35 ! *Zeyher* ! hills of Addo, *Ecklon & Zeyher,* 21 ! Koegas
Kopje, *Zeyher,* 5 ! Port Elizabeth Div. ; along the Baakens River, near Port
Elizabeth, *Burchell,* 4358 ! Bathurst Div. ; base of Kleinemont, *Atherstone,* 21 !
near Dixons Bush, *Bennie.* Albany Div. ; Fish River Heights, *Hutton* ! Oatlands,
Grahamstown, *Atherstone,* 21 ; near Grahamstown, *Bolton* ! Curries Kloof,
MacOwan, and without precise locality, *Hutton* ! *Miss Bowker* ! Queenstown Div. ;
mountain sides, Queenstown, 4000 ft., *Galpin,* 1578 ! *Wolley-Dod* ! King
Williamstown Div. ; near King Williamstown, *Tyson, Flanagan,* 2206 ! Kei Road,
Krook, Penther, 228 ; Mount Coke, *Sim.* Komgha Div. ; near Komgha, *Flanagan.*
Eastern Frontier of Cape Colony, *Hutton* ! *Hallack* !
CENTRAL REGION : Somerset Div. ; *Bowker* ! Graaff Reinet Div. ; sides of
Oude Berg, near Graaff Reinet, 3500 ft., *Bolus,* 177 ! Camdeboo Mountain, *Dunn.*

EASTERN REGION : Transkei ; Fort Bowker, *Bowker & Mrs. Barber,* 769, partly !

A very remarkable species with dimorphic flowers, those of the upper half of
the raceme being extended into long filaments, more than twice as long as in the
lower flowers. It has been suggested that the upper flowers are abortive, but the
column is normally developed, and in a fruiting specimen collected by Drège the
capsules are developed right to the apex.

XXI. HUTTONÆA, Harv.

Dorsal sepal free, unguiculate, lanceolate or ovate, erect ; lateral
sepals larger and broader, oblique, spreading. *Petals* unguiculate ;
claw elongated, united to the claw of the dorsal sepal at the base ;
limb broad or suborbicular, concave, much fimbriate. *Lip* sessile,
spreading, broadly dilated, much fimbriate, not spurred at the base.
Column short and broad ; anther-bed broad, erect, its margin
continuous with the side lobes of the rostellum and not distinct

from the connective of the anther. *Anther-cells* adnate, diverging,
incurved and ascending at the apex, almost included within the lobes
of the rostellum; pollinia solitary in the cells, coarsely granular;
caudicles somewhat elongated; glands small, scarcely exserted from
the folds of the rostellum. *Stigma* small, depressed; rostellum
3-lobed; side lobes complicate round the caudicles of the anther.
Capsule oblong.

Terrestrial herbs, with erect leafy stems, and a globose tuber; leaves few, broad
or ovate, decreasing upwards; flowers medium-sized, shortly pedicelled, arranged
in short or lax racemes; bracts lanceolate.

DISTRIB. Species 5, all eastern and endemic.

Lower leaf long petiolate ... ... (1) **fimbriata.**

Lower leaf sessile :
    Lower leaf more than 2 in. long, usually much
    larger ... ... (2) **pulchra.**

Lower leaf less than 2 in. long :
    Spikes many-flowered; appendages of petals clavate (3) **Woodii.**

    Spikes few-flowered; appendages of petals slender :
      Lip about 4 lin. broad ... ... (4) **oreophila.**

      Lip 6–8 lin. broad ... ... (5) **grandiflora.**

1. **H. fimbriata** (Reichb. f. in Flora, 1867, 116); plant erect,
with straight or slightly flexuous stems, distantly 2-leaved; lower
leaf long-petiolate, broadly cordate, abruptly acuminate, very
glabrous, membranous, 3–3¾ in. long, 2½–3 in. broad; petiole
sheathing at the base, free part 2–2½ in. long; upper leaf sessile,
2–2¼ in. long, 1½–1¾ in. broad, otherwise like the lower leaf; raceme
2–4 in. long, rather lax, many-flowered; bracts ovate-lanceolate or
oblong-lanceolate, very acuminate, ¼–½ in. long; pedicels ⅓–½ in.
long; dorsal sepal subsessile, ovate, subacute, fimbriate, reflexed,
about 1 lin. long; lateral sepals obliquely and broadly ovate, subobtuse, fimbriate, shortly unguiculate, about 2½ lin. long; petals
distinctly unguiculate, somewhat diverging, broadly dilated and
flabellate at the apex, shortly and irregularly fimbriate, deeply
concave at the base of the limb, about 2½ in. long; lip broadly
flabellate, with a short broad claw, about 2 lin. long, broader than
long, strongly fimbriate at the margin; column strongly incurved,
scarcely 1 lin. long; capsule oblong, over ⅓ in. long. *Bolus in Journ.
Linn. Soc.* xxv. 188, *and Ic. Orch. Austr.-Afr.* ii. *t.* 97; *Kränzl.
Orch. Gen. et Sp.* i. 595, *and in Ann. Naturhist. Hofmus. Wien,* xx. 2.
H. Hallackii, Bolus in Journ. Linn. Soc. xix. 339. *Hallackia
fimbriata, Harv. Thes. Cap.* ii. 2, *t.* 102.

EASTERN REGION : Griqualand East; Zuurberg Range, 4300 ft., *Schlechter,*
6614 ; Kwenkwe Mountain, 5500 ft., *Bolus,* 8774. Natal ; Dargle Farm,
Mrs. Fannin, 72 ! Ingoma, *Gerrard,* 1547 ! 2171 ; Van Reenen, *Krook, Penther,*
320, and without precise locality, *Hallack,* 6 ! *Mrs. K. Saunders.*

Harvey's drawing shows both the leaves sessile, which is incorrect, even for
Hallack's specimen, the only one cited by him.

2. H. pulchra (Harv. Thes. Cap. ii. 1, t. 101); plant erect, $\frac{3}{4}$–1$\frac{1}{2}$ ft.
high, with rather stout straight or rarely flexuous stems, distantly
2-leaved; leaves sessile or the lower subsessile, broadly ovate or
subcordate-ovate, apiculate or subobtuse, membranous, very glabrous, the lower 2–6 in. long, 1$\frac{1}{2}$–3 in. broad, the upper 1–5 in. long,
$\frac{3}{4}$–2$\frac{1}{2}$ in. broad; racemes 1–6 in. long, usually lax, few- to many-
flowered; bracts ovate-lanceolate, very acuminate, $\frac{1}{3}$–$\frac{3}{4}$ in. long;
pedicels about $\frac{1}{2}$ in. long; dorsal sepal subsessile, elliptic-ovate, subobtuse, entire, reflexed about the middle, 2–2$\frac{1}{4}$ lin. long; lateral sepals
oblique and broadly ovate, subobtuse, entire, recurved at the apex,
$\frac{1}{4}$–$\frac{1}{3}$ in. long; petals long-unguiculate, somewhat diverging, adnate at
the extreme base to the dorsal sepal, dilated at the apex into a
rounded limb, deeply fimbriate, with capitate hairs; limb with a
deep obtuse sac in the centre; limb and claw each about $\frac{1}{4}$ in. long;
lip suborbicular, with a short broad claw, irregularly fimbriate,
with capitate hairs, about 2$\frac{1}{2}$ lin. broad; column strongly incurved,
about 1 lin. long; capsule oblong, $\frac{1}{2}$–$\frac{3}{4}$ in. long. *Reichb. f. in Flora,*
1867, 115; *Bolus in Journ. Linn. Soc.* xix. 339, xxv. 188; *Kränzl.
Orch. Gen. et Sp.* i. 595.

COAST REGION: Stockenstrom Div. ; Kat Berg, 4000 ft., under trees, *Hutton,* 2 !
KALAHARI REGION : Orange River Colony ; without locality, *Cooper,* 1091 !
EASTERN REGION : Transkei ; near Tsomo, *Bowker and Mrs. Barber,* 842 !
Griqualand East ; Zuurberg Range, *Schlechter,* 6612 ! Natal ; Fort Nottingham,
2000–3000 ft., *Buchanan in Herb. Sanderson,* 1049 ! 1056 ! Byrne, 4500 ft.,
Wood, 1871 !

3. H. Woodii (Schlechter in Engl. Jahrb. xxxviii. 144); plant
$\frac{3}{4}$–1 ft. high, erect, with rather stout straight stem, distantly
2-leaved; leaves sessile, elliptic-ovate, subobtuse, glabrous, some-
what fleshy, 1$\frac{1}{4}$–2 in. long, $\frac{1}{3}$ to nearly 1 in. broad; raceme oblong,
2–3$\frac{1}{2}$ in. long, somewhat dense- and many-flowered; bracts ovate-
lanceolate, acuminate, $\frac{1}{3}$–$\frac{1}{2}$ in. long; pedicels 4–5 lin. long; dorsal
sepal reflexed, elliptic-oblong, with recurved acute apex and entire
margin, 1–1$\frac{1}{4}$ lin. long; lateral sepals obliquely and broadly ovate,
subacute, the outer margin subcordate, subcrenulate, 2$\frac{1}{2}$–3 lin. long;
petals unguiculate, somewhat diverging, dilated above, with rather
long capitate appendages, deeply concave at the base of the limb,
2 lin. long, claw about as long as the limb; lip with a short broad
claw and an obcordately dilated limb, about $\frac{1}{3}$ in. broad, irregularly
fimbriate, with subcapitate appendages; column strongly incurved,
1$\frac{1}{2}$ lin. long.

EASTERN REGION : Natal ; Seven Fontein, near Boston, 5000 ft., *Wood,* 5577 !
Dargle Farm, *Mrs. Fannin,* 87 !

4. H. oreophila (Schlechter in Engl. Jahrb. xxiv. 420); plant
erect, 5–12 in. high, with straight or somewhat flexuous stem, distantly 2-leaved; leaves sessile, broadly ovate, apiculate or shortly
acuminate, very glabrous, $\frac{3}{4}$–2$\frac{1}{2}$ in. long, $\frac{1}{3}$–1$\frac{1}{2}$ in. broad; raceme

$\frac{3}{4}$-2$\frac{1}{2}$ in. long, 2–15-flowered; bracts oblong-lanceolate, acute or acuminate, $\frac{1}{3}$ to nearly $\frac{1}{2}$ in. long; flowers with yellowish-green sepals, mauve-purple petals and a white lip (*Allison*); pedicels 3–4 lin. long; dorsal sepal elliptic-lanceolate, acute, denticulate, spreading or reflexed, about 1$\frac{1}{2}$ lin. long; lateral sepals obliquely and broadly ovate, subobtuse, closely fringed at the margin, about 2$\frac{1}{2}$ lin. long; petals long-unguiculate, somewhat diverging, broadly dilated and flabellate at the apex, irregularly fimbriate to the middle, with narrow subcapitate filaments, deeply concave at the base of the limb, $\frac{1}{3}$ in. long; claws connate to the middle; lip broadly flabellate, with a short broad claw, about $\frac{1}{4}$ in. long by $\frac{1}{3}$ in. broad, the limb irregularly fimbriate to nearly the middle, with narrow subcapitate lobes; column strongly incurved, 1$\frac{1}{2}$ lin. long.

EASTERN REGION: Natal; Dargle Farm, *Mrs. Fannin*, 87! Van Reenan, 6900 ft., *Schlechter*, 6931! Oliviers Hoek, sources of Tugela River, 5000 ft., *Allison*, 7!

H. oreophila, var. *grandiflora*, Schlechter, is structurally distinct. See the following species.

5. **H. grandiflora** (Rolfe); plant 6–8 in. high, erect, with some-what slender flexuous stem, distantly 2-leaved; leaves sessile, ovate or cordate-ovate, acute, glabrous, submembranous, $\frac{1}{2}$–1$\frac{1}{4}$ in. long, $\frac{1}{3}$–1 in. broad; raceme very short, 2–4-flowered; bracts ovate, acute, 4–5 lin. long; pedicels about $\frac{1}{4}$ in. long; dorsal sepal ovate-oblong, obtuse, fimbriate, 2 lin. long; lateral sepals oblique and very broadly ovate, obtuse, much fimbriate, 3–4 lin. long; petals unguiculate, somewhat diverging, broadly dilated above, deeply and compoundly fimbriate, 5–6 lin. long by about as broad, obtusely saccate at the base of the limb, claws rather shorter than the limb and connate to about the middle; lip with a short broad claw and a much dilated limb, 6–8 lin. broad by nearly as long, deeply and compoundly fimbriate; column strongly incurved, 2 lin. long. *H. oreophila, var. grandiflora, Schlechter in Engl. Jahrb.* xxiv. 420.

KALAHARI REGION: Orange River Colony; rocky grassy ledges near the summit of Mapedis Peak, Witzies Hoek (Mont aux Sources), 8600 ft., *Thode*, 49!

XXII. PERISTYLUS, Blume.

Sepals and *petals* free, subequal, connivent or subconnivent. *Lip* continuous with the column, free or slightly adnate to it, produced at the base into a short (sometimes very short) spur; limb erect or somewhat spreading, entire or 3-lobed. *Column* short, footless; anther-bed erect, short. *Anther-cells* parallel, apex inferior, short and adnate to the column; pollinia granular, with short caudicles and exserted naked glands; staminodes lateral, auriculate. *Stigma* sessile; rostellum subulate or tooth-like, situated between the anther-cells. *Capsule* ellipsoid or oblong.

Terrestrial herbs with erect, sometimes leafy stems, and oblong or subglobose tubers; leaves radical and cauline, few or numerous, decreasing upwards; flowers rather small, shortly pedicelled, arranged in narrow spikes or racemes; bracts lanceolate.

DISTRIB. Species about 50, the majority Indian and extending into the Malayan and Chinese regions, with six tropical African representatives and the following somewhat anomalous species within our limits.

1. **P. natalensis** (Rolfe); plant 1–1½ ft. high, slender, somewhat flexuous, with leafy stems and a few narrow basal sheaths; leaves 4–5, subsessile, elliptic-ovate, acute or acuminate, membranous, 1–2½ in. long, ½–1 in. broad; racemes 1–3 in. long, lax; bracts narrowly ovate-lanceolate, acuminate, about 3 lin. long; pedicels shorter than the bracts; flowers small; sepals ovate, subobtuse, about 1 lin. long, 1-nerved; petals obliquely ovate-oblong, subobtuse, rather shorter and much narrower than the sepals, 1-nerved; lip about as long as the petals, with a broadly oblong base, 3-lobed to nearly the middle, with oblong or triangular-oblong obtuse subequal fleshy lobes; spur broadly oblong, obtuse, somewhat curved, about half as long as the limb; column very short and broad. *Herminium natalense, Reichb. f. Otia Bot. Hamb.* ii. 108; *Bolus in Journ. Linn. Soc.* xxv. 188; *Kränzl. Orch. Gen. et Sp.* i. 536. *Platanthera natalensis, Schlechter in Engl. Jahrb.* xx. *Beibl.* 50, 6.

EASTERN REGION: Natal; Ingoma, on half-decayed trees, *Gerrard*, 1541!

XXIII. STENOGLOTTIS, Lindl.

Sepals free, subequal, ultimately spreading. *Petals* somewhat narrower than the sepals, suberect. *Lip* continuous with the base of the column, cuneate-oblong, spurless, 3–5-fid at the apex. *Column* very short and broad; anther-bed broad and erect, thickened at the margin. *Anther-cells* parallel; pollinia granular, affixed by a short stipes to a small oblong gland; staminodes lateral, oblong, tuberculate or glandular at the apex. *Stigmatic* processes 2, clavate or capitate, short; rostellum broad and very short. *Capsule* oblong, erect.

Terrestrial herbs, with short stems, and tuberiferous or thickened fleshy fasciculate roots; leaves radical, numerous, rosulate or tufted; flowers small, shortly pedicelled, arranged in loose or sometimes dense somewhat one-sided racemes; bracts small.

DISTRIB. Species 3, two occurring within our limits, the third in British Central Africa.

Leaves usually more or less blotched with brown; lip 3-lobed, rarely with a pair of short lateral teeth   ... (1) **fimbriata.**

Leaves concolorous; lip 5-lobed ...   . (2) **longifolia.**

1. S. fimbriata (Lindl. in Hook. Comp. Bot. Mag. ii. 210); leaves numerous, arranged in a dense rosette, oblong or narrowly lanceolate-oblong, acute, slightly undulate at the margin, all spreading, slightly recurved at the apex, bright green, usually with few or numerous blackish or purple-black blotches, mostly arranged in transverse bars, $1\frac{1}{2}$–$3\frac{1}{2}$ in. long, $\frac{1}{3}$–$1\frac{1}{4}$ in. broad; scapes erect, slender, $\frac{1}{2}$–1 ft. high, with 2 or 3 lanceolate sheaths; racemes $1\frac{1}{2}$–6 in. long, usually lax, many-flowered; bracts lanceolate, acuminate, $\frac{1}{6}$–$\frac{1}{3}$ in. long; pedicels $\frac{1}{4}$–$\frac{1}{3}$ in. long; flowers light purple with a few elongated dark purple blotches on the lip; sepals broadly ovate, obtuse or subacute, $\frac{1}{8}$ in. long, lateral somewhat falcate; petals ovate, subobtuse, obscurely erose at the margin, slightly shorter and broader than the sepals; lip linear-oblong, $\frac{1}{4}$–$\frac{1}{3}$ in. long, 3-fid or 3-partite, rarely with a pair of short additional lateral teeth; column short and broad; capsule narrowly oblong, $\frac{1}{3}$–$\frac{1}{2}$ in. long. *Drège, Zwei Pfl. Documente,* 153; *Harv. Thes. Cap.* i. 36, *t.* 56; *Bot. Mag. t.* 5872; *Gard. Chron.* 1889, vi. 438; 1894, xvi. 563, *in note; Durand & Schinz, Conspect. Fl. Afr.* v. 68; *Schlechter in Engl. Jahrb.* xx. *Beibl.* 50, 30; *Kränzl. Orch. Gen. et Sp.* i. 567, *excl. syn.*

VAR. **saxicola** (Schlechter ex Kränzl. Orch. Gen. et Sp. i. 567); a very dwarf form, about 2 in. high, with all the parts much reduced in size.

SOUTH AFRICA: without locality, *Zeyher,* 3921! *Mrs. Bowker*!
COAST REGION: Stockenstrom Div.; in woods near Stockenstrom, 2000 ft., *Scully in MacOwan & Bolus, Herb. Norm. Austr.-Afr.,* 690! Kat Berg, 2500–3000 ft., *Baur,* 1105! *Hutton*! Albany Div.; near Grahamstown, 2200 ft., *MacOwan,* 515! Howisons Poort, *Hutton*!
KALAHARI REGION: Transvaal; Houtbosch, *Rehmann,* 5855! 5856! *Nelson,* 377! *Schlechter,* 4471; Spitz Kop, near Lydenberg, *Wilms,* 1384!
EASTERN REGION: Transkei; Krielis Country, *Bowker,* 217! Pondoland; between St. John's River and Umsikaba River, under 1000 ft., *Drège,* 4574! Natal; Umcomaas River, *Gerrard,* 1536! ravines of Townhill, Maritzburg, 2500 ft., *Sanderson,* 503! Umtshunga Cutting, 2000 ft., *Sanderson,* 735! Shafton, Howick, *Mrs. Hutton,* 7! Itafamasi, *Wood,* 854! Inanda, *Wood,* 548! 858! Highlands, *Schlechter*! Ingoma, *Gerrard,* 1539! Table Mountain, *Sanderson,* 897! Sterk Spruit, *Sanderson,* 897! and without precise locality, *Gueinzius*! Var. β, Natal; Mount West, *Schlechter,* 6820.

A very variable species. Some of the Natal specimens have unspotted leaves, and at first sight look different from the form figured in the Botanical Magazine. Examination of plants in the field might throw some light on the cause of this variation.

2. S. longifolia (Hook. f. Bot. Mag. t. 7186); leaves numerous, arranged in a dense rosette, ensiform or linear-oblong, acuminate, undulate at the margin, the outer spreading, the inner suberect, all recurved at the apex, uniformly light green, 3–7 in. long, $\frac{1}{3}$–1 in. broad; scapes erect, often stout, $\frac{3}{4}$–$1\frac{3}{4}$ ft. high, with numerous linear-lanceolate somewhat recurved sheaths; racemes 4–10 in. long, many-flowered; bracts lanceolate, acuminate, $\frac{1}{4}$–$\frac{1}{2}$ in. long; pedicels $\frac{1}{3}$–$\frac{3}{4}$ in. long; flowers light purple with a few minute darker dots on the lip, occasionally white; sepals broadly ovate, subobtuse, $\frac{1}{4}$ in. long; petals ovate, subacute, erose or minutely denticulate,

smaller than the sepals; lip linear-oblong, $\frac{1}{3}$–$\frac{1}{2}$ in. long, 5-fid or 5-partite; column short and broad. *Gard. Chron.* 1894, xvi. 563, *fig.* 72; *Durand & Schinz, Conspect. Fl. Afr.* v. 68. *S. fimbriata, N. E. Br. in Gard. Chron.* 1889, vi. 438, *partly, not of Lindl.*

EASTERN REGION: Zululand; Eshowe, *Mrs. K. Saunders,* 2! near Ungoya, 1000–2000 ft., *Wood,* 5650! *and in MacOwan & Bolus, Herb. Norm. Austr.-Afr.,* 1024!

XXIV. HABENARIA, Willd.

Sepals unequal, free; lateral more or less spreading or reflexed. *Petals* simple or deeply bilobed, usually narrower than the dorsal sepal and adpressed to it, forming a galea, or the posticous lobe so adpressed, and the anticous descending, and simulating a lobe of the lip. *Lip* continuous with the column, often shortly adnate to it, produced at the base into a short or much elongated spur; limb spreading or pendulous, narrow or broad, undivided or trilobed; side lobes somewhat fimbriate or pectinate. *Column* short, footless; anther-bed erect, short or scarcely as long as the anther. *Anther-cells* parallel or diverging; apex inferior, short and adnate or free, elongate and horizontal or descending in slender channels on the margins of the side lobes of the rostellum; pollinia granular, with short or elongate caudicles and exserted naked glands; staminodes lateral, small, rounded or auriculate. *Stigma* bilobed or extended into two short or elongate, often clavate, papillose processes; rostellum trilobed; middle lobe subulate or tooth-like, situated between the anther-cells; side lobes much longer, and acting as carriers for the caudicles of the pollinia. *Capsule* ellipsoid or oblong, sometimes beaked.

Terrestrial herbs, with the habit of *Orchis*; tubers ovoid-globose or rarely lobed; flowers small or large, in lax or dense spikes or racemes; bracts mostly narrow.

DISTRIB. Species over 500, widely diffused through tropical and subtropical regions.

Petals simple :
  Spur of lip short, not exceeding 1$\frac{1}{4}$ in. long :
    Leaves radical, 2 or 3 ...   . (1) **arenaria.**
    Leaves more or less cauline : :
      Leaves not numerous, and not gradually passing
        into the bracts :
        Leaves elongate-linear   ... ... (2) **natalensis.**
        Leaves broadly elliptic-lanceolate   ...   ... (3) **microrhynchos.**
      Leaves usually numerous, more or less imbricate
        and gradually passing into the bracts :
        Lip simple :
          Spur of lip clavate or subclavate :
            Stigmas about $\frac{1}{2}$ lin. long ...   . (4) **anguiceps.**
            Stigmas about $\frac{1}{4}$ lin. long .... (5) **Readei.**
          Spur of lip cylindrical ... ... (6) **foliosa.**

Lip 3-partite :
  Leaf-sheaths uniformly green :
    Spikes 1–2 in. long ; spur ¾–1¼ in. long ... (7) **lævigata.**

    Spikes about 8 in. long ; spur 3–3½ lin.
    long ... ... (8) **bicolor.**

  Leaf-sheaths more or less transversely barred
  with black ... ... (9) **ciliosa.**

Spur of lip 4 in. or more long ... ... (10) **Schlechteri.**

Petals 2-partite :
  *Leaves linear to oblong, cauline or subradical :
    Side lobes of lip very short and tooth-like ... ... (11) **Culveri.**

    Side lobes of lip linear or elongate :
    †Stigmas short, under 2½ lin. long, or if longer
    leaves radical or subradical :
    Stigmas under 1 lin. long :
      Leaves linear or narrowly lanceolate :
      Leaves cauline, narrowly lanceolate :
        Upper lobe of petals over twice as long
        as the lower lobe ... ... (12) **tridens.**

        Lobes of petals subequal ... ... (13) **Galpini.**

      Leaves radical, linear ... ... (14) **Woodii.**

      Leaves broadly lanceolate or oblong-lanceolate (15) **malacophylla.**

    Stigmas 1 lin. or more long :
    Leaves oblong or broad ; inflorescence lax :
      Leaves cauline :
        Lobes of petals linear ... ... (16) **transvaalensis.**
        Upper lobe of petals over 3 times as
        broad as the lower ... ... (17) **Barbertoni.**

      Leaves radical or subradical :
        Spur of lip under 1 in. long ... ... (18) **insignis.**

        Spur of lip 1¼–1½ in. long ... ... (19) **polypodantha.**

    Leaves linear or linear-oblong ; inflorescence
    many-flowered :
    Leaves cauline :
    Spur of lip slender or subclavate :
      Stigmas about 1 lin. long :
        Side lobes of lip and lower lobe of
        petals linear-oblong or narrow (20) **orangana.**

        Side lobes of lip and lower lobe of
        petals ovate-lanceolate or broad (21) **dives.**

      Stigmas about 2 lin. long :
        Spur of lip ⅔ in. long or less :
        Lobes of petals simple ... ... (22) **Rehmanni.**

        Upper lobe of petals bifid (petals
        3-lobed) ... ... (23) **tetrapetaloides.**

        Spur of lip 1 in. or more long :
        Lower lobe of petals broadly ovate,
        over 3 times as broad as the
        upper lobe ... ... (24) **falcicornis.**

        Lobes of petals narrow and sub-
        equal ... ... (25) **tetrapetala.**

      Spur of lip strongly clavate at the apex (26) **cornuta.**

Leaves radical or subradical :
  Pedicels of flowers 1 in. or more long ;
    bracts shorter than the pedicels :
      Spur of lip straight, clavate ...   ... (27) **involuta.**

      Spur of lip much curved, slightly
        thickened at the apex   ...   ... (28) **incurva.**

  Pedicels of flowers 1½ in. long ; bracts as
    long as or longer than the pedicels   (29) **trachychila.**

††Stigmas 3 lin. or more long ; leaves not radical or
  subradical :
  Spur of lip longer than the lateral sepals :
    Spur of lip about 8 lin. long ...   . (30) **umvotensis.**

    Spur of lip over 1 in. long :
      Stem leaves reduced to membranous sheaths (31) **porrecta.**

      Stem leafy ...   . (32) **clavata.**

    Spur of lip shorter than the lateral sepals   ... (33) **stenorhynchos.**

**Leaves cordate-ovate or suborbicular, radical :
  Flowers small and dense :
    Lower lobe of petals much shorter than the upper ;
      lobes of lip about half as long as the spur   ... (34) **dregeana.**

    Lower lobe of petals usually longer than the upper ;
      lobes of lip about as long as the spur...   ... (35) **macowaniana.**

  Flowers larger and somewhat lax :
    Lower lobe of petal about as long as the dorsal
      sepal, slightly hispidulous   ...   . (36) **Tysoni.**

    Lower lobe of petals 3 or 4 times as long as the
      dorsal sepal, strongly hispid ...   . (37) **krænzliniana.**

1. **H. arenaria** (Lindl. Gen. & Sp. Orch. 317) ; plant ¾–1½ ft.
high, with oblong tubers and 2 or 3 radical or subradical leaves ;
leaves subsessile, spreading or suberect, membranous, elliptic-oblong,
obtuse or subacute, 2½–6 in. long, 1–2¼ in. broad ; scapes ¾–1½ ft.
high, with several lanceolate sheaths ; racemes elongate, lax, 3–7 in.
long, many-flowered ; bracts lanceolate or oblong-lanceolate, acuminate, 5–8 lin. long ; pedicels 7–10 lin. long ; flowers small, green ;
dorsal sepal broadly ovate, obtuse, 1½–2 lin. long ; lateral sepals
obliquely ovate, somewhat acuminate, 2–2½ lin. long ; petals simple,
lanceolate-oblong, acute, nearly as long as the dorsal sepal ; lip
3-partite ; front lobe oblong or linear-oblong, obtuse, 2–2½ lin. long ;
side lobes spreading, filiform, nearly as long as the front lobe ; spur
½–1¼ in. long, slender below, somewhat thickened at the apex ;
column ¾ lin. long, obtuse ; side lobes of rostellum ¼ lin. long ;
stigmas oblong, about ⅓ lin. long ; capsule oblong, about ½ in. long.
Harv. Thes. Cap. i. 35, *t.* 55 ; *Bolus in Journ. Linn. Soc.* xxv. 190,
and Ic. Orch. Austr.-Afr. ii. *t.* 42 ; *Durand & Schinz, Conspect. Fl.
Afr.* v. 73 ; *Kränzl. in Engl. Jahrb.* xvi. 172, *and Orch. Gen. et
Sp.* i. 371. *H. micrantha, Reichb. f. in Flora,* 1865, 180. *Bonatea
micrantha, Lindl. Gen. & Sp. Orch.* 329 ; *Sond. in Linnæa,* xx. 218.

SOUTH AFRICA : without locality, *Bowie,* 622 ! 630 ! *Prior* ! *Ecklon & Zeyher,*
35 ! *Mrs. Bowker* !

COAST REGION : Swellendam Div. ; rocks near Swellendam, *Bowie*, 5 ! Knysna Div. ; sand hills near the west end of Groene Vallei, *Burchell*, 5654 ! woods near Knysna, *Bowie* ! *Miss Newdigate.* Albany Div. ; Howisons Poort, *Zeyher*, 1188 ! 3922 ! *Pappe*, 70 ! *Hutton* ! Grahamstown, 2650 ft., *Schlechter*, 2763 ! *MacOwan* ! Featherstone Kloof, *MacOwan*, 273 ! and without precise locality, *Miss Bowker* ! *Atherstone*, 17 ! Stockenstrom Div. ; Kat Berg, *Hutton* ! Komgha Div. ; Kei River, *Bowker*, 569 ! *Flanagan*, 823. British Kaffraria ; without precise locality, *Cooper*, 1803 !
CENTRAL REGION : Somerset Div. ; Bosch Berg, near Somerset East, *MacOwan.* Graaff Reinet Div.; Graff Reinet, under rocks, *Bolus*, 83 !
KALAHARI REGION : Transvaal ; Barberton, *Culver*, 37 ! 38.
EASTERN REGION : Natal, Springbok, *McKen in Herb. Sanderson* !

2. **H. natalensis** (Reichb. f. Otia Bot. Hamb. ii. 97) ; plant 1¼–2 ft. high, with slender leafy stems ; leaves 2 or 3 on the lower part of the stem, linear or lanceolate-linear, acute or acuminate, 2–6 in. long, 1½–3 lin. broad, reduced to narrow sheaths above ; racemes oblong or elongate, 2–6 in. long, lax, many-flowered ; bracts lanceolate, very acuminate, ½–1 in. long ; pedicels 5–6 lin. long ; flowers small, white (*Gerrard*) or green (*Wood*) ; dorsal sepal ovate, obtuse, about 1⅓ lin. long ; lateral sepals obliquely oblong, obtuse, over 1½ lin. long ; petals simple, linear-oblong, subfalcate, subacute, as long as the dorsal sepal ; lip 3-partite, about ¼ in. long ; lobes subequal and slightly thickened towards the apex ; spur 7–9 lin. long, curved, cylindrical ; column ¾ lin. long, obtuse ; side lobes of rostellum about half as long as the column ; stigmas oblong, ¾ lin. long. *Bolus in Journ. Linn. Soc.* xxv. 191 ; *Durand & Schinz, Conspect. Fl. Afr.* v. 82 ; *Kränzl. in Engl. Jahrb.* xvi. 146, *and Orch. Gen. et Sp.* i. 334 ; *Schlechter in Engl. Jahrb.* xx. *Beibl.* 50, 35. *H. wilmsiana, Kränzl. Orch. Gen. et Sp.* i. 464.

KALAHARI REGION : Transvaal ; in marshes near Houtbosch (Woodbush), 6500 ft., *Schlechter*, 4465 ! near Lydenburg, *Wilms*, 1371 !
EASTERN REGION : Natal ; near Glencoe Junction, 4000–5000 ft., *Wood*, 4822 ! *Wylie in Herb. Wood*, 7601 ! Zululand, *Gerrard*, 1552 !

I cannot find a character to separate *H. wilmsiana*, Kränzl.

3. **H. microrhynchos** (Schlechter in Engl. Jahrb. xx. Beibl. 50, 36) ; plant ¾–1¼ ft. high, with slender leafy stems and a few reduced sheaths at the base ; leaves sessile, elliptic-oblong, acute, membranous, 1½–2¼ in. long, the upper somewhat smaller ; racemes slender, lax, about 3 lin. long, many-flowered ; bracts lanceolate, acuminate, ¼–⅓ in. long ; pedicels about ¼ in. long ; flowers very small, green ; dorsal sepal broadly ovate-oblong, obtuse, 1 lin. long ; lateral sepals broadly ovate, apiculate, slightly longer than the dorsal ; petals simple, obovate-oblong, obtuse, as long as the dorsal sepal ; lip deeply 3-lobed ; front lobe linear-oblong, obtuse, 1¼ lin. long ; side lobes diverging, triangular-linear, subacute, longer than the front lobe ; spur ellipsoid-oblong, obtuse, ¾ lin. long ; column ½ lin. long, obtuse ; side lobes of rostellum minute ; stigmas oblong, obtuse, very short. *Kränzl. Orch. Gen. et Sp.* i. 388.

KALAHARI REGION : in woods at Houtbosch (Woodbush), 6000 ft., *Schlechter*, 4468 !

4. H. anguiceps (Bolus in Journ. Linn. Soc. xxv. 164, 165, fig. 2) ;
plant $\frac{1}{2}$–$\frac{3}{4}$ ft. high, with large oblong-ovoid tubers and leafy stems ;
leaves sessile, oblong-lanceolate, acute, somewhat fleshy, 1–1$\frac{1}{4}$ in.
long, decreasing upwards into the bracts; racemes oblong, about
3 in. long, dense, many-flowered ; bracts ovate-lanceolate, acute,
$\frac{1}{2}$–$\frac{3}{4}$ in. long; pedicels about 5. lin. long; flowers medium-sized ;
dorsal sepal broadly ovate, obtuse, concave, about 2$\frac{1}{2}$ lin. long ;
lateral sepals obliquely semiovate-oblong, obtuse, about 3 lin. long ;
petals simple, semiovate-oblong, obtuse, nearly as long as the dorsal
sepal; lip simple, linear-oblong, obtuse, fleshy, about 4 lin. long ;
spur 3$\frac{1}{2}$–4 lin. long, rather stout, subclavate at the apex ; column
about 1 lin. long, obtuse ; side lobes of rostellum short ; stigmas
$\frac{1}{2}$ lin. long, oblong, obtuse. *Durand & Schinz, Conspect. Fl. Afr.* v.
73 ; *Kränzl. in Engl. Jahrb.* xvi. 216, *Orch. Gen. et Sp.* i. 438, *and
in Ann. Naturhist. Hofmus. Wien*, xx. 3.

COAST REGION : Uitenhage Div. ; Van Stadens River, *Mackie.* Albany Div. ;
Brookhuizens Poort, Grahamstown, 2250 ft., *Bolus*, 7312 !
KALAHARI REGION : Orange River Colony ; near Harrismith, *Krook, Penther*, 123.
EASTERN REGION : Natal ; Van Reenen, *Penther*, 87.

5. H. Readei (Harv. MSS.) ; plant about $\frac{3}{4}$ ft. high, with stout
very leafy stem ; leaves sessile, ovate-oblong, acute or acuminate,
somewhat fleshy, 1–2 in. long, imbricate ; raceme somewhat elongate,
5 in. long, dense, many-flowered ; bracts ovate, acute, $\frac{1}{2}$–$\frac{3}{4}$ in. long,
somewhat fleshy ; pedicels $\frac{1}{3}$–$\frac{1}{2}$ in. long ; flowers medium-sized ; dorsal
sepal broadly ovate, obtuse, very concave, 2$\frac{1}{4}$ lin. long ; lateral sepals
narrowly and obliquely ovate-oblong, obtuse, 2$\frac{1}{2}$ lin. long ; petals
simple, narrowly and obliquely ovate-oblong, obtuse, about 2 lin.
long ; lip simple, oblong-linear, obtuse, rather broader at the base,
fleshy, about 3 lin. long ; spur about 4 lin. long, clavate from a
narrow base, somewhat curved ; column very short and broad ; side
lobes of rostellum short ; stigmas oblong, about $\frac{1}{4}$ lin. long.

COAST REGION : Albany Div. ; Fullers, *Reade*, 130 !

6. H. foliosa (Reichb. f. in Flora, 1865, 180) ; plant $\frac{1}{2}$–1$\frac{1}{2}$ ft. high,
with large oblong tubers and stout densely leafy stems ; leaves
sessile, ovate-oblong or oblong, obtuse or subacute, rather fleshy,
imbricate, 1–3 in. long, $\frac{1}{3}$–1 in. broad, gradually decreasing
upwards ; racemes oblong, 2–6 in. long, very dense, many-flowered ;
bracts ovate or oblong-lanceolate, subacute, $\frac{1}{2}$–1 in. long; pedicels
1–1$\frac{1}{2}$ in. long ; flowers rather large, green or green and white ;
dorsal sepal elliptic-ovate, obtuse, cucullate, 4$\frac{1}{2}$–6 lin. long ; lateral
sepals obliquely semiovate, obtuse, concave, 4$\frac{1}{2}$–6 lin. long ; petals
simple, very broadly ovate or orbicular-ovate, obtuse, about as long
as the sepals, sometimes broader than long ; lip linear-oblong, obtuse,
very fleshy, revolute at the margin, 5–7 lin. long, with a pair of
short slender lateral teeth near the base; spur 1–1$\frac{1}{4}$ in. long,
slender below, clavate at the apex ; column about 2$\frac{1}{2}$ lin. long,
obtuse ; side lobes of rostellum short and broad; stigmas oblong,

very broad and short. *Bolus in Journ. Linn. Soc.* xxv. 191. *Orchis foliosa, Sw. in Vet. Acad. Handl. Stockh.* 1800, 206. *Bonatea foliosa, Lindl. Gen. & Sp. Orch.* 329 ; *Sond. in Linnæa,* xix. 82 ; *Durand & Schinz, Conspect. Fl. Afr.* v. 89. *H. polyphylla, Kränzl. in Engl. Jahrb.* xvi. 214, *and Orch. Gen. et Sp.* i. 436 ; *Schlechter in Engl. Jahrb.* xx. *Beibl.* 50, 37 ; *Bolus, Ic. Orch. Austr.-Afr.* ii. *t.* 46.

SOUTH AFRICA : without locality, *Mrs. Bowker* !
COAST REGION : Swellendam Div. ; Kloof of Kaga Hoogte, *Bowie* ! Uitenhage Div. ; Van Stadens River, *Burchell,* 4657 ! Port Elizabeth Div. ; around Krakakamma, *Burchell,* 4563 ! *Ecklon & Zeyher,* Emerald Hill, *Bolus.* Bathurst Div. ; Tharfield, *Atherstone,* 6 ! Dugmore and near the mouth of the Kowie River, *MacOwan,* 543 ! Albany Div. ; without precise locality, *Miss Bowker* ! Eastern Frontier of Cape Colony, *Hutton* !
CENTRAL REGION : Somerset Div. ; Somerset East, *Bowker* !
KALAHARI REGION : Orange River Colony ; Leeuw Spruit and Vredefort, *Barrett-Hamilton* ! Witzies Hoek, *Thode* ! Transvaal ; near Lydenburg, *Roe in Herb. Bolus,* 2650 ! *Wilms,* 1362 ! near Barberton, 2800 ft., *Galpin,* 793. Waterval Boven, *Burtt-Davy,* 1428 ! near Potchefstroom, *Burtt-Davy,* 1488 ; near Pietersburg, 4000 ft., *Bolus,* 10882 ; near Ermelo, *Miss Leendertz,* 3129 !
EASTERN REGION : Transkei ; Tsomo River, *Mrs. Barber,* 833 ! Krielis Country, *Bowker* ! near Kentani, *Miss Pegler,* 864. Tembuland ; Umtata District, at Engocasi, 3500 ft., *Bolus,* 10287. Griqualand East ; near Kokstad, *Schlechter* ! by the Tsitsa River, *Schlechter* ! near Clydesdale, 2200 ft., *Tyson,* 2054. Natal ; Attercliffe, 800 ft., and Bishopstowe, *Sanderson,* 499 ! Itafamasi, *Wood,* 724 ! 1018 ! Oliviers Hoek, sources of Tugela River, 5000 ft., *Allison,* 37 ! near Durban, *Gueinzius* ! *Gerrard,* 643 ! 2183 ! 2185 ! hill near Mooi River, 4000–5000 ft., *Wood,* 5245 ! near Ellesmere, *Rudatis,* 553 ! and without precise locality, *Mrs. K. Saunders.*

7. **H. lævigata** (Lindl. in Ann. Nat. Hist. ser. i. iv. 315) ; plant $\frac{3}{4}$–$1\frac{1}{2}$ ft. high, with rather stout leafy stems ; leaves sessile, suberect, oblong or oblong-lanceolate, acute, somewhat fleshy, 1–2 in. long, $\frac{1}{4}$–$\frac{1}{2}$ in. broad ; racemes oblong or elongate, $1\frac{1}{2}$–4 lin. long, dense, many-flowered ; bracts ovate or oblong-ovate, acuminate, $\frac{1}{2}$–$\frac{3}{4}$ in. long ; pedicels about $\frac{1}{2}$ in. long ; flowers green or greenish-yellow ; dorsal sepal ovate-oblong, obtuse, about 3 lin. long ; lateral sepals obliquely semiovate-oblong, obtuse, about $3\frac{1}{2}$ lin. long ; petals simple, obliquely semiovate-oblong, obtuse, nearly as long as the dorsal sepal ; lip 3-lobed, with oblong obtuse fleshy lobes ; middle lobe 2 lin. long ; side lobes rather shorter and diverging ; spur $\frac{3}{4}$–$1\frac{1}{4}$ in. long, very slender ; column 1 lin. long, obtuse ; side lobes of rostellum very short ; stigmas oblong, about $\frac{1}{2}$ lin. long. *Drège, Zwei Pfl. Documente,* 147 ; *Bolus in Journ. Linn. Soc.* xxv. 190 ; *Durand & Schinz, Conspect. Fl. Afr.* v. 80 ; *Kränzl. in Engl. Jahrb.* xvi. 171, *Orch. Gen. et Sp.* i. 369, *and in Ann. Naturhist. Hofmus. Wien,* xx. 3 ; *Schlechter in Engl. Jahrb.* xx. *Beibl.* 50, 36. *H. ornithopoda, Reichb. f. in Linnæa,* xx. 696. *H. anguiceps, Schlechter in Engl. Jahrb.* xx. *Beibl.* 50, 37, *not of Bolus.*

COAST REGION : Albany Div. ; Howisons Poort, *Hutton* ! Div. ? Ruyterbosch, *Mund & Maire* !
KALAHARI REGION : Orange River Colony ; Harrismith, 7500 ft., *Sankey,* 260 ! Transvaal ; near Bergendal, 6200 ft., *Schlechter,* 4012 !
EASTERN REGION : Transkei ; Tsomo River, *Mrs. Barber,* 832 ! Tembuland ;

Morley, 1000–2000 ft., *Drège,* 8287! Griqualand East; Mount Currie, near
Kokstad, 5000–6000 ft., *Tyson,* 30! 1075! near Tsitsa River, 3500 ft., *Schlechter,*
6367! Zuurbergen, 4800 ft., *Schlechter,* 6579! near Nalogha, *Krook, Penther,*
123.

8. **H. bicolor** (Conrath & Kränzl. in Vierteljahrsschr. Nat. Ges.
Zürich, li. 131); plant 1½–2 ft. high, with leafy stems; leaves
subamplexicaul, linear, acuminate, 2½–4 in. long, the upper somewhat shorter; racemes very long and lax, 8 in. or more long, many-
flowered; bracts broadly lanceolate, acuminate, about ¾ in. long,
longer than the flowers; pedicels 3–3½ lin. long; flowers minute,
greenish-yellow, with brown inside (*Conrath*); dorsal sepal broadly
ovate, obtuse, concave, 1½ lin. long; lateral sepals 1¼ lin. long;
petals simple, narrower than the dorsal sepal and agglutinated to it,
the anterior side somewhat thickened; lip 3-partite, 1½ lin. long;
side lobes ligulate, obtuse; front lobe ovate, obtuse, as long as the
side lobes, but twice as broad; spur 3–3½ lin. long, slightly thickened
near the apex; column rather large; side lobes of rostellum short;
stigmas long, subparallel, obtuse, concave at the apex.

KALAHARI REGION: Transvaal; Modder Fontein, *Conrath,* 1083.

Only known to me from description.

9. **H. ciliosa** (Lindl. in Ann. Nat. Hist. ser. i. iv. 314); plant
1¼–1¾ ft. high, with leafy stems; leaves sessile, suberect, linear-
oblong or oblong-lanceolate, acute, 2–4 in. long, 3–4 lin. broad,
somewhat fleshy, the lower reduced to narrow sheaths, which are
transversely barred with black, gradually reduced upwards to
lanceolate sheaths; racemes oblong, 2–5 in. long, dense, many-
flowered; bracts ovate-lanceolate, acuminate, ¾–1¼ in. long;
pedicels 5–7 lin. long; flowers green; dorsal sepal broadly ovate or
ovate-oblong, obtuse, hispidulous and ciliate, 1½–3 lin. long; lateral
sepals obliquely ovate or ovate-oblong, obtuse, hispidulous and
ciliate, 2–3½ lin. long; petals simple, obliquely semiovate-oblong,
obtuse, glabrous, 1½–3 lin. long; lip 3-partite, with fleshy lobes;
front lobe oblong or linear, obtuse, 1½–4 lin. long; side lobes linear-
oblong or filiform, obtuse, rather shorter than the front lobe; spur
½–¾ in. long, slender below, somewhat thickened and obtuse at the
apex; column ¾ lin. long, obtuse; side lobes of rostellum about half
as long as the column; stigmas oblong, about ½ lin. long. *Drège,
Zwei Pfl. Documente,* 148; *Bolus in Journ. Linn. Soc.* xxv. 190;
Durand & Schinz, Conspect. Fl. Afr. v. 75; *Kränzl. in Engl. Jahrb.*
xvi. 169, *Orch. Gen. et Sp.* i. 367, *and in Ann. Naturhist. Hofmus.
Wien,* xx. 3.

EASTERN REGION: Transkei; in damp situations at Fort Bowker, *Bowker,* 623!
Tembuland; between Morley and Umtata River, 1000–2000 ft., *Drège,* 4573!
Griqualand East; near Clydesdale, 2500–2700 ft., *Tyson,* 2044! Natal; Dumisa,
Alexandra District, 2000 ft., *Rudatis,* 223! N'kandhla, 4000–5000 ft., *Wylie in
Herb. Wood,* 11822! Colenso, *Krook, Penther,* 274, and without precise locality,
Buchanan in Herb. Sanderson, 1059!

10. **H. Schlechteri** (Kränzl. ex Schlechter in Engl. Jahrb. xx. Beibl. 50, 35); plant $1\frac{1}{2}$–2 ft. high, with stout leafy stem, turning black in drying; leaves erect, sheathing, adpressed to the stem, narrowly linear-oblong, acuminate, 3–4 in. long, the lower sheath-like, the upper passing into the bracts; racemes about 4 in. long, 6–8-flowered; bracts oblong-lanceolate, acuminate, $1\frac{1}{2}$–2 in. long, suberect; pedicels slender, $1\frac{1}{2}$–$2\frac{1}{4}$ in. long; flowers white, somewhat fleshy; dorsal sepal broadly elliptic-ovate, apiculate, about 7 lin. long; lateral sepals oblong-lanceolate, acuminate, about 7 lin. long; petals simple, oblong-lanceolate, apiculate, nearly as long as the dorsal sepal; lip deeply 3-lobed, about $\frac{3}{4}$ in. long; front lobe oblong-lanceolate, subacute; side lobes slightly broader than the front lobe and strongly pectinate; spur very slender, elongate, 4–5 in. long, often included within the bract; column 4 lin. long, curved, apiculate; side lobes of rostellum much curved, about 4 lin. long; stigmas 3 lin. long, clavate-capitate and very broad at the apex. *Kränzl. in Reichb. f. Xen. Orch.* iii. 148, *t.* 286, *fig.* 5–9, *and Orch. Gen. et Sp.* i. 244; *Schlechter in Engl. Jahrb.* xx. *Beibl.* 50, 35.

KALAHARI REGION : Transvaal ; near Wilge River, 4600 ft., *Schlechter,* 4121 ! Belfast, 6500 ft., *Doidge,* 4794 !

A striking species, closely resembling *H. occultans,* Welw., in habit, but differing in having simple petals.

11. **H. Culveri** (Schlechter in Engl. Jahrb. xx. Beibl. 50, 14, 32); plant $1\frac{1}{2}$–$1\frac{3}{4}$ ft. high, with rather stout leafy stems ; leaves 4 or 5, subsessile, elliptic or lanceolate-oblong, subacute, very membranous, $2\frac{1}{2}$–$4\frac{1}{2}$ in. long, 1–$1\frac{3}{4}$ in. broad ; racemes $4\frac{1}{2}$–6 in. long, lax, many-flowered ; bracts lanceolate, acuminate, $\frac{1}{4}$–$\frac{1}{2}$ in. long; pedicels 5–6 lin. long ; flowers green (*Culver*); dorsal sepal broadly ovate, obtuse, somewhat concave, 2 lin. long ; lateral sepals obliquely ovate, obtuse, $2\frac{1}{2}$ lin. long ; petals 2-partite ; upper lobe broadly linear, subobtuse, as long as the dorsal sepal ; lower lobe narrowly linear or subfili-form, half as long and a quarter as broad as the upper lobe ; lip 3-lobed ; front lobe narrowly lanceolate or linear-lanceolate, sub-acute, about 3 lin. long ; side lobes diverging at right angles, linear, acute, about $\frac{1}{2}$ lin. long ; spur about 4 lin. long, cylindrical, some-what curved ; column about 1 lin. long, broad, obtuse ; side lobes of rostellum about $\frac{1}{3}$ lin. long ; stigmas clavate-oblong, 1 lin. long *Kränzl. Orch. Gen. et Sp.* i. 269.

KALAHARI REGION : Transvaal ; Rimers Creek, Barberton, 3000 ft., *Culver,* 47 !

12. **H. tridens** (Lindl. in Hook. Comp. Bot. Mag. ii. 208); plant $\frac{1}{3}$–$1\frac{1}{4}$ ft. high, with rather slender leafy stems ; leaves 3–6, sessile, oblong-lanceolate, acute, somewhat fleshy, suberect, 1–4 in. long, 2–6 lin. broad, with a few reduced sheaths above ; racemes 1–4 in. long, usually lax, few- to many-flowered ; bracts lanceolate or oblong-lanceolate, acuminate, 3–6 lin. long ; pedicels 5–6 lin. long ;

flowers sweet-scented (*Miss Pegler*); dorsal sepal broadly ovate or
orbicular-ovate, obtuse, about 1½–2 lin. long; lateral sepals broadly
oblong, obtuse, 2½–3 lin. long; petals 2-partite; upper lobe falcate-
oblong, obtuse, 1½–2 lin. long; lower lobe falcate-linear, subacute,
rather shorter than the upper lobe and about a third as broad; lip
3-partite; lobes linear, subobtuse, subequal or the lateral pair
shorter, equalling or rather shorter than the lateral sepals; spur
4–7 lin. long, nearly straight, slender below, somewhat thickened
above; column ½ lin. long, obtuse; side lobes of rostellum less than
½ lin. long; stigmas ½ lin. long, clavate and truncate; capsule
oblong, about 4 lin. long. *Drège, Zwei Pfl. Documente,* 149; *Bolus
in Journ. Linn. Soc.* xxv. 190; *Durand & Schinz, Conspect. Fl. Afr.*
v. 88; *Kränzl. in Engl. Jahrb.* xvi. 105, *and Orch. Gen. et Sp.* i.
262, 895. *H. Gerrardi, Reichb. f. Otia Bot. Hamb.* ii. 97; *Kränzl.
in Engl. Jahrb.* xvi. 110; *Durand & Schinz, Conspect. Fl. Afr.* v. 78.
H. Barberæ, Schlechter in Engl. Jahrb. xx. *Beibl.* 50, 7; *Kränzl.
Orch. Gen. et Sp.* i. 228.

EASTERN REGION: Transkei; along streams near Kentani, 1200 ft., *Miss Pegler,*
346! Krielis Country, *Mrs. Barber*! Pondoland; between Umtata River and
St. Johns River, *Drège,* 4567! Natal; Highlands, *Gerrard,* 1559! Blue Kranz
River, 3700 ft., *Schlechter,* 6862!

The specimen of *H. Barberæ,* kindly lent by Dr. Schlechter, is in poor
condition, but I am unable to separate it from *H. tridens,* Lindl.

13. **H. Galpini** (Bolus, Ic. Orch. Austr.-Afr. i. t. 17); plant about
8 in. high, with short, rather stout leafy stem; leaves 6–8, sessile,
lanceolate, acuminate, suberect, 3–4 in. long, the upper gradually
decreasing into the bracts; racemes elongate, about 5 in. long,
somewhat lax, many-flowered; bracts lanceolate or linear-lanceolate,
acuminate, ⅓–½ in. long; pedicels ½–¾ in. long; flowers light green;
dorsal sepal ovate-elliptic, subacute, concave, 2½ lin. long; lateral
sepals semiobovate-oblong, obliquely apiculate, 3 lin. long; petals
2-partite; upper lobe linear, as long as the dorsal sepal; lower lobe
lanceolate-linear, about 3½ lin. long; lip 3-partite, with linear or
filiform lobes; front lobe 5 lin. long; side lobes 3½ lin. long; spur
about ¾ in. long, slender below, clavate at the apex; column about
1 lin. long, obtuse; side lobes of rostellum not half as long as the
column; stigmas clavate, about ¾ lin. long. *Kränzl. Orch. Gen. et
Sp.* i. 218.

KALAHARI REGION: Transvaal; rocky places near Johannesburg, 6000 ft.,
Galpin, 392; near Lydenburg, *Wilms,* 1356! 1383!

14. **H. Woodii** (Schlechter in Engl. Jahrb. xxxviii. 149); plant
1½–1¾ ft. high, with slender stems; leaves few, chiefly basal, erect,
narrowly elongate-linear, acute, 3–5 in. long, 1–1½ lin. broad,
reduced to lanceolate acuminate sheaths above; racemes oblong,
2–3½ in. long, rather lax, 6–15-flowered; bracts lanceolate, very
acuminate, 4–6 lin. long; pedicels slender, ¾–1 in. long; flowers
white (*Wylie*); dorsal sepal ovate-elliptic, obtuse, concave, 1½ lin.
long; lateral sepals obliquely semiovate, obtuse or apiculate, about

2 lin. long; petals 2-partite; upper lobe linear, obtuse, about as long as the dorsal sepal; lower lobe lanceolate-linear, obtuse, somewhat falcate, a third longer than the upper lobe; lip 3-partite, with oblong-linear obtuse lobes; front lobe 2 lin. long; lateral pair diverging and rather shorter; spur $\frac{1}{2}-\frac{3}{4}$ in. long, slender below, somewhat thickened at the apex; column 1 lin. long, obtuse; side lobes of rostellum $\frac{1}{2}$ lin. long; stigmas $\frac{3}{4}$ lin. long, clavate-oblong.

EASTERN REGION : Zululand; Ungoya, 1000–2000 ft., *Wylie in Herb. Wood*, 7601, 9468 !

15. **H. malacophylla** (Reichb. f. Otia Bot. Hamb. ii. 97); plant 1–1$\frac{1}{2}$ ft. high, with ovoid tubers and rather stout leafy stems; leaves 5–7, subsessile, elliptic-lanceolate, acuminate, membranous, 2–6 in. long, $\frac{1}{2}$–1$\frac{1}{2}$ in. broad; racemes 2$\frac{1}{2}$–9 in. long, lax; bracts narrowly ovate-lanceolate, acuminate, $\frac{1}{4}$–$\frac{1}{2}$ in. long; pedicels $\frac{1}{3}$–$\frac{1}{2}$ in. long; flowers green; dorsal sepal ovate, obtuse, galeate, about 2$\frac{1}{4}$ lin. long; lateral sepals spreading, obliquely ovate-oblong, obtuse, $\frac{1}{4}$ in. long; petals 2-partite; upper lobe linear-oblong, subacute, 2$\frac{1}{4}$ lin. long; lower lobe subfiliform, curved, about $\frac{1}{4}$ in. long; lip 3-lobed, $\frac{1}{4}$ in. long; lobes subfiliform, the lateral pair spreading and rather shorter than the front lobe; spur rather stout, slightly curved, $\frac{1}{3}$–$\frac{1}{2}$ in. long; column stout, 1 lin. long; stigmas oblong, obtuse, about $\frac{3}{4}$ lin. long; capsule elliptic-oblong, 4–5 lin. long. *Bolus in Journ. Linn. Soc.* xxv. 191; *Durand & Schinz, Conspect. Fl. Afr.* v. 81; *Rolfe in Dyer, Fl. Trop. Afr.* vii. 230; *Kränzl. in Engl. Jahrb.* xvi. 66, *and Orch. Gen. et Sp.* i. 198; *Schlechter in Engl. Jahrb.* xx. *Beibl.* 50, 32.

COAST REGION : Stockenstrom Div. ; Kat Berg, 2000–3000 ft., *Hutton*, 45 !
KALAHARI REGION : Transvaal ; Rimers Creek, near Barberton, 3000 ft., *Culver*, 47 ! Houtbosch (Woodbush) Mountains, 6600 ft., *Schlechter*, 4470 !
EASTERN REGION : Transkei ; Tsomo River, *Barber & Bowker*, 4 ! Bazeia mountain forests, 2500 ft., *Baur*, 812 ! Griqualand East ; in damp woods at Malowe, near Clydesdale, *Tyson*, 2043 ! Natal ; Polela, 4000–5000 ft., *Wood*, 4586 ! Alexandra District, at Ifafa, *Rudatis*, 295 !

Also in East and West Tropical Africa.

16. **H. transvaalensis** (Schlechter in Engl. Jahrb. xx. Beibl. 50, 6, 32); plant about 1$\frac{3}{4}$ ft. high, with stout leafy stem; leaves sessile, oblong or ovate-oblong, acute, membranous, 2–4 in. long, $\frac{3}{4}$–1$\frac{1}{4}$ in. broad; racemes oblong, about 5 in. long, rather lax, about 10-flowered; bracts ovate-lanceolate, acute, $\frac{3}{4}$–1 in. long; pedicels $\frac{1}{2}$–$\frac{3}{4}$ in. long; flowers medium-sized; dorsal sepal ovate-oblong, subobtuse, about 5 lin. long; lateral sepals semiovate, obliquely acute, about $\frac{1}{2}$ in. long; petals 2-partite; upper lobe linear, acute, as long as the dorsal sepal; lower lobe narrowly linear, acuminate, about 7 lin. long; lip 3-partite; middle lobe linear, obtuse, about $\frac{1}{2}$ in. long; side lobes diverging, narrowly linear, acuminate, about 8 lin. long; spur about $\frac{3}{4}$ in. long, somewhat curved, narrow at base and apex, somewhat thickened above the middle; column 2 lin. long, obtuse; side lobes of rostellum 1$\frac{1}{2}$ lin. long; stigmas clavate, truncate, 2 lin. long. *Kränzl. Orch. Gen. et Sp.* i. 208, *and in Ann. Naturhist. Hofmus. Wien*, xx. 2.

KALAHARI REGION: Transvaal; among shrubs near Barberton, 5000 ft., *Thorn-croft*, 466; Houtbosch (Woodbush), *Schlechter*, 4383 !
EASTERN REGION: Natal; Weenen District, near Estcourt, *Krook, Penther*, 131.

17. **H. Barbertoni** (Kränzl. & Schlechter in Kränzl. Orch. Gen. et Sp. i. 199); plant 8–12 in. high, with large ovoid tubers; leaves sessile, cauline, the few lower oblong, obtuse, 1–2 in. long, the rest ovate-lanceolate, acute and gradually ‘decreasing in size up to the bracts; racemes short, 3–10-flowered; bracts ovate-lanceolate, ½–¾ in. long; pedicels about ¾ in. long; flowers green and white (*Culver*); dorsal sepal ovate, subobtuse, galeate, ⅓ in. long; lateral sepals spreading, obliquely ovate-oblong, subobtuse, 3–4 lin. long; petals 2-partite; upper lobe triangular-oblong, subobtuse, ¼ in. long; lower lobe linear-subulate, acute, spreading, about ¼ in. long; lip 3-lobed, about 5 lin. long; lobes subfiliform, the lateral pair spreading and rather narrower than the front lobe; spur filiform, 1¼–1½ in. long, slightly compressed near the apex; stigmas clavate, 1½ lin. long.

KALAHARI REGION: Transvaal; rocky hillsides at Barberton, 3300 ft., *Culver*, 81! Koodoes Poort, near Pretoria, *Reck*, 165 ! Ermelo, *Miss Leendertz*, 3115 !

18. **H. insignis** (Schlechter in Engl. Jahrb. xx. Beibl. 50, 32, not of Rolfe); plant 1–2 ft. high; leaves few, chiefly basal, sub-sessile, oblong, subacute, membranous, 2–3¼ in. long, ½–¾ in. broad, upper reduced to narrow acute sheaths; scapes 1–2 ft. high; racemes about 4 in. long, lax, 3–9-flowered; bracts lanceolate, acuminate, ½–¾ in. long; pedicels about ¾ in. long; flowers medium-sized; dorsal sepal elliptic-lanceolate, acute, cucullate, 5–7 lin. long; lateral sepals obliquely and broadly semiovate, acuminate, curved at the apex, about as long as the dorsal sepal; petals 2-partite; upper lobe falcate-linear, acute, as long as the dorsal sepal; lower lobe filiform, flexuous, about twice as long as the upper lobe; lip 3-partite; front lobe linear, obtuse, with revolute margin, about 8 lin. long; side lobes filiform, flexuous, about as long as the lower lobe of the petals; spur cylindric, ½–¾ in. long, slender below, somewhat clavate above; column 2 lin. long, obtuse; side lobes of rostellum rather short; stigmas linear, rather longer than the anther-channels. *Kränzl. Orch. Gen. et Sp.* i. 204.

KALAHARI REGION: Transvaal; among shrubs on Mailas Kop, 2400 ft., *Schlechter*, 4517 !

The flowers are rather young in the specimens examined, and the parts not fully developed.

19. **H. polypodantha** (Reichb. f. Otia Bot. Hamb. ii. 97); plant ⅓–¾ ft. high, rather slender; leaves radical or subradical, 2 or 3, sessile or subsessile, elliptic-oblong or lanceolate-oblong, acute, membranous, 1½–3 in. long; scapes ⅓–¾ ft. high, with 2 or 3 narrow sheaths; racemes 1½–3 in. long, 3–10-flowered; bracts oblong-

lanceolate, acuminate, $\frac{1}{3}$–$\frac{1}{2}$ in. long; pedicels $\frac{3}{4}$ to nearly 1 in. long; flowers white (*Gerrard*); dorsal sepal elliptic-lanceolate, acuminate, 4–5 lin. long; lateral sepals broadly semiovate, very acuminate, 4–5 lin. long; petals 2-partite; upper lobe narrowly linear, obtuse, rather shorter than the dorsal sepal; lower lobe filiform, curved, nearly 1 in. long; lip 3-partite; lobes filiform, curved, the middle lobe about $\frac{3}{4}$ in. long; the side lobes rather shorter; spur filiform, somewhat thickened near the apex, $1\frac{1}{4}$–$1\frac{1}{2}$ in. long; column 2 lin. long, apiculate; side lobes of rostellum 2 lin. long; stigmas sub-capitate, 3–$3\frac{1}{2}$ lin. long. *Bolus in Journ. Linn. Soc.* xxv. 191; *Durand & Schinz, Conspect. Fl. Afr.* v. 84; *Kränzl. in Engl. Jahrb.* xvi. 70, *and Orch. Gen. et Sp.* i. 203.

SOUTH AFRICA: without locality, *Trimen*!
COAST REGION: Queenstown Div.; in scrub on lower slopes near the junction of Zwart Kei and White Kei Rivers, 2350 ft., *Galpin*, 8180!
EASTERN REGION: Natal; in thorny bush by the Upper Tugela River, *Gerrard*, 1554! at or near Krans Kop, *McKen*, 27!

20. **H. orangana** (Reichb. f. Otia Bot. Hamb. ii. 101); plant $\frac{3}{4}$–1 ft. high, with stout leafy stems and 2–3 short sheaths below; leaves sessile, oblong-lanceolate, acute or subacute, somewhat fleshy, 2–3 in. long, gradually decreasing upwards; racemes $2\frac{1}{2}$–6 in. long, dense, many-flowered; bracts oblong-lanceolate, acuminate, $\frac{1}{2}$–$\frac{3}{4}$ in. long; pedicels $\frac{1}{3}$–$\frac{1}{2}$ in. long; flowers white; dorsal sepal elliptic or obovate-elliptic, obtuse, $2\frac{1}{2}$ lin. long; lateral sepals obliquely and broadly obovate, obtuse, 3–$3\frac{1}{2}$ lin. long; petals 2-partite; upper lobe oblong, obtuse, rather shorter than the dorsal sepal; lower lobe oblong, obtuse, about $\frac{1}{2}$ lin. long; lip 3-partite; front lobe linear, obtuse, $2\frac{1}{2}$–$3\frac{1}{2}$ lin. long; side lobes lanceolate-oblong, obtuse, $1\frac{1}{4}$–$1\frac{3}{4}$ lin. long; spur $\frac{1}{3}$–$\frac{1}{2}$ in. long, slightly curved, narrowly funnel-shaped below, then filiform, slightly thickened near the apex; column $1\frac{1}{4}$ lin. long, obtuse; side lobes of rostellum $\frac{1}{2}$ lin. long; stigmas 1 lin. long, strongly clavate. *Bolus in Journ. Linn. Soc.* xxv. 191; *Durand & Schinz, Conspect. Fl. Afr.* v. 83; *Kränzl. in Engl. Jahrb.* xvi. 85, *and in Orch. Gen. et Sp.* i. 214.

KALAHARI REGION: Orange River Colony; Harrismith, 7500 ft., *Sankey*, 263! and without precise locality, *Cooper*, 1093! 1872! Transvaal; Belfast, *Burtt-Davy*, 1298! between Pilgrims Rest and Sabie Falls, *Burtt-Davy*, 5061! Bamboo Mountain, Drakensberg, *Doidge*, 5573!
EASTERN REGION: Oliviers Hoek, sources of Tugela River, 5000 ft., in wet places, *Allison*, 14!

21. **H. dives** (Reichb. f. in Flora, 1867, 117); plant 1–$1\frac{1}{2}$ ft. high, with stout leafy stems; leaves linear or oblong-linear, acute or subobtuse, somewhat fleshy, $1\frac{1}{2}$–4 in. long, 3–7 lin. broad, decreasing upwards into imbricating sheaths; racemes oblong or somewhat elongate, 2–10 in. long, dense, many-flowered; bracts lanceolate or oblong-lanceolate, acuminate, 4–9 lin. long; pedicels 5–7 lin. long; flowers white or green and white; dorsal sepal oblong-lanceolate, subacute, 2–$2\frac{1}{2}$ lin. long; lateral sepals broadly and obliquely obovate, obtusely apiculate, 2–$2\frac{1}{2}$ lin. long; petals 2-partite; upper

lobe linear, obtuse, shorter than the dorsal sepal; lower lobe lanceolate, subobtuse, rather shorter than the upper lobe; lip 3-partite; lobes linear, subacute; front lobe 2½–3 lin. long, the lateral pair rather shorter and diverging; spur 5–7 lin. long, linear, slightly curved, very slightly thickened near the apex; column 1½ lin. long, obtuse; side lobes of rostellum ¼ lin. long; stigma 1 lin. long, clavate and nearly truncate at the apex; capsule oblong, 3–4 lin. long. *Bolus in Journ. Linn. Soc.* xxv. 191; *Kränzl. in Engl. Jahrb.* xvi. 86, *Orch. Gen. et Sp.* i. 215, *and in Ann. Naturhist. Hofmus. Wien,* xx. 2; *Durand & Schinz, Conspect. Fl. Afr.* v. 77.

EASTERN REGION : Tembuland ; Bazeia, 3500–4000 ft., *Baur,* 632! Natal ; Clairmont, Icarion and Congella Flats, *Sanderson,* 501! Inanda, *Wood,* 437! Clairmont, near Durban, 50 ft., *Wood,* 1015! and in *MacOwan, Herb. Austr.-Afr.,* 1534! near Ixopo, 4000 ft., *Schlechter,* 6658! Mount West, *Schlechter*! Itafamasi, *Wood,* 737! Tongaat, *Mrs. Saunders*! Dumisa, *Rudatis,* 548! Inanda, *Wood,* 437! Umkomanzi River, *Krook, Penther,* 109; near Estcourt, *Krook, Penther,* 133, and without precise locality, *Gerrard,* 4! 1549! *Buchanan,* 14! 15! *Hallack*! Zululand, *Gerrard,* 583!

22. **H. Rehmanni** (Bolus in Journ. Linn. Soc. xxv. 169, fig. 6); plant 1¼–1¾ ft. high, with stout somewhat leafy stems; leaves erect, lanceolate-linear, subacute, somewhat fleshy, 2–4 in. long, ¼–⅓ in. broad; racemes somewhat elongate, 3–5 in. long, somewhat lax; bracts oblong-lanceolate, acuminate, ½–¾ in. long; pedicels about ¾ in. long; flowers green (*Culver*); dorsal sepal ovate-oblong, subobtuse, 2–2½ lin. long; lateral sepals obliquely obovate-oblong, apex lateral and obtusely apiculate, 2½–3 lin. long; petals 2-partite; upper lobe linear, acute, about as long as the dorsal sepal; lower lobe falcately incurved from a broad base, linear, acute and puberulous above, twice. as long as the front lobe; lip 3-partite with linear acute curved lobes; front lobe 4–5 lin. long; the lateral pair diverging and rather shorter; spur about 5 lin. long, curved, clavate above; column 1½ lin. long, obtuse, side lobes of rostellum 1½ lin. long; stigmas 2 lin. long, clavate and truncate above. *Durand & Schinz, Conspect. Fl. Afr.* v. 85; *Kränzl. in Engl. Jahrb.* xvi. 81, *and Orch. Gen. et Sp.* i. 224.

KALAHARI REGION: Transvaal; Houtbosch (Woodbush), *Rehmann,* 5780! swampy places in Lomati Valley, near Barberton, 3500 ft., *Culver,* 74!

23. **H. tetrapetaloides** (Schlechter in Engl. Jahrb. xx. Beibl. 50, 34); plant about 1½ ft. high, with stout leafy stem; leaves erect, linear-lanceolate, acute, somewhat coriaceous, 3–6 in. long, ½–¾ in. broad, decreasing upwards into imbricating sheaths; racemes cylindric, 3–5 in. long, dense, many-flowered; bracts lanceolate or ovate-lanceolate, acuminate, ½–¾ in. long; pedicels ¾–1 in. long; flowers green (*Schlechter*); dorsal sepal ovate-lanceolate, subobtuse, 2½–3 lin. long; lateral sepals obliquely obovate, 3½–4 lin. long, apex lateral, apiculate; petals 2-partite; upper lobe linear, acute, shorter than the dorsal sepal; lower lobe oblong;

unequally 2-lobed in the upper half, 3½–4 lin. long, the upper
segment subulate, acute, the other broader and subobtuse ; lip
3-partite ; lobes linear, acute, the front one 5 lin. long, the lateral
pair shorter and diverging ; spur 8 lin. long, curved, filiform below,
clavate above; column 1½ lin. long, obtuse, side lobes of rostellum
1½ lin. long ; stigmas about 2½ lin. long, broadly clavate. *Kränzl.
Orch. Gen. et Sp.* i. 228.

KALAHARI REGION : Transvaal ; in damp places near Houtbosch (Woodbush),
6600 ft., *Schlechter*, 4464 !

24. **H. falcicornis** (Bolus in Journ. Linn. Soc. xix. 340, *falciformis*,
by error) ; plant 1–3 ft. high, with stout leafy stems ; leaves linear-
oblong or somewhat elongate, acute or subacute, somewhat
coriaceous, 2½–8 in. long, ½–¾ in. broad, gradually decreasing
upwards into imbricating sheaths ; racemes oblong or elongate,
3–9 in. long, dense, many-flowered ; bracts lanceolate or oblong-
lanceolate, acute or acuminate, ½–1 in. long ; pedicels ¾–1 in. long ;
flowers white or green and white ; dorsal sepal elliptic-oblong, obtuse,
3–3½ lin. long ; lateral sepals broadly and obliquely semiovate,
obtusely apiculate, 3–3½ lin. long ; petals 2-partite ; upper lobe
linear or lanceolate-linear, acute or subacute, as long as the dorsal
sepal ; lower lobe broadly ovate, obtuse, three to five times as
broad as the upper lobe ; lip 3-partite ; lobes 4–5 lin. long, the
front lobe linear, straight, the lateral pair broader and diverging ;
spur linear, somewhat curved, 1–1½ in. long, slightly thickened
near the apex ; column 2 lin. long, obtuse ; side lobes of rostellum
1 lin. long ; stigmas 1½ lin. long, broadly clavate ; capsule oblong,
½–¾ in. long. *Durand & Schinz, Conspect. Fl. Afr.* v. 77 ; *Kränzl.
Orch. Gen. et Sp.* i. 214. *H. Bilabrella, Kränzl. in Engl. Jahrb.* xvi.
86. *H. tetrapetala, Reichb. f. in Flora;* 1865, 180 ; *Kränzl. in
Engl. Jahrb.* xvi. 79, *partly* ; *Bolus, Ic. Orch. Austr.-Afr.* i. *t.* 16,
partly, not of Kränzl. H. Mundtii, Kränzl. in Engl. Jahrb. xvi. 79 ;
Durand & Schinz, Conspect. Fl. Afr. v. 82 ; *Kränzl. Orch. Gen. et
Sp.* i. 222. *Bonatea minor, Mund ex Kränzl. in Engl. Jahrb.* xvi. 79.
Orchidea falcicornis, Burch. ex Lindl. Gen. & Sp Orch. 328. *Bila-
brella falcicornis, Lindl. in Bot. Reg. sub t.* 1701. *Bonatea Bila-
brella, Lindl. Gen. & Sp. Orch.* 328. *B. tetrapetala, Lindl. in Hook.
Comp. Bot. Mag.* ii. 208 ; *Sond. in Linnæa,* xix. 81 ; *Drège, Zwei
Pfl. Documente,* 148 ; *Krauss in Flora,* 1845, 307, *and Beitr. Fl.
Cap- und Natal.* 159, *partly.*

COAST REGION : George Div. ; near George, 650 ft., *Moyle Rogers* ! *Schlechter,*
2325 ! Knysna Div. ; near the Keurbooms River, *Burchell,* 5178 ! Ruyterbosch,
Mund and Maire ! Uitenhage Div. ; Van Stadens Berg, *Ecklon & Zeyher* ; Zuur-
berg Range, *Hallack.* Bathurst Div. ; near Bathurst, *Atherstone,* 7 ! Albany Div. ;
Howisons Poort, *Hutton* ! near Grahamstown, *Bolton* ! *MacOwan* ! Coldstream,
Glass, and without precise locality, *Prior* ! Bedford Div. ; summit of Kaga Berg,
3200 ft., *Atherstone* ! Stockenstrom Div. ; Kat Berg, 2000 ft., *Hutton* ! British
Kaffraria ; without locality, *Mrs. Barber,* 859 ! *Brownlee* ! Eastern Frontier of Cape
Colony, *MacOwan,* 511 ! *Brownlee* ! *Hutton* !
CENTRAL REGION : Somerset Div. ; Somerset, *Bowker* !

EASTERN REGION : Transkei ; Fort Bowker, *Bowker*, 624 ! Butterworth, *Bowker*, 325 ! Tembuland ; Bazeia, 2000 ft., *Baur*, 816 ! between Morley and Umtata River, 1000–2000 ft., *Drège*, 4566 ! Pondoland ; near St. Andrews Mission Station, 1000 ft., *Tyson*, 3080. Griqualand East ; near Kokstad, 5000 ft., *Bolus in MacOwan & Bolus, Herb. Norm. Austr.-Afr.*, 477 ! *Tyson*, 1071 ! Natal ; near Murchison, *Wood*, 3088 ! near Durban, 100 ft., *Sanderson* ! *Krauss*, 101, partly ! Clairmont, *Wood*, 696 !

A common and widely diffused species, which includes *H. tetrapetala*, Reichb. f. Under the latter name an allied species has been included ; this now stands as *H. tetrapetala*, Kränzl., which see.

25. **H. tetrapetala** (Kränzl. Orch. Gen. et Sp. i. 221, not of Reichb. f.) ; plant 1–3 ft. high, with stout leafy stems ; leaves linear-oblong or somewhat elongate, acute or subacute, somewhat coriaceous, 3–6 in. long, ½–¾ in. broad, gradually decreasing upwards into imbricating sheaths ; racemes oblong or somewhat elongate, 3–8 in. long, dense, many-flowered ; bracts lanceolate or oblong-lanceolate, acute or acuminate, ½–1 in. long ; pedicels ¾ in. long ; flowers white or green and white ; dorsal sepal elliptic-oblong, obtuse, 3½–4 lin. long ; lateral sepals broadly and obliquely semi-ovate, obtusely apiculate, 3½–4 lin. long ; petals 2-partite ; lobes oblong-lanceolate, acute, subequal, nearly as long as the dorsal sepal ; lip 3-partite ; lobes 5–6 lin. long, the front lobe narrow and obtuse, the lateral pair somewhat broader and diverging ; spur linear, curved, 1–1½ in. long, slightly thickened near the apex ; column 2 lin. long, obtuse ; side lobes of rostellum 1 lin. long ; stigmas 2 lin. long, broadly clavate. *Bolus in Journ. Linn. Soc.* xxv. 191, *partly* ; *Durand & Schinz, Conspect. Fl. Afr.* v. 87, *partly* ; *N. E. Br. in Gard. Chron.* 1885, xxiv. 307, *partly* ; *Kränzl. in Engl. Jahrb.* xvi. 79, *partly, and in Ann. Naturhist. Hofmus. Wien,* xx. 2 ; *Bolus, Ic. Orch. Austr.-Afr.* i. t. 16, *partly, excl. tab., not of Reichb. f.* ; *Schlechter in Engl. Jahrb.* xx. Beibl. 50, 34. *Bonatea tetrapetala, Krauss in Flora,* 1845, 307, *and Beitr. Fl. Cap- und Natal.* 159, *partly.*

VAR. β, **major** (Schlechter in Engl. Jahrb. xx. Beibl. 50, 34) ; taller and stouter, sometimes as much as 4½ ft. high ; flowers larger ; front lobe of petals elongate-lanceolate.

KALAHARI REGION : Transvaal ; hillsides at Johannesburg, 5000 ft., *Mrs. Galpin in Herb. Galpin*, 1400 ! near Barberton, 4900 ft., *Galpin*, 877 ; Houtbosch, 6000 ft., *Schlechter*, near Tsewasso, 1500 ft., *Schlechter* ; Merwe Station, *Rech*, 6861 ! Var. β : Badsloop, near Waterberg, 4300 ft., *Schlechter*, 4779 ; River Limpopo, near Valdisia, 2000 ft., *Schlechter* ; Warm Baths, *Burtt-Davy*, 1713 ! EASTERN REGION : Transkei ; near Kentani, 1200 ft., *Miss Pegler*, 347 ! Griqualand East ; Nalogha, *Penther*, 172 ; foot of Insiswa Range, *Krook, Penther,* 136, 161. Natal ; near Durban, *Gerrard*, 865 ! *Sanderson* ! *Krauss*, 101, partly ! Congella, Seaview and Clairmont Flats, *Sanderson*, 500 ! Oliviers Hoek, sources of Tugela River, 5000 ft., *Allison* ! Ixopo, *Schlechter* ! Klip River District, at Colenso, *Penther*, 342 ; Van Reenen, *Krook, Penther,* 71 ; Dumisa, *Rudatis*, 240 ! and without precise locality, *Gerrard*, 11 ! 583 ! *Buchanan* ! *Allison* !

This species has been completely confused with *H. falcicornis*, Bolus. Dr. Kränzlin has pointed out the difference between the two, but did not discover that *H. tetrapetala*, Reichb. f., was identical with *H. falcicornis*, Bolus. See note under the latter species.

26. **H. cornuta** (Lindl. in Hook. Comp. Bot. Mag. ii. 208); plant 1–1½ ft. high, with stout leafy stems; leaves elliptic-oblong or oblong-lanceolate, acute or subacute, 1½–4 in. long, subcoriaceous, decreasing upwards; racemes 3–5 in. long, dense, many-flowered; bracts ovate or ovate-lanceolate, acute or acuminate, ½–¾ in. long; pedicels 7–8 lin. long; flowers green and white; dorsal sepal elliptic-ovate, obtuse, about ⅓ in. long; lateral sepals spreading, obliquely semiovate, subobtuse, about ⅓ in. long; petals 2-partite; upper lobe narrowly linear, subobtuse, as long as the dorsal sepal; lower lobe strongly incurved, more than twice as long as the upper lobe and tapering upwards from a broader base; lip 3-partite, ⅓ in. long, with tapering lobes, the lateral pair distinctly broader at the base; spur ½ in. long, slender below, strongly clavate above; column over 2 lin. long, subobtuse; side lobes of rostellum 2 lin. long; stigmas 1½ lin. long, very stout and capitate. *Drège, Zwei Pfl. Documente,* 154; *Bolus in Journ. Linn. Soc.* xxv. 190; *Durand & Schinz, Conspect. Fl. Afr.* v. 75; *Kränzl. in Engl. Jahrb.* xvi. 71, *and Orch. Gen. et Sp.* i. 206; *Schlechter in Engl. Jahrb.* xx. *Beibl.* 50, 33.

SOUTH AFRICA: Kaffirland; *Brownlee,* 4! *Trimen*!
KALAHARI REGION: Transvaal: marshes at Houtbosch (Woodbush), 6600 ft., *Schlechter,* 4474! *Rehmann,* 5815!
EASTERN REGION: Transkei; Tsomo, *Barber & Bowker,* 839! 855! Pondoland; Umtentu River, near the mouth, and near Umsikaba River, *Drège,* 4570! Griqualand East; near Kokstad, 5000 ft., *Tyson,* 1076! near Clydesdale, 3500 ft., *Tyson,* 2857! Natal; Izongolweni, *Wood,* 1997! Oliviers Hoek, sources of the Tugela River, 5000 ft., *Allison,* 21! near Charlestown, 5000–6000 ft., *Wood,* 5529!

27. **H. involuta** (Bolus in Journ. Linn. Soc. xxv. 165, 166, fig. 3); plant 2 ft. or more high, with rather stout stems and ovoid tubers; leaves 2–4, basal or subbasal, linear-lanceolate or elongate-linear, ½–1½ ft. long, ¼–⅓ in. broad, upper leaves much reduced; racemes oblong, about 6 in. long, somewhat dense, many-flowered; bracts lanceolate or oblong-lanceolate, acuminate, about ½ lin. long; pedicels about ¾ in. long; flowers light green; dorsal sepal elliptic-oblong, obtuse, 2½ lin. long; lateral sepals obliquely obovate-elliptic, obtusely apiculate, 3–3¼ lin. long; petals 2-partite; upper lobe falcate-linear, subacute, shorter than the dorsal sepal; lower lobe lanceolate-oblong, longer than the upper lobe and about four times as broad; lip 3-partite, about 5 lin. long, with linear curved lobes, the lateral pair shorter and broader than the front lobe; spur about ½ in. long, nearly straight, slender below, subclavate above; column about 1 lin. long, obtuse, side lobes of rostellum curved, 2 lin. long; stigmas 2¼ lin. long, straight, capitate at the apex. *Bolus in Journ. Linn. Soc.* xxv. 192; *Durand & Schinz, Conspect. Fl. Afr.* v. 80; *Kränzl. in Engl. Jahrb.* xvi. 87, *and Orch. Gen. et Sp.* i. 229.

EASTERN REGION: Natal; Durban Bay, near the mouth of the Umhlatuzaan River, *Sanderson,* 833!

Bolus originally described the petals as undivided, but this is a mistake.

28. **H. incurva** (Rolfe); plant over ½ ft. high; leaves cauline,
3–5, lanceolate-linear, acuminate, 2–2½ in. long, decreasing upwards
into the bracts; racemes about 2½ in. long, somewhat lax, many-
flowered; bracts oblong-lanceolate, acuminate, ¼–⅓ in. long;
pedicels ½–¾ in. long, rather slender; flowers rather small;
dorsal sepal broadly elliptic, obtuse, 2 lin. long; lateral sepals
obliquely semiobovate, laterally apiculate, 2½ lin. long; petals
2-partite; upper lobe narrowly falcate-linear, obtuse, as long as the
dorsal sepal; lower lobe linear, obtuse, rather longer and broader
than the upper lobe; lip 3-partite, with subequal filiform lobes,
4–5 lin. long; spur 7–9 lin. long, much curved, filiform below,
somewhat thickened above; anther 1 lin. long, obtuse; side lobes
of rostellum 1¼ lin. long; stigmas about 2 lin. long, subclavate,
recurved at the apex.

KALAHARI REGION: Transvaal: hillsides among rocks at 6000 ft., *Galpin*, 392!

29. **H. trachychila** (Kränzl. in Bull. Herb. Boiss. 2ᵐᵉ sér. iv.
1007); plant about 1⅓ ft. high, with large tubers and leafy stem;
leaves linear from a sheathing base, acuminate, lower 8 in. long by
about 5 lin. broad, upper reduced to sheaths; spikes 10 in. long,
lax, many-flowered; bracts lanceolate, acuminate, longer than the
pedicels; pedicels about 1½ in. long; flowers medium-sized; dorsal
sepal oblong-lanceolate, acute, concave, 5 lin. long; lateral sepals
obliquely ovate or subrhomboid, slender, 3–3½ lin. long; petals
2-partite; upper lobe linear, as long as the dorsal sepal; lower lobe
linear, 6–7 lin. long; lip 3-partite, about 7 lin. long, with narrowly
lanceolate acuminate lobes, very hairy on the upper half; spur ¾ in.
long, filiform below, clavate and somewhat compressed near the
apex; column 4 lin. long; side lobes of rostellum curved; stigmas
deflexed, 2½ lin. long, thickened at the apex.

KALAHARI REGION: Transvaal; *Rehmann*, without number.

Only known to me from description.

30. **H. umvotensis** (Rolfe); plant about 1 ft. high, with rather
stout leafy stem and large oblong tubers; leaves about 6, lanceolate, subobtuse, attenuate below, submembranous, 3–5 in. long;
raceme short, lax, about 5-flowered; bracts ovate or ovate-lanceolate, acute or acuminate, ⅓–½ in. long; pedicels about ¾ in. long;
flowers medium-sized, white and green (*Culver*); dorsal sepal ovate,
concave, shortly acuminate, 5 lin. long; lateral sepals obliquely
ovate-oblong, acuminate, 5 lin. long; petals 2-partite; upper lobe
linear, as long as the dorsal sepal, lower lobe elongate-linear,
acuminate, very narrow, over ¾ lin. long; lip 3-partite, ¾ in.
long; front lobe linear, subobtuse; side lobes elongate-linear,
acuminate, very narrow, longer than the middle lobe; spur about
4 lin. long, rather stout, somewhat curved; column 2 lin. long,

apiculate ; side lobes of rostellum 2 lin. long ; stigmas linear, ¼ in. long, slightly curved.

KALAHARI REGION : Transvaal ; Umvoti Creek, near Barberton, 3000 ft., *Culver*, 30 ! *Galpin*, 954 !

31. **H. porrecta** (Bolus in Journ. Linn. Soc. xxv. 167, 168, fig. 5) ; plant 1¼–2 ft. high, with stout leafy stems and numerous imbricating sheaths above ; leaves narrowly oblong, subobtuse, rather fleshy, those seen 3–4 in. long, ½–¾ in. broad, but all more or less imperfect ; racemes 4–8 in. long, dense, many-flowered ; bracts ovate or ovate-lanceolate, acuminate, submembranous, ½–1 in. long ; pedicels ¾–1¼ in. long ; dorsal sepal elliptic-ovate, acuminate, about ½ in. long ; lateral sepals spreading, obliquely and broadly semi-ovate, shortly caudate-apiculate, about ½ in. long ; petals 2-partite ; upper lobe linear, acute, ½ in. long ; lower lobe filiform, curved, about twice as long as the upper lobe ; lip 3-partite ; lobes subfiliform, curved, somewhat stouter below, ¾–1 in. long ; spur 1¼–1½ in. long, slender below, clavate above ; column ¼ in. long, apiculate, side lobes of rostellum ¼ in. long ; stigmas clavate, ½ in. long. *Bolus in Journ. Linn. Soc.* xxv. 192 ; *Durand & Schinz, Conspect. Fl. Afr.* v. 84 ; *Kränzl. in Engl. Jahrb.* xvi. 71, *and Orch. Gen. et Sp.* i. 207.

EASTERN REGION : Natal ; Green Vale, *Plant*, 52 ! Durban, *McKen & Gerrard*, 2 ! Alexandra District, *Rudatis*, 411 ! and without precise locality, *Gueinzius* ! *Sanderson* ! *Pappe*, 77 ! Zululand ; without precise locality, *Gerrard*, 737 !

The leaves seem to disappear early, being more or less shrivelled or absent in all the above cited specimens.

32. **H. clavata** (Reichb. f. in Flora, 1865, 180) ; plant ¾–2½ ft. high, with stout leafy stems, and 2–3 short sheaths below ; leaves sessile, oblong or elliptic-oblong, subobtuse or acute, submembranous, 2–4½ in. long, gradually decreasing upwards ; racemes 4–8 in. long, somewhat dense and many-flowered ; bracts oblong-lanceolate or ovate-lanceolate, acuminate, 1–1½ in. long ; pedicels 1¼–2 in. long ; flowers green and white or yellow ; dorsal sepal elliptic-ovate, subobtuse, 5–6 lin. long ; lateral sepals obliquely falcate-obovate, apiculate, 6–7 lin. long ; petals 2-partite ; upper lobe linear, 5–6 lin. long ; lower lobe subfiliform, falcately curved, much stouter than the upper lobe and twice as long ; lip 3-partite ; lobes narrowly linear, curved, somewhat stouter below, ½–¾ lin. long ; spur 1¼–1½ in. long, filiform below, clavate above ; column ¼ in. long, apiculate ; side lobes of rostellum 5–6 lin. long ; stigmas 5–6 lin. long, capitate. *Bolus in Journ. Linn. Soc.* xxv. 191, *and Ic. Orch. Austr.-Afr.* ii. *t.* 43 ; *N. E. Br. in Gard. Chron.* 1885, xxiv. 307 ; *Durand & Schinz, Conspect. Fl. Afr.* v. 75 ; *Kränzl. in Engl. Jahrb.* xvi. 72, *Orch. Gen. et Sp.* i. 208, *and in Ann. Naturhist. Hofmus. Wien*, xx. 2 ; *Schlechter in Engl. Jahrb.* xx. *Beibl.* 50, 34. *Bonatea clavata, Lindl. in Hook. Comp. Bot. Mag.* ii. 208 ; *Drège, Zwei Pfl. Documente*, 146, 149.

COAST REGION: Bedford Div. ; Kaga Berg, *Hutton.* Stockenstrom Div. ; Kat
Berg, 2000 ft., *Hutton,* 46 ! King Williamstown Div. ; Dohne Hill, *Sim,* 44.
Komgha Div. ; flat near Kei River Mouth, 200 ft , *Flanagan.* Eastern Frontier
of Cape Colony, *Mrs. Barber* !
   KALAHARI REGION: Orange River Colony ; Harrismith, 6500 ft., *Sankey,* 259 !
Besters Vlei, 5300 ft., *Bolus,* 13459, and without precise locality, *Cooper,* 1097 !
Transvaal ; Houtbosch (Woodbush), 5100 ft., *Bolus,* 10974 ! Wilge River, 6400 ft.,
Schlechter, 4123 ! Belfast, 6400–6650 ft., *Burtt-Davy,* 1297 ! 1396 ! Lake
Chrissie, *Burtt-Davy,* 2354a ! Ermelo, *Burtt-Davy,* 5421 ! Spion Kop, *Burtt-Davy,*
8904 ! Saddleback Mountain, near Barberton, 4600 ft., *Galpin,* 801 ; Rocky
Mountains in Petersburg District, 4600 ft., *Bolus,* 10974 ; Wonder Fontein,
6000 ft., *Bolus,* 12310.
   EASTERN REGION : Transkei ; between Gekau River and Bashee River, 1000–
2000 ft., *Drège,* 4568 ! Tsomo River, *Mrs. Barber,* 829, partly ! Tembuland ;
Bazeia, 2000 ft., *Baur,* 626 ! Pondoland ; between Umtata River and St. Johns
River, *Drège* ! Griqualand East ; near Kokstad, 5000 ft., *Tyson,* 1067 ! and in
McOwan & Bolus, Herb. Norm. Austr.-Afr., 451 ! *Krook, Penther,* 222 ; Nalogha,
Penther, 170 ; Insiswa Range, *Penther,* 145, 146. Natal ; Inanda, *Wood,* 723 !
Dargle Farm, *Mrs. Fannin,* 93 ! Attercliff, 800 ft., *Sanderson,* 489 ! near Fairfield,
Rudatis, 565 ! Zululand ; on plains, *Gerrard,* 1546 !

33. H. stenorhynchos (Schlechter in Engl. Jahrb. xx. Beibl. 50,
33) ; plant about 8 in. high, with slender, somewhat leafy stem ;
leaves few, erect, linear-lanceolate, acute, submembranous, the lower
about 1½ in. long, 2 lin. broad, the upper reduced to short sheaths ;
racemes very short, few-flowered ; bracts oblong-lanceolate, acuminate,
6–7 lin. long ; pedicels about ¾ in. long ; flowers white (*Schlechter*) ;
dorsal sepal elliptic-lanceolate, subobtuse, 4½ lin. long ; lateral sepals
obliquely obovate-oblong, subobtuse, 4½–5 lin. long ; petals 2-partite,
with narrowly linear acute lobes, the lower about as long as the
dorsal sepal, the upper rather shorter and more slender ; lip 3-partite,
with filiform acute lobes ; front lobe about ½ in. long ; side lobes
diverging and only half as long ; spur 3–3½ lin. long, strongly
clavate ; column 1½ lin. long, obtuse ; side lobes of rostellum 3 lin.
long ; stigmas 3½ lin. long, broadly capitate. *Kränzl. Orch. Gen. et
Sp.* i. 205.

   KALAHARI REGION : Transvaal ; Houtbosch (Woodbush) Mountains, 6400 ft.,
in marshes, *Schlechter,* 4416 !

34. H. dregeana (Lindl. in Ann. Nat. Hist. ser. 1, iv. 314) ; plant
½–1¼ ft. high ; leaves 2, radical, sessile, broadly cordate-orbicular,
apiculate, horizontally spreading, fleshy, 1–1¾ in. long, 1–2 in. broad ;
scapes ½–1¼ ft. high, with numerous lanceolate very acuminate free
sheaths ; racemes oblong, 1½–4 in. long, dense, many-flowered ; bracts
lanceolate or oblong-lanceolate, acuminate, ⅓–½ in. long ; pedicels
about ⅓ in. long ; flowers green ; dorsal sepal ovate, subacute, 2–2½ lin.
long ; lateral sepals obliquely semiovate-oblong, acuminate, 2½–3 lin.
long ; petals 2-partite ; upper lobe falcate-oblong, acute, obscurely
serrulate, three-quarters as long as the dorsal sepal ; lower lobe
diverging, linear, subacute, about a third as long as the upper lobe ;
lip 3-partite, with obscurely hispidulous lobes ; front lobe linear-
oblong, subacute, 2–2½ lin. long ; side lobes diverging, linear or
filiform, subobtuse, often shorter than the front lobe ; spur 4–4½ lin.

long, slender below, clavate above ; column very broad, obtuse, $\frac{3}{4}$ lin. long ; side lobes of rostellum $\frac{1}{2}$ lin. long ; stigmas oblong, about 1 lin. long ; capsule oblong, 4–5 lin. long. *Drège, Zwei Pfl. Documente,* 147 ; *Bolus in Journ. Linn. Soc.* xxv. 190 ; *Durand & Schinz, Conspect. Fl. Afr.* v. 77 ; *Kränzl. in Engl. Jahrb.* xvi. 149, *Orch. Gen. et Sp.* i. 338, *excl. syn., and in Ann. Naturhist. Hofmus. Wien,* xx. 3 ; *Bolus, Ic. Orch. Austr.-Afr.* i. *t.* 15 ; *Schlechter in Engl. Jahrb.* xx. *Beibl.* 50, 35.

COAST REGION : Stockenstrom Div. ; Kat Berg, 2000 ft., *Hutton*, 42 !

KALAHARI REGION : Orange River Colony ; Harrismith, *Sankey*, 261 ! *Krook, Penther*, 121. Transvaal ; Lydenberg District ; near Paardeplatze, *Wilms*, 1381 (in fruit) ! Ermelo, *Burtt-Davy*, 7722 ! Belfast, 6500 ft., *Doidge*, 4806 ! and without precise locality, *Schlechter*.

EASTERN REGION : Transkei ; near the mouth of the Bashee River, *Bowker*, 453 ! Kentani, 1200 ft., *Miss Pegler*, 348 ! 1814 ! Tembuland ; on hills at Bazeia, 2000 ft., *Baur*, 815 ! between the Bashee River and Morley, *Drège* ! Griqualand East ; Insiswa Range, *Penther*, 409 ; Newmarket, *Krook, Penther*, 111. Natal ; Sterk Spruit, at Caversham, 4000 ft., *Buchanan in Herb. Wood* and *Herb. Sanderson*, 1058 ! Attercliff to Pinetown, 800–1000 ft., *Sanderson*, 487 ! Dargle Farm, *Mrs. Fannin*, 81 ! Inanda, *Wood*, 815 ! 852 ! Tugela River, *Gerrard*, 1811 ! Oliviers Hoek, source of Tugela River, 5000 ft., *Allison*, 24 ! Alexandra District, *Rudatis*, 258 ! Summit of Amajuba, *Burtt-Davy*, 7762 ! near Estcourt, *Krook, Penther*, 132 ; Van Reenen, *Penther*, 93 ; Colenso, *Krook, Penther*, 275.

35. **H. macowaniana** (Kränzl. in Engl. Jahrb. xvi. 150) ; plant $\frac{1}{2}$–1 ft. high ; leaves 2, radical, sessile, orbicular or broadly cordate-orbicular, apiculate, horizontally spreading, fleshy, 1–2 in. long, 1–2$\frac{3}{4}$ in. broad ; scapes $\frac{1}{2}$–1 ft. high, with numerous narrowly lanceolate very acuminate free sheaths, $\frac{1}{2}$–$\frac{3}{4}$ in. long ; racemes oblong or somewhat elongate, 1$\frac{1}{2}$–6 in. long, dense, many-flowered ; bracts lanceolate or oblong-lanceolate, acuminate, $\frac{1}{3}$–$\frac{1}{2}$ in. long ; pedicels about $\frac{1}{3}$ in. long ; flowers green (*MacOwan*) ; dorsal sepal broadly ovate, apiculate, 2–2$\frac{1}{2}$ lin. long ; lateral sepals obliquely semiovate-oblong, acuminate, 3–3$\frac{1}{2}$ lin. long ; petals 2-partite ; upper lobe narrowly falcate-oblong, acute, ciliate, as long as the dorsal sepal ; lower lobe filiform, acute, hispidulous, rather longer than the upper lobe ; lip 3-partite, with linear or filiform hispidulous lobes, 3–3$\frac{1}{2}$ lin. long, the lateral pair diverging and rather narrower than the front lobe ; spur about 4 lin. long, clavate ; column 1 lin. long, very broad, obtuse ; side lobes of rostellum suberect, about $\frac{3}{4}$ lin. long ; stigmas oblong, apiculate, nearly $\frac{1}{2}$ lin. long ; capsule oblong, about 4 lin. long. *Durand & Schinz, Conspect. Fl. Afr.* v. 80. *H. arachnoidea, MacOwan ex Kränzl. in Engl. Jahrb.* xvi. 150, *not of Thouars.*

COAST REGION : Albany Div. ; tops of hills near Grahamstown, *MacOwan*, 767 ! Bedford Div. ; Kaga Berg, 4000 ft., *MacOwan*, 767 ! *Hutton* !

CENTRAL REGION : Somerset Div. ; Bosch Berg, 4000 ft., *MacOwan*, 767 !

Reduced to *H. dregeana*, Lindl., by Kränzlin, but quite distinct.

36. **H. Tysoni** (Bolus in Journ. Linn. Soc. xxv. 166, 167, fig. 4) ; plant $\frac{3}{4}$–1 ft. high ; leaves 2, radical, sessile, suborbicular or broader

than long, shortly apiculate, spreading, somewhat fleshy, 1–1½ in.
long, 1–2¼ in. broad ; scapes ¾–1 ft. high, with numerous narrowly
lanceolate very acuminate free sheaths, about ¾ in. long; racemes
oblong, about 4½ in. long, lax, many-flowered ; bracts narrowly
lanceolate, very acuminate, ½–1 in. long; pedicels 6–7 lin. long;
dorsal sepal ovate, acuminate, 3½–4 lin. long ; lateral sepals obliquely
semiovate, acuminate, 4–4½ lin. long ; petals 2-partite ; upper lobe
falcate-linear, acute, slightly ciliate, rather shorter than the dorsal
sepal ; lower lobe filiform, curved, hispidulous, about 5 lin. long ;
lip 3-partite, hispidulous ; front lobe linear, acute, curved, 5–6 lin.
long ; side lobes diverging, shorter and narrower than the front
lobe ; spur 6–7 lin. long, slender below, clavate above ; column
oblong, obtuse ; side lobes of rostellum ¾ lin. long ; stigmas clavate-
oblong, subacute, 1 lin. long. *Durand & Schinz, Conspect. Fl. Afr.*
v. 88.

EASTERN REGION: Griqualand East ; among grasses on Mount Currie, near
Kokstad, 6000 ft., *Tyson,* 1068 ! Natal ; without precise locality, *Sanderson,* 2 !

Flowers much larger than in *H. dregeana,* Lindl., to which species it is reduced
by Kränzlin.

37. **H. krænzliniana** (Schlechter in Engl. Jahrb. xx. Beibl. 50, 35) ;
plant ¾–1¼ ft. high ; leaves radical, sessile, 2, spreading, broadly
ovate-orbicular, shortly and abruptly apiculate, somewhat fleshy,
1¼–1¾ in. long, 1½–2 in. broad ; scapes ¾–1¼ ft. high, with numerous
lanceolate-linear acuminate free sheaths ¾–1 in. long ; racemes
oblong or somewhat elongate, 2–5 in. long, dense, many-flowered ;
bracts oblong-lanceolate, very acuminate, ½–¾ in. long ; flowers
yellow-green (*Wood*) ; dorsal sepal elliptic-ovate, acute, cucullate,
3 lin. long ; lateral sepals obliquely semiovate, acuminate, 3½ lin.
long ; petals 2-partite ; upper lobe lanceolate-oblong, obliquely
acuminate, nearly as long as the dorsal sepal, minutely hispidulous ;
lower lobe filiform, elongate, flexuous, hispid, three or four times as
long as the upper lobe ; lip 3-partite ; front lobe lanceolate, acuminate,
about 3 lin. long ; side lobes filiform, elongate, flexuous, hispid, about
¾ in. long ; spur 1–1¼ in. long, slender below, subclavate above ;
column 1¼ lin. long, obtuse ; side lobes of rostellum 1–1¼ lin. long ;
stigmas clavate, 1 lin. long. *Kränzl. Orch. Gen. et Sp.* i. 340.

VAR. β, **natalensis** (Rolfe) ; flowers rather larger ; spur 1½–1¾ in. long and
more clavate at the apex ; dorsal sepal more obtuse ; lower lobes of petals and side
lobes of lip longer and more strongly hispid.

KALAHARI REGION : Transvaal ; in wet places, among grasses, near Sandloop,
between Pietersburg and Houtbosch, 4600 ft., *Schlechter,* 4369 !
EASTERN REGION : Var. β : Natal ; near Colenso, 3400 ft., *Schlechter,* 6889 !
stony hill near Ladysmith, 3400 ft., *Wood,* 5528 !

XXV. BONATEA, Willd.

Sepals unequal, free; dorsal erect, lateral oblique, reflexed.
Petals deeply bipartite; posterior lobe erect, adpressed to the
margin of the dorsal sepal; anterior lobe descending and simulating a lobe of the lip. *Lip* continuous with the column, produced
at the base into a more or less elongated spur; limb spreading,
base produced into a narrow claw, adnate to the base of the lateral
sepals, the anterior lobe of the petals, and the stigmatic processes,
tripartite above with narrow lobes. *Column* short, footless; anther-
bed erect, as long as the anther. *Anther-cells* somewhat diverging;
apex inferior, prolonged in front into channels, confluent with the
side lobes of the rostellum; pollinia granular, with elongate curved
caudicles and exserted naked glands; staminodes lateral, small,
auriculate. *Stigma* bipartite, extended in front into a pair of
elongate subclavate processes; rostellum trilobed; middle lobe
cucullate, apiculate; side lobes linear, elongate. *Capsule* oblong.

Terrestrial herbs, with the habit of a large *Habenaria*; flowers rather large, in
more or less elongated racemes; bracts ovate.

DISTRIB. Species about 18, exclusively Continental African, and 7 occurring
within our limits. It is distinguished from *Habenaria* by the cucullate middle
lobe of the rostellum, and by the union of the base of the lateral sepals, the
anterior lobe of the petals and the base of the stigmatic processes with the claw of
the lip.

Spur cylindrical or subcylindrical, ¾–1 in. long:
  Side lobes of lip broadly elliptic-lanceolate, under ½ in.
    long ...   .   (1) **cassidea.**
  Side lobes of lip lanceolate, acuminate, about ¾ in. long   (2) **Saundersiæ.**
  Side lobes of lip linear, about 1¼ in. long   ...   ...   (3) **saundersioides.**
Spur somewhat elongate at the apex, 1¼–1½ in. long:
  Side lobes of lip linear or lanceolate-linear:
    Dorsal sepal about 7 lin. long; stigmas about ½ in.
    long   ...   .   (4) **Boltoni.**
    Dorsal sepal ¾–1 in. long; stigmas about ¾ in. long   (5) **speciosa.**
  Side lobes of lip filiform:
    Flowers rather small; spur under 1 in. long   ...   (6) **densiflora.**
    Flowers large; spur over 1 in. long ...   .   (7) **antennifera.**

1. **B. cassidea** (Sond. in Linnæa, xix. 81); plant 1–1½ ft. high,
with somewhat slender leafy stems; leaves linear or oblong-linear,
acute, membranous, 3–6 in. long, ⅓–½ in. broad, the upper suddenly
reduced to lanceolate sheaths; racemes oblong or somewhat elongate,
2½–6 in. long, dense, many-flowered; bracts lanceolate or oblong-
lanceolate, acuminate, ½–¾ in. long; pedicels ½–¾ in. long; flowers
white and light green; dorsal sepal elliptic-oblong, apiculate or
subacute, somewhat galeate, 3–4½ lin. long; lateral sepals oblong-
lanceolate, obliquely acuminate or apiculate, 4–5 lin. long; petals
2-partite; upper lobe narrowly lanceolate, acuminate, as long as

the dorsal sepal; lower lobe broadly lanceolate, acute, a third longer than the upper lobe; lip 3-partite, 6–8 lin. long; front lobe elongate-linear; side lobes broadly lanceolate, acute; spur 8–10 lin. long, cylindrical or slightly thickened near the apex, somewhat curved; column 2 lin. long, apiculate; rostellum cucullate, 2 lin. long; side lobes 1½ lin. long; stigmas 2½ lin. long, clavate, somewhat curved at the apex. *B. . Darwinii, Weale in Journ. Linn. Soc.* x. 470, 473, *with fig. Habenaria cassidea, Reichb. f. in Walp. Ann.* i. 797; *Bolus in Journ. Linn. Soc.* xxv. 191; *Durand & Schinz, Conspect. Fl. Afr.* v. 74; *Kränzl. in Engl. Jahrb.* xvi. 80, and *Orch. Gen. et Sp.* i. 224.

SOUTH AFRICA: without precise locality, *Mrs. Bowker*!
COAST REGION: Uitenhage Div. ; Olifants Hoek, near Bosjesmans River, *Ecklon & Zeyher*. Bathurst Div. ; woods near the mouth of the Riet River, *Atherstone*! Albany Div. ; Lushington Valley, 3800 ft., *MacOwan*, 1529! and without precise locality, *Cooper*, 1865! *Hutton*! Bedford Div. ; Kaga Berg, 4000 ft., *MacOwan*, 1529! Bedford, in damp hollows, *Weale*. Stockenstrom Div. ; Kat Berg, *Hutton*! East London Div. ; Second Creek, East London, *Dodd*!
CENTRAL REGION: Somerset Div. ; *Bowker*!
EASTERN REGION: Transkei ; Fort Bowker, in woods, *Bowker*, 587!

This has the cucullate rostellum and other characters of *Bonatea*, and is out of place in *Habenaria* § *Replicatæ*, where Kränzlin places it. It is intimately allied to *B. Saundersiæ*, but has smaller and more numerous flowers.

2. **B. Saundersiæ** (Durand & Schinz, Conspect. Fl. Afr. v. 89, excl. syn. Weale); plant 1–1½ ft. high, with rather stout leafy stems; leaves 5–6, sessile, amplexicaul, linear or oblong-linear, acute or acuminate, membranous, 3–6½ in. long, ¼–1 in. broad, the upper reduced to acuminate sheaths; racemes 4–6 in. long, usually lax, many-flowered; bracts ovate-lanceolate, very acuminate, ½–1 in. long; pedicels ¾–1¼ in. long; flowers large, white; dorsal sepal cucullate, elliptic-oblong, acuminate, 6–7 lin. long; lateral sepals obliquely semiovate, apiculate, 6–7 lin. long; petals 2-partite, upper lobe linear or lanceolate-linear, acuminate, about as long as the dorsal sepal; lower lobe falcate-lanceolate, acuminate, 9–10 lin. long; lip 3-partite, about 1 in. long; front lobe narrowly linear, acute; side lobes falcate-lanceolate, acuminate, about as long as the front lobe; spur ¾–1 in. long, cylindrical, somewhat curved; column 2½ lin. long, obtuse; rostellum cucullate, acuminate, front lobe 1½ lin. long; side lobes about twice as long; stigmas clavate, somewhat recurved at the apex, 5 lin. long, the lower half united to the combined base of the lip, petals and lateral sepals. *Habenaria Saundersiæ, Harv. Thes. Cap.* ii. 29, *t.* 147; *Bolus in Journ. Linn. Soc.* xxv. 191; *Kränzl. in Engl. Jahrb.* xvi. 57, *excl. syn. Weale, Orch. Gen. et Sp.* i. 181, *excl. MacOwan*, 1529, *and in Ann. Naturhist. Hofmus. Wien*, xx. 2.

COAST REGION: Mossel Bay Div. ; Mossel Bay, *Penther*, 333.
EASTERN REGION: Natal ; Umbilo River, near Durban, 5000 ft., *Sanderson*, 822! *Miss Wheelwright*! and without precise locality, *Mrs. Saunders*, 2!

3. **B. saundersioides** (Rolfe); plant with large oblong tubers, 2 in. long, and leafy stems about 16 in. high; leaves 6–8, linear-lanceolate, acuminate, about 6 in. long, ½ in. broad, not decreasing upwards into bracts; scapes with a few bract-like sheaths; racemes lax, few-flowered; bracts ovate, acuminate, pallid, nearly half as long as the pedicelled ovary; flowers rather large, with green sepals and white petals and lip; dorsal sepal cucullate, acuminate, 5 lin. long; lateral sepals deflexed, obliquely ovate, acute, rather longer than the dorsal; petals 2-partite; upper lobe linear, as long as the dorsal sepal and united to its margins; lower lobe linear-lanceolate, falcate, very acuminate, 1 in. long; lip 3-partite, with linear lobes; side lobes about 1¼ in. long; front lobe a third to a half shorter; spur cylindrical, somewhat thickened, about ¾ in. long, its mouth nearly closed by the presence of a small tooth (*Bolus*); column apiculate; rostellum large, cucullate, side lobes scarcely half as long as the stigmas; stigmas elongated, free from the lip, somewhat curved at the apex. *Habenaria saundersioides, Kränzl. & Schlechter in Kränzl. Orch. Gen. et Sp.* i. 181.

KALAHARI REGION : Transvaal; Umvoti Creek, Barberton, 3300 ft., *Culver*, 30.
Only known to me from the description.

4. **B. Boltoni** (Bolus in Journ. Linn. Soc. xix. 340); plant ¾–1¼ ft. high, with large oblong tubers and leafy stems; leaves 5–7, sessile, linear-oblong to ovate, subacute, somewhat fleshy, 1–5 in. long, ½–1¼ in. broad; racemes broadly oblong, rarely subcapitate, 2–4 in. long, dense, many-flowered; bracts ovate-lanceolate, acute or acuminate, ¾–1¼ in. long; pedicels about 1½ in. long; flowers large, green and white; dorsal sepal cucullate, elliptic-ovate, acute, 6–7 lin. long; lateral sepals obliquely semiovate, acute, with an acute tooth on the inner margin near the apex, about ½ in. long; petals 2-partite; upper lobe linear-lanceolate, acute, as long as the dorsal sepal; lower lobe linear-lanceolate, somewhat falcate, 7–8 lin. long; lip 3-partite, ¾ in. long, with linear acute lobes; front lobe bent about the middle; side lobes rather narrower and more acute; spur 1¼–1½ in. long, cylindrical, somewhat clavate near the apex; column ¼ in. long, apiculate; rostellum cucullate, ¼ in. long; front lobe apiculate; side lobes about 2½ lin. long; stigmas ½ in. long, clavate at the apex, united to the combined base of the lip, petals and lateral sepals. *Durand & Schinz, Conspect. Fl. Afr.* v. 89. *Habenaria Boltoni, Harv. Thes. Cap.* i. 55, *t.* 88; *Bolus in Journ. Linn. Soc.* xxv. 191; *Kränzl. in Engl. Jahrb.* xvi. 142. *H. Bonatea, Schlechter in Engl. Jahrb.* xx. *Beibl.* 50, 32, *not of Willd. H. Bonatea, var. Boltoni, Bolus, Ic. Orch. Austr.-Afr.* ii. *t.* 45.

COAST REGION : Uitenhage Div.; Karroo grounds of Uitenhage, *Bowie*! Albany Div.; near Grahamstown, *Bolton*! *Glass in Herb. Bolus*, 6238; woods near Howisons Poort, *MacOwan*. British Kaffraria, *Mrs. Hutton*! Eastern Frontier of Cape Colony, *Mrs. Barber*, 829, partly!
KALAHARI REGION : Transvaal; Barberton, *Culver*, 18! Houtbosch, among shrubs, *Schlechter*, 4381!

EASTERN REGION : Natal ; near the Tugela River, 500 ft., *Buchanan* ! Noodsberg, *Buchanan in Herb. Wood* ! Drakensberg, near Newcastle, *Wilms,* 2284 !
Inanda, *Wood,* 270 ! grassy hill above Pinetown, 2200 ft., *Wood,* 946 ! and without
precise locality, *Cooper,* 3618 ! *Sanderson,* 172 ! *Lange,* 100 !

Quite distinct from *B. speciosa,* Willd. (*Habenaria Bonatea,* Reichb. f.), to which
Kränzlin has reduced it.

5. B. speciosa (Willd. Sp. Pl. iv. 43) ; plant 1–3 ft. high, with
stout leafy stems ; leaves 8–10, sessile, amplexicaul, elliptic or
elliptic-oblong to ovate, subobtuse or apiculate, 2–6 in. long, 1–2¼ in.
broad, somewhat fleshy, spreading or recurved, the upper reduced
to bract-like sheaths ; racemes oblong, rarely somewhat elongate,
4–8 in. long, dense, many-flowered ; bracts ovate or ovate-lanceolate,
acuminate, ¾–1½ in. long ; pedicels 1¾–2 in. long ; flowers large,
green, except the stigmas and lower lobe of the petals, which are
white ; dorsal sepal cucullate, elliptic-oblong, acute, 9–11 lin. long ;
lateral sepals obliquely semiovate, acute, 12–14 lin. long ; the inner
margin adnate to the lip for ½ in., then revolute ; petals 2-partite ;
upper lobe falcate-linear, acute ; lower lobe lanceolate-linear, acute,
1½ lin. long, curved at the apex ; lip 3-lobed to below the middle,
1¾ in. long, with linear acute lobes, reflexed at the apex ; side
lobes diverging, longer than the front lobe ; spur 1¼–1½ in. long,
slender below, clavate and compressed at the apex ; column 4 lin.
long, apiculate ; rostellum cucullate ; front apiculate, recurved ; side
lobes ½ in. long ; stigmas 1 in. long, spathulate at the apex, the
lower half adnate to the combined base of the lip, front lobe of
petals and lateral sepals ; capsule oblong, 1½–1¾ in. long. *Lindl. Gen.
& Sp. Orch.* 327 ; *Bot. Mag. t.* 2926 ; *Lodd. Bot. Cab. t.* 284 ; *Bauer,
Ill. Orch. Gen. tt.* 12, 13 ; *Drège, Zwei Pfl. Documente,* 125 ; *R. Trimen
in Journ. Linn. Soc.* ix. 156, *t.* 1 ; *Durand & Schinz, Conspect. Fl. Afr.*
v. 89. *Orchis speciosa, Linn. f. Suppl.* 401 ; *Thunb. Prodr.* 4, *and
Fl. Cap. ed. Schult.* 5 ; *Jacq. Hort. Schœnbr. t.* 451. *Habenaria Bonatea,
Reichb. f. Otia Bot. Hamb.* ii. 101 ; *Bolus in Journ. Linn. Soc.* xxv.
191, *and Ic. Orch. Austr.-Afr.* ii. *t.* 44 ; *Kränzl. in Engl. Jahrb.* xvi. 57,
Orch. Gen. et Sp. i. 179, *excl. syn. Harv.,* and in *Ann. Naturhist. Hofmus.
Wien,* xx. 2. *H. robusta, N. E. Br. in Gard. Chron.* 1885, xxiv. 307.

SOUTH AFRICA : without locality, *Masson* ! *Thom,* 32 ! 96 ! *Moyle Rogers* !
COAST REGION : Malmesbury Div. ; near Darling, *Peringuey in Herb. Bolus,*
13458. Cape Div. ; Table Mountain, *Miss Kensit.* Mossel Bay Div. ; near the
landing-place at Mossel Bay, *Burchell,* 6231 ! thickets near Mossel Bay, *Bowie* !
Knysna Div. ; sand hills, Plettenbergs Bay, *Burchell,* 5306 ! Groene Vallei, under
500 ft., *Drège,* 8297 ! Uitenhage Div. ; forests near the Koega River, *Zeyher,* 6 !
near Uitenhage, *Tredgold* ! Port Elizabeth Div. ; Algoa Bay, *Rabjohn* ! *Wilson* !
Albany Div. ; Tharfield, near Grahamstown, *Atherstone,* and without precise
locality, *Hutton* ! *Miss Bowker* ! Bedford Div. ; without precise locality, *Hutton* !
Fort Beaufort Div. ; Fort Beaufort, and the Kunap River, *Atherstone* ! Stockenstrom Div. ; near Kat Berg, *Scully* ! Komgha Div. ; rocks near the Kei River
mouth, 1500 ft., *Flanagan,* 647. British Kaffraria ; without precise locality,
Cooper, 1802 ! *Mrs. Hutton* !
KALAHARI REGION : Transvaal ; near Potchefstroom, *McLea in Herb. Bolus,*
3038 ; near Rustenberg, 4400 ft., *Miss Nation,* 184.
EASTERN REGION : Transkei ; near the Coast and the Bashee River, *Mrs. Barber,*

565! Fort Bowker, *Mrs. Barber*, 779! Krielis Country, *Bowker*! Griqualand East; near Newmarket, *Krook, Penther*, 112. Natal; near Durban, *Sanderson,* 172! Wentworth Bluff, *Sanderson*, 490! 1037! near the mouth of the Umlwalumi River, *Gerrard*, 1848! Inanda, *Wood*, 969! Weenen District, at Colenso, *Krook, Penther*, 272.

6. **B. densiflora** (Sond. in Linnæa, xix. 80); plant 1½ ft. high, with stout leafy stem; leaves oblong-lanceolate, acuminate, sheathing at the base, the lower 1 in. broad, the rest narrower; racemes elongated, ½ ft. long, dense, many-flowered; bracts membranous, very acuminate, longer than the pedicels; dorsal sepal galeate, elliptic-oblong, canaliculate, acuminate, incurved, 6–7 lin. long; lateral sepals oblong-lanceolate, acuminate, reflexed, 7 lin. long; petals 2-partite, the upper lobe linear, acute; lower lobe elongate-linear, acute, curved, rather shorter than the sepals; lip 3-partite; lobes elongate, narrowly linear, the lateral pair rather shorter and acuminate at the apex; spur elongate, slightly thickened above, subobtuse, about 1 in. long; column nearly ¼ in. long; rostellum acuminate, its side lobes 2½ lin. long; stigmas clavate, 5 lin. long; capsule fusiform-oblong, 1¼ in. long. *Habenaria densiflora, Reichb. f. in Walp. Ann.* i. 797; *Bolus in Journ. Linn. Soc.* xxv. 191; *Durand & Schinz, Conspect. Fl. Afr.* v. 76; *Kränzl. in Engl. Jahrb.* xvi. 68, *and Orch. Gen. et Sp.* i. 200.

COAST REGION: Stockenstrom Div.; Kat River, *Balfour, in Herb. Ecklon & Zeyher*!

There is a single leaf and flower of the type in Dr. Lindley's Herbarium.

7. **B. antennifera** (Rolfe in Gard. Chron. 1905, xxxviii. 450); plant about 3 ft. high, with stout somewhat glaucous leafy stem; leaves sessile, amplexicaul, oblong, subacute, spreading, 3–6 in. long, decreasing upwards into the bracts; racemes about 9 in. long, lax, many-flowered; bracts lanceolate, acuminate, convolute round the pedicels at the base, 1–2 in. long; pedicels 2½ in. long; flowers large, green and white; dorsal sepal cucullate, elliptic, acuminate, ¾ to nearly 1 in. long; lateral sepals obliquely oblong, acuminate, recurved at the apex, ¾–1 in. long, with an acute tooth on the inner margin; petals 2-partite; upper lobe linear, acute, as long as the dorsal sepal; lower lobe filiform, curved, 1¾ in. long; lip 3-partite; side lobes filiform, 2 in. or more long; front lobe linear, recurved, sharply bent about the middle, 1¼ in. long; spur 1¾ in. long, cylindrical, somewhat clavate near the apex; column ⅓ in. long, apiculate; rostellum cucullate; front lobe acuminate, ⅓ in. long; side lobes rather shorter than the front lobe; stigmas clavate, 10 lin. long, united to the combined base of the lip, petals and lateral sepals. *Orch. Rev.* 1906, 38.

KALAHARI REGION: Transvaal; Potchefstroom, *McLea in Herb. Bolus*, 3028! Koodoo's Poort, Pretoria, *Rech*, 38! 58! Groene Kloof, *Burtt-Davy*, 2430! near Potgieters Rust, *Burtt-Davy*, 5631!

The type is from Rhodesia, Tropical Africa, but the above specimens appear to be identical.

XXVI. CYNORCHIS, Thouars.

Sepals subequal; dorsal erect, concave; lateral oblique and spreading. *Petals* entire, narrower than the dorsal sepal, and adpressed to its margin, forming a hood. *Lip* inferior, continuous with the column, spreading, usually broad, 3–5-lobed, produced at the base into a short or elongated spur. *Column* very short and broad; anther-bed erect or somewhat reclinate. *Anther-cells* diverging; apex inferior, more or less elongated and adnate to the side lobes of the rostellum, forming slender channels; pollinia granular, extended into slender caudicles, terminating in small exserted naked glands. *Stigma* bilobed, extended into a pair of oblong or clavate papillose processes; rostellum trilobed; middle lobe concave or complicate, often large; side lobes elongated, forming channels for the caudicles of the pollinia. *Capsules* erect, oblong or fusiform, sometimes shortly beaked.

Terrestrial herbs, with fasciculate fleshy roots, sometimes thickened into tubers; leaves radical, few or solitary, oblong or elongated; flowers small or medium-sized, pedicellate, arranged in lax, short or elongated racemes; bracts small or narrow.

DISTRIB. Species about 40, the majority natives of the Mascarene Islands, with a few Tropical African; 3 others in China and Japan, and one representative in extra-tropical South Africa.

1. **C. compacta** (Rolfe in Bot. Mag. t. 8053); plant $\frac{1}{4}$–$\frac{1}{2}$ ft. high, with ovoid or fusiform tubers, $\frac{1}{2}$–1 in. long; leaf solitary, suberect, ovate-oblong, shortly acuminate, 1$\frac{1}{2}$–3 in. long, with 1 or 2 short imbricate basal sheaths; scape erect, 4–7 in. high, glabrous; raceme many-flowered, compact or sometimes lax, 1–3 in. long; bracts lanceolate or ovate-lanceolate, acuminate, 2–6 lin. long; pedicels 4–9 lin. long; flowers small, white, with a few red-purple dots on the disc; dorsal sepal erect, ovate, obtuse, 2 lin. long; lateral sepals somewhat spreading, oblique, 2$\frac{1}{4}$ lin. long; petals ovate-oblong, obtuse, slightly oblique, incurved beneath the dorsal sepal, 2 lin. long; lip spreading, 4–5 lin. long, lobes obovate-oblong, somewhat crenulate, disc with a thin green line at the base; spur oblong or subclavate, somewhat curved, 1$\frac{1}{2}$ lin. long; column very short, obtuse; anther with parallel cells; rostellum minutely 3-lobed; stigmas very short, included within the cavity of the rostellum. *Cynosorchis compacta, Reichb. f. in Flora*, 1888, 149; *Kränzl. Orch. Gen. et Sp.* i. 495.

EASTERN REGION: Natal; rocks near Emberton, 2100 ft., *Wood, 5765*! and without precise locality, *Sanderson*!

XXVII. SATYRIDIUM, Lindl.

Sepals free, subequal, spreading or reflexed. *Petals* free, rather narrower than the lateral sepals, reflexed. *Lip* superior, sessile at the base of the column, erect, cucullate, acuminate or rostrate, produced behind into a pair of short oblong descending sacs. *Column* erect under the lip, rather short, divided at the apex into two lobes, the upper subcapitate, bearing the elliptic or umbonate stigma on its anterior surface, the lower anticous and forming the short ridge-like rostellum. *Anther* hanging vertically and nearly free in front of the column ; cells parallel ; pollinia granular, each with a slender stipes, united to a single orbicular gland. *Ovary* and *pedicel* not twisted.

An erect glabrous herb, with a few thickened roots ; leaves few and subbasal, somewhat spreading, with a few sheaths on the lower part of the stem, reduced upwards into the bracts ; spikes somewhat dense, many-flowered ; flowers rather small ; bracts membranous, medium-sized, reflexed after flowering.

DISTRIB. A single species, confined to the south-western corner of Cape Colony.

1. **S. rostratum** (Lindl. Gen. & Sp. Orch. 345) ; plant $\frac{3}{4}$–1$\frac{3}{4}$ ft. high, with rather slender stem ; leaves 2–4, basal or nearly so, suberect or somewhat spreading, lanceolate or oblong, subacute, 1–2$\frac{1}{2}$ in. long, submembranous ; scapes $\frac{3}{4}$–1$\frac{3}{4}$ ft. high, with numerous lanceolate or spathaceous sheaths ; spikes 1–6 in. long, somewhat dense, many-flowered ; bracts ovate-lanceolate, subacute, $\frac{1}{3}$–$\frac{1}{2}$ in. long, reflexed when the flower expands ; pedicels 4–5 lin. long, much compressed ; flowers medium-sized, lilac, spotted with purple inside the lip and spurs, and the anther carmine ; odd sepal oblong, obtuse, about $\frac{1}{4}$ in. long ; lateral sepals rather longer than the odd one and somewhat falcate and spreading ; petals elliptic-oblong, rather larger than the odd sepal, sometimes minutely toothed at the apex ; lip scarcely galeate, ovate at the base, produced into an oblong obtuse somewhat incurved beak above, about 2$\frac{1}{2}$ lin. long ; spurs oblong, very stout, obtuse, about $\frac{1}{4}$ in. long ; column straight, 2 lin. long ; stigma elliptic, obtuse, $\frac{1}{2}$ lin. long ; rostellum very minute. *Drège, Zwei Pfl. Documente,* 82 ; *Krauss in Flora,* 1845, 307, *and Beitr. Fl. Cap- und Natal.* 159 ; *Harv. Thes. Cap.* i. 55, *t.* 87. *Satyrium rhynchanthum, Bolus in Trans. S. Afr. Phil. Soc.* v. 133, *t.* 25, *Orch. Cap. Penins.* 133, *t.* 25, *and in Journ. Linn. Soc.* xxv. 195 ; *Durand & Schinz, Conspect. Fl. Afr.* v. 98 ; *Kränzl. Orch. Gen. et Sp.* i. 657 ; *Schlechter in Engl. Jahrb.* xxxi. 193.

SOUTH AFRICA : without locality, *Villett* !

COAST REGION : Worcester Div. ; Dutoits Kloof, 3000–4000 ft., *Drège,* 1245 ! Hex River, *Ecklon & Zeyher.* Paarl Div. ; French Hoek Pass, in a wet spongy bog, *Harvey* ! *Schlechter,* 9301 ! Cape Div. ; Stein Berg, near Muizenberg, 1100 ft., *Bodkin in MacOwan & Bolus, Herb. Norm. Austr.-Afr.,* 331 ! *Bolus,* 4946, 4999 ; Table Mountain, 3000 ft., *Harvey* ! *Prior* ! Waai Vley, *Wolley-Dod,* 2121 ! marshes on Constantia Berg, 1500 ft., *Schlechter,* 207 ; near Cape Town, *Wilms,* 646. Caledon Div. ; near Villiersdorp, 1300 ft., *Bolus,* 5277 !

XXVIII. SATYRIUM, Sw.

Sepals subequal or the lateral broader, spreading or reflexed. *Petals* more or less united to the sepals at the base, similar to or narrower than the odd or front sepal. *Lip* superior, sessile at the base of the column, erect, galeate or cucullate, base more or less united to the lateral sepals, produced behind into a pair of descending spurs or sacs, rarely without a sac. *Column* erect under the lip, short or somewhat elongated, divided at the apex into two lobes, the upper convex or rarely concave, bearing the pulvinate stigma on its anterior surface, the lower anticous and forming the rostellum. *Anther* hanging under the rostellum or horizontal; cells nearly parallel; pollinia granular, each with a slender stipes and distinct gland. *Ovary* and *pedicel* not twisted. *Bracts* erect or often reflexed.

Terrestrial herbs with ovoid or globose tubers; leaves few and basal, sometimes prostrate on the ground, or more numerous and cauline, decreasing upwards into the bracts; spikes usually dense or many-flowered; flowers small or medium-sized; bracts membranous, sometimes large, often reflexed after flowering.

DISTRIB. Species about 100, most numerous in extra-tropical South Africa, but copiously represented in Tropical Africa, with a few representatives in the Mascarene Islands and two or three others in India and South China.

*Lips with 2 short broad sacs at the base :
  Bracts usually broad, not elongated, green; plants
    $\frac{1}{2}$-$\frac{3}{4}$ ft. high :
    Lip broadly concave, slightly arched at the apex;
      flowers few; bracts not deflexed .. ... (1) **striatum.**

    Lip more or less distinctly galeate; flowers usually
      numerous; bracts more or less deflexed :
    Lip 1–1$\frac{1}{2}$ lin. long; flowers not striped :
      Apex of lip not deflexed :
        Flowers small; sepals under 1 lin. long ... (2) **debile.**

        Flowers larger; sepals 1$\frac{1}{2}$ lin. or more long :
          Spurs broadly saccate, under $\frac{1}{2}$ lin. long ... (3) **lindleyanum.**

          Spurs saccate-oblong, about 1 lin. long ... (4) **retusum.**

      Apex of lip strongly deflexed in front ... ... (5) **bicallosum.**

    Lip 2 lin. or more long; flowers distinctly striped :
    Bracts ample, broad, strongly acuminate, $\frac{1}{2}$-$\frac{3}{4}$ in.
      long :
      Lower leaves much larger than the upper ... (6) **cordifolium.**

      Leaves regularly decreasing upwards into the
        bracts ... ... (7) **Fanniniæ.**

    Bracts smaller, acute, 3–6 lin. long :
      Plant suberect or decumbent; spike rather
        lax and few-flowered ... ... (8) **saxicolum.**

      Plant erect; spikes dense and many-flowered :
        Nerves of lip distinctly papillate :
          Lip as broad as or broader than long ... (9) **lineatum.**

        Lip longer than broad :
          Lower leaves ovate, half as broad as
            long ... ... (10) **bracteatum.**

          Lower leaves oblong or ovate-oblong,
            not half as broad as long ... ... (11) **Dregei.**

Nerves of lip not distinctly papillate :
   Leaves lax ... ... (12) **Bowiei.**

   Leaves somewhat crowded ... ... (13) **pictum.**

Bracts usually lanceolate or elongated, whitish, often
   much longer than the flowers ; plant 1–2 ft. high (14) **Atherstonei.**

****Lip with two slender often elongated spurs at the base,
   or the spurs rarely subobsolete :**
†Radical leaves more or less ascending or not adpressed
   to the ground :
   Spurs subobsolete... ... ... (15) **microrrhyn-
                                   chum.**

   Spurs distinct, more or less elongated :
      Lip subglobose ; flowers green :
         Lip 2 or rarely 3 lin. long :
            Leaves present at flowering time :
               Spurs 2–3 lin. long ... ... (16) **parviflorum.**

               Spurs 7–9 lin. long ... ... (17) **wilmsianum.**

            Leaves not present at flowering time ... ... (18) **aphyllum.**

         Lip 4 lin. or more long ... ... (19) **odorum.**

      Lip ovate or ovate-oblong, white, yellow or rose-red :
        Leaves on short shoots, separate from and at the
         base of the flowering stems :
          Spurs scarcely ¾ in. long :
            Flowers small, in long cylindrical spikes ... (20) **neglectum.**

            Flowers medium-sized, in short or oblong
              spikes ... ... (21) **Woodii.**

          Spurs 1–1¼ in. long ... ... (22) **longicauda.**

        Leaves at the base of the flowering stems :
         Bracts persistently erect :
          Spurs 8–10 lin. long ; sepals and petals linear-
           oblong ... ... (23) **stenopetalum.**

          Spurs 5–6 lin. long ; sepals and petals oblong (24) **marginatum.**

         Bracts spreading or deflexed after flowering :
          Flowers yellow, ochreous or brown :
           Lip about ½ in. long ; flowers yellow or
            flame-colour ... ... (25) **coriifolium.**

           Lip about ¼ in. long ; flowers ochreous-
            yellow or brown :
              Spurs about ½ in. long ... ... (26) **lupulinum.**

              Spurs about 1 in. long ... ... (27) **foliosum**

          Flowers white, rose or spotted :
           Spurs ½ in. or less long :
            Lip 2 lin. or less long ... ... (28) **rupestre.**

           Lip 3–4 lin. long :
            Scapes very stout ; spurs of lip about
             twice as long as the limb... ... (29) **Hallackii.**

           Scapes moderately stout ; spurs of lip
            about as long as the limb :
             Stigma narrowly oblong ; spikes
              rather lax ... ... (30) **ligulatum.**

             Stigma quadrate ; spikes dense ... (31) **cristatum.**

          Lip 5–7 lin. long ... ... (32) **sphærocarpum.**

Spurs ¾ in. or more long :
  Leaves ⅓–1 ft. long, 2½–5 in. broad ... (33) **macrophyllum**.

  Leaves 3–6 in. long, 1–2½ in. broad ... (34) **ocellatum**.

††Radical leaves broad and adpressed to the ground :
  Spurs exceedingly short or subobsolete :
    Plant 1–1½ ft. high ; lip ½ in. or more long ... (35) **muticum**.

    Plant ⅓–½ ft. high ; lip ⅓ in. or less long ... ... (36) **paludicola**.

Spurs more or less elongated and cylindrical :
  ‡Flowers green or yellowish ; limb of the lip under
    ⅓ in. long :
    Spurs ⅛–⅓ in. long, not or scarcely exceeding
      the ovary :
      Lip without a distinct apical lobe :
        Leaves ovate or ovate-oblong :
          Lip ovate-oblong, with a very broad oblong
            apex ... ... (37) **pygmæum**.

          Lip broadly ovate, apex not half as broad
            as the limb ... ... (38) **Schlechteri**.

          Leaves broadly ovate-cordate ... ... (39) **pentherianum**.

        Lip with a reflexed acuminate apex ... ... (40) **emarcidum**.

    Spurs ⅓–¾ in. long or about twice as long as the
      ovary :
      Stem-leaves large, tubular, subobtuse ... (41) **bicorne**.

      Stem-leaves small, ovate, acute :
        Leaves small, seldom 1¼ in. broad :
          Stigma broadly oblong ; limb of lip very
            broad and obtuse ... ... (42) **outeniquense**.

          Stigma narrowly oblong ; limb of lip
            narrow and subacute ... ... (43) **humile**.

        Leaves ample, usually 2 in. or more broad (44) **ochroleucum**.

  ‡‡Flowers white, rose or spotted ; limb of lip usually
    ½ lin. or more long :
    Flowers white, unspotted :
      Spurs over ½ in. long :
        Stigmas oblong ... ... (45) **acuminatum**.

        Stigma broader than long ... ... (46) **candidum**.

      Spurs about ¼ in. long ... ... (47) **Guthriei**.

    Flowers white or rose, with purple spots :
      Spurs 1–1¼ in. long ... ... (48) **maculatum**.

      Spurs under ½ in. long... ... ... (49) **erectum**.

    Flowers uniformly rose-coloured, unspotted :
      Spurs over twice as long as the limb... ... (50) **membranaceum**.

      Spurs not much exceeding the limb :
        Stigma broadly obovate-oblong, as broad as
          long ... ... (51) **princeps**.

        Stigma linear-oblong, much longer than
          broad ... ... (52) **carneum**.

1. S. striatum (Thunb. Prodr. 6) ; plant ⅓–½ ft. high ; stem
slender ; leaves cauline, 3–4, ovate or ovate-orbicular, subobtuse,
⅓–½ in. long, somewhat fleshy ; scapes ⅓–½ ft. high ; spikes oblong,
½–1½ in. long, dense, usually few-flowered ; bracts broadly ovate or

suborbicular-ovate, acute or apiculate, $\frac{1}{3}$–$\frac{1}{2}$ in. long; pedicels about $\frac{1}{4}$ in. long; flowers small, yellow; odd sepal broadly ovate, obtuse, reflexed at the sides, nearly 2 lin. long; lateral sepals broadly elliptic-oblong, obtuse, 3 lin. long, spreading; petals oblong, obtuse, 1$\frac{1}{2}$ lin. long; lip broadly oblong, not galeate, 3 lin. long, truncate or emarginate, with 3 to 5 prominent nerves; spurs broadly saccate, $\frac{1}{2}$ lin. long; column curved, 2 lin. long; stigma oblong, shortly bilobed, 1 lin. long; rostellum triangular, apiculate, about as long as the stigma. *Thunb. Fl. Cap. ed. Schult.* 19; *Lindl. Gen. & Sp. Orch.* 345; *Krauss in Flora,* 1845, 307, *and Beitr. Fl. Cap- und Natal.* 159; *Bolus in Trans. S. Afr. Phil. Soc.* v. 132, *t.* 33, *and in Journ. Linn. Soc.* xx. 475; xxv. 195; *Durand & Schinz, Conspect. Fl. Afr.* v. 99; *Kränzl. Orch. Gen. et Sp.* i. 718; *Schlechter in Engl. Jahrb.* xxxi. 185.

COAST REGION: Piquetberg Div.; Piquet Berg, *Thunberg*! Malmesbury Div.; near Hopefield, 150 ft., *Schlechter*, 5307. Cape Div.; sandy places on Steen Berg, 1200 ft., *Bolus*, 4946! and in *MacOwan & Bolus, Herb. Norm. Austr.-Afr.*, 3171! *Bodkin*. Stellenbosch Div.; Vlagge Berg, near Stellenbosch, *Miss de Waal in Herb. Bolus*, 6090!

2. **S. debile** (Bolus in Journ. Linn. Soc. xxii. 67); plant 4–8 in. high; stem slender; leaves 2, spreading or suberect, ovate or ovate-oblong, apiculate or subacute, 1–2 in. long, 4–7 lin. broad, somewhat fleshy; scapes 4–8 in. high, with about 3 narrow spathaceous sheaths; the lower sometimes leaf-like; spikes 1–2 in. long, narrow, somewhat dense, many-flowered; bracts ovate, acute, 1$\frac{1}{2}$–2 lin. long; pedicels about 1$\frac{1}{2}$ lin. long; flowers very small, green tinged with red; odd sepal oblong or linear-oblong, subacute, $\frac{3}{4}$ lin. long; lateral sepals oblong, subobtuse, 1 lin. long, spreading; petals ovate, subacute, $\frac{3}{4}$ lin. long; lip somewhat galeate, broadly ovate-suborbicular, 1 lin. long, obtuse or truncate, broader than long; spurs minutely saccate or subobsolete; column slightly curved, $\frac{3}{4}$ lin. long; stigma suborbicular, obtuse or emarginate, $\frac{1}{4}$ lin. long; rostellum minute, triangular; capsule subglobose, about 2 lin. long. *Bolus, Ic. Orch. Austr.-Afr.* i. *t.* 24, *and in Journ. Linn. Soc.* xxv. 195; *Durand & Schinz, Conspect. Fl. Afr.* v. 95; *Kränzl. Orch. Gen. et Sp.* i. 667; *Schlechter in Engl. Jahrb.* xxxi. 186.

COAST REGION: Tulbagh Div.; slopes of Great Winterhoek Mountain, at Klein Poort, 3000 ft., *Bodkin in Herb. Bolus*, 5907! and in *MacOwan & Bolus, Herb. Norm. Austr.-Afr.* 318!

3. **S. lindleyanum** (Bolus in Journ. Linn. Soc. xx. 474); plant 3–10 in. high; stem rather stout; leaves several, cauline, ovate or ovate-oblong, subobtuse or acute, $\frac{3}{4}$–2$\frac{1}{4}$ in. long, decreasing upwards into the bracts; scapes 3–10 in. high; spikes 1–4 in. long, dense, many-flowered; bracts ovate or ovate-lanceolate, acute or acuminate, 3–7 lin. long; pedicels 1$\frac{1}{2}$ lin. long; flowers small, yellowish-white, sacs and apex of lip tinted with red-brown; odd sepal ovate-oblong, subobtuse, 2 lin. long; lateral sepals oblong,

obtuse, spreading, about 2½ lin. long ; petals ovate-oblong, subacute, rather smaller than the odd sepal ; lip galeate, 1½ lin. long, broader than long, with a narrow inflexed margin ; spurs saccate, ½ lin. long ; column 1¼ lin. long ; stigma obovate, pulvinate, about ½ lin. long ; rostellum deflexed, obtuse, very short, with 2 tubercles at the base ; capsule subglobose, 1½ lin. long. *Bolus in Trans. S. Afr. Phil. Soc.* v. 130, *t.* 30, *and in Journ. Linn. Soc.* xxv. 195 ; *Durand & Schinz, Conspect. Fl. Afr.* v. 96 ; *Kränzl. Orch. Gen. et Sp.* i. 668 ; *Schlechter in Engl. Jahrb.* xxxi. 188. *S. bracteatum, Lindl. Gen. & Sp. Orch.* 342, *not of Thunb.* ; *Drège, Zwei Pfl. Documente,* 82 ; *Kränzl. Orch. Gen. et Sp.* i. 666.

COAST REGION : Worcester Div. ; Dutoits Kloof, 3000–4000 ft., *Drège,* 1251d ! *Zeyher* ! Cape Div. ; near Simons Bay, *Wright,* 134 ! on Table Mountain, *Burchell,* 650 ! Klaver Vley, near Simons Town, 800 ft., *Bolus,* 4828 ! 7024 ! and in *MacOwan & Bolus, Herb. Norm. Austr.-Afr.* 404 ! rocks over Waai Vley, *Wolley-Dod,* 2184 !

4. S. retusum (Lindl. Gen. & Sp. Orch. 343) ; plant 3–7 in. high ; stem rather stout ; leaves several, the two lower ovate, subobtuse, 1–1¼ in. long, ¾–1 in. broad, the others decreasing upwards into the bracts ; scapes 3–7 in. long ; spikes 1–4 in. long, dense, many-flowered ; bracts ovate, acute, ¼–½ in. long ; pedicels 1½ lin. long ; flowers small, white, smelling of old cheese (*Bolus*) ; odd sepal oblong or linear-oblong, obtuse, 1½ lin. long ; lateral sepals oblong or ovate-oblong, obtuse, spreading, longer than the odd sepal ; petals linear, obtuse, about as long as the odd sepal ; lip galeate, broadly ovoid, obtuse, 1½ lin. long, with a broad inflexed apex and margin ; spurs broadly saccate-oblong, about 1 lin. long ; column 1 lin. long ; stigma semiorbicular, very obtuse, scarcely ½ lin. long ; rostellum cuneate, 3-lobed at the apex, very short ; capsule broadly oblong, 3 lin. long. *Drège, Zwei Pfl. Documente,* 124 ; *Bolus in Journ. Linn. Soc.* xxv. 195 ; *Durand & Schinz, Conspect. Fl. Afr.* v. 98 ; *Kränzl. Orch. Gen. et Sp.* i. 666 ; *Schlechter in Engl. Jahrb.* xxxi. 188.

COAST REGION : Cape Div. ; Table Mountain, 3000 ft., *Schlechter.* Swellendam Div. ; Zuurbraak, 800 ft., *Galpin,* 4593a ! 2000ft., *Schlechter,* 5669 ! Langeberg Range, near Swellendam, *Schlechter* ! *Mund.* Riversdale Div. ; Langeberg Range, at 1000 ft., *Schlechter* ! Knysna Div. ; near Knysna, *Mund* ! *Forcade, Newdigate* ; Ruigte Valley, under 500 ft., *Drège,* 3578 ! Port Elizabeth Div. ; sandy places at Algoa Bay, *Hallack* !

5. S. bicallosum (Thunb. Prodr. 6) ; plant 3–10 in. high ; stem rather stout ; leaves several, cauline, ovate or ovate-oblong, sub-obtuse or acute, ¾–1½ in. long, decreasing upwards into the bracts ; scapes 3–10 in. high ; spikes 1–4 in. long, dense, many-flowered ; bracts ovate or ovate-lanceolate, acute or acuminate, ¼–½ in. long ; pedicels 1½ lin. long ; flowers small, white or yellowish-white ; odd sepal ovate-oblong, obtuse, about 1 lin. long ; lateral sepals obliquely ovate-oblong, obtuse, about 1½ lin. long, much broader than the odd sepal ; petals ovate-oblong, obtuse, about 1 lin. long ; lip galeate,

about 1¼ in. long, apex broadly oblong, obtuse and deflexed in front,
leaving two lateral openings to the flower ; spurs broadly saccate,
½ lin. long ; column about ⅓ lin. long ; stigma minute, semiorbicular ;
rostellum triangular, very short, with 2 large tubercles at the base ;
capsule subglobose, ½ lin. long. *Sw. in Vet. Acad. Handl. Stockh.*
1800, 216 ; *Lindl. Gen. & Sp. Orch.* 343 ; *Drège, Zwei Pfl. Documente,* 83, 100, 105, 114, 115 ; *Bolus in Trans. S. Afr. Phil. Soc.*
v. 128, *t.* 31, *fig.* 3, *and in Journ. Linn. Soc.* xxv. 195 ; *Durand &
Schinz, Conspect. Fl. Afr.* v. 93 ; *Kränzl. Orch. Gen. et Sp.* i. 669 ;
Schlechter in Engl. Jahrb. xxxi. 187.

VAR. β, **ocellatum** (Bolus in Trans. S. Afr. Phil. Soc. v. 129, t. 31, excl. fig. 3) ;
upper bracts shorter than the flowers ; deflexed point of the galea very long,
reaching nearly to the base and forming two circular entrances to the flower.
Bolus, Orch. Cap. Penins. 129, *t.* 31, *excl. fig.* 3.

SOUTH AFRICA : without locality, *R. Brown, Harvey,* 241 ! *Rogers* !
COAST REGION : Tulbagh Div. ; Tulbagh, *Pappe* ! near Tulbagh Waterfall,
1000–3000 ft., *Ecklon & Zeyher, Thom,* 1180 ! Worcester Div. ; mountains near
De Liefde, 2000–3000 ft., *Drège,* 1251e ! Paarl Div. ; Klein Drakenstein
Mountains, under 1000 ft., *Drège,* 1251a ! Drakenstein Mountains, 3000–4000 ft.,
Drège. Cape Div. ; Paarde Berg, *Thunberg* ! Simons Bay, *Wright,* 148 ! Devils
Mountain and Table Mountain, 1000–2000 ft., *Bolus,* 4554 ! and in *MacOwan
& Bolus, Herb. Norm. Austr.-Afr.* 335 ! *Zeyher* ! *Wilson* ; Wynberg, *Schlechter* !
Ecklon & Zeyher, Kässner ; by Slangkop River, *Wolley-Dod,* 1543 ! Constantia
Berg, near Houts Bay, *Schlechter.* Caledon Div. ; Bavians Kloof, near Genadendal,
1000–2000 ft., *Drège* ; stony hill near Houw Hoek, 1000 ft., *Schlechter,* 5442 ;
Caledon, *Prior* ! Swellendam Div. ; Zuurbraak, 800 ft., *Galpin,* 4597 ! George
Div. ; Oakford, near George, *Rehmann,* 583.

6. S. cordifolium (Lindl. Gen. & Sp. Orch. 344) ; plant 4–6 in.
high ; stem stout ; leaves 2, sessile, basal, broadly ovate or cordate-
orbicular, obtuse, 2–2½ in. long, 1½–1¾ in. broad, sometimes with
1 or 2 smaller cauline leaves ; scapes 4–6 in. long ; spikes 1¼–3 in.
long, dense, many-flowered ; bracts ovate, very acuminate, ½–¾ in.
long ; pedicels about ¼ in. long ; flowers medium-sized ; sepals and
petals united to about the middle ; odd sepal oblong, obtuse, about
2½ lin. long ; lateral sepals falcate-oblong, obtuse, much larger than
the odd sepal, not spreading ; petals oblong, subobtuse, as long as
the odd sepal ; lip cucullate, broadly ovate, ½ in. long, with an
obtuse slightly reflexed apex ; spurs broadly saccate, 1 lin. long ;
column curved, 4½ lin. long ; stigma ovate-oblong, 1½ lin. long,
acutely 2-dentate at the apex ; rostellum 3-lobed from a cuneate
base, about half as long as the stigma. *Drège, Zwei Pfl. Documente,*
45 ; *Bolus in Journ. Linn. Soc.* xxv. 195 ; *Durand & Schinz, Conspect.
Fl. Afr.* v. 94 ; *Kränzl. Orch. Gen. et Sp.* i. 673 ; *Schlechter in Engl.
Jahrb.* xxxi. 189.

COAST REGION : Stockenstrom Div. : Kat Berg, above the forests, 4000–5000 ft.,
Drège, 8286 ! *Scully in Herb. Bolus,* 5912 !
EASTERN REGION : Transkei ; mountains near Bazeia, 3500 ft., *Baur,* 551 !

7. S. Fanniniæ (Rolfe) ; plant 2–4 in. high ; stem stout ; leaves
2–4, spreading, sessile, ovate, subacute, 1–1¼ in. long, somewhat

fleshy, gradually passing into the bracts, the upper somewhat reflexed at the apex; scapes 2–4 in. high; spikes $\frac{3}{4}$–1$\frac{1}{2}$ in. long, dense, few- to many-flowered; bracts ovate, very acuminate, $\frac{1}{2}$–$\frac{3}{4}$ in. long; pedicels about $\frac{1}{4}$ in. long; flowers medium-sized, purple in the dried state; odd sepal oblong, obtuse, about 2 lin. long; lateral sepals falcate-oblong, obtuse, much larger than the odd sepal, not spreading; petals oblong, obtuse, about 2 lin. long; lip galeate, very broadly ovate, about 5 lin. long, with an obtuse slightly reflexed apex; spurs very broadly saccate, 1$\frac{1}{2}$ lin. long; column curved, 4$\frac{1}{2}$ lin. long; stigma oblong, minutely 2-dentate at the apex, 1$\frac{1}{2}$ lin. long; rostellum 3-lobed, half as long as the stigma.

EASTERN REGION · Natal; Darglo Farm, *Mrs. Funnin*!

8. **S. saxicolum.** (Bolus in Journ. Linn. Soc. xx. 474); plant 2–6 in. high; stems moderately stout, suberect or somewhat decumbent; leaves 2–4, ovate to elliptic-oblong, subobtuse or apiculate, $\frac{3}{4}$–2$\frac{1}{2}$ in. long, $\frac{1}{2}$–1$\frac{1}{4}$ in. broad, submembranous, the lower spreading, upper much smaller; scapes 2–6 in. high; spikes $\frac{1}{2}$–1$\frac{1}{2}$ in. long, usually lax and few-flowered; bracts broadly ovate, acute or shortly acuminate, $\frac{1}{3}$–$\frac{3}{4}$ in. long, the lower somewhat reflexed; pedicels about 2 lin. long, with broad denticulate ridges; flowers rather small, brownish-yellow lined with red; sepals and petals united to nearly the middle, each with a strong central nerve; odd sepal oblong, obtuse, about 1$\frac{1}{2}$ lin. long; lateral sepals falcate-oblong, obtuse, rather longer than the odd sepal; petals ovate, obtuse, about 1$\frac{1}{2}$ lin. long; lip cucullate, 3–3$\frac{1}{2}$ lin. long, with about 9 strong nerves, apex narrowed to a broad reflexed point, margin and nerves outside slightly scaberulous; spurs broadly saccate, very short; column curved, 2 lin. long; stigma ovate-oblong, subobtuse, about $\frac{1}{2}$ lin. long, broader than long; rostellum broad, 3-lobed at the apex, nearly as long as the stigma. *Bolus in Trans. S. Afr. Phil. Soc.* v. 131, *t. 4, and Journ. Linn. Soc.* xxv. 195; *Kränzl. Orch. Gen. et Sp.* i. 673. *S. bracteatum, var. saxicola, Schlechter in Engl. Jahrb.* xxxi. 191.

COAST REGION: Cape Div.; Table Mountain, 1000–3500 ft., *Harvey*, 117! wet slopes on Constantia Berg, *Bodkin, Schlechter*, 1478; shady rocks on Devils Mountain, near the Cataract, 1200 ft., *Bolus*, 3855! *Schlechter*; Muizenberg Mountains, 900 ft., *Bolus*, 3855! and in *MacOwan & Bolus, Herb. Norm. Austr.-Afr.* 156! near Simons Bay, *Wright*, 136! Caledon Div.; Klein Houw Hoek, *Zeyher*! *Schlechter*. Riversdale Div.; Garcias Pass, 900 ft., *Galpin*, 4592!

Near *S. bracteatum*, Thunb., in floral character, but markedly different in habit. Bolus remarks that it is found in the clefts of steep shaded rocks, while *S. bracteatum* affects open sunny flats and heathy mountain sides. Schlechter reduces it to the rank of a variety of *S. bracteatum.*

9. **S. lineatum** (Lindl. Gen. & Sp. Orch. 343, partly, and excl. var. β); plant 4–6 in. high; stem moderately stout; leaves several, the 2 lower ovate or elliptic-ovate, subobtuse, spreading, 1–1$\frac{3}{4}$ in. long, the upper narrowly ovate, suberect, subacute, decreasing upwards into the bracts; scapes 4–6 in. high; spikes 1–2$\frac{1}{2}$ in.

long, somewhat lax, few- to many-flowered ; bracts ovate, acute,
3–5 lin. long, reflexed after flowering ; pedicels 1½ lin. long ; flowers
small ; sepals and petals united to the middle, each with a
slender nerve ; odd sepal ovate-oblong, obtuse, 1 lin. long ; lateral
sepals falcate-oblong, obtuse, longer than the odd sepal ; petals oblong,
obtuse, 1 lin. long ; lip galeate, broadly ovate, 2 lin. long, with
7 somewhat prominent nerves, apex and margin incurved, papillate,
nerves papillate ; spurs broadly saccate, very short ; column curved,
1½ lin. long ; stigma obovate-oblong, obtuse, ½ lin. long ; rostellum
3-dentate in front, rather shorter than the stigma. *Drège, Zwei Pfl.
Documente,* 78.

COAST REGION : Tulbagh Div. ; New Kloof, under 1000 ft., *Drège,* 1259b !

10. **S. bracteatum** (Thunb. Prodr. 6, partly) ; plant 3–9 in. high ;
stems stout ; leaves cauline, 4–6, ovate or ovate-oblong, acute,
¾–2 in. long, somewhat fleshy ; scapes 3–9 in. long ; spikes broadly
oblong, 1–3 in. long, dense, many-flowered ; bracts broadly ovate,
acute, spreading or the lower somewhat reflexed, 3–6 lin. long ;
pedicels about 2 lin. long ; flowers small, ochre-yellow or brown
with red stripes ; sepals and petals united to about the middle,
each with a strong central nerve ; odd sepal oblong, obtuse, 1¼–1½
lin. long ; lateral sepals falcate-oblong, obtuse, rather longer than
the odd sepal and somewhat spreading ; petals oblong or ovate-
oblong, obtuse, 1¼–1½ lin. long ; lip cucullate, 2½–3 lin. long, with
about 9 strong nerves, apex broad and somewhat reflexed, margin
and nerves outside minutely scaberulous ; spurs broadly saccate, very
short ; column curved, 2 lin. long ; stigma oblong, obtuse, 1 lin. long ;
rostellum broad, 3-toothed in front, lateral teeth spreading ; capsule
broadly oblong, 2½ lin. long. *Thunb. Fl. Cap. ed. Schult.* 18, *partly ;
Sw. in Vet. Acad. Handl. Stockh.* 1800, 216 ; *Ker in Quart. Journ.
Sci. & Arts,* viii. 221, *t.* 3, *fig.* 1 ; *N. E. Br. in Gard. Chron.* 1885, xxiv.
331 ; *Bolus in Journ. Linn. Soc.* xxv. 195, *partly ; Durand & Schinz,*
Conspect. Fl. Afr. v. 93, *partly ; Kranzl. Orch. Gen. et Sp.* i. 666,
partly ; Schlechter in Engl. Jahrb. xxxi. 190, *partly. S. bracteatum,*
var. glandulosum, Sond. in Linnæa, xix. 89. *S. bracteatum, var.*
latebracteatum, Sond. in Linnæa, xix. 89, *excl. syn. ; Durand &*
Schinz, Conspect. Fl. Afr. v. 94. *S. bracteatum, var. lineatum, Sond.*
in Linnæa, xix. 88, *partly, excl. syn. ; Bolus in Trans. S. Afr. Phil.
Soc.* v. 130, *t.* 32, *fig.* A, *partly. S. pictum, Sond. in Linnæa,* xix.
89, *partly, not of Lindl. Ophrys bracteata, Linn. f. Suppl.* 403 ;
Murr. Syst. Veg. 814. *Diplecthrum bracteatum, Pers. Syn.* ii. 509.

VAR. β, **nanum** (Bolus in Trans. S. Afr. Phil. Soc. v. 131, t. 32, fig. B) ; plant
2 to 3 in. high, smaller in all its parts than the type ; spike proportionately
longer ; bracts more reflexed ; stigma broader than long. *Durand & Schinz,
Conspect. Fl. Afr.* v. 94, *excl. syn. ; Schlechter in Engl. Jahrb.* xxxi. 191.

SOUTH AFRICA : without locality, *Sparrman ! Bergius, Mund, Masson ! Brown !
Leibold, Wallich, Trimen ! Rogers ! Prior ! Harvey,* 120 ! *Ecklon !*
COAST REGION : Piquetberg Div. ; Piquet Berg, *Thunberg !* Malmesbury Div. ;
Riebecks Castle, *Thunberg !* Tulbagh Div. ; near Tulbagh Waterfall, *Ecklon &*

Zeyher. Cape Div. ; Claremont, below 100 ft., *Bolus,* 3932 ! *Dümmer,* 274 !
275 ! Muizenberg Mountains, near Kalk Bay, 1300 ft., *Bolus,* 4904 ! near
Wynberg, 100 ft., *Bolus in MacOwan & Bolus, Herb. Norm. Austr.-Afr.* 1094 !
Zeyher, 1562 ! *Prior* ! *Schlechter,* 1549 ! *Kässner* ; Cape Flats, *Pappe* ! between
Devils Mountain and Table Mountain, *Zeyher* ! Flats near Kenilworth Racecourse,
Wolley-Dod, 499 ! near Slangkop River, *Wolley-Dod,* 1545 ! 3211 ! Caledon Div. ;
near Houw Hoek, *Schlechter.*   Riversdale Div. ; near Riversdale, 300 ft.,
Schlechter, 2029 ! Humansdorp Div. ; Zitzikama, 200–300 ft., *Schlechter.*   Var. *β* :
Cape Div. ; near streams, in Klaver Vley, behind Simons Town, 800 ft., *Bolus,*
4820 ! marsh at source of Slang Kop River, *Wolley-Dod,* 3211 ! and without
precise locality, *Thunberg* !

11. **S. Dregei** (Rolfe); plant 6–9 in. high; stem stout; leaves
several, suberect, rather lax, the 2 or 3 lower ovate-oblong or ovate-
lanceolate, acute, 1–1½ in. long, the upper linear and much smaller,
somewhat fleshy; scapes 6–9 in. high; spikes 1½–3 in. long, dense,
many-flowered; bracts ovate, acute, 3–5 lin. long, reflexed after
flowering; pedicels 1½ lin. long; flowers small; sepals and petals
united for one-third of their length, each with a slender nerve; odd
sepal oblong, obtuse, 1 lin. long; lateral sepals falcate-oblong,
obtuse, longer than the odd sepal; petals oblong, obtuse, 1 lin.
long; lip galeate, 2¼ lin. long, with 7 prominent nerves, apex
subobtuse and reflexed, keel distinctly papillate, margin entire;
spurs saccate, short, not broader than long; column curved, 1¾ lin.
long; stigma oblong, obtuse, ¾ lin. long; rostellum 3-lobed in front,
half as long as the stigma; capsule broadly oblong, ¼ in. long.
S. lineatum, Lindl. Gen. & Sp. Orch. 344, *excl. vars.   S. lineatum,
Drège, Zwei Pfl. Documente,* 76.   *S. bracteatum, var. pictum, Schlechter
in Engl. Jahrb.* xxxi. 191, *excl. syn.*

Coast Region : Piquetberg Div. ; Piquet Berg, 2000–3000 ft., *Drège,* 1259a !

12. **S. Bowiei** (Rolfe) ; plant about 7–9 in. high ; stem moderately
stout; leaves several, suberect, rather lax, ovate, subacute, the
lower about 1¼ in. long, the rest decreasing upwards, somewhat
fleshy; scapes 7–9 in. long; spikes 1½–2½ in. long, dense, many-
flowered ; bracts ovate, acute, 3½–5 lin. long, reflexed after
flowering; pedicels 2 lin. long; flowers small; sepals and petals
united to nearly the middle ; each with a slender nerve; odd sepal
ovate-oblong, obtuse, 1¼ lin. long; lateral sepals falcate-oblong,
obtuse, longer and broader than the odd sepal; petals oblong,
obtuse, 1¼ lin. long; lip galeate, ovate-oblong, 2½ lin. long, with 7
prominent nerves, apex broad, obtuse, margin slightly scaberulous,
nerves not scaberulous; spurs broadly saccate, very short; column
curved, 1¾ lin. long; stigma oblong, obtuse, ½ lin. long; rostellum
3-dentate in front, about half as long as the stigma.

Coast Region : George Div. ; plains of George, *Bowie* !

13. **S. pictum** (Lindl. Gen. & Sp. Orch. 344) ; plant 6–8 in. high ;
stem rather stout ; leaves several, suberect, rather crowded, ovate,
subacute, the 2 or 3 lower 1–1½ in. long, the rest decreasing

upwards into the bracts, somewhat fleshy ; scapes 6–8 in. long ;
spikes cylindrical, 1–3 in. long, dense, many-flowered ; bracts
ovate, acute or acuminate, 3–5 lin. long, reflexed after flowering ;
pedicels 1½ lin. long ; flowers small ; sepals and petals united for
about a third of their length, each with a prominent nerve ; odd
sepal ovate-oblong, obtuse, 1½ lin. long ; lateral sepals falcate-oblong,
obtuse, broader than the odd sepal ; petals oblong, obtuse, 1½ lin.
long ; lip galeate, 3 lin. long, with 9 prominent nerves, apex subacute, margin slightly scaberulous, nerves not scaberulous ; spurs
broadly saccate, very short ; column curved, 1¾ lin. long ; stigma
suborbicular-oblong, obtuse, ¾ lin. long ; rostellum 3-dentate in
front, not half as long as the stigma. *Drège, Zwei Pfl. Documente,*
124.

COAST REGION : Knysna Div. ; Ruigte Vallei, under 500 ft., *Drège,* 3579 !

14. S. Atherstonei (Reichb. f. in Flora, 1881, 328) ; plant
1–2 ft. high ; stems stout ; leaves cauline, 2–4, suberect, oblong or
linear-oblong, acute or subacute, 3–6 in. long, ¾–1¼ in. broad,
somewhat fleshy ; scapes 1–2 ft. high, with two or three narrow
sheaths above ; spikes 1–4 in. long, very dense, many-flowered ;
bracts lanceolate to ovate-lanceolate, acute or acuminate, ⅓–1 in.
long ; pedicels 1–2 lin. long ; flowers very small, white or white
and yellow ; sepals and petals united to the middle ; odd sepal
oblong, obtuse, 1½–2 lin. long ; lateral sepals longer and broader
than the odd sepal, somewhat spreading ; petals oblong, obtuse,
1½–2 lin. long ; lip cucullate, ovoid-globose, 2–2½ lin. long, subobtuse, not or scarcely reflexed at the apex ; spurs rather stout,
somewhat curved, 1¼–2 lin. long ; column curved, 1½–2 lin. long ;
stigma broadly rhomboid-ovate, ¾ lin. long, broader than long ;
rostellum rhomboid-oblong, minutely 3-lobed at the apex, scarcely
half as long as the stigma. *Bolus in Journ. Linn. Soc.* xxv. 194 ;
Durand & Schinz, Conspect. Fl. Afr. v. 93 ; *Kränzl. Orch. Gen. et
Sp.* i. 660. *S. triphyllum, Kränzl. Orch. Gen. et Sp.* i. 660.
S. monopetalum, Kränzl. Orch. Gen. et Sp. i. 662. *S. trinerve,
Schlechter in Engl. Jahrb.* xxxi. 182, *partly, not of Lindl.*

COAST REGION : Albany Div. ; Howisons Poort, *Hutton* !
KALAHARI REGION : Transvaal ; near Lydenburg, *Atherstone* ! *Wilms*, 1358,
1359 ; Mac Mac, *Mudd* ! Houtbosch, *Rehmann*, 5824 ! *Schlechter* ; Belfast, *Burtt-
Davy*, 1305 ! *Doidge*, 4790 ! swamps of Nyl River, *Nelson*, 108 ! Klein Oliphants
River, 5000 ft., *Schlechter*, 4038 ; Witbank, near Middelburg, 5300 ft., *Gilfillan
in Herb. Galpin*, 7243 ! Merwe Station, *Rech*, 6293 !
EASTERN REGION : Pondoland ; summit of West Gate, Port St. John, 1200 ft.,
Galpin, 3420 ! Natal ; Clairmont Flat, *Sanderson*, 478 ! *Wood*, 1599 ! 1716 !
Newcastle, 3800 ft., *Wood*, 6657 ! Oliviers Hoek, sources of Tugela River,
5000 ft., *Allison*, 28 ! Highlands, *Gerrard*, 1562 ! Fairfield, *Rudatis*, 572 ! and
without precise locality, *Mrs. Fannin*, 4 ! *Mrs. K. Saunders* ! *Gerrard*, 2181 !

15. S. microrrhynchum (Schlechter in Engl. Jahrb. xx. Beibl.
50, 14) ; plant ¾ ft. high, with straight stem, sheathed at the base ;
leaves radical, nearly opposite, suberect, broadly ovate, shortly
acuminate, 2–2½ in. long, ½–¾ in. broad ; scapes ¾ ft. high, with 2

or 3 spathaceous sheaths ; spikes about 3 in. long, dense, many-
flowered ; bracts ovate or ovate-lanceolate, acute or acuminate,
spreading, deflexed at the apex, $\frac{1}{4}$–$\frac{1}{3}$ in. long ; pedicels about 2 lin.
long ; flowers very small ; sepals and petals somewhat pilose at the
base ; odd sepal linear-oblong, obtuse, about 2$\frac{1}{2}$ lin. long ; lateral
sepals more acute than the odd sepal, not spreading ; petals linear-
oblong, subobtuse, shorter than the odd sepal'; lip galeate, about
3 lin. long, with a short obtuse, somewhat reflexed apex ; spurs
very short or nearly obsolete ; column about 1$\frac{1}{2}$ lin. long ; stigma
rounded-oblong, short ; rostellum shortly triangular, minutely 3-
lobed at the apex. *Kränzl. Orch. Gen. et Sp.* i. 671 ; *Schlechter in
Engl. Jahrb.* xxxi. 192, *t.* 2, *fig.* A–G.

KALAHARI REGION: Orange River Colony ; in turf at the summit of Mont-aux-
Sources, 11000 ft., *Thode* !

16. **S. parviflorum** (Sw. in Vet. Acad. Handl Stockh. 1800, 216) ;
plant 1–2$\frac{1}{2}$ ft. high ; stems stout ; leaves 2–4, suberect or somewhat
spreading, broadly ovate to elliptic-oblong, obtuse or subacute,
3–8 in. long, 1–4$\frac{1}{2}$ in. broad, somewhat fleshy ; scapes 1–2$\frac{1}{2}$ ft. high,
with several spathaceous sheaths, the lower often somewhat
foliaceous ; spikes 4–8 in. long, usually dense, many-flowered ;
bracts lanceolate to ovate-lanceolate, acuminate, $\frac{1}{3}$–$\frac{3}{4}$ in. long,
usually reflexed ; pedicels about $\frac{1}{3}$ in. long ; flowers small, green or
yellowish-green ; sepals and petals united to nearly the middle ;
odd sepal spathulate-oblong, obtuse, 1$\frac{1}{2}$–2 lin. long ; lateral sepals
falcate-oblong, obtuse, longer and broader than the odd sepal ;
petals spathulate-oblong, obtuse, 1$\frac{1}{2}$–2 lin. long ; lip cucullate,
broadly ovoid or obovate-globose, 2$\frac{1}{2}$–3 lin. long, with the apex and
margin reflexed and crenulate ; spurs curved, sometimes diverging,
3–5 lin. long ; column curved, 2–2$\frac{1}{2}$ lin. long ; stigma obovate-
oblong, obtuse, 1–1$\frac{1}{4}$ lin. long ; rostellum trilobed, with broadly
rhomboid spathulate concave front lobe, rather shorter than the
stigma ; capsule oblong, about $\frac{1}{2}$ in. long. *Krauss in Flora*, 1845,
307, *and Beitr. Fl. Cap- und Natal.* 159 ; *Bolus in Journ. Linn. Soc.*
xxv. 194; *Durand & Schinz, Conspect. Fl. Afr.* v. 98 ; *Kränzl.
Orch. Gen. et Sp.* i. 686 ; *Schlechter in Engl. Jahrb.* xxxi. 175. *S.
densiflorum, Lindl. Gen. & Sp. Orch.* 340 ; *Drège, Zwei Pfl. Documente,* 127. *S. cassideum, Lindl. Gen. & Sp. Orch.* 341 ; *Drège,
Zwei Pfl. Documente,* 83, 98, 125. *S. eriostomum, Lindl. Gen. & Sp.
Orch.* 342 ; *Drège, Zwei Pfl. Documente,* 45, 143 ; *Kränzl. Orch. Gen.
et Sp.* i. 686. *S. lydenburgense, Reichb. f. in Flora,* 1881, 328.
S. tenuifolium, Kränzl. in Engl. Jahrb. xxiv. 505, *and Orch. Gen. et
Sp.* i. 684. *S. parviflorum, var. Schimperi, Schlechter in Engl. Jahrb.*
xxxi. 176, *partly. Diplecthrum parviflorum, Pers. Syn.* ii. 509.

COAST REGION : Paarl Div. ; between Paarl and Lady Grey Railway Bridge,
under 1000 ft., *Drège,* 8294a ! George Div. ; between Malgaten and Great Brak
Rivers, *Burchell,* 6134 ! Knysna Div. ; Groene Valley, under 500 ft., *Drège,*
8294d ! Port Elizabeth Div. ; near Port Elizabeth, *Hallack* ! and in *Herb. Bolus,*
5928 ! Uitenhage Div. ; Van Stadens River, under 200 ft., *Drège,* 8285 ! near
Uitenhage, *Zeyher,* 1563 ! *Tredgold* ! Bathurst Div. ; source of the Kasuga River,

Burchell, 3905! between Port Alfred and Kaffir Drift, *Burchell* 3854! Fish
River Flats, 500–1500 ft., *Atherstone*, 4! between Bathurst and the Kowie River,
Atherstone, 5! Albany Div.; Sidbury, *Burke*; Featherstone Kloof, 2000 ft.,
MacOwan, 127! *Galpin*, 301! *Schönland*, and without precise locality, *Hutton*!
Williamson! Stockenstrom Div.; Kat Berg, 2000–4000 ft., *Drège*, *Hutton*!
swampy ground on Benholm Mountains, 4000 ft., *Scully*, 373! Queenstown
Div.; summit of Long Hill, 4400 ft., *Galpin*, 1502! Stutterheim Div.; Dohne
Hill, 300 ft., *Sim*, 35. King Williamstown Div.; Perie Forest, *Sim*, 17, 34.
Komgha Div.; beyond Kei, in forests, *Mrs. Barber*, 853! near the mouth of the
Kei River, 100 ft., *Flanagan*, 262.

CENTRAL REGION: Somerset Div.; Bosch Berg, near Somerset, *Bowker*!

KALAHARI REGION: Orange River Colony; Laai Spruit, *Burke*! Harrismith,
Sankey, 266! 272! Basutoland; Mametsana, *Dieterlen*, 494! Transvaal; near
Lydenburg, *Atherstone*! near Spitzkop, *Wilms*, 1370; Houtbosch, 5700–7000 ft.,
Bolus, 10979! *Schlechter*; Olifants River, *Schlechter*, 4141! Little Lomati Valley,
3500–4500 ft., *Culver*, 13, 1890; Belfast, *Burtt-Davy*, 1310! Bamboo Mountain,
Drakensberg Range, 6000 ft., *Doidge*, 5574! Belfast, 6500 ft., *Doidge*, 4792!
Barberton, *Thorncroft*, 2780! Merwe Station, *Rech*, 6294! Witbank, Middelburg
District, 5300 ft., *Gilfillan in Herb. Galpin*, 7242!

EASTERN REGION: Transkei; Krielis Country, *Bowker*! Griqualand East;
Mount Currie, 4800–6000 ft., *Tyson*, 1069! Natal; Oliviers Hoek, sources of
Tugela River, 5000 ft., *Allison*! Pine Town, 600–1000 ft., *Sanderson*, 486!
Inanda, *Wood*, 1185! near Byrne, 4000 ft., *Wood*, 1872! Ingoma, *Gerrard*,
1537! near Umkomaas River, 4000–5000 ft., *Wood*; Dargle Farm, *Mrs. Fannin*,
76! and without precise locality, *Cooper*, 3619! *Gerrard*, 7!

A• common and variable plant. The Tropical African *S. Schimperi*, Hochst.,
and *S. zombense*, Rolfe, which are referred here by Schlechter, are quite distinct,
and *S. wilmsianum*, Kränzl., is a more slender plant, with a much longer spur,
and very different texture.

17. **S. wilmsianum** (Kränzl. in Engl. Jahrb. xxiv. 505); plant
4–6 in. high; stems slender; leaves 2, suberect, elliptic-oblong,
subacute, $1\frac{1}{2}$–2 in. long, $\frac{3}{4}$–1 in. broad, somewhat membranous;
scapes 4–6 in. long, with about 2 membranous sheaths; spikes
1–$1\frac{1}{2}$ in. long, lax, 5–7-flowered; bracts ovate, acute, about $\frac{1}{3}$ in.
long; pedicels $\frac{1}{3}$ in. long; flowers small; sepals and petals united
to the middle; odd sepal subspathulate-oblong, obtuse, $1\frac{1}{2}$ lin. long;
lateral sepals oblong, obtuse, twice as large as the odd sepal; petals
subspathulate-oblong, obtuse, rather broader than the odd sepal;
lip cucullate, broadly ovoid or obovate-globose, $2\frac{1}{2}$–3 lin. long, with
the apex and margin reflexed and crenulate; spurs slender, curved,
elongate, 7–8 lin. long; column curved, 2 lin. long; stigma obovate-
orbicular, obtuse, crenulate, 1 lin. long; rostellum broadly oblong,
obtuse, over half as long as the stigma. *Kränzl. Orch. Gen. et Sp.* i.
684.

KALAHARI REGION: Transvaal; Devils Knuckles, near Spitz Kop, Lydenburg
District, *Wilms*, 1380!

18. **S. aphyllum** (Schlechter in Engl. Jahrb. xxiv. 424); plant
$\frac{3}{4}$–2 ft. high; stems rather stout; leaves 2, produced on lateral buds
at the base of the stem, spreading, ovate-oblong, subacute, $2\frac{1}{4}$ in.
long, $1\frac{1}{4}$ in. broad, shrivelling before flowering time; scapes $\frac{3}{4}$–2 ft.
high, with several spathaceous imbricate sheaths; spikes 4–8 in.
long, somewhat lax, many-flowered; bracts lanceolate or ovate-
lanceolate, acuminate, $\frac{1}{3}$–$\frac{1}{2}$ in. long; pedicels about $\frac{1}{4}$ in. long;

flowers small, green or greenish; sepals and petals united to the
middle; odd sepal linear-oblong, obtuse, 1½ lin. long; lateral sepals
longer and broader than the odd sepal, somewhat spreading; petals
elliptic-oblong, obtuse, somewhat crenulate, shorter than the odd
sepal; lip cucullate, ovoid-globose, 2 lin. long, with a narrow subentire reflexed margin; spurs curved, somewhat stout, about 3½ lin.
long; column curved, 2 lin. long; stigma oblong, obtuse, ¾ lin. long;
rostellum broadly pandurate-oblong, obtuse, about as long as the
stigma. *Schlechter in Engl. Jahrb.* xxxi. 178, *t.* 1, *fig.* G-M.

COAST REGION: Albany Div.; Coldspring, near Grahamstown, *Glass*, 490!
Queenstown Div.; Shepstone Berg, 5500 ft., *Galpin*, 1896!
CENTRAL REGION: Craddock Div.; without precise locality, *Cooper*, 1320!
KALAHARI REGION: Transvaal; marshes near Donker Hoek, 4900 ft.,
Schlechter, 3723!
EASTERN REGION: Tembuland; Umnyolo and Mount Bazeia, 2000–3000 ft.,
Baur, 306! 535! 739! 810! Griqualand East; Zuurberg Range, 5000 ft.,
Schlechter, 6599! Natal; near Inanda, *Wood*, 660! Dargle Farm, *Mrs. Fannin*,
27! near Emberton, 1800 ft., *Schlechter*, 3234! Zululand; without precise
locality, *Gerrard*, 358!

19. **S. odorum** (Sond. in Linnæa, xix. 86); plant ½–2 ft. high;
stems stout; leaves 3–5, suberect, ovate to elliptic-oblong, subobtuse, 3–8 in. long, 1¼–3 in. broad, somewhat fleshy; scapes 1½–2
ft. high, with several spathaceous sheaths, the lower often somewhat
foliaceous; spikes 3–6 in. long, usually somewhat lax, many-
flowered; bracts ovate, acute, ½–¾ in. long; pedicels about ½ in.
long; flowers medium-sized, light green, sometimes with dull purple
tips to the segments, carnation-scented; sepals oblong, obtuse,
3–3½ lin. long, the lateral spreading; petals spathulate-oblong,
obtuse, narrower and rather shorter than the sepals; lip cucullate,
broadly elliptic-oblong, 4–4½ lin. long, with a broad and obtuse
reflexed apex; spurs slender, curved, 7–9 lin. long; column curved,
3½ lin. long; stigma oblong, obtuse or truncate, 2–2½ lin. long;
rostellum broadly oblong, with an obtuse crenulate apex, half as
long as the stigma; capsule broadly oblong, 7–8 lin. long. *Bolus
in Trans. S. Afr. Phil. Soc.* v. 125, *and in Journ. Linn. Soc.* xxv.
194; *Durand & Schinz, Conspect. Fl. Afr.* v. 97; *Kränzl. Orch. Gen.
et Sp.* i. 710; *Schlechter in Engl. Jahrb.* xxxi. 175.

SOUTH AFRICA: without locality, *Bergius, Mund, Drège*, 8291a! *Zeyher*,
1557! *Harvey*, 135! *Prior*! *Rogers*!
COAST REGION: Cape Div.; Simons Bay, *Wright*, 138! Devils Mountain,
Wolley-Dod, 393! near Rondebosch, 300 ft., *Bolus*, 4559! Cape Peninsula,
100 ft., *Bolus in MacOwan & Bolus, Herb. Norm. Austr.-Afr.* 157! *Schlechter*, 57!
Muizenberg, *Schlechter*, 1555! Wynberg, *Ecklon*! Cape Flats, *Pappe*! 68!
Ecklon!

20. **S. neglectum** (Schlechter in Engl. Jahrb. xx. Beibl. 50, 39);
plant 1¼–2½ ft. high; stems usually very stout; leaves 2, subradical,
usually on short branches at the base of the flowering stem, suberect,
oblong or elliptic-oblong, subobtuse, ½–1 ft. long, 1½–2¾ in. broad,
with 2 or 3 loose sheaths below; scapes 1¼–2½ ft. long, with

numerous imbricate spathaceous sheaths; spikes 6–11 in. long, dense, many-flowered; bracts lanceolate or ovate-lanceolate, acute, ½–1 in. long; pedicels about ½ in. long; flowers rather small, white and pink or carmine-rose; odd sepal linear-oblong, obtuse, 2½ lin. long; lateral sepals twice as broad as the odd sepal and spreading, 3½ lin. long; petals elliptic-oblong, obtuse, 2½ lin. long; lip cucullate, broadly elliptic-ovate, 3–3½ lin. long, with a broadly orbicular-oblong crenulate reflexed apex; spurs slender, curved, 7–8 lin. long; column curved, 3 lin. long; stigma obovate-orbicular, 1 lin. long; rostellum subquadrate, shortly 3-lobed at the apex, much shorter than the stigma. *Kränzl. Orch. Gen. et Sp.* i. 706; *Schlechter in Engl. Jahrb.* xxxi. 157.

KALAHARI REGION: Orange River Colony; Harrismith, *Sankey*, 271! Transvaal; Houtbosch Range, 5100 ft., *Bolus*, 10976! *Rehmann*, 5831! *Nelson*, 379! Lomati Valley, near Barberton, 3500–4500 ft., *Culver*, 58, *Galpin*, 718; summit of Mount Mpome, near Houtbosch, 6400 ft., *Schlechter*.

EASTERN REGION: Transkei; Tsomo mountain ridges, *Mrs. Barber*, 858! and without precise locality, *Mrs. Barber*, 521! Griqualand East; near Clydesdale, 3500 ft., *Tyson*, 2696! Mount Currie, 3500 ft., *Tyson*, 1887. Natal; Oliviers Hoek, sources of the Tugela River, 5000 ft., *Allison*! slopes of the Drakensberg, *Wood*, 3418! and without precise locality, *Mrs. K. Saunders*!

21. **S. Woodii** (Schlechter in Engl. Jahrb. xx. Beibl. 50, 16); plant 1¼–1¾ ft. high; stems stout; leaves 2, radical, subsessile, on short branches at the foot of the flowering stem, suberect, oblong or elliptic, obtuse or subacute, 4–7 in. long, 1¼–4 in. broad, with about 2 basal sheaths; scapes 1¼–1¾ ft. high, with numerous imbricate spathaceous sheaths; spikes 3–6 in. long, dense, many-flowered; bracts ovate or ovate-lanceolate, acute, ¾–1 in. long; pedicels about ½ in. long; flowers medium-sized, pink or orange-red; odd sepal linear-oblong, obtuse, 4–4½ lin. long; lateral sepals twice as broad as the odd sepal and spreading; petals elliptic-oblong, obtuse, 4–4½ lin. long; lip cucullate, broadly elliptic-ovate, ½ in. long, with a reflexed apiculate apex; spurs somewhat slender, curved, 7–8 lin. long; column curved, 4–5 lin. long; stigma obovate or orbicular, obtuse or emarginate, 1½ lin. long; rostellum subquadrate, 3-lobed, with short side lobes and an acute tooth-like front lobe, about as long as the stigma. *Kränzl. Orch. Gen. et Sp.* i. 699; *Schlechter in Engl. Jahrb.* xxxi. 157.

KALAHARI REGION: Orange River Colony; in marshy places, *Cooper*, 1099! 1874!

EASTERN REGION: Pondoland; near Fort William, 2500 ft., *Tyson*, 2813! Natal; Bothas Hill, *Wood*, 427! Puff-adder Hill, Umzimkulu River, *Wood*, 1411! Inanda, *Wood*, 1435! Town Bush, near Maritzburg, 2800–3000 ft., *Adlam*, 6; Dumisa, *Rudatis*, 482! and without precise locality, *Sanderson*!

22. **S. longicauda** (Lindl. Gen. & Sp. Orch. 337); plant 1–1¾ ft. high; stems stout; leaves 2, radical, subsessile, suberect or somewhat spreading, oblong or elliptic-oblong, rarely lanceolate, acute or subacute, 2½–8 in. long, ¾–2½ in. broad, usually produced on short

branches at the base of the flowering stem; scapes 1–1¾ ft. high,
with very numerous imbricate spathaceous sheaths; spikes 3–8 in.
long, dense, many-flowered; bracts ovate or ovate-lanceolate, acute,
¾–1¼ in. long, usually reflexed after flowering; pedicels ½–¾ in.
long; flowers medium-sized, white or white and pink, fragrant; odd
sepal oblong or linear-oblong, obtuse, about 4 lin. long; lateral
sepals rather broader than the odd sepal and spreading; petals
oblong or elliptic-oblong, obtuse, rather shorter than the sepals; lip
cucullate, elliptic-ovate, about 5 lin. long, with a suborbicular or
obcordate denticulate reflexed apex; spurs filiform, somewhat
curved, 1–1¼ in. long; column curved, 3–3½ lin. long; stigma
suborbicular or obovate, 1½ lin. long, broader than long; rostellum
spathulate, obtuse or 3-denticulate, complicate at the base, not half
as long as the stigma. *Drège, Zwei Pfl. Documente*, 143; *Krauss
in Flora*, 1845, 307, *and Beitr. Fl. Cap- und Natal.* 159; *Bolus in
Journ. Linn. Soc.* xxv. 194; *Durand & Schinz, Conspect. Fl. Afr.* v.
96; *Kränzl. Orch. Gen. et Sp.* i. 706; *Schlechter in Engl. Jahrb.*
xxxi. 155.

SOUTH AFRICA: without locality, *Trimen*!
COAST REGION: Albany Div.; Howisons Poort, *Hutton*! Stockenstrom Div.!
slopes of Great Kat Berg, 4500 ft., *Scully*, 394! and in *MacOwan & Bolus, Herb.
Norm. Austr.-Afr.* 1372! Kat Berg, *Zeyher*. Queenstown Div.; Hangklip
Mountain, 5500 ft., *Galpin*, 1805! Komgha Div.; between Zandplaat and
Komgha, *Drège*! near Komgha, 2000 ft., *Flanagan*, 526. British Kaffraria,
Brownlee, 2, partly! Eastern Frontier of Cape Colony, *MacOwan*, 471!
CENTRAL REGION: Somerset Div.; Bosch Berg, 4500 ft., *MacOwan*, 1952!
KALAHARI REGION: Orange River Colony; Harrismith, *Sankey*, 268! 269!
and without precise locality, *Cooper*, 1094! 1098! 3614! Basutoland; Mametsana,
Dieterlen, 485! Lekuesha, *Dieterlen*, 244! Transvaal; near Barberton, 4000–5000 ft.,
Culver, 11! *Galpin*, 824! Spitzkop Goldmine, Lydenberg, *Wilms*, 1367! Hout-
bosch, *Rehmann*, 5832! *Schlechter*, 4472; Belfast, *Burtt-Davy*, 1300! 1307!
Miss Leendertz, 2859! near Donkerhoek, 4500 ft., *Schlechter*.
EASTERN REGION: Tembuland; Umlenze, near Bazeia, 3000 ft., *Baur*, 590!
Pondoland; summit of West Gate, Port St. John, 1200 ft., *Galpin*, 3419!
Griqualand East; near Mount Frere, 4300 ft., *Schlechter*, 6410. Natal; Inanda,
Wood, 443! near Ixopo, *Mrs. Clarke*! summit of Mawahqui Mountain, 6000–
7000 ft., *Wood*, 4617; near Charlestown, 5000–6000 ft., *Wood*, 5540; Great
Noodsberg, *Buchanan*! Oliviers Hoek, 5000 ft., *Allison*, Q! U! Bothas Hill,
2500 ft., *Schlechter*, 3253; Dumisa, 2000 ft., *Rudatis*, 444! 501! and without
precise locality, *Buchanan*, 8! *Sanderson*, 477! *Krauss*! *Mrs. K. Saunders*!
Zululand; coast to Isandhlwana, *Mrs. McKenzie*! and without precise locality,
Gerrard, 1550! 1551!

23. **S. stenopetalum** (Lindl. Gen. & Sp. Orch. 336); plant ½–1¼ ft.
high; stems stout; leaves 3 or 4, erect or suberect, oblong or ovate-
oblong, subobtuse, 1½–3 in. long, fleshy, reduced upwards into the
sheaths; scapes ½–1¼ ft. high, with several spathaceous sheaths;
spikes 2–7 in. long, somewhat dense, many-flowered; bracts ovate
or elliptic-ovate, acute, ½–¾ in. long, not reflexed after flowering;
pedicels about ½ in. long; flowers medium-sized, white; odd sepal
linear, subobtuse, 4½ lin. long; lateral sepals linear-lanceolate,
rather longer than the odd sepal and spreading; petals linear-
lanceolate, subobtuse, 4½ lin. long; lip cucullate, elliptic-ovate,

5 lin. long, with a long linear acuminate reflexed apex; spurs
slender, somewhat curved, about 8–10 lin. long; column curved,
2½ lin. long; stigma broadly obcordate-oblong, obtuse or emarginate,
1¼ lin. long; rostellum spathulate, with a broad limb, much constricted in the middle, three-quarters as long as the stigma; capsule
oblong, ½ in. long. *Bolus, Ic. Orch. Austr.-Afr.* i. *t.* 71, *and in
Journ. Linn. Soc.* xxv. 194; *Durand & Schinz, Conspect. Fl. Afr.* v.
99; *Kränzl. Orch. Gen. et Sp.* i. 712, *excl. var.*; *Schlechter in Engl.
Jahrb.* xxxi. 162, *excl. var.*

SOUTH AFRICA : without locality, *Masson*!
COAST REGION : Riversdale Div. ; between Little Vet River and Garcias Pass,
Burchell, 6856! 6880! Knysna Div.; near Knysna, *Newdigate.* Humansdorp
Div. ; Storms River, 200 ft., *Schlechter*, 5995! *Penther.*

24. S. marginatum (Bolus in Journ. Linn. Soc. xx. 476); plant
½–1¾ ft. high; stems stout; leaves 3 or 4, erect or suberect, oblong
or ovate-oblong, subobtuse or apiculate, 2–4 (rarely 5) in. long,
fleshy, reduced upwards into the sheaths; scapes ½–1¾ ft. high,
with several spathaceous sheaths; spikes 2–6 in. long, dense, many-
flowered; bracts ovate or elliptic-ovate, acute, ½–¾ in. long, not
reflexed after flowering; pedicels about ½ in. long; flowers medium-
sized, white or light pink; odd sepal linear-oblong, obtuse, 3 lin.
long; lateral sepals rather longer than the odd sepal and spreading;
petals linear-oblong, obtuse, 3 lin. long; lip cucullate, broadly
elliptic-ovate, 4 lin. long, with a broad triangular-oblong obtuse
reflexed apex; spurs slender, curved, 5–6 lin. long; column curved,
2 lin. long; stigma broadly obovate-oblong, obtuse, about 1 lin.
long; rostellum spathulate, with a broad limb, much constricted
in the middle, half as long as the stigma; capsule oblong, ½ in.
long. *Bolus in Trans. S. Afr. Phil. Soc.* v. 127; *Durand & Schinz,
Conspect. Fl. Afr.* v. 97; *Kränzl. Orch. Gen. et Sp.* i. 714. *S.
parviflorum, Lindl. Gen. & Sp. Orch.* 336, *not of Sw.*; *Drège, Zwei
Pfl. Documente*, 78, 100, 120; *Sond. in Linnæa*, xix. 83, *partly.
S. stenopetalum, var. brevicalcaratum, Bolus, Ic. Orch. Austr.-Afr.* i.
t. 72. *S. stenopetalum, var. parviflorum, Schlechter in Engl. Jahrb.*
xxxi. 162.

SOUTH AFRICA : without locality, *Ecklon*! *Grisebach*, 6! *Lehmann*! *Leibold,
Ludwig, Trimen*!
COAST REGION : Tulbagh Div. ; valley near the waterfall, 1200 ft., *Bolus*, 5551!
Pappe. Worcester Div. ; Mord Kuil, by the Doorn River, Boschesfeld Range,
under 1000 ft., *Drège*, 1260c! and without precise locality, *Zeyher*! *Cooper*,
1613! 1684! Paarl Div. ; banks of the Berg River, *Drège*, 1260a! Cape Div. ;
Cape Flats, *Zeyher*, 1561! *Pappe*, 66! near Cape Town, below 100 ft., *Bolus*,
4550! near Wynberg, 100 ft., *Bolus in MacOwan & Bolus, Herb. Norm. Austr.-Afr.*,
691! *Schlechter*; Klaver Vley, *Wolley-Dod*, 2149! by Orange Kloof Swamp,
Wolley-Dod, 3600! Stellenbosch Div. ; plateau above Lowrys Pass, 1500 ft.,
Galpin, 4595! Caledon Div.; Appels Kraal, by the River Zondereinde, *Zeyher*,
3913!

A common western plant, closely allied to *S. stenopetalum*, Lindl., to which it
has been referred as a variety, but differing in its much shorter spurs, shorter and
broader sepals and petals, and the quite differently shaped lip.

25. S. coriifolium (Sw. in Vet. Acad. Handl. Stockh. 1800, 216);
plant $\frac{3}{4}$–2 ft. high; stems stout; leaves 2–4, suberect, elliptic-ovate
to oblong, acute or subacute, 2–6 in. long, 1–2$\frac{1}{4}$ in. broad, fleshy;
scapes $\frac{3}{4}$–2 ft. high, with numerous spathaceous imbricate sheaths;
spikes 3–6 in. long, dense, many-flowered; bracts ovate, acute,
$\frac{1}{2}$–1$\frac{1}{4}$ in. long; pedicels about $\frac{1}{2}$ in. long; flowers rather large,
deep yellow to orange or flame-coloured, sometimes spotted; odd
sepal linear-oblong, obtuse, 7–8 lin. long; lateral sepals much
broader than the odd sepal and spreading; petals linear or linear-
oblong, obtuse, 7–8 lin. long; lip cucullate, broadly ovate, 6–7 lin.
long, with a very broad obtuse reflexed apex; spurs linear-oblong,
straight, stout, about 5 lin. long; column curved, 5 lin. long;
stigma broadly ovate-oblong, obtuse or emarginate, about 2 lin.
long, broader than long; rostellum very broad, shortly 3-lobed in
front, about as long as the stigma. *Bot. Mag. t.* 2172; *Bot. Reg.
t.* 703; *Lindl. Gen. & Sp. Orch.* 341; *Sweet, Fl. Gard. ser.* 2, *t.* 3;
Drège, Zwei Pfl. Documente, 101; *Krauss in Flora,* 1845, 307, *and
Beitr. Fl. Cap- und Natal.* 159; *Bolus in Trans. S. Afr. Phil. Soc.*
v. 124, *and in Journ. Linn. Soc.* xxv. 193; *Durand & Schinz,
Conspect. Fl. Afr.* v. 94; *Kränzl. Orch. Gen. et Sp.* i. 710; *Schlechter
in Engl. Jahrb.* xxxi. 158; *var. maculatum, Hook. f. in Bot. Mag.
t.* 7289. *S. cucullatum, Lodd. Bot. Cab. t.* 104, *not of Sw. S. erectum,
Lindl. Gen. & Sp. Orch.* 340; *Drège, Zwei Pfl. Documente,* 100;
Maund, Bot. iii. *t.* 117, *not of Sw. S. aureum, Paxt. Mag.' Bot.*
xv. 31, *with plate. Diplecthrum coriifolium, Pers. Syn.* ii. 509.
Orchis bicornis, Linn. Sp. Pl. ed. ii. 1330. *O. cornuta, Houtt.
Handl.* xii. 456, *t.* 80, *fig.* 2.—*O. luteo caule purpureo, Buxb. Pl.
Min. Cogn. Cent.* iii. 7, *t.* 10.

SOUTH AFRICA: without locality, *Masson*! *Bergius, Brown*! *Mund, Liebold,
Ludwig, Lichtenstein, Bowie, Harvey,* 123! 136! *Stuart*! *Wallich*! *Rogers*! *Pappe*!
Forbes, 86!
COAST REGION: Malmesbury Div.; near Groene Kloof, *Thunberg*! Paarl Div.;
Great and Little Drakenstein Mountains, under 100 ft., *Drège,* 1256a! 1256b!
Div.? between Tulbagh and Hantam, *Thom,* 1298! Cape Div.; Simons Bay,
Wright, 147! Cape Flats, 100 ft., *Bolus,* 4557! *Bowie,* 4! *Prior*! *Taylor*!
Kässner, Schlechter, 1553; Table Mountain, 600–700 ft., *Bolus, Kassner, Schlechter*;
Red Hill, *Mrs. Jameson*! Raapenburg, 100 ft., *Guthrie in Herb. Bolus,* 4878!
Stellenbosch Div.; Stellenbosch, *Lloyd,* 12! (in *Herb. Sanderson,* 939!) Caledon
Div.; near Caledon, *Zeyher,* 1555! between Bot River and the Zwart Berg,
Ecklon & Zeyher! Swellendam Div.; Swellendam, *Bowie.* George Div.; near
George, on the plain, *Burchell,* 6067! Knysna Div.; near Knysna, *Burchell,*
5547! 5551! 5746! *Forcade, Newdigate.* Port Elizabeth Div.; Algoa Bay,
Forbes, 80! Albany Div.; Grahamstown, *Atherstone,* 1241!

26. S. lupulinum (Lindl. Gen. & Sp. Orch. 338); plant $\frac{1}{2}$–1$\frac{1}{2}$ ft.
high; stems stout; leaves 2 or 3, suberect, subsessile, ovate or
ovate-oblong, subacute, 1–2$\frac{1}{4}$ in. long, somewhat fleshy; scapes
$\frac{1}{2}$–1$\frac{1}{2}$ ft. high, with several spathaceous imbricate sheaths; spikes
3–6 in. long, dense, many-flowered; bracts ovate, acute or acumi-
nate, $\frac{1}{2}$–$\frac{3}{4}$ in. long, reflexed after flowering; pedicels $\frac{1}{3}$–$\frac{1}{2}$ in. long;
flowers rather small, tawny yellow or dark brown; odd sepal
linear-oblong, obtuse, 3 lin. long; lateral sepals rather longer

than the dorsal sepal and somewhat spreading; petals oblong,
obtuse, somewhat falcate above the middle and crenulate, 3 lin.
long; lip cucullate, broadly elliptic-ovate, 3 lin. long, with a
broadly triangular-ovate obtuse reflexed apex; spurs slender,
somewhat curved, about ½ in. long; column curved, 2½ lin. long;
stigma oblong or obovate-oblong, obtuse or truncate, 1½ lin. long;
rostellum oblong, truncate, about half as long as the stigma.
Drège, Zwei Pfl. Documente, 105; *Bolus in Trans. S. Afr. Phil.
Soc.* v. 126, *Ic. Orch. Austr.-Afr.* i. *t.* 73, *and in Journ. Linn. Soc.*
xxv. 194; *Durand & Schinz, Conspect. Fl. Afr.* v. 96; *Kränzl.
Orch. Gen. et Sp.* i. 664; *Schlechter in Engl. Jahrb.* xxxi. 159.
S. pallidiflorum, Schlechter in Engl. Jahrb. xx. *Beibl.* 50, 15; *Kränzl.
Orch. Gen. et Sp.* i. 720.

SOUTH AFRICA : without locality, *Brown*! *Zeyher*, 1560! *Rogers*!
COAST REGION: Tulbagh Div.; Tulbagh Waterfall, *Bolus*. Cape Div. ; Table
Mountain and Devils Mountain, under 100 ft., *Drège*, 8292! Table Mountain,
800 ft., *Bolus*, 4553! Cape Flats, *Bolus*! Devils Mountain, 500 ft., *Bolus in
MacOwan & Bolus, Herb. Norm. Austr.-Afr.*, 1373! near Wynberg, 80 ft., *Bolus,
Schlechter*, 1554 ; Slang Kop, *Wolley-Dod*, 3083! hill west of Simons Town,
Wolley-Dod, 1523! mountain slopes near Vlagge Berg, *Mund*. Caledon Div. :
Houw Hoek Mountains, *Bolus* ; near Steenbrass River, 1000 ft., *Schlechter*, 5405.
Swellendam Div. : plains near Swellendam, *Bowie* ! Riversdale Div. ; Langeberg
Range, near Riversdale, 1500 ft., *Schlechter*. George Div. ; Outeniqua Mountains,
above Montague Pass, 4000 ft., *Schlechter*, 5854. Port Elizabeth Div. ; Algoa
Bay, *Forbes*, 80 !

27. S. foliosum (Sw. in Vet. Acad. Handl. Stockh. 1800, 216);
plant ½-1¼ ft. high; stems stout; leaves 2 or 3, suberect, ovate
or ovate-oblong, subobtuse, 1½-2¼ in. long, 1-1½ in. broad, somewhat fleshy ; scapes ½-1¼ ft. high, with several spathaceous sheaths ;
spikes 2-4 in. long, rather dense, many-flowered ; bracts ovate or
elliptic-ovate, acute, ½-¾ in. long, reflexed after flowering; pedicels
about ½ in. long ; flowers rather small, pale ochre-yellow ; odd sepal
subspathulate-oblong, obtuse, 2½-3 lin. long ; lateral sepals oblong,
obtuse, spreading, 3 lin. long ; petals oblong, obtuse, rather shorter
than the sepals ; lip cucullate, broadly ovate, 2¾ lin. long, with a
suborbicular or broadly oblong slightly crenulate reflexed apex;
spurs slender, nearly straight, 1 in. or rather more long ; column
curved, 2 lin. long ; stigma quadrate, about 1 lin. long, broader
than long ; rostellum broad, 3-denticulate at the apex, half as long
as the stigma, tubercled at the base.   *Thunb. Fl. Cap. ed. Schult.* 18 ;
Krauss in Flora, 1845, 307, *and Beitr. Fl. Cap- und Natal.* 159 ;
Bolus in Trans. S. Afr. Phil. Soc. v. 126, *and in Journ. Linn. Soc.*
xxv. 194 ; *Durand & Schinz, Conspect. Fl. Afr.* v. 95 ; *Kränzl. Orch.
Gen. et Sp.* i. 713 ; *Schlechter in Engl. Jahrb.* xxxi. 166.

COAST REGION : Cape Div. ; Table Mountain, 3000-3500 ft., *Thunberg* ! *Mund
& Maire, Bodkin in Herb. Bolus*, 4858 ! *Bolus in MacOwan & Bolus, Herb.
Norm. Austr.-Afr.* 155 ! *Schlechter*, 305.

A rare species, only found at present near the summit of Table Mountain.
Lindley's plant of this name belongs to *S. Hallackii*, Bolus.

28. S. rupestre (Schlechter in Engl. Jahrb. xxiv. 422); plant $\frac{1}{2}$–1$\frac{1}{2}$ ft. high; stem stout; leaves 2 or 3, subsessile, suberect or spreading, oblong or ovate-oblong, subobtuse, 1$\frac{1}{2}$–4 in. long, 1–2 in. broad, submembranous; scapes $\frac{1}{2}$–1$\frac{1}{2}$ ft. high, with several spathaceous sheaths,; spikes 1$\frac{1}{4}$–6 in. long, lax, many-flowered; bracts lanceolate or ovate-lanceolate, acute, $\frac{1}{3}$–$\frac{1}{2}$ in. long; pedicels $\frac{1}{4}$–$\frac{1}{3}$ in. long; flowers small, white; odd sepal oblong-lanceolate, subacute, 1–1$\frac{1}{4}$ in. long; lateral sepals broader and somewhat spreading, with an oblique acute apex; petals obovate-oblong, subobtuse, 1$\frac{1}{4}$–1$\frac{1}{2}$ lin. long; lip cucullate, broadly elliptic-ovate, 1$\frac{1}{2}$–1$\frac{3}{4}$ lin. long, with a suberect triangular subacute apex; spurs slender, nearly straight, 1$\frac{3}{4}$ lin. long; column somewhat curved, 1$\frac{1}{4}$ lin. long; stigma oblong, truncate or obtuse, $\frac{3}{4}$ lin. long; rostellum quadrate-oblong, obscurely trilobed at the apex, scarcely one-third as long as the stigma. *Bolus, Ic. Orch. Austr.-Afr.* i. *t.* 69; *Kränzl. Orch. Gen. et Sp.* i. 692; *Schlechter in Engl. Jahrb.* xxxi. 161.

COAST REGION : George Div. ; Silver River, 400 ft., *Schlechter,* 5870 ! *Penther,* 219! Humansdorp Div. ; Storms River, 200 ft., *Schlechter,* 5980 ! *Penther, Krook.*

29. S. Hallackii (Bolus in Journ. Linn. Soc. xx. 476); plant 1–2 ft. high; stems very stout; leaves 4–5, suberect, subcauline, narrowly oblong to elliptic-oblong, acute or subacute, 4–8 in. long, 1$\frac{1}{4}$–2$\frac{1}{4}$ in. broad, somewhat fleshy, passing upwards into the sheaths; scapes 1–2 ft. high, with several spathaceous sheaths; spikes 3–8 in. long, very dense, many-flowered; bracts lanceolate or ovate-lanceolate, acute or acuminate, $\frac{1}{2}$–1 in. long, reflexed after flowering; pedicels about $\frac{1}{2}$ in. long; flowers medium-sized, pink; odd sepal linear, obtuse, 3$\frac{1}{2}$–4 lin. long; lateral sepals linear oblong, obtuse, twice as broad as the odd sepal and spreading; petals linear, obtuse, 3$\frac{1}{2}$–4 lin. long; lip cucullate, broadly elliptic-ovate, 3 lin. long, with a broadly oblong obtuse reflexed apex; spurs slender, somewhat curved, about $\frac{1}{2}$ in. long; column curved, 2$\frac{1}{2}$ lin. long; stigma broadly deltoid-ovate, obtuse, $\frac{3}{4}$ lin. long, broader than long; rostellum 3-lobed, with an oblong subacute front lobe, longer than the stigma. *Bolus in Trans. S. Afr. Phil. Soc.* v. 128, *t.* 29, *and Journ. Linn. Soc.* xxv. 194; *Durand & Schinz, Conspect. Fl. Afr.* v. 96; *Kränzl. Orch. Gen. et Sp.* i. 704; *Schlechter in Engl. Jahrb.* xxxi. 170. *S. foliosum, and var. helonioides, Lindl. Gen. & Sp. Orch.* 336; *Drège, Zwei Pfl. Documente,* 127.

SOUTH AFRICA : without locality, *Bergius, Harvey*!
COAST REGION : Cape Div. ; near Vlagge Berg and Steen Berg, *Mund* ; Cape Flats, near Zeekoe Valley, *Pappe,* 65 ! *Zeyher,* 1556! Hout Bay, 50–100 ft., *Bodkin in MacOwan & Bolus, Herb. Norm. Austr.-Afr.,* 692 ! Knysna Div. ; near Knysna, *Pappe* ! Uitenhage Div. ; Cape Recife, *Burchell,* 4379! 4387 ! near Sand Fontein and Matjes Fontein, under 500 ft., *Drège,* 4772 ! Uitenhage, *Tredgold* ! Port Elizabeth Div. ; near Port Elizabeth, *Burchell,* 4372 ! *Miss West,* 26 ! *Hallack in Herb. Bolus,* 6092 ! and in *MacOwan & Bolus, Herb. Norm. Austr.-Afr.,* 948 ! Walmer, *MacOwan,* 948 ! Eastern Frontier of Cape Colony, *Prior* !
EASTERN REGION : Transkei ; Krielis Country, *Bowker* ! Kentani, 1200 ft., *Miss Pegler,* 1817 !

M 2

30. **S. ligulatum** (Lindl. Gen. & Sp. Orch. 342); plant 1–1¾ ft.
high ; stems stout ; leaves 2–5, subsessile, suberect, oblong or ovate-
oblong, subobtuse, 2–5½ in. long, ¾–1¾ in. broad, submembranous ;
scapes 1–1¾ ft. high, with numerous imbricate spathaceous sheaths ;
spikes 2–6 in. long, rather dense, many-flowered ; bracts ovate or
ovate-lanceolate, acute, ½–¾ in. long, reflexed after flowering ;
pedicels ⅓–½ in. long; flowers rather small, white ;. odd sepal
lanceolate-oblong, subacute, 4 lin. long ; lateral sepals lanceolate,
with a long acuminate twisted and obliquely ascending apex,
4½–5 lin. long; petals similar to the lateral sepals but rather
shorter ; lip cucullate, broadly elliptic-ovate, 4 lin. long, with a
broadly triangular acuminate reflexed apex ; spurs rather stout
below, acuminate, curved, 3 lin. long ; column slightly curved,
2½ lin. long ; stigma narrowly oblong, obtuse, 1½ lin. long ; rostellum broadly oblong, truncate, scarcely one-third as long as the
stigma, with 2 basal tubercles ; capsule oblong, ½ in. long. *Drège,
Zwei Pfl. Documente,* 76, 78 ; *Bolus in Trans. S. Afr. Phil. Soc.* v.
122, *t.* 28, *and in Journ. Linn. Soc.* xxv. 194 ; *Durand & Schinz,
Conspect. Fl. Afr.* v. 96 ; *Kränzl. Orch. Gen. et Sp.* i. 713 ; *Schlechter
in Engl. Jahrb.* xxxi. 160.

SOUTH AFRICA : without locality, *Masson* ! *Zeyher*, 3913 !
COAST REGION : Piquetberg Div. : Piquet Berg, 2000–3000 ft., *Drège.*
Tulbagh Div. ; New Kloof, under 1000 ft., *Drège*, 1258a ! Cape Div. ; Table
Mountain, 2500 ft., *Bolus*, 4583 ! and in *MacOwan & Bolus, Herb. Norm. Austr.-Afr.,*
332 ! *Harvey*, 245 ! Tiger Berg, *Mund* ; Orange Kloof Swamp, *Wolley-Dod*, 3460 !
Caledon Div. ; near Caledon, 850 ft., *Schlechter*, 5605 ! foot of Zwart Berg, near
Caledon, 800 ft., *Zeyher*, 3910 ! Swellendam Div. ; Zondereinde River, *Zeyher*, 3910 !
near Swellendam, 600 ft., *Galpin*, 4604 ! George Div. ; George, 600 ft., *Schlechter*,
5866 ! Knysna Div. ; Plettenbergs Bay, 200 ft., *Schlechter.* Albany Div. ;
Howisons Poort, 1800 ft., *MacOwan*, 693 ! Fort Beaufort Div. ;
Winter Berg. *Mrs. Barber*, 535 ! Stockenstrom Div. ; Kat Berg, 5000–5300 ft.,
Galpin, 1687. Queenstown Div. ; river side at Bongolo Nek, 4000 ft., *Galpin*,
7974 !

31. **S. cristatum** (Sond. in Linnæa, xix. 84) ; plant ¾–1½ ft. high ;
stems stout ; leaves 2–3, suberect, ovate-oblong to broadly elliptic,
subacute, 3–6 in. long, somewhat fleshy ; scapes ¾–1½ ft. high, with
several somewhat imbricate spathaceous sheaths ; spikes 2–6 in.
long, dense, many-flowered ; bracts ovate to ovate-lanceolate, acute
or acuminate, ½–¾ in. long, reflexed after flowering ; pedicels ⅓–½ in.
long ; flowers medium-sized, white, streaked with red (*Allison,
Wood*) ; odd sepal oblong or elliptic-oblong, obtuse, 2½–3½ lin. long ;
lateral sepals rather broader and spreading ; petals oblong or
elliptic-oblong, obtuse, 2½–3½ lin. long ; lip cucullate, broadly
elliptic-oblong, 3–5 lin. long, with a broadly oblong obtuse crenulate
reflexed apex ; spurs rather stout below, more or less acuminate
above, 3½–5 lin. long, somewhat curved ; column curved, 2½–4 lin.
long ; stigma subquadrate, 1½ lin. long, broader than long ; rostellum shortly tridentate, rather shorter than the stigma. *Bolus in
Journ. Linn. Soc.* xxv. 194 ; *Durand & Schinz, Conspect. Fl. Afr.* v.
95 ; *Kränzl. Orch. Gen. et Sp.* i. 717 ; *Schlechter in Engl. Jahrb.*

xxxi. 169, *partly, excl. syn. Reichb. f.   S. pentadactylum, Kränzl. in Engl. Jahrb.* xxiv. 506, *and Orch. Gen. et Sp.* i. 716.

SOUTH AFRICA : without locality, *Scott-Elliot* !
COAST REGION : Port Elizabeth Div. ; Port Elizabeth, *Miss West* ! Bedford Div. ; Kaga Berg, 2000 ft., *MacOwan,* 1105 ! Stockenstrom Div. ; Kat Berg, 2000 ft., *Hutton* ! *Zeyher* ! *Scully,* 414 ! Elands Berg, 6000 ft., *Scully,* 407 ! Willsdale, 2500 ft., *Scully,* 414 ! Queenstown Div. ; Madeira Hill, 4000–4500 ft., *Galpin,* 2035 ! British Kaffraria, *Brownlee,* 2, partly ! Eastern Frontier of Cape Colony, *MacOwan,* 513 !
CENTRAL REGION : Somerset Div. ; summit of Bosch Berg, 4800 ft., *MacOwan,* 1527 ! 1900 ! and in *Herb. Bolus,* 6126 !
KALAHARI REGION : Orange River Colony, *Cooper,* 1096 ! Transvaal ; Houtbosch, *Rehmann,* 5836 ! 5837 ! *Schlechter,* 4414 ; Spitzkop, near Lydenburg, *Wilms,* 1369 ; near Barberton, 4000–5000 ft., *Galpin,* 724 ! *Culver,* 48, *Thorncroft,* 456 ; Belfast, 6500 ft., *Burtt-Davy,* 1306 ! 4795 ! Ermelo, *Miss Leendertz,* 3127 ! Middelburg, *Hewitt,* 8041 !
EASTERN REGION : Tembuland ; Bazeia, 2000–4000 ft., *Baur,* 140 ! 633 ! Griqualand East ; near Kokstad, 6000 ft., *Tyson,* 1082. Pondoland ; Fakus Territory, *Sutherland* ! Natal ; Sevenfontein, 3000–4000 ft., *Wylie* ; Attercliff, *Sanderson,* 477 ! near Durban, *Sanderson,* 236 ! *Gerrard,* 736 ! Oliviers Hoek, 4000 ft., *Allison* ! Inanda, *Wood,* 336 ! 1582 ! near Van Reenan, 5000–6000 ft., *Wood,* 5533 ! Table Land, 2000 ft., *Buchanan* ! Highlands, 5000 ft., *Schlechter,* 6851 ! *Gerrard,* 1548 ! Dumisa, 2300 ft., *Rudatis,* 571 ! and without precise locality, *Buchanan* ! *Mrs. Fannin* ! *Mrs. K. Saunders* ! Zululand ; without precise locality, *Mrs. McKenzie* !

A very variable species, if all the above are really identical. I have not seen the type of *S. pentadactylum,* Kränzl., but the Angolan *S. Ivantalæ,* Reichb. f., which is referred here by Schlechter, is clearly different.

32. **S. sphærocarpum** (Lindl. Gen. & Sp. Orch. 337) ; plant 1–1¾ ft. high ; stems stout ; leaves 2–4, suberect or rarely somewhat spreading, ovate-oblong or elliptic-oblong, acute or subacute, 2–6 in. long, 1–2 in. broad, somewhat fleshy ; scapes 1–1¾ ft. high, with a few large spathaceous sheaths ; spikes 2–8 in. long, dense and many-flowered, sometimes lax and few-flowered ; bracts ovate or ovate-lanceolate, acute, ¾–1 in. long ; pedicels ½–¾ in. long ; flowers large, white, variously blotched and sometimes suffused with red ; sepals and petals united to nearly the middle, lanceolate-oblong, subobtuse, 6–8 lin. long, more or less recurved, lateral sepals broader than the other segments ; lip cucullate, broadly elliptic-ovate, 5–7 lin. long, with a short obtuse or apiculate reflexed apex ; spurs curved, rather stout, 4–6 lin. long ; column curved, 4–5 lin. long ; stigma broadly suborbicular, 1 lin. long ; rostellum oblong, 3-denticulate at the apex, half as long as the stigma. *Hook. f. in Bot. Mag. t.* 7295 ; *Bolus in Journ. Linn. Soc.* xxv. 194, *and Ic. Orch. Austr.-Afr.* i. *t.* 75 ; *N. E. Br. in Gard. Chron.* 1885, xxiv. 331 ; *Durand & Schinz, Conspect. Fl. Afr.* v. 99 ; *Kränzl. Orch. Gen. et Sp.* i. 702 ; *Schlechter in Engl. Jahrb.* xxxi. 167. *S. militare, Lindl. Gen. & Sp. Orch.* 342 ; *Drège, Zwei Pfl. Documente,* 143, 146. *S. beyrichianum, Kränzl. in Engl. Jahrb.* xxiv. 508, *and Orch. Gen. et Sp.* i. 705.

COAST REGION : Bathurst Div. ; shores near Port Alfred, *MacOwan,* 173, *Schönland* ; Flats of Lower Albany, *Hutton* ! Albany Div. ; Grahamstown, *MacOwan,* 175 ! Howisons Port, *Hutton* ! Coldstream, *Glass,* and without precise locality, *Cooper,* 1873 ! Komgha Div. ; between Zandplaat and Komgha, 1000–

2000 ft., *Drège* ; marshy plains at Keimouth, 200 ft., *Flanagan*, 648. East London Div. ; East London, *Galpin*, 3138 ! Eastern Region of Cape Colony, *Prior* !

EASTERN REGION : Transkei ; between Gekau River and Basche River, 1000–2000 ft., *Drège*, 4563 ! Krielis County, *Bowker*, 13 ! Kentani, 1200 ft., *Miss Pegler*, 289 ! Tembuland ; slopes of Bazeia Mountain, 3000 ft., *Baur*, 589 ! various localities, *Bolus*. Pondoland ; Emagushen, *Tyson*, 2840 ; between Umtamouma and Fort William, 3000 ft., *Tyson*, 2840 ! near Umtata, 3500 ft., *Schlechter*, 5344, and without precise locality, *Beyrich*, 374 ! Griqualand East ; near Kokstad, 4800 ft., *Tyson*, 1081 ; Mount Ayliffe, 4300 ft., *Schlechter*, 6305 ! Natal ; dry rocks on Great Noodsberg, *Buchanan* ! near Durban, *Krauss*, 172 ! *Sanderson*, 189 ! 479 ! *Gerrard*, 8, *Wood*, 204 ! and in *MacOwan Herb. Austr.-Afr.*, 1536 ! Clairmont, below 400 ft., *Wood*, 1093 ! 1291 ! Dumisa, 2000 ft., *Rudatis*, 547 ! Delagoa Bay ; *Forbes*, 67 ! *Mrs. Monteiro* ! *Junod*.

33. S. macrophyllum (Lindl. Gen. & Sp. Orch. 338) ; plant 1¼–2½ ft. high ; stem stout ; leaves 2–4, suberect, subsessile, ovate-oblong to broadly elliptic, obtuse or subacute, ⅓–1 ft. long, 2½–5 in. broad, somewhat fleshy ; scapes 1¼–2½ ft. high, with several somewhat imbricate sheaths ; spikes 4–10 in. long, rather dense, many-flowered ; bracts ovate or lanceolate-ovate, acute or acuminate, ½–¾ in. long, reflexed after flowering ; pedicels 6–7 lin. long ; flowers rather small, light or dark pink ; odd sepal linear, obtuse, 4–5 lin. long ; lateral sepals shorter and broader than the odd sepal and spreading ; petals linear, obtuse, almost as long as the lateral sepals ; lip cucullate, broadly ovate, 3 lin. long, with an apiculate reflexed apex ; spurs slender, somewhat curved, ¾–1 in. long, adnate at the base to the pedicel ; column curved, 2½ lin. long ; stigma broadly obovate-oblong, obtuse, 1 lin. long ; rostellum broadly oblong or suborbicular, 3-denticulate at the apex, rather shorter than the stigma. *Drège, Zwei Pfl. Documente*, 151 ; *Krauss in Flora*, 1845, 307, *and Beitr. Fl. Cap- und Natal.* 159 ; *Bolus, Ic. Orch. Austr.-Afr.* i. t. 74, *and in Journ. Linn. Soc.* xxv. 194 ; *Durand & Schinz, Conspect. Fl. Afr.* v. 97 ; *Kränzl. Orch. Gen. et Sp.* i. 695 ; *Schlechter in Engl. Jahrb.* xxxi. 168.

COAST REGION : Komgha Div. ; near Komgha, *Flanagan*, 646.

EASTERN REGION : Transkei ; valleys and slopes near Kentani, 1200 ft., *Miss Pegler*, 395 ! Pondoland ; between St. Johns River and Umsikaba River, 1000–2000 ft., *Drège* ! Griqualand East ; near Kokstad, 6000 ft., *Tyson*, 1089 ; mountains near Emyembe, 5000 ft., *Tyson*, 2087 ! mountains near Clydesdale, 3500 ft., *Tyson*, 2735 ! Natal ; Umzinto' River, *McKen*, 1535 ! near Richmond, 3000 ft., *Wood*, 1848 ! Inanda, 2200 ft., *Wood*, 3585 ! Manderston, 2800 ft., *Wood* ! Attercliff, 800 ft., *Sanderson*, 477 ! between Umgeni Falls and Pietermaritzburg, 3700 ft., *Schlechter*, 7035 ; Dumisa, 2300 ft., *Rudatis*, 623 ! and without precise locality, *Sanderson* ! *Mrs. K. Saunders* !

34. S. ocellatum (Bolus, Ic. Orch. Austr.-Afr. i. t. 23, partly) ; plant 1–2½ ft. high ; stems stout ; leaves 3 or 4, suberect, oblong-lanceolate, acute, 3–6 in. long, somewhat fleshy ; scapes 1–2½ ft. high, with several spathaceous sheaths ; spikes 3–6 in. long, dense or somewhat lax, many-flowered ; bracts ovate or ovate-lanceolate, acute, ½–1 in. long ; pedicels about ½ in. long ; sepals and petals united to nearly the middle ; flowers medium-sized, white or pink ;

odd sepal oblong or linear-oblong, obtuse, 3½–4 lin. long; lateral
sepals longer and twice as broad as the odd sepal, somewhat spreading;
petals linear-oblong, obtuse, about as long as the odd sepal; lip
cucullate, obovate-elliptic, 4–4½ lin. long, with a narrow apiculate
reflexed apex; spurs slender, curved, 10–12 lin. long; column
curved, about 3½ lin. long; stigma broadly suborbicular, ¾ lin. long,
broader than long; rostellum triangular-oblong, subacute, as long
as the stigma. *Kränzl. Orch. Gen. et Sp.* i. 707; *Schlechter in Engl.
Jahrb.* xxxi. 172, *partly.* *S. rostratum, Krauss in Flora*, 1845, 307,
and Beitr. Fl. Cap- und Natal. 159, *not of Lindl.* *S. nutans, Kränzl.
in Engl. Jahrb.* xxiv. 507, *and Orch. Gen. et Sp.* i. 704.

COAST REGION: Fort Beaufort Div.; Winterberg Range, *Mrs. Barber*, 519!
Stockenstrom Div.; Chumie (Tyumie) Peak, *Mrs. Barber*, 22! Komgha Div.;
near Komgha River, 1100 ft., *Flanagan*, 527.
KALAHARI REGION: Orange River Colony; Harrismith, *Sankey*, 270! Basuto-
land; Mametsana, *Dieterlen*, 486! Transvaal; near Sterk Spruit, Lydenburg,
Wilms, 1357! 1360! marshes on Houtbosch Mountains, 6500 ft., *Schlechter*,
4386! Koodoes Poort, near Pretoria, *Rech*, 47b! Ermelo, *Burtt-Davy*, 985!
Belfast, 6500 ft., *Burtt-Davy*, 1357! 1464! 4743! near Pretoria, 4700 ft.,
McLea; near Johannesburg, *Hall*; near Barberton, *Culver*, 36, and without
precise locality, *Hallack*, 21!
EASTERN REGION: Griqualand East; Mount Currie, 4500–5000 ft., *Tyson*,
1091! Natal; near Sterk Spruit, *Wood*, 3415! near Umlaas River, *Krauss*, 12!
Oliviers Hoek, 4000–5000 ft., *Allison*, C! 29! and without precise locality,
Mrs. Fannin, B.

35. **S. muticum** (Lindl. Gen. & Sp. Orch. 344); plant 1–1½ ft.
high; stems stout; leaves 2, basal, subsessile, cordate-orbicular or
broadly ovate-orbicular, subobtuse, somewhat fleshy, spreading,
2¼–5 in. long, 1½–4 in. broad; scapes 1–1½ ft. high, with numerous
tubular or spathaceous imbricating sheaths; spikes oblong, 2–3½ in.
long, somewhat dense, 8–12-flowered; bracts ovate or ovate-oblong,
acute or subobtuse, ¾–1¼ in. long, reflexed after flowering; pedicels
about ½ in. long; flowers rather large, pink with carmine tips to
the segments; sepals and petals united for ¼ in. at the base, reflexed,
about 7 lin. long; odd sepal oblong, obtuse; lateral pair rather
broader; petals oblong-spathulate, obtuse, somewhat crisped and
lacerate, rather broader than the sepals; lip galeate, broadly ovate,
obtuse, about 7 lin. long, reflexed and crisped-undulate at the apex,
united to the margin of the lateral sepals at the base; spurs subobsolete or shortly saccate, sometimes nearly 1 lin. long; column
curved, 4 lin. long; stigma suborbicular, somewhat concave; rostellum rhomboid, tridentate, shorter than the stigma. *Drège, Zwei
Pfl. Documente*, 123; *Bolus in Journ. Linn. Soc.* xxv. 195, *and Ic
Orch. Austr.-Afr.* i. *t.* 22; *Durand & Schinz, Conspect. Fl. Afr.* v
97; *Kränzl. Orch. Gen. et Sp.* i. 672; *Schlechter in Engl. Jahrb.*
xxxi. 143.

SOUTH AFRICA: without locality, *Mund*!
COAST REGION: Mossel Bay Div.; Driefontein, under 500 ft., *Drège*, 1758b!
Knysna Div.; in sandy soil near Knysna, *Bolus*, 6227.

36. S. paludicola (Schlechter in Engl. Jahrb. xx. Beibl. 50, 37);
plant about ½ ft. high, slender; leaves 2, radical, spreading, broadly
ovate or orbicular, shortly acuminate, 1¼–2¾ in. long, 1¼–1¾ in.
broad; scape about ½ ft. high, with 2 or 3 lanceolate subacute
sheaths; spike 2–2¼ in. long, dense, many-flowered; bracts ovate,
acute, ⅓–½ in. long, spreading; pedicels ¼–⅓ in. long; flowers small;
odd sepal lanceolate, obtuse; lateral sepals oblique, falcate-ovate,
subacute, entire, 2 lin. long; petals oblong, subacute, entire, shorter
than the sepals and united to them for nearly half their length; lip
cucullate, ovate, keeled behind, 3 lin. long, with revolute obtuse
apex; spurs saccate, short; column curved, slender; stigma oblong,
bifid at the apex; rostellum 3-lobed, middle lobe tooth-like, side
lobes much shorter. *Kränzl. Orch. Gen. et Sp.* i. 669; *Schlechter
in Engl. Jahrb.* xxxi. 142, *t.* 1, *fig.* A–F (*S. muticum on plate*).

KALAHARI REGION: Transvaal; Little Olifant River, 5100 ft., *Schlechter*, 4047.

37. S. pygmæum (Sond. in Linnæa, xix. 86); plant 3–7 in. high,
with ovoid tubers and rather slender stem; leaves 2, radical,
sessile, spreading, ovate, subacute or apiculate, ¾–1¾ in. long, ½–1 in.
broad, somewhat membranous; scapes 3–7 in. high, with numerous
ovate amplexicaul acute sheaths; spikes 1–3 in. long, lax, many-
flowered; bracts ovate, acute, 3–4 lin. long; pedicels 2½–3 lin.
long; flowers small, greenish-yellow with the spurs and tips of the
segments red; odd sepal oblong, obtuse, 1½ lin. long; lateral sepals
nearly twice as broad; petals ovate-oblong, obtuse, 1½ lin. long;
lip cucullate, broadly ovate, obtuse, slightly crenulate, 3 lin. long;
spurs rather stout, curved, 3–3½ lin. long; column curved, 1¾ in.
long; stigma oblong, obtuse, 1 lin. long; rostellum subquadrate,
apiculate, not half as long as the stigma. *Reichb. f. in Walp. Ann.*
i. 799; *Bolus, Ic. Orch. Austr.-Afr.* i. *t.* 20, *and in Journ. Linn. Soc.*
xxv. 195; *Durand & Schinz, Conspect. Fl. Afr.* v. 98; *Kränzl. Orch.
Gen. et Sp.* i. 711; *Schlechter in Engl. Jahrb.* xxxi. 151, *partly.*

COAST REGION: Tulbagh Div.; Winterhoek Mountains, 3000–4000 ft., *Ecklon
& Zeyher*! Tygers Kloof, *Pappe*! Ceres Div.; Skurfdeberg Range, near Ceres,
1800 ft., *Bolus*, 7327! and in *MacOwan & Bolus, Herb. Norm. Austr.-Afr.* 1095!
Swellendam Div.; Grootvaders Bosch, *Zeyher*, 3914! Zuurbraak, 800 ft., *Galpin*,
4593a, partly! Riversdale Div.; Garcias Pass, 1200 ft., *Galpin*, 4593! George
Div.; Outeniqua Mountains, above Montagu Pass, 2500 ft., *Schlechter, Penther*.

38. S. Schlechteri (Rolfe); plant 4–6 in. high; stem moderately
stout; leaves 2, radical, sessile, spreading, ovate or ovate-oblong,
obtuse or apiculate, 1¼–1½ in. long, submembranous; scapes 4–6 in.
high, with 2 to 4 spathaceous sheaths; spikes 1½–3 in. long, somewhat lax, many-flowered; bracts ovate or ovate-lanceolate, subacute,
⅓–½ in. long; pedicels 3–4 lin. long; flowers rather small; odd
sepal oblong, obtuse, 2 lin. long; lateral sepals much broader than
the odd sepal, somewhat spreading; petals oblong, obtuse, 2 lin.
long; lip cucullate, broadly ovate, 2½–3 lin. long, with a rather
small reflexed apex; spurs somewhat curved, 4 lin. long; column

curved, 1½ lin. long; stigma ovate or rounded, ½ lin. long; rostellum 3-lobed with acute lobes, as long as the stigma. *S. pygmæum, Schlechter in Engl. Jahrb.* xxxi. 151, *partly.*

COAST REGION: Caledon Div. ; Zwart Berg, 2000 ft., *Schlechter!* Swellendam Div.; Zuurbraak, 600 ft., *Galpin,* 4598! Riversdale Div.; Langeberg Range, 1500 ft., *Schlechter,* 2030!

This has been included under *S. pygmæum,* Sond., by Schlechter.

39. S. pentherianum (Kränzl. Orch. Gen. et Sp. i. 944); plant 3-8 in. high, leafy, papillose; leaves broadly ovate-cordate, acute, about 1¼ in. long, ¾ in. broad, withering soon after flowering; cauline oblong, acute, margin and nerves beneath papillose; spikes elongate, many-flowered; bracts broadly ovate, acute, 4-5 lin. long, 3-4 lin. broad, deflexed after flowering; pedicels 3 lin. long; flowers small; odd sepal ligulate, subobtuse, 2½-3 lin. long; lateral oblong, obtuse; petals ligulate, subobtuse, about as long as the odd sepal; lip galeate, ovate, with a broad mouth, acute, without an apical lobe, 2½ lin. long; spurs cylindrical, acute, 3 lin. long; stigma subquadrate, subobtuse; rostellum deeply bilobed.

COAST REGION: Knysna Div. ; Elands River, *Penther,* 292.

Only known to me from the description. Kränzlin places it next to *S. emarcidum,* Bolus, to which it seems most allied in floral structure.

40. S. emarcidum (Bolus in Journ. Linn. Soc. xxii. 67); plant ⅓-¾ ft. high; stems stout; leaves several, amplexicaul, lowest one spreading, cordate-ovate or orbicular, subobtuse, 1¾-2¼ in. long, the rest gradually decreasing upwards into the bracts; spikes oblong, 1-2½ in. long, dense, many-flowered; bracts ovate or ovate-lanceolate, acute, ½-¾ in. long, reflexed after flowering; pedicels about ¼ in. long; flowers small, dirty yellowish-white; odd sepal linear, subobtuse, 2¼ lin. long, recurved; lateral sepals lanceolate-linear, acute, 3 in. long, suberect; petals oblong-lanceolate, acute, rather shorter than the odd sepal, spreading; lip galeate, ovate, acuminate, subobtuse, 3-3¼ lin. long; spurs subfiliform, somewhat curved, 3½-4 lin. long; column curved, 1¾ lin. long; stigma oblong, submarginate; rostellum subclavate at the apex, about as long as the stigma; capsule oblong, about ½ lin. long. *Bolus in Trans. S. Afr. Phil. Soc.* v. 121, *t.* 27, *and in Journ. Linn. Soc.* xxv. 193; *Durand & Schinz, Conspect. Fl. Afr.* v. 95; *Kränzl. Orch. Gen. et Sp.* i. 665; *Schlechter in Engl. Jahrb.* xxxi. 141.

COAST REGION: Cape Div. ; sandy flat ground in Fish Hoek, north shores of False Bay, below 50 ft., *Bolus,* 4847, and in *MacOwan & Bolus, Herb. Norm. Austr.-Afr.,* 159! *Fair in Herb. Wolley-Dod,* 1756! sand dunes between Retreat Station and Muizenberg, *Schlechter,* 1480!

41. S. bicorne (Thunb. Prodr. 6); plant ¾-1½ ft. high; stems stout; leaves 2, radical, sessile, spreading, cordate-orbicular or cordate-ovate, subobtuse, ciliate, 2-3½ in. long, 1¼-3 in. broad, fleshy; scapes ¾-1½ ft. high, with several broad tubular or spathaceous bracts; spikes 2–4 in. long, somewhat lax, many-flowered; bracts ovate-oblong or elliptic, acute or subobtuse, ½–¾ in. long, reflexed after flowering; pedicels about ⅓ in. long; flowers dull ochre-yellow, scented; odd sepal oblong, obtuse, 2½ lin. long, reflexed; lateral sepals rather larger and spreading; petals oblong, obtuse, about as large as the odd sepal, spreading; lip galeate, broadly ovate, obtuse, 3½ lin. long, basal third united to the margin of the lateral sepals; spurs cylindrical, slender and somewhat curved, 7–9 lin. long; column curved, 2½ lin. long; stigma oblong, obtuse, 1½ lin. long; rostellum oblong, with a short tooth in front, deflexed, much shorter than the stigma. *Bolus in Trans. S. Afr. Phil. Soc.* v. 122, *and in Journ. Linn. Soc.* xxv. 192; *Durand & Schinz, Conspect. Fl. Afr.* v. 93; *Kränzl. Orch. Gen. et Sp.* i. 675; *Schlechter in Engl. Jahrb.* xxxi. 144. *S. cucullatum, Sw. in Vet. Acad. Handl. Stockh.* 1800, 216, *t.* 3, *fig.* C; *Thunb. Fl. Cap. ed. Schult.* 17, *partly*; *Drège, Zwei Pfl. Documente,* 69, 102, 121. *Orchis bicornis, Linn. Sp. Pl. ed.* ii. 1330; *Houtt. Handl.* xii. 455, *t.* 86, *fig.* 1.—*Orchis lutea, caule geniculato, Buxb. Pl. Min. Cogn. Cent.* iii. 6, *t.* 8.

SOUTH AFRICA: without locality, *Thunberg*! *Brown*! *Thom,* 715! *Harvey,* 119! 133! *Lehmann*! *Oldenburg,* 54!
COAST REGION: Malmesbury Div.; *Laaus Kloof, under 1000 ft., Drège,* 8294c! Cape Div.; near Simons Town, *Wolley-Dod,* 2986! Doorn Hoogte, *Zeyher,* 1559! between Kirstenbosch and Wynberg, *Zeyher,* 4679! Simons Bay, *Wright,* 139! Lions Head, *Wolley-Dod,* 376! Cape Flats, *Pappe,* 63! heathy places up to 2000 ft., *Bolus,* 4556. Stellenbosch Div.; Lowrys Pass, 1500 ft., *MacOwan,* 3128! and *Herb. Austr.-Afr.* 1758! Stellenbosch, *Lloyd,* 13! (*in Herb. Sanderson,* 940!) Swellendan Div.; near Appels Kraal, *Zeyher*! Zuurbraak, on the ruggens, 800 ft., *Galpin,* 4602! Uniondale Div.; between Keurbooms River and Kromme River, 2000–3000 ft., *Drège,* 8295d!
WESTERN REGION: Little Namaqualand; near Ezelsfontein and Roode Berg, 3000–4000 ft., *Drège,* 8295a!

42. **S. outeniquense** (Schlechter in Engl. Jahrb. xxiv. 421); plant ⅓–½ ft. high; stem rather slender; leaves basal, 2, spreading, broadly ovate, subacute, rather fleshy, ¾–1¼ in. long, ½–1 in. broad; cauline reduced to ovate-lanceolate sheaths, ⅓–½ in. long; spikes subcylindrical, 1½–2¾ in. long, rather lax, few- to many-flowered; bracts ovate or ovate-lanceolate, acute, ⅓–½ in. long, reflexed after flowering; pedicels ¼–⅓ in. long; flowers small, pale ochre-yellow; odd sepal oblong, subobtuse, 2 lin. long, reflexed; lateral sepals oblong-ligulate, subobtuse, 2½ lin. long, spreading; petals oblong, subobtuse, as long as the odd sepal, spreading; lip galeate, broadly elliptic-oblong, obtuse and crisped-undulate at the apex, 2½ lin. long; spurs cylindrical, slender, curved, 5 lin. long; column curved, 1½ lin. long; stigma obovate-oblong or orbicular; rostellum clavate, rather shorter than the stigma. *Bolus, Ic. Orch. Austr.-Afr.* i. *t.* 68; *Kränzl. Orch. Gen. et Sp.* i. 703; *Schlechter in Engl. Jahrb.* xxxi. 141.

COAST REGION: George Div.; Montague Pass, 2000 ft., *Schlechter,* 5792!

43. S. humile (Lindl. Gen. & Sp. Orch. 339); plant ½–¾ ft. high, with rather stout stems; leaves 2, radical, sessile, spreading, ovate or suborbicular, obtuse or apiculate, about 1¼ in. long, somewhat fleshy; scapes ½–¾ ft. high, with several spathaceous sheaths; spikes about 2 in. long, lax, many-flowered; bracts ovate or elliptic-ovate, acute, ⅓–½ in. long; pedicels ⅓–½ in. long; flowers rather small; odd sepal linear-oblong, obtuse, 3 lin. long; lateral sepals broader and somewhat spreading; petals linear-oblong, obtuse, as long as the odd sepal; lip cucullate, ovate-oblong, 3–3½ lin. long, with a very broad obtuse reflexed apex; spurs slender, somewhat curved, 6–7 lin. long; column curved, 2¼ lin. long; stigma oblong or elliptic-oblong, obtuse, 1 lin. long; rostellum spathulate-ovate, somewhat concavo, half as long as the stigma. *Drège, Zwei Pfl. Documente,* 82; *Sond. in Linnæa,* xix. 86; *Bolus in Journ. Linn. Soc.* xxv. 193; *Durand & Schinz, Conspect. Fl. Afr.* v. 96; *Kränzl. Orch. Gen. et Sp.* i. 676, *partly*; *Schlechter in Engl. Jahrb.* xxxi. 152, *partly.*

COAST REGION: Worcester Div.; Dutoits Kloof, 3000–4000 ft., *Drège,* 8295b! 8295bb!

Only known from Drège's original specimen. Lindley remarked: "This looks like a starved state of *S. cucullatum* (i.e. *S. bicorne*), but the form and texture of the flowers are different." *S. ochroleucum,* Bolus, has since been considered identical, but in that the stigma is much shorter and broader. The origin of the confusion is probably that Lindley himself wrote "*S. humile,* m.," on a sheet in Sir William Hooker's Herbarium. This sheet contains two small specimens of *S. ochroleucum,* Bolus, localised "Cape," and without any collector's name, but they are quite different from the two specimens collected by Drège on which Lindley founded the species, and which are preserved in his own Herbarium.

44. S. ochroleucum (Bolus in Journ. Linn. Soc. xxii. 66); plant ½–1½ ft. high; stems stout; leaves 2, radical, subsessile, spreading, broadly ovate or ovate-orbicular, obtuse, 1–6 in. long, ¾–4 in. broad, membranous; scapes ½–1½ ft. high, with numerous spathaceous sheaths, the lower often larger; spikes 2–6 in. long, somewhat lax, many-flowered; bracts ovate or elliptic-ovate, acute, ½–¾ in. long; pedicels ⅓–½ in. long; flowers rather small, pale yellow; odd sepal linear-oblong, obtuse, 2½–3 lin. long; lateral sepals nearly twice as broad as the odd sepal, spreading; petals linear-oblong, obtuse, 2½–3 lin. long; lip cucullate, ovate, 3–3½ lin. long, with a broad obtuse crenulate apex; spurs slender, curved, about 5 lin. long; column curved, 2 lin. long; stigma broadly ovate, obtuse, about ¾ lin. long; rostellum broadly oblong, tridentate at the apex, shorter than the stigma; capsule oblong, about ½ in. long. *Bolus in Trans. S. Afr. Phil. Soc.* v. 123, *t. 26, and in Journ. Linn. Soc.* xxv. 193; *Durand & Schinz, Conspect. Fl. Afr.* v. 97. *S. candidum,* Lindl. Bot. Reg. 1838, Misc. 82, *and Gen. & Sp. Orch.* 339, *partly*; *Drège, Zwei Pfl. Documente,* 103, 115. *S. humile, Kränzl. Orch. Gen. et Sp.* i. 676, *partly*; *Schlechter in Engl. Jahrb.* xxxi. 152, *partly. Orchis bicornis,* Jacq. Hort. Schœnbr. ii. *t.* 179, *not of Linn.*

45. **S. acuminatum** (Lindl. Gen. & Sp. Orch. 339) ; plant $\frac{3}{4}$–2$\frac{1}{4}$ ft.
high, usually stout ; leaves 2, radical, sessile, spreading, broadly
ovate-orbicular, subobtuse, 1–4$\frac{1}{2}$ in. long, by about as broad, somewhat fleshy ; scapes $\frac{3}{4}$–2$\frac{1}{4}$ ft. high, with several ovate or spathaceous
sheaths ; spikes $\frac{1}{4}$–1 ft. long, dense, many-flowered ; bracts ovate or
ovate-oblong, acute, $\frac{1}{2}$–1 in. long, reflexed after flowering ; pedicels
$\frac{1}{3}$–$\frac{1}{2}$ in. long ; flowers white ; odd sepal oblong, obtuse, $\frac{1}{4}$ in. long ;
lateral sepals rather broader ; petals elliptic-oblong, obtuse, $\frac{1}{4}$ in.
long ; lip cucullate, ovate-orbicular, $\frac{1}{4}$ in. long, with a broad obtuse
somewhat crenulate apex ; spurs slender, curved, 7–9 lin. long ;
column curved, 2$\frac{1}{4}$ lin. long ; stigma oblong, obtuse, 1$\frac{1}{2}$ lin. long ;
rostellum half as long as the stigma, constricted above the middle,
dilated and anchor-shaped at the apex. *Sond. in Linnæa*, xix. 86 ;
Bolus in Journ. Linn. Soc. xxv. 193 ; *Durand & Schinz, Conspect. Fl.
Afr.* v. 92 ; *Kränzl. Orch. Gen. et Sp.* i. 675 ; *Schlechter in Engl.
Jahrb.* xxxi. 145. *S. humile, Schlechter in Engl. Jahrb.* xxxi. 152,
partly.

46. **S. candidum** (Lindl. Bot. Reg. 1838, Misc. 82, partly) ; plant
$\frac{3}{4}$–1$\frac{1}{2}$ ft. high ; stems stout ; leaves 2, sessile, spreading, ovate or cordate-ovate, subobtuse, somewhat fleshy, 1$\frac{1}{2}$–4 in. long, 1–4 in. broad ;
scapes $\frac{3}{4}$–1$\frac{1}{2}$ ft. high, with several spathaceous or tubular sheaths ;
spikes 2–6 in. long, dense, many-flowered ; bracts ovate or elliptic-
ovate, acute or subacute ; pedicels about $\frac{1}{2}$ in. long ; flowers medium-
sized, white or faintly suffused with pink, fragrant ; odd sepal
linear-oblong, obtuse, 5–6 lin. long ; lateral sepals broader, shorter
and spreading ; petals linear-oblong, obtuse, 5–6 lin. long ; lip
cucullate, broadly elliptic-ovate, 5–6 lin. long, with a broad obtuse
crenulate reflexed apex ; spurs slender, curved, about $\frac{3}{4}$ in. long ;
column curved, 3$\frac{1}{2}$ lin. long ; stigma ovate-reniform or obovate,
1$\frac{1}{4}$ lin. long, broader than long ; rostellum short and broad, 3-dentate
at the apex. *Lindl. Gen. & Sp. Orch.* 339, excl. *Drège sp.* ; *Bolus
in Journ. Linn. Soc.* xxv. 194, *and in Trans. S. Afr. Phil. Soc.* v.
121 ; *Durand & Schinz, Conspect. Fl. Afr.* v. 94 ; *Kränzl. Orch. Gen.
et Sp.* i. 678 ; *Schlechter in Engl. Jahrb.* xxxi. 151. *S. utriculatum,
Sond. in Linnæa,* xix. 84.

SOUTH AFRICA : without locality, *Brown* ! *Thom* ! *Zeyher*, 1558 ! *Rogers* ! *Harvey*, 144 !

COAST REGION : Cape Div. ; Claremont, near Capetown, under 100 ft., *Bolus*, 4331 ! and in *MacOwan & Bolus, Herb. Norm. Austr.-Afr.* 158 ! Vygeskraal Farm, *Wolley-Dod*, 2022 ! Cape Flats, *Pappe*, 69 ! *Ecklon* ! *Fielden* ! Wynberg Flats, *Prior* ! Simons Bay, *Wright* ! Caledon Div. ; Genadendal, *Pappe* !

47. S. Guthriei (Bolus, Ic. Orch. Austr.-Afr. i. t. 21) ; plant about 6 in. high ; stem stout ; leaves 2, radical, the lower ovate, obtuse, spreading, 3 in. long, about 2 in. broad, somewhat fleshy, the upper much smaller and suberect ; scapes about 6 in. high, with 2 inflated sheaths ; spikes 1 in. long, somewhat lax, about 7-flowered ; bracts ovate-lanceolate, acute, $\frac{1}{2}$–$\frac{1}{2}$ in. long ; pedicels about $\frac{1}{4}$ in. long ; flowers medium-sized, very pale pink with a few darker markings ; odd sepal linear-oblong, obtuse, about $3\frac{1}{2}$ lin. long ; lateral sepals larger than the odd sepal, somewhat spreading ; petals linear, obtuse, rather shorter than the odd sepal, minutely denticu-late at the margin ; lip cucullate, globose-inflated, about 3 lin. long, with a triangular acute somewhat reflexed apex ; spurs slightly curved, slender, about $\frac{1}{4}$ in. long ; column curved, about $2\frac{1}{2}$ lin. long ; stigma oblong, 2-lobed at the apex, short ; rostellum broadly triangular, emarginate in front, longer than the stigma. *Kränzl. Orch. Gen. et Sp.* i. 719 ; *Schlechter in Engl. Jahrb.* xxxi. 195 ; *Rolfe in Orch. Rev.* i. 269.

COAST REGION : Cape Div. ; in burnt-off places on the Cape Flats at Tokai, 100 ft., *Guthrie in Herb. Bolus*, 7095.

Described and figured from a single specimen, which was found growing with *S. candidum*, Lindl. Bolus remarks : " The column resembles in some degree that of *S. bicallosum*, Thunb., while both are, in this respect, very different from that of any other *Satyrium* known. In every other respect this differs greatly from *S. bicallosum*, and I doubt very much if it is a natural hybrid." But it appears to me to combine the characters of *S. candidum* and *S. bicallosum*, which grow intermixed.

48. S. maculatum (Burch. ex Lindl. Gen. & Sp. Orch. 337) ; plant $\frac{1}{2}$–$1\frac{1}{2}$ ft. high ; stems stout ; leaves 2, radical, sessile, spreading, ovate or ovate-orbicular, obtuse, somewhat fleshy, $1\frac{1}{4}$–3 in. long, nearly as broad as long ; scapes $\frac{1}{2}$–$1\frac{1}{2}$ ft. high, with numerous spathaceous imbricate sheaths ; spikes 2–6 in. long, many flowered, dense or rarely somewhat lax ; bracts ovate or ovate-oblong, acute, reflexed after flowering ; pedicels 5–7 lin. long ; flowers medium-sized, white or pale pink, spotted with purple ; sepals oblong, obtuse, 4–5 lin. long ; petals rather shorter and narrower than the sepals ; lip cucullate, oblong or ovate-oblong, about $\frac{1}{2}$ in. long, with broad obtuse somewhat reflexed apex ; spurs slender, curved, 1–$1\frac{1}{4}$ in. long ; column curved, about 4 lin. long ; stigma obovate-oblong, obtuse, 1 lin. long ; rostellum rhomboid, tridentate at the apex, rather shorter than the stigma. *Drège, Zwei Pfl. Documente*, 123, 134, 137 ; *Bolus in Journ. Linn. Soc.* xxv. 193, and *Ic. Orch. Austr.-Afr.* i. t. 19 ; *Durand & Schinz, Conspect. Fl. Afr.* v. 97 ; *Kränzl. Orch. Gen. et Sp.* i. 679 ; *Schlechter in Engl.*

Jahrb. xxxi. 150. *S. longicolle, Lindl. Gen. & Sp. Orch.* 335. *S. longicollum, Drège, Zwei Pfl. Documente,* 139.

SOUTH AFRICA: without locality, *Thom,* 90 ! *Zeyher,* 1565, partly ! *Oldenburg,* 873 ! *Burke* ! *Mund, Krebs,* 350.
COAST REGION : Riversdale Div. ; between Zoetemelks River and Little Vet River, *Burchell,* 6854 ! Muiskraal Ridge, near Garcias Pass, 1500 ft., *Galpin,* 4601 ! Mossel Bay Div. ; Driefontein and Mossel Bay, under 500 ft., *Drège,* 2206a ! George Div. ; Langekloof, near George, 5300 ft., *Young.* Knysna Div. ; near Keurebooms River, *Schlechter,* 5940 ! Plettenbergs Bay, *Bergius.* Uitenhage Div. ; Enon, 1000–2000 ft., *Drège,* 2206b ! Zuurberg Range, 2500–3500 ft., *Drège* ; Uitenhage, *Tredgold* ! Port Elizabeth Div. ; near Port Elizabeth, *Hallack in Herb. Bolus,* 6096 ! Albany Div. ; near Assegai Bosch and Botran, 1000–2000 ft., *Drège* ! Grahamstown, *MacOwan* ! *Tuck* ! *Galpin,* 30, 303 ! *Schönland, Glass* ; Fullers, *Reade,* 103 ! Eastern part of Cape Colony, *Prior* ! *MacOwan,* 383 ! *Hutton* !

49. S. erectum (Sw. in Vet. Acad. Handl. Stockh. 1800, 216);

plant $\frac{3}{4}$–1$\frac{1}{2}$ ft. high; stems stout; leaves 2, radical, subsessile, spreading, ovate or ovate-orbicular, obtuse, somewhat fleshy, 1$\frac{1}{2}$–5 in. long, 1$\frac{1}{2}$–4$\frac{1}{2}$ in. broad ; scapes $\frac{3}{4}$–1$\frac{1}{2}$ ft. high, with numerous imbricate spathaceous sheaths; spikes 3–8 in. long, dense, many-flowered ; bracts ovate-lanceolate, acute, $\frac{3}{4}$–1 in. long, reflexed after flowering; pedicels about $\frac{1}{2}$ in. long; flowers medium-sized, rose (*Bolus*) or paler and spotted with red at the base of the segments; sepals oblong, obtuse, somewhat narrowed at the base, 4–5 lin. long ; petals oblong, obtuse, 4–5 lin. long; lip cucullate, ovate-oblong, 5–6 lin. long, with very broad obtuse slightly crenulate reflexed apex ; spurs linear, slightly curved, 4–5 lin. long ; column curved, 4–5 lin. long; stigma suborbicular, obtuse, 1 lin. long; rostellum rhomboid, bifid at the apex, rather shorter than the stigma ; capsule oblong, 6–8 lin. long. *Thunb. Fl. Cap. ed. Schult.* 18, *excl. syn.* ; *Bolus in Journ. Linn. Soc.* xxv. 193 ; *Durand & Schinz, Conspect. Fl. Afr.* v. 95 ; *Kränzl. Orch. Gen. et Sp.* i. 696 ; *Schlechter in Engl. Jahrb.* xxxi. 148. *S. pustulatum, Lindl. Bot. Reg.* 1840, *t.* 18. *S. papillosum, Lindl. Gen. & Sp. Orch.* 341 ; *Drège, Zwei Pfl. Documente,* 68, 113, 114, 118, 136. *S. Herscheliæ, Harv. Gen. S. Afr. Pl. ed.* i. 326, *name only. Diplecthrum erectum, Pers. Syn. Pl.* ii. 509.

SOUTH AFRICA: without locality, *Thom,* 980 ! *Masson* !
COAST REGION : Clanwilliam Div. ; near Piqueniers Kloof, 1000 ft., *Schlechter.* Malmesbury Div. ; Zwartland, *Thunberg* ! between Groene Kloof and Saldanha Bay, under 500 ft., *Drège* ! Tulbagh Div. ; near Tulbagh, *Pappe,* 23 ! Tulbagh Kloof, 500 ft., *Bolus,* 5443 ! Worcester Div. ; on mountains near De Liefde, 1000–2000 ft., *Drège* ! Stellenbosch Div. ; Hottentots Holland, *Herschel* ! Caledon Div. ; Donker Hoek and Eseljagt mountains, 1000–2000 ft., *Drège* ! Genadendal, *Prior* ! Swellendam Div. ; near Riet Kuil, *Zeyher,* 3911 ! Riversdale Div. ; near Riversdale, *Rust,* 18. George Div.; near George, *Prior* ! Uitenhage Div. ; Zuurberg Range, 2000–3000 ft., *Drège.* Albany Div. ; without precise locality, *Atherstone,* 60 ! Eastern district of Cape Colony, *Hutton* !
WESTERN REGION : Little Namaqualand ; between Pedros Kloof and Lily Fontein, *Drège.*

50. S. membranaceum (Sw. in Vet. Acad. Handl. Stockh. 1800,

216); plant $\frac{3}{4}$–1$\frac{1}{2}$ ft. high; stems stout; leaves 2, radical, sessile,

spreading, orbicular or broadly ovate-orbicular, obtuse, 1½–4 in.
long, somewhat fleshy; scapes ¾–1½ ft. high, with numerous
imbricate membranous sheaths; spikes 3–8 in. long, dense, many-
flowered; bracts ovate or ovate-elliptic, acute, ½–¾ in. long; pedicels
about ½ in. long; flowers medium-sized, pink; odd sepal oblong,
apiculate, 3½–4 lin. long; lateral sepals spreading and rather
broader; petals oblong, obtuse, with much fringed margin, rather
shorter than the sepals; lip broadly cucullate, 4–4½ lin. long, with
reflexed oblong fimbriate apex; spurs slender, curved, ¾–1 in. long;
column curved, 3 lin. long; stigma broadly oblong, 1 lin. long;
rostellum 3-fid, with broad acute lobes, nearly as long as the stigma.
Lindl. Gen. & Sp. Orch. 335; *Drège, Zwei Pfl. Documente,* 136;
Bolus in Journ. Linn. Soc. xxv. 193; *Durand & Schinz, Conspect. Fl.
Afr.* v. 97; *N. E. Br. in Gard. Chron.* 1889, v. 136, *partly*; *Kränzl.
Orch. Gen. et Sp.* i. 678, *partly?* *Schlechter in Engl. Jahrb.* xxxi.
146. *S. cucullatum, Thunb. Fl. Cap. ed.* i. 88; *ed. Schult.* 17, *partly.*

SOUTH AFRICA: without locality, *Bowie!* *Prior!*
COAST REGION: Riversdale Div.; near Zoetemelks River, *Burchell,* 6738/² !
Mossel Bay Div.; between Zout River and Duyker River, *Burchell,* 6345 !
George Div.; Long Kloof, 5300 ft., *Young in Herb. Bolus,* 2471b ! Uitenhage
Div.; Zuurberg Range, 2000–3000 ft., *Drège,* 2207 ! Port Elizabeth Div.; Algoa
Bay, *Forbes!* Port Elizabeth, *Miss West!* *Bolus,* 5930 ! Bathurst Div.; between
Riet Fontein and the source of the Kasuga River, *Burchell,* 4153 ! Albany Div.;
between Howisons Poort and Tharfield, *Atherstone,* 9 ! near Grahamstown, 2000 ft.,
MacOwan, 87 ! *Bolton!* *Galpin,* 304 ! *Williamson!* Stockenstrom Div.; Katberg,
Scully, 114 ! and in *Herb. Bolus,* 5913 ! 5914 ! Queenstown Div.; northern slopes
of Kat Berg, 5000–5300 ft., *Galpin,* 1686 ! British Kaffraria; without precise
locality, *Cooper,* 384 ! Eastern Frontier, *Hutton!*
CENTRAL REGION: Somerset Div.; Somerset, *Bowker!*

51. **S. princeps** (Bolus in Hook. Ic. Pl. t. 1729); plant 1½–2¾ ft.
high, very stout; leaves 2, radical, sessile, spreading, ovate-
orbicular or cordate-orbicular, subobtuse, 4–8½ in. long, 3–7 in.
broad, fleshy; scapes 1½–2¾ ft. high, with several broad spathaceous
imbricate sheaths; spikes oblong, 4–10 in. long, dense, many-
flowered; bracts ovate or ovate-oblong, subobtuse, ¾–1¼ in. long,
reflexed after flowering; pedicels 6–8 lin. long; flowers large,
bright carmine or rose with carmine tips to the segments; sepals
and petals united at the base; odd sepal oblong, obtuse, about ½
in. long; lateral sepals spreading, elliptic-oblong, longer than the odd
sepal; petals oblong, subacute, nearly as long as the odd sepal;
lip cucullate, broadly orbicular-ovate, 6–7 lin. long, with apiculate
reflexed apex; spurs elongate, curved, slender, 8–10 lin. long;
column curved, 5–6 lin. long; stigma subquadrate, emarginately
2-lobed, 2 lin. long; rostellum triangular, subacute, nearly as long
as the stigma. *Bolus in Journ. Linn. Soc.* xxv. 193; *Durand &
Schinz, Conspect. Fl. Afr.* v. 98; *Schlechter in Engl. Jahrb.* xxxi. 147.
S. membranaceum, N. E. Br. in Gard. Chron. 1889, v. 136, *partly*; *Bot.
Mag. t.* 7104; *Kränzl. Orch. Gen. et Sp.* i. 678, *partly, not of Sw.*

SOUTH AFRICA: without locality, *Mrs. Holland,* 15 ! 16 !
COAST REGION: Knysna Div.; among shrubs on the coast near Knysna,

Schlechter, 5906.   Port Elizabeth Div.; sandy dunes on the coast near Port
Elizabeth, *Hallack in Herb. Bolus*, 5929!

A very robust species, bearing some resemblance to *S. carneum*, R. Br., but
markedly different in the shape of the stigma. I have not seen the Knysna
specimen.

52. **S. carneum** (R. Br. in Ait. Hort. Kew. ed. 2, v. 196); plant
1–2 ft. high; stems stout; leaves 2, radical, subsessile, spreading,
ovate or orbicular-ovate, subobtuse, 3–8 in. long, fleshy; scapes
1–2 ft. high, with several ovate or spathaceous sheaths, the lower
sometimes subfoliaceous; spikes oblong, 3–8 in. long, dense, many-
flowered; bracts ovate or ovate-oblong, obtuse or subobtuse, $\frac{1}{2}$–1 in.
long; pedicels about $\frac{1}{2}$ in. long; flowers large, pink or rose-coloured;
sepals and petals united at the base; odd sepal lanceolate-oblong,
obtuse, 6–8 lin. long; lateral sepals spreading, rather larger than
the odd sepal; petals elliptic-oblong, obtuse, as long as the odd
sepal; lip cucullate, broadly ovate, obtuse, reflexed at the apex,
7–9 lin. long; spurs slender, somewhat curved, 7–9 lin. long;
column curved, 6–7 lin. long; stigma oblong, obtuse, somewhat
curved, $2\frac{1}{2}$ lin. long; rostellum ovate-oblong, subacute, over half as
long as the stigma. *Bot. Mag. t.* 1512; *Lindl. Gen. & Sp. Orch.*
336; *Fl. des Ser. t.* 329; *Reichb. Fl. Exot.* iv. 32, *t.* 266; *Drège,
Zwei Pfl. Documente*, 125; *Bolus in Journ. Linn. Soc.* xxv. 193, *and
in Trans. S. Afr. Phil. Soc.* v. 120; *Durand & Schinz, Conspect. Fl.
Afr.* v. 94; *Gard. Chron.* 1885, xxiv. 331; 1888, iv. 696, 697, *fig.*
98; *Kränzl. Orch. Gen. et Sp.* i. 677; *Schlechter in Engl. Jahrb.* xxxi.
145. *S. cucullatum, Thunb. Fl. Cap. ed.* i. 88, *and ed. Schult.* 17,
partly.   *Orchis carnea, Dryand. in Ait. Hort. Kew. ed.* 1, iii. 294.

SOUTH AFRICA: without locality, *Thunberg*! *Brown*! *Wright*, 130! *Grey*!
COAST REGION: Cape Div.; Simons Bay, *Wright*, 143! sandy plains near
Cape Town, *Zeyher*, 4991! flats between Wynberg and Muizenberg, *Bolus*, 4881!
near Fish Hoek and neighbouring hills up to 800 ft., *Bolus*, 4881; Red Hill,
Wolley-Dod, 1808! Knysna Div.; Groene Vallei, under 500 ft., *Drège*, 8293!

XXIX. AVICEPS, Lindl.

Sepals and *petals* united into a broadly oblong spreading papillose
limb almost to the apex. *Lip* superior, sessile at the base of the
column, erect, cucullate, broadly ovate, acute, produced behind into
a pair of saccate or broadly oblong descending spurs. *Column* erect
under the lip, somewhat curved, divided at the apex into two lobes,
the upper oblong, bearing the papillose stigma on its anterior
surface, the lower anticous and forming the rostellum. *Anther*
hanging under the rostellum; cells parallel; pollinia granular, each
with a short stipes and distinct gland. *Ovary* and *pedicel* not
twisted.

A dwarf glabrous herb, with elliptic-ovoid tubers; leaves in a basal tuft,
spreading; spikes very short, subcapitate, and almost included within the tuft

of leaves, generally few-flowered; flowers relatively large; bracts leaf-like, spreading.

DISTRIB. A single species, almost confined to the western and south-western part of Cape Colony.

1. **A. pumila** (Lindl. Gen. & Sp. Orch. 346); plant 1¼–4 in. high; stem stout; leaves 4–6 in a basal tuft, the lower somewhat spreading, ovate, acute, ¾–1¾ in. long, ½–1 in. broad, somewhat fleshy, passing upwards into the bracts; scapes 1¼–4 in. high, stout; spikes congested, very short, usually few-flowered; bracts leaf-like, ovate to ovate-lanceolate, acute or acuminate, ¾–1¼ in. long; pedicels 3–4 lin. long; flowers rather large, brown and yellow, with a strong putrid odour like a *Stapelia;* sepals and petals united into a broadly oblong limb almost to the apex; limb about ½ in. long, over ¼ in. broad, fleshy, sepia-brown, with numerous whitish papillæ, distinctly saccate at the base, upper margin of sac formed by a prominent transverse crenulate ridge, from which a prominent central keel extends upwards; free lobes scarcely 1 lin. long; sepals broad, obtuse, lateral pair falcate; petals linear; lip galeate, elliptic-ovate, acute, about ½ in. long, with slightly papillate nerves, yellow, with numerous transverse brownish streaks and spots; spurs very broadly saccate, about 1 lin. long; column curved, 5 lin. long; stigma oblong or elliptic-oblong, 1½ lin. long; rostellum subulate, not half as long as the stigma; capsule oblong, 4–5 lin. long. *Drège, Zwei Pfl. Documente,* 68, 69. *Satyrium pumilum, Thunb. Prodr.* 6, *and Fl. Cap. ed. Schult.* 19; *Sw. in Vet. Acad. Handl. Stockh.* 1800, 216; *Bolus, Ic. Orch. Austr.-Afr.* i. t. 25, *and in Journ. Linn. Soc.* xxv. 195; *Durand & Schinz, Conspect. Fl. Afr.* v. 98; *Kränzl. Orch. Gen. et Sp.* i. 656; *Schlechter in Engl. Jahrb.* xxxi. 194. *Diplecthrum pumilum, Pers. Syn.* ii. 509.

COAST REGION : Piquetberg Div. ; Piquet Berg, near streams, *Thunberg* ! Tulbagh Div. ; mountains near Tulbagh, *Zeyher.* Worcester Div. ; mountains near Hex River, *Tyson* !
CENTRAL REGION : Ceres Div. ; abundant on moist ground near Ceres, 1500 ft., *Bolus,* 7347 ! and in *MacOwan & Bolus, Herb. Norm. Austr.-Afr.* 1096 !
WESTERN REGION : Little Namaqualand ; near Ezels Fontein and Roode Berg, 3500–4000 ft., *Drège,* 2996a ! near Lily Fontein, at the foot of Ezels Kop, 4000–5000 ft., *Drège* ! 2996b !

Remarkable for the Stapelia-like flowers and the united sepals and petals with a distinct sac at the base ; the latter character apparently overlooked in previous descriptions.

XXX. PACHITES, Lindl.

Sepals subequal, free, spreading. *Petals* similar to the sepals or rather smaller. *Lip* superior, erect, undivided or lobed, without spur. *Column* erect, cylindrical below ; anther-bed obliquely directed towards the summit of the column ; anther-cells pendulous, separate or approximate at the base ; pollinia 2, in separate cells, granular, with ascending caudicles ; glands terminal or deflexed in front, somewhat distant. *Rostellum* fleshy, pyramidal or horseshoe-shaped, with two horn-like appendages; stigma pendulous between the anther-cells. *Ovary* and *pedicel* not twisted.

Terrestrial glabrous herbs, resembling a slender *Disa* in habit, with somewhat thickened roots; leaves linear, cauline, erect; spikes cylindrical or short, few- or many-flowered; flowers small, pink; bracts lanceolate.

DISTRIB. Species 2, limited to the south-western corner of Cape Colony.

An anomalous genus, which was long known from a solitary individual collected by Burchell, but of which a few additional individuals have since been discovered, also a second species. Dr. Bolus remarks that it seems to connect *Disa* with *Satyrium*, but has very distinct characters.

Sepals and petals elliptic-oblong, obtuse; lip entire   ... (1) **appressa.**

Sepals and petals narrow, acute or acuminate; lip
    3-lobed ...   . (2) **Bodkini.**

1. **P. appressa** (Lindl. Gen. & Sp. Orch. 301); plant $\frac{3}{4}$–1$\frac{1}{2}$ ft. high; stem rather stout; leaves cauline, 5–14, erect, linear, acuminate, the lower 3–5 in. long, upper decreasing into the bracts, broader and sheathing at the base; scapes $\frac{3}{4}$–1$\frac{1}{2}$ ft. high; spikes oblong or elongate, 3–7 in. long, lax or somewhat dense; bracts lanceolate, acuminate, $\frac{1}{2}$–1 in. long; pedicels 5–7 lin. long; flowers lilac, with two yellow stripes on the lip, and the column yellow with the outer margin of the rostellum red; sepals suberect, elliptic-oblong, obtuse, about 4 lin. long, with incurved margin; petals elliptic-oblong, obtuse, nearly as long as the sepals, with incurved margins; lip entire, elliptic lanceolate, subacute, rather shorter than the sepals, straight, erect, about 2 lin. long, furnished with two horn-like obtuse processes (apparently staminodes), about 2 lin. long; rostellum terminal, erect, triangular, obtuse, thick, broader than long; anther-cells placed one on each side of the stigma, parallel; glands affixed to the base of the rostellum, distant; stigma large, oblong, with a prominent margin; capsule oblong, 6–7 lin. long. *Krauss in Flora,* 1845, 306, *and Beitr. Fl. Cap- und Natal.* 158; *Bolus, Ic. Orch. Austr.-Afr.* i. *t.* 76, *and in Journ. Linn. Soc.* xxv. 196; *Durand & Schinz, Conspect. Fl. Afr.* v. 99; *Kränzl. Orch. Gen. et Sp.* i. 721; *Schlechter in Engl. Jahrb.* xxxi. 198.

COAST REGION: Caledon Div.; Baviaans Kloof, *Krauss*; Swellendam Div.; summit of the Langeberg Range, near Swellendam, *Burchell,* 7356! slopes near Zuurbraak, 2000–4000 ft., *Schlechter,* 2157! Riversdale Div.; Garcias Pass, near Riversdale, 1500–2000 ft., *Schlechter!*

Burchell's original specimen is much stouter and has far more numerous leaves and flowers than those subsequently gathered, but the structure seems substantially identical.

2. **P. Bodkini** (Bolus, Ic. Orch. Austr.-Afr. i. t. 26); plant 5–8 in. high; stem erect, moderately stout; leaves cauline, about 5, linear, acuminate, with a broad membranous sheathing base, erect or suberect, 1–3 in. long, gradually decreasing upwards; scapes 5–8 in. high; spikes 1$\frac{1}{4}$–2$\frac{1}{2}$ in. long, lax, few- to many-flowered; bracts

linear, acuminate, suberect, $\frac{1}{2}$–$\frac{3}{4}$ in. long; pedicels 4–5 lin. long; flowers lilac-purple, with the anther and side lobes of the lip purple; sepals lanceolate, acuminate, somewhat spreading, 5–6 lin. long 1-nerved; petals linear, acute, 4–5 lin. long, margins slightly incurved, 1-nerved; lip erect, 3-lobed, as long as the petals; front lobe linear, acuminate, 2–2$\frac{1}{2}$ lin. long; side lobes broadly oblong, obtuse, under 1 lin. long, clasping the column, furnished at the base beneath with a cluster of minute yellow tubercles; column straight, slender, about 3$\frac{1}{2}$ lin. long; anther-cells curved, approximate at the base; glands capping the arms of the rostellum; rostellum horseshoe-shaped, with erect arms; stigma horseshoe-shaped, cushion-like. *Kränzl. Orch. Gen. et Sp.* i. 722; *Schlechter in Engl. Jahrb.* xxxi. 199.

COAST REGION : Cape Div. ; in a moist place on Muizen Berg, 1300 ft., rare, *Bodkin in Herb. Bolus*, 7071. Caledon Div. ; Zwart Berg, 2000 ft., *Bodkin in Herb. Bolus*, 6970 ! Houw Hoek Mountains, *Bodkin*. Bredasdorp Div. ; mountains near Koude River, 1000 ft., *Schlechter*, 9624 !

XXXI. ORTHOPENTHEA, Rolfe.

Sepals free, odd one inferior, expanded and horizontal or hooded and reflexed, concave, saccate or with a short spur. *Petals* reflexed, small or somewhat exserted, more or less adnate to the column at the base. *Lip* superior, sessile, oblong, narrow, or trowel-shaped or broadly elliptical, without a spur. *Column* short. *Anther* horizontal or reflexed, 2-celled ; cells distinct and parallel ; pollinia solitary in each cell, granular, attached by short caudicles to two distinct glands, seated at the apex of the rostellum. *Ovary* straight, or twisted in *O. elegans* ; rostellum erect, emarginate, obovate, bifid or trifid, with short side arms ; stigma cushion-like or somewhat concave in *O. fasciata*. *Capsule* cylindrical or oblong.

Terrestrial herbs, with simple sessile tubers ; leaves cauline or radical, the upper often reduced and sheath-like ; flowers medium-sized, usually produced in corymbs or short spikes, rarely subsolitary ; bracts usually narrow, sheathing in *O. fasciata*.

DISTRIB. Species 10, almost entirely south-western.

Much like *Penthea*, Lindl., in habit, but differing in having inverted flowers as in *Pachites*, Lindl., and a differently-shaped dorsal sepal. It includes *Disa* section *Vaginaria*, Lindl., and section *Orthocarpa*, Bolus. Lindley also included some of the species in his genus *Penthea*.

Odd sepal more or less concave or subsaccate :
     Leaves narrow, not broadened upwards, more or less
         elongate and acute :
         Leaves 1–3 in. long ; flowers medium-sized and
             usually numerous :
             Lateral sepals ovate-oblong, obtuse, flat and
                 white   ...   . (1) **bivalvata.**

Lateral sepals elliptic-lanceolate, shortly apiculate,
  the adjacent inner halves black, the outer
  white   ...   . (2) **atricapilla.**

Leaves 3–5 in. long; flowers rather larger and few
  in number:
  Bracts ovate; pedicels ½ in. long, not twisted;
    flowers green and brown   ...   . (3) **Bodkini.**

Bracts lanceolate; pedicels ¾–1 in. long, spirally
  twisted; flowers white   ...   . (4) **elegans.**

Leaves lanceolate or oblong, rarely subspathulate or
  somewhat broader upwards:
  Flowers about ¼ in. long:
    Dorsal sepal entire ...   . (5) **minor.**
    Dorsal sepal 3-lobed   ...   . (6) **triloba.**

Flowers ⅛ to over ⅓ in. long:
  Leaves short, ¼–⅓ in. broad:
    Flowers usually several, subcorymbose, under
      ½ in. long   ...   . (7) **richardiana.**

    Flowers 1 or 2, over ½ in. long ...   . (8) **schizodioides.**

  Leaves longer, usually over ½ in. broad ...   ... (9) **rosea.**

Odd sepal with a short acute spur:
  Flowers yellow, with long aristate sepals   ...   ... (10) **Telipogonis.**

  Flowers white, with obtuse sepals   ...   . (11) **fasciata.**

1. O. bivalvata (Rolfe); plant ⅓–1 ft. high; stem moderately
stout; leaves radical and cauline, numerous, spreading or suberect,
linear or lanceolate-linear, acute, somewhat rigid, 1–3 in. long,
reduced upwards into the bracts; scapes ⅓–1 ft. high, with narrow
sheaths above; spikes subcorymbose, 1–2 in. broad, many-flowered;
bracts lanceolate or linear-lanceolate, acuminate, ½–¾ in. long;
pedicels ½–¾ in. long; flowers medium-sized, white, slightly veined
with green on the dorsal sepal, and the petals and lip deep brown
or nearly black; dorsal sepal galeate, obovate from a narrow base,
obtuse, concave or somewhat conduplicate, subsaccate above the
middle, 5–6 lin. long; lateral sepals oblique, ovate-oblong, obtuse,
flat, with a fleshy keel near the apex, 5–6 lin. long; petals oblique,
broadly oblong, truncate, minutely denticulate, fleshy, about 3 lin.
long; lip rhomboid-oblong, subacute, fleshy, 3½ lin. long; column
2½ lin. long; anther reflexed; rostellum erect, trifid at the apex,
concave, with oblong bidentate side lobes; stigma pulvinate. *Ophrys
bivalvata, Linn. f. Suppl.* 403; *Murr. Syst. Veg.* xiv. 814. *Serapias
melaleuca, Thunb. Prodr.* 3. *Disa melaleuca, Sw. in Vet. Acad.
Handl. Stockh.* 1800, 213; *Thunb. Fl. Cap. ed. Schult.* 16, *partly*;
Harv. Thes. Cap. i. 53, *t.* 84; *Bolus in Trans. S. Afr. Phil. Soc.* v.
166, *Orch. Cap. Penins.* 166, *and Journ. Linn. Soc.* xxv. 201. *D.
bivalvata, Durand & Schinz, Conspect. Fl. Afr.* v. 100; *Kränzl. Orch.
Gen. et Sp.* i. 763; *Schlechter in Engl. Jahrb.* xxxi. 279. *Penthea
melaleuca, Lindl. Gen. & Sp. Orch.* 361; `*Drège, Zwei Pfl. Documente,*
73, 82; *Krauss in Flora,* 1845, 306, *and Beitr. Fl. Cap- und Natal.*
158; *Sond. in Linnæa,* xx. 220.

SOUTH AFRICA : without locality, *Brown* ! *Masson* ! *Sieber* ! *Villett* ! *Thom*, 276 ! *Harvey*, 246 !

COAST REGION : Clanwilliam Div. ; Blue Berg, 4000-5000 ft., *Drège*, 1247b ! Worcester Div. ; Dutoits Kloof, 3000-4000 ft., *Drège* ! 1247a ! Paarl Div. ; Bains Kloof, *Rehmann*, 2287 ! Cape Div. ; Table Mountain, 1200-3500 ft., *Burchell*, 651 ! *Thunberg* ! *Ludwig*, *Harvey* ! *Bolus*, 4208 ! *Pappe*, *Zeyher*, *Rehmann*, 568 ! *Prior* ! *Kässner*, *Schlechter* ! *Wolley-Dod*, 2212 ! Cape Flats, near Doorn Hoogte, *Ecklon* ; Cape Point, *Herschel* ! Muizen Berg, 1400 ft., *Bolus*, 4208 ! *Schlechter*, 148 ; hill west of Simonstown, *Wolley-Dod*, 2095 ! Steenberg, 900 ft., *Dümmer*, 973 ! Red Hill, Mrs. *Jameson* ! Riversdale Div. ; between Little Vet River and Garcias Pass, *Burchell*, 6855 ! George Div. ; Cradock Berg, near George, *Mund & Maire*, Outeniqua Mountains, above Montagu Pass, 3000 ft., *Schlechter*, *Penther*, *Krook*. Knysna Div. ; Elands River, *Penther*, 296 ! Humansdorp Div. ; Storms River, 200 ft., *Schlechter*, *Penther*, *Krook*.

CENTRAL REGION : Ceres Div. ; Witzenberg and Skurfdeberg Rangen, *Zeyher*.

2. O. atricapilla (Rolfe) ; plant $\frac{1}{3}$–1 ft. high ; stem moderately stout ; leaves radical and cauline, rather numerous, spreading or suberect, linear or lanceolate-linear, acute, somewhat rigid, 1–2$\frac{1}{2}$ in. long, reduced upwards into the bracts ; scapes $\frac{1}{3}$–1 ft. high, with narrow sheaths above ; spikes subcorymbose, ultimately subracemose, 1–2 in. broad, many-flowered or rarely few-flowered ; bracts lanceolate or linear-lanceolate from a broader base, $\frac{1}{2}$–$\frac{3}{4}$ in. long ; pedicels $\frac{1}{2}$–$\frac{3}{4}$ in. long ; flowers medium-sized ; dorsal sepal white, slightly veined with green, lateral sepals blackish-purple on the adjacent inner halves, white on the outer, petals and lip green, dotted with purple ; dorsal sepal galeate, obovate from a narrow base, obtuse, the upper half with inflexed margin, about $\frac{1}{2}$ in. long, obtusely saccate above the middle ; lateral sepals oblique, elliptic-lanceolate, obtuse, with a short obtuse apiculus behind the apex, and a rounded basal lobe in front, conduplicate, about $\frac{1}{2}$ in. long ; petals ovate-oblong, obliquely acute, denticulate at the apex, with an ample rounded basal lobe in front, 3 lin. long ; lip lanceolate-linear, obtuse, about 3 lin. long ; column about 3 lin. long ; anther reflexed ; rostellum erect, bifid at the apex, with oblong bidentate side lobes ; stigma pulvinate ; capsule oblong, about $\frac{3}{4}$ in. long. *Penthea atricapilla*, *Harv. in Hook. Lond. Journ. Bot.* i. 17. *P. melaleuca*, var. *atricapilla*, *Sond. in Linnæa*, xx. 220. *Disa atricapilla*, *Bolus in Journ. Linn. Soc.* xix. 344, xxv. 202, *Trans. S. Afr. Phil. Soc.* v. 166, t. 10, *and Orch. Cap. Penins.* 166, t. 10 ; *Kränzl. Orch. Gen. et Sp.* i. 764. *D. bivalvata*, var. *atricapilla*, *Schlechter in Engl. Jahrb.* xxxi. 280. *D. melaleuca*, *Thunb. Fl. Cap. ed. Schult.* 16, *partly.*

SOUTH AFRICA : without locality, *Oldenburg*, 1050 ! *Scott-Elliot* ! COAST REGION : Tulbagh Div. ; mountains near Tulbagh Kloof, *Pappe* ! Worcester Div. ; Slang Hoek, *Cooper*, 1686 ! and without precise locality, *Cooper*, 1614 ! 3597 ! Cape Div. ; Cape Flats at Doorn Hoogte, *Zeyher* ! Muizenberg, 1400-2000 ft., *Bolus*, 4638 ! *Harvey* ! Table Mountain, 2000-2500 ft., *Harvey*, *Schlechter*, 91, *Dümmer*, 2136 ! Stellenbosch Div : Hottentots Holland Mountains, *Masson* ! *Bowie* ! Caledon Div. ; Genadendal, *Prior* ! Baviaans Kloof, *Krauss* ! Swellendam Div. ; by the River Zondereinde, near Appels Kraal, *Zeyher*, 1579 ! *Pappe* ! 3 ! CENTRAL REGION : Ceres Div. ; Cold Bokkeveld, near Gydouw, 3000 ft., *Bolus*, 4638 ; near Ceres, 1500 ft., *Bolus in MacOwan & Bolus. Herb. Norm. Austr.-Afr.* 409 !

3. O. Bodkini (Rolfe); plant 3–10 in. high; stem stout; leaves cauline, 6–8, erect or suberect, linear from a broad sheathing base, acuminate, 3–5 in. long; scapes 3–10 in. high, with a few leaf-like sheaths above; spikes subcorymbose, up to 1½ in. broad, 4–8-flowered, occasionally reduced and 1–2-flowered, after flowering somewhat racemose; bracts broadly ovate, acute or shortly acuminate, ½–¾ in. long; pedicels about ½ in. long; flowers rather large, dull red or greenish with a red suffusion, petals and lip deep purple with yellow tips; dorsal sepal galeate, broadly elliptic, subobtuse, much incurved, 7–8 lin. long, very concave but scarcely saccate behind; lateral sepals oblique, ovate-elliptic, subobtuse, 7–8 lin. long; petals obliquely ovate, obtuse, 3 lin. long; lip elliptic-ovate, obtuse, concave, about 3½ lin. long; column about 3 lin. long; anther reflexed; rostellum erect, tall, shortly bifid; stigma pulvinate. *Disa Bodkini, Bolus in Journ. Linn. Soc. xxii. 74, xxv. 202, Trans. S. Afr. Phil. Soc. v. 165, t. 13, and Orch. Cap. Penins. 165, t. 13; Durand & Schinz, Conspect. Fl. Afr. v. 100; Kränzl. Orch. Gen. et Sp. i. 763; Schlechter in Engl. Jahrb. xxxi. 281.*

SOUTH AFRICA : without locality, *Mund.*
COAST REGION : Cape Div.; moist places on Table Mountain, 2300–3000 ft., *Bodkin in Herb. Bolus,* 4968! and in *MacOwan & Bolus, Herb. Norm. Austr.-Afr.,* 333 ! Muizen Berg, *Upjohn* !

4. O. elegans (Rolfe); plant ½–1½ ft. high; stem moderately stout; leaves cauline, 5–8, suberect, lanceolate or linear-lanceolate, acute or acuminate, 3–5 in. long, reduced upwards into the bracts; scapes ½–1½ ft. high, with several narrow sheaths above; spikes subcorymbose or shortly racemose, 2–6-flowered; bracts lanceolate, acuminate, ¾–1 in. long; pedicels ¾–1 in. long; flowers rather large, white, petals and lip purple, tipped with yellow; dorsal sepal galeate, broadly elliptic, obtuse, somewhat curved, concave but scarcely saccate behind, 6–8 lin. long; lateral sepals broadly elliptic, obtuse, 7–8 lin. long; petals spathulate-oblong, obtuse, falcately curved, 3½–4 lin. long; lip rhomboid or rhomboid-lanceolate, subobtuse, 4–4½ lin. long; column 3 lin. long; anther reflexed; rostellum erect, emarginate, with short divaricate side lobes; stigma pulvinate. *Penthea elegans, Sond. in Linnæa, xx. 220. Disa elegans, Reichb. f. in Flora, 1865, 182; Bolus in Journ. Linn. Soc. xxv. 200, and Ic. Orch. Austr.-Afr. i. t. 35; Durand & Schinz, Conspect. Fl. Afr. v. 102; Schlechter in Engl. Jahrb. xxxi. 281.*

COAST REGION : Caledon Div.; mountains near the Zondereinde River, *Zeyher,* 3934! *Pappe,* 5 ! Swellendam Div.; mountains above Zuurbraak, 4000 ft., *Schlechter.*
CENTRAL REGION : Ceres Div.: Skurfdeberg Range, near Ceres, 3200 ft., *Bolus,* 7372 ! Prince Albert Div.; summit of the Zwartberg Range, 5300 ft., *Bolus,* 12328 !

This species differs from the others in having a complete twist to the ovary, which brings the lip into the same position.

5. O. minor (Rolfe); plant 3–5 in. high; stem rather slender; leaves radical, 4–6, spreading or suberect, linear-spathulate, subobtuse, attenuate at the base, $\frac{3}{4}$–1 in. long; scapes 3–5 in. high, with one or two lanceolate sheaths above; spikes very short, several-flowered; bracts oblong-lanceolate, subacute, 3–5 lin. long; pedicels about 4 lin. long; flowers small, golden-yellow; dorsal sepal galeate, subglobose, very obtuse, subsaccate behind, about 3 lin. long; lateral sepals broadly obovate-elliptic, obtuse, concave, about $3\frac{1}{2}$ lin. long; petals suberect, oblong, obtuse, concave, about $1\frac{1}{2}$ lin. long; lip linear-spathulate, very obtuse, 2 lin. long; column short; anther reflexed; rostellum erect, shortly 3-lobed, with linear side lobes; stigma pulvinate. *Penthea minor, Sond. in Linnæa,* xix. 104. *Disa minor, Reichb. f. in Flora,* 1865, 182; *Bolus in Journ. Linn. Soc.* xxv. 202; *Durand & Schinz, Conspect. Fl. Afr.* v. 104; *Schlechter in Engl. Jahrb.* xxxi. 278; *Kränzl. Orch. Gen. et Sp.* i. 800.

COAST REGION: Tulbagh Div. ; Great Winterhoek, near Tulbagh, 4000–5000 ft., *Zeyher* ! *Bolus.*

6. O. triloba (Rolfe); plant $2\frac{1}{2}$–$5\frac{1}{2}$ in. high; stem slender; leaves radical, 3–5, subspathulate or oblanceolate, subobtuse, attenuate into a sheathed petiole at the base, $\frac{3}{4}$–$1\frac{3}{4}$ in. long; scapes $2\frac{1}{2}$–$5\frac{1}{2}$ in. long, with several narrow acute sheaths above; spikes subcorymbose, 2–4-flowered; bracts lanceolate, acuminate, 4–5 lin. long; pedicels about 3 lin. long; flowers rather small, white with yellow petals, lip and column; dorsal sepal galeate, horizontal, 3-lobed, about 3 lin. long; side lobes ample, rounded; front lobe triangular-oblong, obtuse, concave; spur reduced to a concave gibbosity; lateral sepals broadly elliptic-oblong, obtuse, 3–$3\frac{1}{2}$ lin. long; petals broadly elliptic-oblong, obtuse, about 2 lin. long; lip linear, obtuse, 2 lin. long; column short; anther reflexed; rostellum erect, 3-lobed, with very short side lobes; stigma pulvinate. *Penthea triloba, Sond. in Linnæa,* xix. 104. *Disa oligantha, Reichb. f. in Flora,* 1865, 182; *Bolus in Journ. Linn. Soc.* xxv. 202, *and Ic. Orch. Austr.-Afr.* ii. t. 87; *Durand & Schinz, Conspect. Fl. Afr.* v. 105; *Kränzl. Orch. Gen. et Sp.* i. 800; *Schlechter in Engl. Jahrb.* xxxi. 298. *D. parvilabris, Bolus in Journ. Linn. Soc.* xix. 344.

COAST REGION : Worcester Div. ; mountains near Hex River, *Ecklon & Zeyher.* Cape Div. ; Upper Plateau of Table Mountain, near Maclears Beacon, 3600 ft., *Bodkin in Herb. Bolus, Wolley-Dod,* 2183 ! 2338 ! *Marloth in Herb. Bolus,* 7984. Stellenbosch Div. ; Jonkers Hoek, *Marloth.*

7. O. richardiana (Rolfe); plant 3–6 in. high; stem rather slender; leaves radical, 6–8, spreading or suberect, spathulate-lanceolate, subacute or obtuse, attenuate below, $\frac{3}{4}$–$1\frac{1}{4}$ in. long; scapes 3–6 in. long, with several narrow sheaths above; spikes short, subcorymbose, $\frac{3}{4}$–1 in. broad; bracts oblong-lanceolate, subobtuse, 5–7 lin. long; pedicels about 1 in. long; flowers rather

small, with white sepals and golden-yellow petals and lip; dorsal
sepal galeate, subglobose, very obtuse, with inflexed margin, subsaccate behind, about 5 lin. long; lateral sepals broadly ovate, very
obtuse, spreading, subconcave at the apex, about 5 lin. long; petals
obovate-oblong, obliquely apiculate, denticulate, about 2½ lin. long,
included within the dorsal sepal; lip oblong or obovate, obtuse or
truncate, about 1½ lin. long; column short; anther reflexed;
rostellum erect, 3-lobed, with short linear side lobes; stigma pulvinate. *Disa richardiana, Lehm. ex Lindl. Gen. & Sp. Orch.* 361;
N. E. Br. in Gard. Chron. 1885, xxiv. 232; *Bolus in Trans. S.
Afr. Phil. Soc.* v. 164, *Orch. Cap. Penins.* 164, *Journ. Linn. Soc.* xxv.
202, *and Ic. Orch. Austr.-Afr.* i. t. 36; *Durand & Schinz, Conspect.
Fl. Afr.* v. 107; *Kränzl. Orch. Gen. et Sp.* i. 762; *Schlechter in
Engl. Jahrb.* xxxi. 278. *Penthea obtusa, Lindl. Gen. & Sp. Orch.* 361.

SOUTH AFRICA : without locality, *Bergius, Mund.*
COAST REGION : Cape Div. ; moist banks on the eastern side of Table Mountain,
near the summit, *Brown*! *Harvey*, 121! *Bolus*, 4846! rocky clefts on the lower
plateau, 2500 ft., *Bolus*, and in *MacOwan & Bolus, Herb. Norm. Austr.-Afr.* 168!
Wolley-Dod, 1787! near Wynberg Reservoir, *Wolley-Dod*, 3176! Caledon Div. ;
Genadendal, 5000 ft., *Galpin*, 4621!

8. O. schizodioides (Rolfe); plant 5–8 in. high; stem rather
slender; leaves radical, rosulate, 4–8, spreading, lanceolate to
ovate-lanceolate, acute or apiculate, narrowed at the base, ½–¾ in.
long; scapes 5–8 in. high, with several narrow acuminate sheaths
above; spikes short, 1–2-flowered; bracts lanceolate, acuminate,
6–8 lin. long; pedicels slender, ¾–1 in. long; flowers medium-sized,
submembranous, white with dull purple tips to the petals and lip;
dorsal sepal galeate, broadly ovate, subobtuse or acute, 5–6 lin.
long, obtusely saccate behind; lateral sepals elliptic-ovate, subacute,
about 7 lin. long; petals oblong, obtuse, somewhat incurved and
concave at the apex, 2½ lin. long; lip obovate-oblong, obtusely
apiculate, about 3 lin. long; column short; anther reflexed;
rostellum erect, rhomboid or obovate, rounded at the apex; stigma
pulvinate. *Disa schizodioides, Sond. in Linnæa,* xix. 92; *Bolus in
Journ. Linn. Soc.* xxv. 202, *and Ic. Orch. Austr.-Afr.* i. t. 85; *Durand
& Schinz, Conspect. Fl. Afr.* v. 107; *Kränzl. Orch. Gen. et Sp.* i. 790;
Schlechter in Engl. Jahrb. xxxi. 276.

COAST REGION : Swellendam Div. ; on a mountain peak near Swellendam,
Burchell, 7323! mountains of Puspus Valley, *Ecklon & Zeyher*, rocky clefts on the
Langeberg Range, near Zuurbraak, 3300 ft., *Schlechter*, 2045!

9. O. rosea (Rolfe); plant 4–8 in. high; stem rather stout; leaves
radical and cauline, 3–4, spreading, spathulate or elliptic-oblong,
obtuse or subacute, narrowed below, 1½–4½ in. long; scapes 4–8 in.
long, with several narrow acute sheaths above; spikes broadly
oblong, 1–2 in. long, somewhat dense, few- to many-flowered; bracts
lanceolate or oblong-lanceolate, acuminate, ⅓–¾ in. long; pedicels
½–¾ in. long; flowers medium-sized, light rose-coloured, delicate

flesh coloured or sometimes nearly white; dorsal sepal galeate, broadly ovate, obtuse or subacute, obtusely saccate behind, 4–6 lin. long; lateral sepals broadly ovate-elliptic, obtuse or subacute, 4–6 lin. long; petals 2-lobed, about 2½ lin. long, front lobe broadly ovate, obtuse, back lobe triangular, thickened and clavate at the apex, and extending backwards into the spur; lip oblong or ovate-oblong, obtuse, 3½–4½ lin. long; column short; anther reflexed; rostellum suberect, emarginate, with very short side lobes; stigma pulvinate. *Disa rosea, Lindl. Gen. & Sp. Orch.* 350; *Bolus in Trans. S. Afr. Phil. Soc.* v. 164, *Orch. Cap. Penins.* 164, *Journ. Linn. Soc.* xxv. 202, *and Ic. Orch. Austr.-Afr.* ii. *t.* 64; *Durand & Schinz, Conspect. Fl. Afr.* v. 107; *Kränzl. Orch. Gen. et Sp.* i. 762; *Schlechter in Engl. Jahrb.* xxxi. 277.

SOUTH AFRICA : without locality, *Bergius, Mund, Hesse* ! *Villet* !
COAST REGION : Cape Div. ; moist banks, ridges and clefts of rock on Table Mountain, 1500–3200 ft., *Brown* ! *Harvey* ! *Bolus,* 4562 ! and in *MacOwan & Bolus, Herb. Norm. Austr.-Afr.* 319 ! *Schlechter,* 158 ; Muizenberg, 1400 ft., *Bolus* ; Orange Kloof, *Wolley-Dod,* 392 ! Caledon Div. ; Houw Hoek Mountains, *Bolus* ; Genadendal, 3200 ft., *Galpin,* 4623 !

10. O. Telipogonis (Rolfe); plant 2–5½ in. high ; stem somewhat stout ; leaves radical and cauline, suberect or arcuate, linear, acuminate, 1–4 in. long, reduced upwards into the bracts; scapes 2–5½ in. long, with short leaf-like sheaths above ; spikes subcapitate, ½–1 in. long, dense, many-flowered ; bracts lanceolate or linear-lanceolate, acuminate, 3–6 in. long; pedicels 3–4 lin. long; flowers small, bright yellow ; dorsal sepal galeate, broadly ovate, about 2½ lin. long, abruptly narrowed into a long slender bristle at the apex, about half as long as the limb, incurved at the margin ; spur conical, obtuse, about half as long as the limb ; lateral sepals ovate, acute, about 3 lin. long, with a long slender bristle at or sometimes behind the apex ; petals oblong, unequally bilobed at the apex, with the upper lobe acuminate, 1¼ lin. long; lip linear-lanceolate, acuminate, 2 lin. long; column short; anther reflexed ; rostellum suberect, 3-lobed, with tooth-like side lobes, and a much longer triangular concave front lobe ; stigma pulvinate. *Disa Telipogonis, Reichb. f. in Linnæa,* xx. 689 ; *Bolus in Journ. Linn. Soc.* xxv. 204, *and Ic. Orch. Austr.-Afr.* ii. *t.* 70 ; *Durand & Schinz, Conspect. Fl. Afr.* v. 108 ; *Kränzl. Orch. Gen. et Sp.* i. 779 ; *Schlechter in Engl. Jahrb.* xxxi. 267, *t.* 5, *fig.* A–E.

COAST REGION : Paarl or Worcester Div. ; mountains about Bains Kloof, Wellington District, 2300 ft., *Schlechter,* 9165 ! Cape Div. ; summit of Table Mountain, in rock crevices, 3500 ft., *Bergius* ! *Miss Kensit in Herb. Bolus,* 9355 !

11. O. fasciata (Rolfe); plant 3–6 in. high ; stem rather stout ; leaves cauline, 4–6, spreading, ovate from a sheathing base, acute or acuminate, limb ⅓–½ in. long, decreasing upwards into the bracts, leaf-sheaths barred with red ; scapes 3–6 in. high, with short ovate sheaths above ; spikes corymbose, 1–6-flowered ; bracts elliptic-

ovate, acute, 6–9 in. long, margined with red, sheathing the
pedicels; pedicels about 4 lin. long; flowers rather large, white,
with a few purple dots in the centre; dorsal sepal subgaleate,
obovate, expanded and obtuse above, somewhat concave below,
5–7 lin. long; spur narrowly conical, obtuse, about a quarter as long
as the limb; lateral sepals broadly elliptic-oblong, obtuse, 5–7 lin.
long, spreading; petals oblique, auriculate, acute, $2\frac{1}{2}$ lin. long,
with an oblong basal lobe in front; lip broadly elliptic-spathulate,
obtuse, 4–5 lin. long, spreading; column short; anther reflexed;
rostellum erect, deeply emarginate; stigma pulvinate. *Disa
fasciata, Lindl. Gen. & Sp. Orch.* 350; *Harv. Thes. Cap.* i. 54, *t.* 85;
Bolus in Trans. S. Afr. Phil. Soc. v. 167, *t.* 36, *Orch. Cap. Penins.*
167, *t.* 36, *and Journ. Linn. Soc.* xxv. 202; *N. E. Br. in Gard.
Chron.* 1885, xxiv. 231; *Durand & Schinz, Conspect. Fl. Afr.* v. 102;
Kränzl. Orch. Gen. et Sp. i. 798; *Schlechter in Engl. Jahrb.* xxxi. 272.

SOUTH AFRICA : without locality, *Masson* ! *Ecklon, Leibold.*
COAST REGION : Cape Div.; summit of Table Mountain, *Harvey*; Constantia
Berg, 2700–3000 ft., *Bodkin in Herb. Bolus,* 4965 ! and in *MacOwan & Bolus,
Herb. Norm. Austr.-Afr.* 320 ! mountains south of Simons Town, *Miller.*
Caledon Div.; Zwart Berg, near Caledon, *Ecklon* ! *Pappe,* 2 ! Houw Hoek,
2500 ft., *Bolus, Schlechter.* Stellenbosch Div.; Hottentots Holland, *Alexander* !
Lowrys Pass, 1500–2000 ft., *Schlechter,* 5378, *Penther, Krook.* Riversdale Div.;
Langeberg Range, near Riversdale, 1000–1500 ft., *Schlechter,* 2028 ; Garcias
Pass, 1200 ft., *Galpin,* 4618 ! George Div.; plains near George, *Bowie* !
Outeniqua Mountains, above Montagu Pass, 3000 ft., *Schlechter, Penther, Krook.*

XXXII. MONADENIA, Lindl.

Sepals free, the odd one superior, hood-shaped, with an oblong or
cylindrical spur, lateral pair spreading. *Petals* erect, included or
half-exserted, entire or bilobed. *Lip* anticous, sessile, usually small
and narrow, without a spur. *Column* short. *Anther* erect or
reclinate, 2-celled; cells distinct and parallel; pollinia solitary in
each cell, granular, attached by short or long caudicles to a single
gland seated near the apex of the rostellum. *Ovary* twisted;
rostellum erect, broad, subentire, emarginate or 3-lobed with
spreading arms, occasionally minute; stigma cushion-like, seated in
front of the rostellum. *Capsule* oblong or elliptic-oblong.

Terrestrial herbs with simple sessile tubers; leaves usually cauline and gradually
reduced upwards or the lower subradical and the upper reduced to short sheaths;
flowers small or medium-sized, usually in dense cylindrical spikes, sometimes
more lax or in short spikes; bracts usually narrow or narrowed from a broader base.

DISTRIB. Species 19, limited to South Africa, and the majority western.

Readily separated from *Disa* by the single gland of the pollinia, and from
Herschelia and *Amphigena* by habit and floral structure.

Spur of dorsal sepal descending :
   Flowers small; limb of dorsal sepal 1–$2\frac{1}{2}$ lin. long :
     Petals falcate-oblong or ovate :
      Spur of dorsal sepal shorter than the limb :
       Spur reduced to a gibbous prominence ...   ... (1) **conferta.**

Spur saccate or shortly oblong :
  Spike generally as long as the basal part of
    the scape :
      Leaves 2–3 lin. broad ; spikes dense   ... (2) **multiflora.**
      Leaves 4–6 lin. broad ; spikes rather lax ... (3) **macrostachya.**
  Spike much shorter than the basal part of
    the scape :
      Petals falcate-oblong ; lip oblong ...   ... (4) **auriculata.**
      Petals ovate ; lip subspathulate   ...   ... (5) **leydenberg-**
                              ensis.
Spur of dorsal sepal rather longer than the limb ... (6) **micrantha.**
Petals oblong with a short ovate limb at the base ... (7) **pygmæa.**
Flowers medium-sized ; dorsal sepal 3–9 lin. long :
  Spur of dorsal sepal shorter than the limb :
    Dorsal sepal 3½ lin. long ; spur oblong, obtuse ... (8) **densiflora.**
    Dorsal sepal 5–6 lin. long ; spur obovoidly inflated (9) **physodes.**
  Spur of dorsal sepal as long as or longer than the
    limb :
    Plant not suffused with red-purple :
      Spike elongate or somewhat lax :
        Spur of dorsal sepal oblong, obtuse :
          Dorsal sepal broadly elliptic-oblong, 3–4
            lin. long ...   . (10) **brevicornis.**
          Dorsal sepal obovate, about 5 lin. long ... (11) **prasinata.**
        Spur of dorsal sepal filiform, more or less
          elongate :
          Lower leaves usually lanceolate or oblong-
            lanceolate, acute ; spur under 9 lin.
            long :
            Petals oblong, scarcely broadened at the
              base ; bracts reticulate   ...   ... (12) **reticulata.**
            Petals ovate-oblong, distinctly broadened
              at the base ; bracts not reticulate :
              Petals as long as broad ; lip ovate-
                oblong   ...   . (13) **macrocera.**
              Petals half as long as broad ; lip
                elliptic-spathulate   ...   ... (14) **bolusiana.**
          Lower leaves oblong or broad and obtuse ;
            spur 10–12 lin. long   ...   . (15) **comosa.**
      Spike short, broad and dense   ...   . (16) **sabulosa.**
    Plant more or less suffused with red-purple :
      Petals obliquely ovate-oblong, emarginate or
        bidentate ; rostellum longitudinally keeled (17) **ophrydea.**
      Petals obliquely ovate, subobtuse ; rostellum
        not longitudinally keeled   ...   . (18) **atrorubens.**
Spur of dorsal sepal ascending   ...   . (19) **Basutorum.**

1. **M. conferta** (Kränzl. in Orch. Gen. et Sp. i. 810) ; plant ⅓–½ ft.
high ; stem rather stout ; leaves cauline, 4–6, linear or lanceolate-
linear, acute or acuminate, suberect, 1½–3 in. long ; scapes ⅓–½ ft.
high ; spikes 1–3 in. long, very dense, many-flowered ; bracts
lanceolate, acuminate, 1½–2 lin. long ; pedicels 1½–2 lin. long ;

flowers very small, dingy yellow with red petals; dorsal sepal
broadly elliptic-oblong, obtuse, somewhat concave, 1–1¼ lin. long;
spur reduced to a gibbous prominence; lateral sepals ovate-oblong,
obtuse, somewhat reflexed, 1¼–1½ lin. long; petals oblong, obtuse,
slightly concave, ¾–1 lin. long; lip linear, obtuse, 1¼ lin. long;
column very short; anther reflexed; connective shorter than the
cells; rostellum erect, subconcave, broadly oblong, emarginate at
the apex; stigma pulvinate, rounded in front. *Disa conferta, Bolus,*
Ic. Orch. Austr.-Afr. i. t. 28; *Schlechter in Engl. Jahrb.* xxxi. 212.

COAST REGION: Cape Div.; near Raapenburg, 50 ft., *Guthrie in Herb. Bolus,*
7079! Bredasdorp Div.; Koude River, 1000 ft., *Schlechter,* 9618! Caledon Div.;
near Houw Hoek, *Bodkin in Herb. Bolus,* 623.

2. M. multiflora (Sond. in Linnæa, xix. 101); plant ⅓–1 ft. high;

stem stout; leaves cauline, suberect, 6–8, linear or linear-lanceolate,
acute or acuminate, involute, sheathing at the base, 2–3 in.
long; scapes ⅓–1 ft. high; spikes 2–6 in. long, dense, many-
flowered; bracts lanceolate or ovate-lanceolate, acuminate, ½–¾ in.
long; pedicels 3–4 lin. long; flowers small, pale yellow or greenish,
with the dorsal sepal dull red; dorsal sepal ovate-oblong, obtuse or
apiculate, cucullate and incurved at the apex, 2½ lin. long; spur
1½–2 lin. long, oblong, obtuse; lateral sepals oblong or ovate-
oblong, obtuse, spreading, 2¼ lin. long; petals 2 lin. long, obliquely
falcate-oblong, obtuse, somewhat fleshy; lip subspathulate-linear,
obtuse, somewhat fleshy, 2 lin. long; column short; anther reflexed;
rostellum erect, tall, with subentire side lobes; stigma pulvinate,
large. *Kränzl. Orch. Gen. et Sp.* i. 811. *Disa multiflora, Bolus*
in Trans. S. Afr. Phil. Soc. v. 140, *Orch. Cap. Penins.* 140, *and*
in Journ. Linn. Soc. xxv. 196; *Schlechter in Engl. Jahrb.* xxxi. 215.
Disa bracteata, Sw. in Vet. Acad. Handl. Stockh. 1800, 211; *Willd.*
Sp. Pl. iv. 48; *Bolus in Trans. S. Afr. Phil. Soc.* v. 154, *in*
note, and in Journ. Linn. Soc. xxv. 196, *not of Lindl.*; *Schlechter*
in Engl. Jahrb. xxxi. 298.

SOUTH AFRICA: without locality, *Sparmann.*
COAST REGION: Cape Div.; Cape Flats at Doorn Hoogte, *Zeyher,* 1564! *Ecklon!*
and without precise locality, *Pappe,* 57! Table Mountain, above Klassenbosch,
Bolus, 4885! *Dümmer,* 554a! Hout Bay or Camps Bay, *Marloth in Herb. Bolus,*
4972! Riet Vallei, *Pappe,* 58! Lions Head, towards Camps Bay, *Bergius, Wolley-*
Dod, 3587! near summit of Twelve Apostles, *Wolley-Dod,* 3601! between
Retreat Station and Muizenberg, *Schlechter,* 1479!

The history of *Disa bracteata,* Sw., has been much confused, and the original
specimen cannot be traced, but I believe it belongs here, for the phrase "labello
lineari apice latiore; bracteis erectis floribus longioribus" is quite in agreement.
For *Disa bracteata,* Lindl., see *D. cylindrica,* Sw.

3. M. macrostachya (Lindl. Gen. & Sp. Orch. 357); plant

about 1 ft. high; stem stout; leaves in a radical tuft, suberect,
about 8, oblong-lanceolate, acute, fleshy, 3–3½ in. long, ⅓–½ in.
broad; scapes about 1 ft. high; spikes about 8 in. long, somewhat
lax, many-flowered; bracts lanceolate or ovate-lanceolate, acuminate,

$\frac{1}{2}$–1 in. long; pedicels about $\frac{1}{2}$ in. long; flowers medium-sized; dorsal sepal galeate, obovate-oblong, obtusely apiculate, 4 lin. long, 2 lin. broad; spur somewhat stout, straight, obtuse, 4 lin. long; lateral sepals oblong, obtuse, 3$\frac{1}{2}$–4 lin. long; petals obliquely ovate, suddenly narrowed about the middle, obtuse, somewhat fleshy, 3 lin. long; lip oblong, obtuse, somewhat fleshy, 3 lin. long; column short; anther reflexed, connective broad; rostellum erect, broad, with 2 acute incurved arms; stigma pulvinate. *Drège, Zwei Pfl. Documente*, 69; *Kränzl. Orch. Gen. et Sp.* i. 812, *partly.* *Disa macrostachya, Bolus in Journ. Linn. Soc.* xxv. 197; *Schlechter in Engl. Jahrb.* xxxi. 209, *partly.*

WESTERN REGION : Little Namaqualand ; near Ezols Fontein and Roode Berg, *Drège*, 8289 !

Only known from Drège's original specimen, for those subsequently referred to the species are quite distinct.

4. **M. auriculata** (Rolfe); plant 5–14 in. high; stem stout; leaves cauline, suberect, lanceolate-linear, acuminate, somewhat fleshy, 2–4 in. long; scapes 5–14 in. high; spikes 3–5 in. long, cylindrical, dense, many-flowered; bracts narrowly lanceolate, acuminate, $\frac{1}{3}$–$\frac{3}{4}$ in. long; pedicels 3–4 lin. long; flowers small, pale yellow, sometimes tinged with red-purple; dorsal sepal broadly ovate-elliptic, obtuse, concave, 1$\frac{3}{4}$–2 lin. long; spur broadly oblong, obtuse, 1 lin. long; lateral sepals ovate-oblong, obtuse, spreading, 1$\frac{1}{2}$ lin. long; petals obliquely falcate-oblong, obtuse, somewhat concave, 1$\frac{1}{2}$ lin. long; lip oblong, obtuse, fleshy, 1$\frac{1}{2}$–1$\frac{3}{4}$ lin. long; column short; anther reflexed; rostellum erect, transverse, emarginate, auricled at each side; stigma pulvinate. *M. macrostachya, Kränzl. Orch. Gen. et Sp.* i. 812, *partly, not of Lindl.* *Disa auriculata, Bolus, Ic. Orch. Austr.-Afr.* i. t. 77 ; *Schlechter in Engl. Jahrb.* xxxi. 214.

COAST REGION : Cape Div. ; on the Steenberg Range, 1200 ft., *Guthrie.* Humansdorp Div. ; grassy places near Storms River, 300 ft., *Schlechter*, 5958 !

5. **M. leydenbergensis** (Kränzl. Orch. Gen. et Sp. i. 811); plant 1–1$\frac{1}{4}$ ft. high ; stem stout ; leaves cauline, suberect, 6–8, narrowly oblong-lanceolate, acuminate, conduplicate, sheathing at the base, 1$\frac{1}{2}$–2$\frac{1}{4}$ in. long; scapes 1–1$\frac{1}{4}$ ft. high, with a few leaf-like sheaths above ; spikes oblong, 2$\frac{1}{2}$–3$\frac{1}{2}$ in. long, dense, many-flowered; bracts ovate-lanceolate, very acuminate, 6–10 lin. long; pedicels $\frac{1}{3}$–$\frac{1}{2}$ in. long; flowers small, rather fleshy ; dorsal sepal broadly ovate-oblong, very obtuse, deeply concave, about 2 lin. long; spur oblong, very obtuse, straight, more than half as long as the limb; lateral sepals ovate-oblong, obtuse, about 2$\frac{1}{2}$ lin. long; petals obliquely ovate, obtuse and slightly twisted at the apex, about 1 lin. long; lip subspathulate, obtuse, fleshy, 1$\frac{1}{2}$ lin. long, limb broadly elliptic, obtuse, with a thickened median nerve, minutely papillose ; column short ; anther reflexed ; rostellum 3-lobed, with rounded divaricate side lobes and short triangular plicate front lobe; stigma

nearly trapezoid, with thickened margin. *Disa nervosa, Schlechter in Engl. Jahrb.* xxxi. 260, *partly, not of Lindl.*

KALAHARI REGION : Transvaal; Crocodile River, Lydenburg District, *Wilms,* 1364!

Cited as a synonym of *Disa nervosa,* Lindl., by Schlechter, but very different in structure. The Kew specimen is not well preserved.

6. **M. micrantha** (Lindl. Gen. & Sp. Orch. 357); plant $\frac{1}{3}$–1 ft. high ; stem stout ; leaves cauline, suberect, usually numerous, linear or lanceolate-linear, acute or acuminate, involute, sheathing at the base, 2–6 in. long; scapes $\frac{1}{3}$–1 ft. high; spikes 2–5 in. long, densely many-flowered ; bracts ovate or ovate-lanceolate, long-acuminate, $\frac{1}{2}$–$\frac{3}{4}$ in. long; pedicels 3–4 lin. long; flowers small, yellow, with the dorsal sepal and tips of the lateral sepals red ; dorsal sepal oblong, obtuse or apiculate, cucullate, inflexed at the apex, 2–2$\frac{1}{4}$ lin. long; spur somewhat slender, obtuse, about as long as the limb ; lateral sepals oblong or ovate-oblong, obtuse, spreading, 1$\frac{1}{2}$–1$\frac{3}{4}$ lin. long; petals obliquely ovate, subobtuse, concave at the base, about 1$\frac{1}{4}$ lin. long; lip linear-oblong, obtuse, somewhat fleshy, 1$\frac{1}{4}$–1$\frac{1}{2}$ lin. long; column stout ; anther reflexed ; connective narrow, nearly as long as the cells; rostellum erect, with short obtuse arms; stigma pulvinate; capsule elliptic-oblong, 5–6 lin. long. *Kränzl. Orch. Gen. et Sp.* i. 818. *Disa micrantha, Bolus in Trans. S. Afr. Phil. Soc.* v. 142, *Orch. Cap. Penins.* 142, *and in Journ. Linn. Soc.* xxv. 196; *Schlechter in Engl. Jahrb.* xxxi. 213.

SOUTH AFRICA : without locality ; *Bergius, Masson* ! *Brown* ! *Mund & Maire, Harvey,* 141 ! *Rogers* ! *Pappe* ! *Mrs. Holland* ! *Liebold.*
COAST REGION : Tulbagh Div. ; New Kloof, 1000–2000 ft., *Drège,* 1261b!
Paarl Div. ; Paarl Mountain, 1000–2000 ft., *Drège,* 1261a. Cape Div. ; near Wynberg, 50–100 ft., *Kässner, Schlechter.* Table Mountain, *Harvey* ! *Ecklon,* 247 ! *Pappe,* 56 ! *Wolley-Dod,* 1845 ! near Cape Town, under 100 ft., *Bolus,* 3859 ! near summit of Twelve Apostles, *Wolley-Dod,* 3602 ! Claremont, *Schlechter* ! Vyges Kraal, *Wolley-Dod,* 394 ! Cape Flats, *Zeyher,* 4680 ! *Dümmer,* 443 ! Simons Bay, *Wright,* 135 ! Stellenbosch Div.; at Stellenbosch, *Lloyd in Herb. Sanderson,* 938 ! Caledon Div. ; Genadendal, *Ecklon & Zeyher* ; Caledon, *Alexander* ! Swellendam Div. ; Voormans Bosch, *Zeyher,* 1564, partly ! Swellendam Div. ; Zuurbraak, 600 ft., *Galpin,* 4606 ! Riversdale Div. ; Langeberg Range, near Riversdale, 1000 ft , *Schlechter,* 2031 ; Garcias Pass, 1200 ft., *Galpin,* 4607 ! George Div. ; between Malgaten River and Great Brak River, *Burchell,* 6139 ! Knysna Div. ; Ruigte Valley, under 500 ft., *Drège,* 1261c ! Humansdorp Div. ; near Storms River, 200 ft., *Schlechter,* 5965 ! *Penther, Krook.* Port Elizabeth Div. ; Port Elizabeth, *Hallack* ! Albany Div. ; mountains near Grahamstown, 2000–2200 ft., *MacOwan,* 381 ! Bathurst Div. ; near Port Alfred, *Burchell,* 4015 !

7. **M. pygmæa** (Durand & Schinz, Conspect. Fl. Afr. v. 112); plant 2–4 in. high ; stem rather slender; leaves cauline, 3–5, ovate or lanceolate, acute or acuminate, suberect, faintly 3–5-nerved, $\frac{1}{2}$–1 in. long, with 1 or 2 short basal sheaths ; scapes 2–4 in. high ; spikes 1–2$\frac{1}{2}$ in. long, dense, many-flowered ; bracts ovate or ovate-lanceolate, acute or acuminate, $\frac{1}{4}$–$\frac{1}{2}$ in. long ; pedicels 2–3 lin. long ; flowers small, dull red, with pale yellow petals and lip ; dorsal sepal ovate, obtuse or apiculate, concave, 2 lin. long ; spur terete, obtuse,

curved, about half as long as the limb ; lateral sepals oblong, obtuse, somewhat recurved, 1½ lin. long ; petals oblong, obtuse, 1½ lin. long, with a short ovate lobe at the posterior basal angle ; lip linear-oblong, obtuse, ¾ lin. long ; column short ; anther reflexed ; rostellum erect, with short truncate arms ; stigma pulvinate. *Kränzl. Orch. Gen. et Sp.* i. 813. *Disa pygmæa, Bolus in Journ. Linn. Soc.* xxii. 72, xxv. 196, *and in Trans. S. Afr. Phil. Soc.* v. 140, *t.* 17 ; *Schlechter in Engl. Jahrb.* xxxi. 213.

COAST REGION : Cape Div. ; sandy places on the Steenberg Range, 1400–1500 ft., *Bodkin in Herb. Bolus*, 4970 ! between Constantia Berg and Steen Berg, *Wolley-Dod*, 3636 !

8. M. densiflora (Lindl. Gen. & Sp. Orch. 357, partly) ; plant about 1 ft. high ; stem rather stout ; leaves cauline, lanceolate-oblong, subacute, somewhat fleshy, 2½–3 in. long ; scapes about 1 ft. high ; spikes 4 in. long, dense, many-flowered ; bracts lanceolate or lanceolate-ovate, acuminate, 6–9 lin. long ; pedicels about 4 lin. long ; flowers rather small ; dorsal sepal somewhat galeate, elliptic-oblong, obtuse, 3½ lin. long ; spur oblong, obtuse, two-thirds as long as the limb ; lateral sepals obliquely ovate-oblong, obtuse, 2½–3 lin. long ; petals obliquely ovate-oblong, obtuse, 2 lin. long ; lip linear-oblong, obtuse, 2½ lin. long ; column short ; anther reflexed ; rostellum with short obtuse arms ; stigma pulvinate. *Durand & Schinz, Conspect. Fl. Afr.* v. 112 ; *Kränzl. Orch. Gen. et Sp.* i. 814. *Disa densiflora, Bolus in Journ. Linn. Soc.* xxv. 197. *D. cernua, Schlechter in Engl. Jahrb.* xxxi. 210, *partly, not of Sw.*

SOUTH AFRICA : without locality, *Thom*, 732 !

Only known from the original specimen in Lindley's Herbarium. Drège's specimen cited by Lindley belongs to *M. physodes*, Reichb. f. Both are distinct from *M. prasinata*, Lindl., with which they have been confused.

9. M. physodes (Reichb. f. in Flora, 1883, 461) ; plant 1–1¼ ft. high ; stem very stout ; leaves cauline, lanceolate or oblong-lanceolate, acute or acuminate, somewhat fleshy, 3–5 in. long ; scapes 1–1¼ ft. high ; spikes 4–7 in. long, dense, many-flowered ; bracts ovate-lanceolate, acuminate, ¾–1¼ in. long ; pedicels 5–7 lin. long ; flowers medium-sized, dusky purple, with dark purple petals and lip ; dorsal sepal galeate, ovate-oblong, obtuse, 5–6 lin. long ; spur ovoidly inflated, two-thirds as long as the limb ; lateral sepals oblong, obtuse, spreading, 4½–5 lin. long ; petals obliquely ovate-oblong, obtuse, 3½–4 lin. long ; lip linear-oblong, obtuse, 4 lin. long ; column short ; anther reflexed ; rostellum with short obtuse arms ; stigma pulvinate. *M. densiflora, Lindl. Gen. & Sp. Orch.* 357, *partly* ; *Drège, Zwei Pfl. Documente*, 98 ; *Durand & Schinz, Conspect. Fl. Afr.* v. 112, *partly. Disa physodes, Sw. in Vet. Acad. Handl. Stockh.* 1800, 211 ; *Thunb. Fl. Cap. ed. Schult.* 12 ; *Lindl. Gen. & Sp. Orch.* 356 ; *Bolus in Journ. Linn. Soc.* xxv. 197, *partly* ; *Kränzl. Orch. Gen. et Sp.* i. 788. *D. cernua, Schlechter in Engl. Jahrb.* xxxi. 210, *partly, not of Sw.*

SOUTH AFRICA: without locality, *Thunberg*! *Rogers*!
COAST REGION: Malmesbury Div.; hills near Malmesbury, 300 ft., *Bolus*, 4336! Paarl Div.; between Paarl and Lady Grey Railway Bridge, under 1000 ft., *Drège*, 8288! Stellenbosch Div.; Hottentots Holland, *Pappe*, 50! Caledon Div.; Caledon, *Prior*!

10. **M. brevicornis** (Lindl. Gen. & Sp. Orch. 357); plant 1–1½ ft. high; stem stout; leaves cauline, suberect, lanceolate or linear-lanceolate, acute or acuminate, somewhat fleshy, 2–4 in. long; scapes 1–1½ ft. high; spikes 3–7 in. long, dense, many-flowered; bracts ovate-lanceolate, acuminate, ½–¾ in. long; pedicels ⅓–½ in. long; flowers moderate-sized, green, with reddish or purple petals and lip; dorsal sepal broadly elliptic-oblong, obtuse or minutely apiculate, concave, 3–4 lin. long; spur oblong, obtuse, about as long as the limb; lateral sepals obliquely oblong, apiculate, spreading, 3–4 lin. long; petals obliquely ovate-oblong, obtuse, fleshy, 2½–3 lin. long; lip linear-oblong, obtuse, 2–2½ lin. long; column short; anther reflexed, connective as long as the cells; rostellum emarginate, with erect acute side lobes; stigma pulvinate. *Sond. in Linnæa*, xix. 101; *Durand & Schinz, Conspect. Fl. Afr.* v. 111; *Kränzl. Orch. Gen. et Sp.* i. 816. *Disa brevicornis, Bolus in Journ. Linn. Soc.* xxv. 196; *Schlechter in Engl. Jahrb.* xxxi. 211.

SOUTH AFRICA: without locality, *Mund*!
COAST REGION: George Div.; without precise locality, *Bowie*! Humansdorp Div.; near Storms River, 200 ft., *Schlechter*, 5994! *Penther, Krook.* Port Elizabeth Div.; near Port Elizabeth, *Hallack in Herb. Bolus*, 6093! *Hall*! Bathurst Div.; near Bathurst, *Atherstone, Schönland.* Albany Div.; grassy slopes about Fullers, near Grahamstown, 2200 ft., *MacOwan*, 679! and without precise locality, *Atherstone*! Alexandria Div.; Zuurberg Range, *Mrs. Barber*, 445! Eastern Frontier, *Prior*!
KALAHARI REGION: Transvaal; Mount Ingram, Houtbosch Berg, 6600 ft., *Schlechter*, 4713! and without precise locality, *Buchanan*!
EASTERN REGION: Transkei; Fort Bowker, *Bowker*, 609! Tembuland; mountains near Bazeia, 3500 ft., *Baur.* Griqualand East; 6000 ft., *Schlechter, Krook.* Natal; Oliviers Hoek, 5000 ft., *Allison*, 25! near Pine Town, 1000 ft., *Sanderson*, 481! Inanda, *Wood*, from the Qaunda, *Sanderson*, 894! Highlands, *Gerrard*, 1544! and without precise locality, *Mrs. Fannin*, 33! *Buchanan*! *Mrs. K. Saunders.*

11. **M. prasinata** (Lindl. Gen. & Sp. Orch. 358); plant ¾–1¼ ft. high; stem very stout; leaves cauline, lanceolate or oblong-lanceolate, acuminate, fleshy, 2–4½ in. long; scape ¾–1 ft. high; spikes 4–7 in. long, dense, many-flowered; bracts ovate-lanceolate, acuminate, ¾–1¼ in. long; pedicels 5–8 lin. long; flowers medium-sized, green, with 2 or 3 longitudinal red bands on the dorsal sepal and petals; dorsal sepal galeate, obovate, obtuse, about 5 lin. long; spur oblong, obtuse, spreading, about as long as the limb; lateral sepals oblong, obtuse, spreading, 5 lin. long; petals obliquely ovate-oblong, narrowed upwards, obtuse, 3–3½ lin. long; lip linear-oblong, obtuse, 3½–4 lin. long; column short; anther reflexed; rostellum ascending, with obtuse arms; stigma pulvinate. *M. inflata, Sond. in Linnæa*, xx. 219. *M. cernua, Durand*

& *Schinz, Conspect. Fl. Afr.* v. 111 ; *Kränzl. Orch. Gen. et Sp.* i. 815.
Disa cernua, Sw. in Vet. Acad. Handl. Stockh. 1800, 211 ; *Thunb.*
Fl. Cap. ed. Schult. 12 ; *Lindl. Gen. & Sp. Orch.* 356 ; *Bolus in Trans.*
S. Afr. Phil. Soc. v. 141, *Orch. Cap. Penins.* 141, *Journ. Linn.*
Soc. xxv. 196, *and Ic. Orch. Austr.-Afr.* ii. *t.* 91, *excl. several syns.* ;
Schlechter in Engl. Jahrb. xxxi. 210, *partly.* *D. prasinata, Ker in*
Bot. Reg. t. 210.

SOUTH AFRICA : without locality, *Thunberg* ! *Masson* !
COAST REGION : Cape Div. ; lower slopes of the Lions Head, 200 ft., *Bolus,*
4973 ! Cape Flats, near Riet Vallei, *Zeyher,*|1569 ; *Pappe,* 49 ! *Ecklon and Zeyher* !
sandy plains between Cape Town and Stellenbosch, *Villet* ! by the Kommetjes,
Chapmans Bay, *Wolley-Dod,* 3635 ! George Div. ; Montagu Pass, 1200 ft., *Young*
in Herb. Bolus, 5531 !

Quite distinct from *M. densiflora,* Lindl., and *M. physodes,* Reichb. f., which
are referred here by Schlechter.

12. **M. reticulata** (Durand & Schinz, Conspect. Fl. Afr. v. 112) ;
plant ½–1 ft. high ; stem stout ; leaves cauline, 5–6, erect, linear-
lanceolate, acute or acuminate, fleshy, 2–3 in. long, about ¼ in. broad ;
scapes ½–1 ft. high ; spikes 2½–6 in. long, rather dense, many-
flowered ; bracts ovate-lanceolate, acute or acuminate, ½–¾ in. long ;
pedicels ⅓–½ in. long ; flowers medium-sized, yellowish, with dull
red on the dorsal sepal and lip, and sometimes on the whole flower ;
dorsal sepal galeate, broadly elliptic-oblong, apiculate or obscurely
3-denticulate, 3 lin. long, about 2 lin. broad ; spur slender, slightly
curved, slightly narrowed towards the apex, 5–6 lin. long ; lateral
sepals ovate-oblong, subobtuse, 2½–3 lin. long ; petals obliquely
ovate-oblong, obtuse, fleshy, about 2½ lin. long ; lip oblong, obtuse,
somewhat fleshy, 2½ lin. long ; column short ; anther reflexed ;
connective broad ; rostellum erect, broader than long, tubercled on
each side, with a shallow cleft between it and the short obtuse
stigma. *Kränzl. Orch. Gen. et Sp.* i. 816. *Disa reticulata, Bolus*
in Journ. Linn. Soc. xxii. 73, xxv. 196, *Trans. S. Afr. Phil. Soc.* v.
143, *t.* 16, *and Orch. Cap. Penins.* 143, *t.* 16. *D. macrostachya,*
Schlechter in Engl. Jahrb. xxxi. 209, *partly, not of Bolus.*

COAST REGION : Cape Div. ; slopes of Table Mountain, 2800 ft., *Bolus,* 4987 !
wet places on Constantia Berg, behind Tokay, 2000 ft., *Bodkin in Herb. Bolus,*
4988 ! George Div. ; Outeniqua Mountains, near Montagu Pass, 2000 ft.,
Schlechter. Knysna Div. ; Zitzikamma River, *Penther,* 815 !

13. **M. macrocera** (Lindl. Gen. & Sp. Orch. 358) ; plant ½–1½ ft.
high ; stem usually stout ; leaves cauline, 3–5, suberect, lanceolate
or oblong-lanceolate, acute, 1½–3 in. long, ¼–¾ in. broad, decreasing
upwards into the bracts ; scapes ½–1½ ft. long ; spikes 3–8 in. long,
usually somewhat lax, many-flowered ; bracts ovate-lanceolate,
acute or acuminate, ½–1½ in. long ; pedicels ½–¾ in. long ; flowers
medium-sized, sepals yellowish, petals and lip dark red ; dorsal
sepal galeate, oblong, obtuse, 4 lin. long ; spur descending, filiform,
acute, slightly curved, 7–9 lin. long ; lateral sepals oblong, obtuse,
slightly falcate, 3–3½ lin. long ; petals obliquely oblong, emarginate,

somewhat broadened at the base, fleshy, 3–3½ lin. long ; lip oblong
or narrowly ovate-oblong, obtuse, fleshy, 3–3½ lin. long; column
short; anther reflexed, connective about as long as the cells;
rostellum erect, semilunate, with acute arms, as long as the broad
pulvinate stigma. *M. leptostachya, Sond. in Linnæa,* xix. 101, xx.
219. *M. rufescens, Durand & Schinz, Conspect. Fl. Afr.* v. 112;
Kränzl. Orch. Gen. et Sp. i. 818. *Disa rufescens, Sw. in Vet. Acad.
Handl. Stockh.* 1800, 210 ; *Thunb. Fl. Cap. ed. Schult.* 13 ; *Bolus in
Trans. S. Afr. Phil. Soc.* v. 144, *partly, Orch. Cap. Penins.* 144,
partly, and Journ. Linn. Soc. xxv. 196 ; *Schlechter in Engl. Jahrb.*
xxxi. 208.

SOUTH AFRICA : without locality, *Thom* ! *Rogers* ! *Masson* ! *Trimen* !
COAST REGION : Malmesbury Div. ; near Groene Kloof, *Thunberg* ! Cape Div. ;
wet fields near Claremont, under 100 ft., *Bolus,* 4551 ! near Kenilworth, 70 ft.,
Bolus, 4551B ! lower plain of Table Mountain, 2500 ft., *Bolus,* 4969, *Harvey* !
marsh at the source of Slang Kop River, *Wolley-Dod,* 2992 ! Cape Flats, *Ecklon* !
Pappe, 53 ! *Bowie* ! near Wynberg, 100 ft., *Zeyher,* 1570 ! *Schlechter,* 1550 !
Wolley-Dod, 3066 ! Swellendam Div. ; near Swellendam, *Bowie* ! Riversdale Div. ;
Garcias Pass, 1200 ft., *Galpin,* 4609 ! George Div. ; near George, *Bowie* !

14. **M. bolusiana** (Rolfe); plant 6–10 in. high; stem stout;
leaves basal or nearly so, erect or suberect, 3–4, lanceolate or
oblong-lanceolate, subacute, fleshy, 2–3 in. long, ⅓–½ in. broad;
scapes 6–10 in. high, with numerous lanceolate somewhat imbricate
sheaths ; spikes 2–3½ in. long, somewhat lax, often many-flowered ;
bracts lanceolate or oblong-lanceolate, acute or subacute, ½–¾ in.
long; pedicels 6–7 lin. long ; flowers rather large, greenish-yellow,
sepals and lip tipped with red ; dorsal sepal galeate, oblong, obtuse,
¾ lin. long, with 3 distinct nerves; spur descending, 6–7 lin. long,
slightly thickened near the apex, subobtuse ; lateral sepals spreading or deflexed, oblong, obtuse, ¾ lin. long, with 3 distinct nerves ;
petals suberect, obliquely ovate, obtuse or slightly emarginate,
somewhat fleshy, 3–3½ lin. long; lip deflexed or somewhat recurved,
elliptic-oblong or obovate-oblong, obtuse, somewhat fleshy, nearly as
long as the lateral sepals ; column very short ; anther reflexed,
connective as long as the cells ; rostellum erect, broad, emarginate ;
stigma suberect, scarcely exceeding the rostellum. *Disa bolusiana,
Schlechter in Engl. Jahrb.* xxiv. 426, *and* xxxi. 207 ; *Kränzl.
Orch. Gen. et Sp.* i. 820. *D. rufescens, Bolus in Trans. S. Afr.
Phil. Soc.* v. 144, *partly, and Orch. Cap. Penins.* 144, *partly.*

COAST REGION : Cape Div. ; grassy spots below Maclears Beacon on Table
Mountain, 3500 ft., *Bolus,* 4903 ! Swellendam Div. ; above Zuurbraak, 3100 ft.,
Schlechter.

15. **M. comosa** (Reichb. f. in Linnæa, xx. 687); plant ½–1½ ft.
high ; stem stout, green throughout ; leaves 2–3, suberect, lanceolate-
oblong to elliptic-oblong, obtuse or subacute, 3–8 in. long, ½–1¾ in.
broad, submembranous ; scapes ½–1½ ft. high, with several spathaceous acute sheaths ; spikes 2–7 in. long, usually lax, many-flowered ;
bracts ovate or broadly oblong-lanceolate, acute, ½–1 in. long;

pedicels $\frac{3}{4}$–1 in. long; flowers rather large, with sulphur-yellow sepals and deep yellow petals and lip; dorsal sepal galeate, oblong, obtuse, submembranous, 5 lin. long; spur deflexed, slender, somewhat curved, 10–12 lin. long; lateral sepals oblong, subobtuse, submembranous, 4$\frac{1}{2}$–5 lin. long; petals obliquely ovate at the base, falcate-oblong above, usually emarginate, somewhat fleshy, 3$\frac{1}{2}$–4 lin. long; lip broadly oblong or elliptic-oblong, obtuse or subobtuse, fleshy, 3$\frac{1}{2}$–4 lin. long; column short; anther reflexed; rostellum erect, with reflexed arms; stigma prominent, rounded. *Kränzl. Orch. Gen. et Sp.* i. 812. *M. rufescens, Lindl. Gen. & Sp. Orch.* 356, *partly, not of Sw.*; *Drège, Zwei Pfl. Documente,* 71, 73, 83, 115. *Disa affinis, N. E. Br. in Gard. Chron.* 1885, xxiv. 402; *Bolus in Trans. S. Afr. Phil. Soc.* v. 143, *Orch. Cap. Penins.* 143, *and Journ. Linn. Soc.* xxv. 196. *D. comosa, Schlechter in Engl. Jahrb.* xxxi. 206.

SOUTH AFRICA : without locality, *Mund, Bergius, Grey*! *Gueinzius*!
COAST REGION : Vanrhynsdorp Div. ; Gift Berg, 1500–2500 ft., *Drège.* Clanwilliam Div. ; near Honig Valei and Koude Berg, 3000–4000 ft., *Drège.* Paarl Div. ; Drakenstein Mountains, 2000–5000 ft., *Drège,* 1252 a! Cape Div. ; Table Mountain, 1400–2500 ft., *Bolus,* 4555! and in *MacOwan & Bolus, Herb. Norm. Austr.-Afr.* 170! *Pappe,* 52! *Dümmer,* 934! *Prior*! *Harvey*! *Brown*! top of Kloof near Klassenbosch, *Wolley-Dod,* 1788! Caledon Div. ; Genadendal, 2000–3000 ft., *Drège,* 1252 d! Swellendam Div. ; near Swellendam, 1000–4000 ft., *Zeyher,* 3925! *Burchell,* 7321! 7357! Zuurbraak, 2800 ft., *Galpin,* 4605! George Div. ; Outeniqua Mountains above Montagu Pass, *Schlechter.*

16. **M. sabulosa** (Kränzl. Orch. Gen. et Sp. i. 814); plant 3–8 in. high; stem stout; leaves 4–5, suberect, linear-oblong, acute or acuminate, 1–2 in. long, somewhat fleshy, mostly withered at the flowering period, reduced upwards into narrow spathaceous sheaths; scapes 3–8 in. high; spikes broadly oblong, 1$\frac{1}{2}$–3 in. long, very dense, many-flowered; bracts ovate-lanceolate, acute or acuminate, $\frac{1}{2}$–$\frac{3}{4}$ in. long; pedicels about $\frac{1}{2}$ in long; flowers rather large, pale yellow, sometimes with red tips to the sepals, petals and lip rather brighter yellow; dorsal sepal galeate, obovate, retuse or apiculate, 4 lin. long; spur slender, subacute, abruptly curved below the middle, 4 lin. long; lateral sepals oblong, apiculate, deflexed, 3$\frac{1}{2}$–4 lin. long; petals broadly oblong, obtuse, somewhat dilated at the base, fleshy, 3–3$\frac{1}{2}$ lin. long, deeply bilobed, with rounded obtuse lobes; lip linear-oblong, obtuse, 2$\frac{1}{2}$–3 lin. long; column short; anther reflexed, connective broad; rostellum erect, with oblong reflexed side lobes; stigma pulvinate. *Disa sabulosa, Bolus, Ic. Orch. Austr.-Afr.* i. *t.* 27; *Schlechter in Engl. Jahrb.* xxxi. 207.

COAST REGION : Cape Div. ; sandy flats and heathy places at Kenilworth Race Course, near Wynberg, *Bolus,* 7104! and in *MacOwan & Bolus, Herb. Norm. Austr.-Afr.* 1374! *Schlechter* ; Cape Flats, *Harvey*!

17. **M. ophrydea** (Lindl. Gen. & Sp. Orch. 358); plant $\frac{1}{2}$–1$\frac{1}{2}$ ft. high; stem stout, purple throughout; leaves 2–4, mostly basal, suberect, linear-oblong, acute, 2–4 in. long, $\frac{1}{2}$–$\frac{3}{4}$ in. broad, somewhat fleshy, more or less involute at the margin; scapes $\frac{1}{2}$–1$\frac{1}{2}$ ft. high,

with several lanceolate acute sheaths; spikes 3–6 in. long, lax, few-
to many-flowered; bracts lanceolate or oblong-lanceolate, acute,
$\frac{3}{4}$–1 in. long; pedicels 9–10 lin. long; flowers rather large, deep
red, with darker petals and lip; dorsal sepal galeate, oblong, obtuse,
$3\frac{1}{2}$–$4\frac{1}{2}$ lin. long; spur filiform, slightly curved, 8–9 lin. long;
lateral sepals ovate-oblong, obtuse or subobtuse, 4–5 lin. long;
petals obliquely ovate-oblong, emarginate or shortly bidentate,
somewhat fleshy, 3–4 lin. long; lip oblong, obtuse, 4–5 lin. long;
column short; anther reflexed, connective narrow, as long as the
cells; rostellum erect; arms obtuse, longitudinally keeled, much
longer than the stigma; stigma umbonate, prominent. *Drège, Zwei*
Pfl. Documente, 83; *Durand & Schinz, Conspect. Fl. Afr.* v. 112;
Kränzl. Orch. Gen. et Sp. i. 817, *partly. M. lancifolia, Sond. in*
Linnæa, xix. 100. *Disa ophrydea, Bolus in Trans. S. Afr. Phil.*
Soc. v. 142, *Orch. Cap. Penins.* 142, *and Journ. Linn. Soc.* xxv.
196; *Schlechter in Engl. Jahrb.* xxxi. 204.

18. **M. atrorubens** (Rolfe); plant 8–13 in. high; stem stout,
leafy; leaves suberect, linear-lanceolate, acute, the lower about 3 in.
long, 5 lin. broad, upper decreasing into the bracts; scapes 8–13 in.
long; spikes somewhat 1-sided, $2\frac{1}{2}$–8 in. long, 6- to many-flowered;
bracts ovate-lanceolate, acuminate, erect, $\frac{3}{4}$–1 in. long; pedicels
7–9 lin. long; flowers rather large, suberect, reddish, with dark
red petals and lip; dorsal sepal galeate, broadly oblong, obtuse,
somewhat involute at the apex, 3 lin. long; spur deflexed, somewhat
incurved, filiform, subacute, 8–10 lin. long; lateral sepals ovate-
oblong, obtuse, $3\frac{1}{2}$ lin. long, $1\frac{1}{2}$ lin. broad, spreading or deflexed;
petals erect, fleshy, obliquely ovate, obtuse, $2\frac{1}{2}$ lin. long, inner face
verrucose about the middle; lip deflexed, linear-oblong, obtuse,
fleshy, as long as the lateral sepals; anther reflexed, connective
broad, scarcely shorter than the cells; rostellum emarginate, side
lobes erect, longer than the stigma. *M. ophrydea, Kränzl. Orch.*
Gen. et Sp. i. 817, *partly. Disa atrorubens, Schlechter in Engl. Jahrb.*
xxiv. 427, *and* xxxi. 205.

19. **M. Basutorum** (Rolfe); plant 3–5 in. high; stem rather
slender; leaves radical, 2, suberect, ovate, acute, $\frac{3}{4}$–1 in. long,

4–6 lin. broad, with sheathing base ; scapes 3–5 in. high, with 3 or 4 narrowly lanceolate acuminate sheaths ; spike about 1 in. long, lax, 6–14-flowered ; bracts narrowly lanceolate, acuminate, ⅓–½ in. long ; pedicels 3–4 lin. long ; flowers small, brownish-green ; dorsal sepal galeate, ovate or ovate-oblong, obtuse, with somewhat incurved margin, 2½ lin. long ; spur spreading or reflexed, filiform, obtuse, rather longer than the limb ; lateral sepals obliquely ovate, obtuse, dorsally and minutely apiculate, 2 lin. long ; petals erect, broadly ovate, obtuse, with inflexed apex, over half as long as the sepals ; lip linear-oblong, subobtuse, slightly dilated at the apex, 1½–2 lin. long ; column very short ; anther somewhat reflexed, deeply notched at the apex ; rostellum minute ; stigma rather large. *Disa Basutorum, Schlechter in Engl. Jahrb.* xx. *Beibl.* 50, 17, *and in Engl. Jahrb.* xxxi. 204, *t.* 3, *fig. E–L.*

KALAHARI REGION : Basutoland ; heathy summits of the Drakensberg Range, 10000 ft., *Thode* !

XXXIII. AMPHIGENA, Rolfe.

Sepals free, the odd one superior, galeate, ascending, with a conical obtuse spur. *Petals* erect, included within the galea, somewhat oblique, serrulate on the front margin. *Lip* anticous, sessile, small and narrow, entire, without a spur. *Column* short. *Anther* much reflexed ; cells distinct and parallel, with a broad connective ; pollinia granular, attached by short caudicles to a single large nearly square gland, situated at the apex of the rostellum. *Ovary* twisted ; rostellum dwarf, destitute of side appendages ; stigma cushion-shaped, seated in front of the rostellum. *Capsule* narrow.

Terrestrial herbs, with rather large irregular tubers, and narrow grass-like leaves springing from a basal sheath, appearing before the flowers and soon withering ; flowers small, borne in narrow cylindrical spikes on slender wiry scapes ; bracts small and narrow.

DISTRIB. Species 2, limited to the south-western corner of Cape Colony.

A peculiar little genus, closely resembling *Herschelia* in habit, and *Monadenia* in the single pollinary gland, but distinct from both in the unappendaged rostellum, and not agreeing with any other member of the *Disa* group.

Dorsal sepal oblong or elliptic-oblong with entire margin,
　　apex long aristate-apiculate ... ... (1) **tenuis.**

Dorsal sepal broadly ovate with crenulate margin, apex
　　shortly apiculate ... ... (2) **leptostachya.**

1. **A. tenuis** (Rolfe) ; plant ½–1¼ ft. high ; stem slender ; leaves radical, few, disappearing very early (not seen), with several narrow basal sheaths ; scapes ½–1¼ ft. high, with several lanceolate acuminate sheaths ; spikes 2–4 in. long, slender, somewhat lax, many-flowered ; bracts lanceolate, long acuminate-aristate, about 3 lin. long ; pedicels 2–3 lin. long ; flowers very small, greenish,

with occasional purple spots (*Bolus*); dorsal sepal broadly elliptic-oblong, abruptly narrowed into an aristate apiculus one-third as long as the limb, limb somewhat concave, with entire margin, about 2 lin. long; spur diverging, conical, obtuse, about $\frac{1}{4}$ lin. long; lateral sepals oblong, slightly concave, distinctly acuminate-aristate, $2\frac{1}{4}$ lin. long; petals oblong, obtuse, somewhat oblique at the apex, serrulate on the front margin, $1\frac{1}{4}$ lin. long; lip linear-oblong, subobtuse, $1\frac{1}{4}$ lin. long; column broad, 1 lin. long; anther reflexed, connective broad; rostellum shortly subquadrate, gland of pollinia, single, nearly square; stigma pulvinate. *Disa tenuis, Lindl. Gen. & Sp. Orch.* 354; *Drège, Zwei Pfl. Documente*, 111; *Bolus in Journ. Linn. Soc.* xx. 484, *excl. syn.*, xxv. 204; *Trans. S. Afr. Phil. Soc.* v. 173, *and Orch. Cap. Penins.* 173; *Durand & Schinz, Conspect. Fl. Afr.* v. 108; *Schlechter in Engl. Jahrb.* xxxi. 293, *partly. Monadenia tenuis, Kränzl. Orch. Gen. et Sp.* i. 819, *partly.*

COAST REGION : Cape Div. ; Wynberg Flats, under 100 ft., *Drège*, 8274 ! *Prior* ! Cape Flats, near Claremont, *MacOwan*, 2566 ! *Bolus, Kässner, Schlechter* ; Devils Mountain, between the Blockhouse and the Waterfall, 1200–1400 ft., *Bodkin in Herb. Bolus*, 4874 ! *Pappe* ; Steen Berg, *Wolley-Dod*, 2555 ! near the source of Silvermine River, *Wolley-Dod*, 1276 ! base of Table Mountain, *Harvey* ! near the Race Course, *Bolus*.

2. **A. leptostachya** (Rolfe); plant $\frac{1}{2}$–$1\frac{1}{4}$ ft. high; stem slender; leaves radical, 2–4, filiform, slightly broader above, acute, flexuous, 3–5 in. long, with several narrow basal sheaths; scapes $\frac{1}{2}$–$1\frac{1}{4}$ ft. high, with several lanceolate acuminate sheaths; spikes 2–6 in. long, slender, usually lax, many-flowered; bracts lanceolate, long acuminate-aristate, 3–4 lin. long; pedicels about $\frac{1}{4}$ in. long; flowers very small, greenish, pale purple or rosy, with darker spots (*Bolus*); dorsal sepal somewhat galeate, broadly ovate, abruptly narrowed into an aristate apiculus, about a quarter as long as the limb; limb very concave, with crenulate margin; spur broadly conical, obtuse, about $\frac{1}{2}$ lin. long; lateral sepals ovate-oblong, somewhat concave, minutely apiculate, $1\frac{1}{2}$–2 lin. long; petals ovate-oblong, obtuse, somewhat oblique at the apex, serrulate or denticulate on the front margin, about 1 lin. long; lip oblong, obtuse, 1 lin. long; column broad, 1 lin. long; anther reflexed, connective broad; rostellum shortly subquadrate, gland of pollinia single, nearly square; stigma pulvinate. *Disa leptostachya, Sond. in Linnæa*, xix. 98. *D. tenuis, Bolus in Journ. Linn. Soc.* xx. 484, *partly, not of Lindl.* ; *Schlechter in Engl. Jahrb.* xxxi. 293. *partly.*

COAST REGION : Caledon Div. ; near Grietjes Gat, *Ecklon & Zeyher* ; Houw Hoek Mountains, 1400 ft., *Bolus*, 5352 ! *Schlechter*, 7550 !

Confused by Bolus and Schlechter with the preceding species.

XXXIV. HERSCHELIA, Lindl.

Sepals free, odd one superior, galeate, ascending, with a short broadly conical spur. *Petals* erect, included within the galea, unequally and deeply bilobed. *Lip* anticous, sessile or long-stalked, deeply fringed, crenate or subentire, or dilated and variously lobed at the apex. *Column* short. *Anther* much reflexed, cells distinct and parallel, connective broad ; pollinia granular, attached by short caudicles to a large single gland or in *H. lugens,* according to Bolus, occasionally with two glands, situated at the apex of the rostellum. *Ovary* twisted ; rostellum broad, sometimes 3-toothed at the apex, and the middle tooth situated behind and applied to the back of the gland, sometimes with two short arms ; stigma cushion-shaped, situated in front of the rostellum. *Capsule* narrowly oblong or clavate.

Terrestrial herbs, with large ovoid irregular tubers, and narrow grass-like radical leaves, with a few sheaths at the base ; flowers medium-sized or large, usually blue, purple or white, borne in short or somewhat elongated racemes on slender wiry scapes ; bracts usually narrow.

DISTRIB. Species about 14, South African, with a single outlying representative in British Central Africa.

Bolus reduced *Herschelia,* with other of Lindley's genera, to a section of *Disa.*

Lip sessile or subsessile :
  Lip crenulate, waved or entire :
    Spur shortly 2-toothed at the apex ...   . (1) **forcipata.**
    Spur entire at the apex :
      Spike 5–6 in. long, rather dense, many-flowered... (2) **excelsa.**
      Spike short, lax, few-flowered :
        Petals curved ; lip ovate, incurved at the
          margins ; gland broader than long   ... (3) **purpurascens.**
        Petals sharply bent above the middle ; lip oblong,
        reflexed ; gland longer than broad   ... (4) **cœlestis.**
  Lip more or less deeply lacerate :
    Spur of dorsal sepal 1½–2 in. long, narrowed at
      the apex :
      Flowers 2–4 ; sepals whitish, with pale blue lines (5) **barbata.**
      Flowers 4–15 ; sepals blue :
        Lip white, shorter than the lateral sepals   ... (6) **venusta.**
        Lip light green, longer than the lateral sepals ... (7) **lugens.**
    Spur of dorsal sepal 3–4 lin. long, cylindrical   ... (8) **Baurii.**
Lip long-stalked :
  Limb of lip trilobed or cordately trilobed :
    Claw of lip 2½–5 lin. long :
      Lip trilobed, with narrow acuminate side lobes ... (9) **tripartita.**
    Lip cordately trilobed, with broad or rounded side
      lobes ...   . (10) **atropurpurea.**
    Claw of lip ½–1¼ in. long   :..   ...   . (11) **spathulata.**
  Limb of lip broken up into numerous slender filaments :
    Claw of lip ½ in. long ...   . (12) **multifida.**
    Claw of lip 1¾–2½ in. long   ...   . (13) **charpentieriana.**

1. **H. forcipata** (Kränzl. Orch. Gen. et Sp. i. 807); plant about
1 ft. high; stem rather slender; leaves unknown; scapes about
1 ft. high, with several lanceolate acuminate sheaths; spikes 5–8
in. long, somewhat lax, many-flowered; bracts ovate-lanceolate,
setaceous-acuminate, 6–8 lin. long; pedicels about 8 lin. long;
flowers large, greenish-yellow; dorsal sepal galeate, broadly ovate,
acute, nearly ¾ in. long; spur filiform, minutely 2-dentate at the
apex, 1½–2 lin. long; lateral sepals spreading, broadly ovate-oblong,
acute, concave, 6–7 lin. long; petals decumbent, linear-lanceolate,
unequally 2-lobed, 4–5 lin. long; posterior lobe obliquely oblong,
obtuse; anterior lobe falcate-oblong, twice as long as the posterior;
lip lanceolate, acute, entire, over half as long as the lateral sepals;
column broad, short; anther reflexed; connective narrow, shorter
than the cells; rostellum erect, 3-fid, with acute lobes; stigma
pulvinate. *Disa forcipata, Schlechter in Engl. Jahrb.* xxiv. 428,
xxxi. 292, *t.* 6, *fig. E–K.*

COAST REGION : without precise locality, *Trimen.*

2. **H. excelsa** (Rolfe); plant about 1¾ ft. high; stem stout;
leaves not seen; scapes about 1¾ ft. high, with numerous narrow
acuminate sheaths above; spikes 5–6 in. long, somewhat dense,
many-flowered; bracts ovate-lanceolate, acuminate, ⅓–½ in. long;
pedicels 7–8 lin. long; flowers medium-sized, purplish; dorsal sepal
galeate, broadly ovate, acute, about 5 lin. long; spur broadly
conical, obtuse, about a fifth as long as the limb; lateral sepals
broadly elliptic-oblong, obtuse, about 5 lin. long; petals strongly
falcate, acute or acuminate, somewhat crenulate on the outer margin,
or sometimes bifid, with a large rounded auricle at the base in
front, about 3 lin. long; lip broadly elliptic-oblong, subobtuse,
entire or obscurely crenulate close to the apex; column short;
anther reflexed; rostellum erect, shortly 3-lobed; stigma pulvinate.
Disa excelsa, Thunb. Fl. Cap. ed. i. 78, *partly, not of Sw., and
ed. Schult.* 14; *Lindl. Gen. & Sp. Orch.* 356; *Bolus in Journ.
Linn. Soc.* xxv. 203; *Durand & Schinz, Conspect. Fl. Afr.* v. 102;
Kränzl. Orch. Gen. et Sp. i. 800; *Schlechter in Engl. Jahrb.* xxxi.
292.

COAST REGION : George Div.; Outeniqua Mountains, near Lange Kloof and
Kromme River, *Thunberg!*

3. **H. purpurascens** (Kränzl. Orch. Gen. et Sp. i. 803); plant
1–1½ ft. high; stems slender; leaves radical, 6–8, narrowly linear,
acute, more or less recurved above, 6–8 lin. long; scapes 1–1½ ft.
high, with several narrow acuminate sheaths; spikes short, lax,
1–4-flowered; bracts ovate or ovate-lanceolate, acuminate, 6–8 lin.
long; pedicels ½–¾ in. long; flowers large, blue-purple, except the
upper limb of the petals and the spur, which are pale green; dorsal
sepal galeate, broadly ovate, apiculate, with very broad open
mouth, 7–9 lin. long; spur broadly conical below, upper part

abruptly narrowed, oblong, about 2 lin. long ; lateral sepals elliptic-
oblong or ovate-oblong, subacute, 6–7 lin. long ; petals somewhat
curved, about 4 lin. long, dilated at the base in front into a small
rounded auricle, apex obliquely dilated into a broad crenulate
limb ; lip broadly ovate, subobtuse, with crenulate margin, about
6 lin. long ; column broad, about 2 lin. long ; anther reflexed ;
rostellum erect, shortly 3-lobed ; stigma pulvinate. *Disa purpura-*
scens, Bolus in Journ. Linn. Soc. xx. 482, xxv. 203, *Trans. S. Afr.*
Phil. Soc. v. 169, *Orch. Cap. Penins.* 169, *and Ic. Orch. Austr.-*
Afr. i. *t.* 86 ; *Durand & Schinz, Conspect. Fl. Afr.* v. 106 ; *Schlechter*
in Engl. Jahrb. xxxi. 291.

SOUTH AFRICA : without locality, *Grey* !
COAST REGION : Cape Div. ; by a mountain stream in Farmer Pecks Valley
Muizen Berg, 1100 ft., *Bolus*, 4893 ! hills near Simonstown, 800 ft., *Bolus* ! *Prior* !
McKellars Farm, near Cape Point, *Marloth* ; Red Hill, *Mrs. Jameson* ! Klaver
Vley, *Wolley-Dod*, 2005 !

4. **H. cœlestis** (Lindl. Gen. & Sp. Orch. 363) ; plant 1–2 ft. high ;
stem slender ; leaves radical, 4–6, narrowly linear, acute, usually
recurved above, 8–12 in. long ; scapes 1–2 ft. high, with several
narrow acuminate sheaths ; spikes 2–6 in. long, 2–7-flowered ; bracts
ovate-lanceolate, acuminate, 4–8 in. long ; pedicels 6–8 lin. long ;
flowers large, bright blue, with the apex of the petals green, and
the lip pale blue or white with a purple margin ; dorsal sepal
galeate, broadly ovate, acute, 7–9 lin. long, mouth broad and open ;
spur broadly conical below, upper part abruptly narrowed, oblong,
obtuse, about 1½ lin. long ; lateral sepals ovate-oblong, subobtuse,
6–7 lin. long ; petals abruptly bent nearly at a right angle above
the middle, about 5 lin. long, dilated at the base in front into a
rounded auricle, apex rounded and cuneate ; lip elliptic-oblong,
obtuse, 6 lin. long, margin slightly undulate, entire ; column
broad, about 2 lin. long ; anther reflexed ; rostellum erect, 3-fid,
with linear lobes ; stigma pulvinate. *Reichb. f. in Reichb. Fl.*
Germ. xiii. *p.* viii. *t.* 354, *fig.* 18–20 ; *Walp. Ann.* iii. 590.
H. graminifolia, Durand & Schinz, Conspect. Fl. Afr. v. 111 ;
Kränzl. Orch. Gen. et Sp. i. 802. *Disa graminifolia, Ker in Quart.*
Journ. Sci. & Arts, vi. 44, *t.* 1, *fig.* 2 ; *Bolus in Journ. Linn. Soc.*
xix. 234, *fig.* 3, xxv. 203, *Trans. S. Afr. Phil. Soc.* v. 168, *Orch.*
Cap. Penins. 168, *and Ic. Orch. Austr.-Afr.* i. *t.* 37 ; *Warn. Orch.*
Alb. ix. *t.* 399 ; *Schlechter in Engl. Jahrb.* xxxi. 290.

SOUTH AFRICA : without locality, *Masson* ! *Mund, Bergius, Alexander* ! *Prior.*
COAST REGION : Cape Div. ; Table Mountain, 1000–1500 ft., *MacOwan*, 1045 !
Harvey ! *Bolus* ! and in *MacOwan & Bolus, Herb. Norm. Austr.-Afr.* 167 ! *Trimen* !
Kässner, Schlechter, 481 ; Orange Kloof, *Wolley-Dod*, 840 ! about Skeleton Ravine,
Wolley-Dod, 885 ! near Wynberg, 80 ft., *Zeyher* ; Lisbeck River, *Mund.* Caledon
Div. ; near Genadendal, 1200–2000 ft., *Bolus, Bowie* ! *Roser*, 43 ! Bredasdorp
Div. ; near Zoetendals Vley, *Joubert.* Swellendam Div. ; grassy slopes on the
Langeberg Range, 900 ft., *Schlechter*, 2061.

5. **H. barbata** (Bolus in Journ. Linn. Soc. xix. 236) ; plant 1–2 ft.
high ; stem slender ; leaves radical, 3–7, suberect, somewhat

flexuous, narrowly elongate-linear, acute, 6–14 in. long; scapes
1–2 ft. high, with several acuminate sheaths; spikes 1–5-flowered;
bracts ovate, acute or acuminate, ⅓–½ in. long; pedicels 6–8 lin.
long; flowers large, nearly white, with some blue lines on the
dorsal sepal, and the petals and lip pale yellow; dorsal sepal
galeate, broadly ovate, acute, ¾–1 in. long; spur broadly conical at
the base, narrow above, 3–4 lin. long; lateral sepals elliptic-oblong,
acute, 8–10 lin. long; petals falcately bent in the middle, 3–4 lin.
long, with a rounded basal lobe in front, dilated, somewhat
2-lobed and crenulate in front; lip' sessile, deflexed, elliptic-oblong,
5–8 lin. long, lacerate-multifid in front with inflexed segments;
column broad, very short; anther reflexed, retuse; pollinia with
very broad gland; rostellum erect, broader than long, 3-dentate;
stigma pulvinate. *Durand & Schinz, Conspect. Fl. Afr.* v. 111;
Kränzl. Orch. Gen. et Sp. i. 804. *Orchis barbata, Linn. f. Suppl.*
399. *Satyrium barbatum, Thunb. Prodr.* 5. *Disa barbata, Sw. in*
Vet. Acad. Handl. Stockh. 1800, 212; *Thunb. Fl. Cap. ed. Schult.* 11;
Lindl. Gen. & Sp. Orch. 354; *Sond. in Linnæa,* xix. 97; *N. E. Br.*
in Gard. Chron. 1885, xxiv. 231; *Bolus in Trans. S. Afr. Phil. Soc.*
v. 170, *t.* 8, *Orch. Cap. Penins.* 170, *t.* 8, *and Journ. Linn. Soc.* xxv.
202; *Schlechter in Engl. Jahrb.* xxxi. 286.

South Africa: without locality, *Bergius, Krebs, Leibold, Harvey*! *Rogers*!
Trimen, Yorke!
Coast Region: Cape Div.; Cape Flats, 50–100 ft., *Zeyher,* 1567! *Bolus,*
3810! 4566 B! 4857! *Pappe,* 38! *Hooker,* 371! *MacOwan & Bolus, Herb. Norm.*
Austr.-Afr. 166! *Wolley-Dod,* 359! Caledon Div.; mountains of Baviaans Kloof,
near Genadendal, *Burchell,* 7801! Houw Hoek Mountains, *Burchell,* 8084!

6. **H. venusta** (Kränzl. Orch. Gen. et Sp. i. 805); plant 1¼–2½ ft.
high; stem slender; leaves radical, 6–12, elongate-linear, very
narrow, acute, conduplicate, ½–1¼ ft. high; scapes 1¼–2½ ft. high,
with several lanceolate acuminate sheaths; spikes 4–8 in. long, lax,
few- to many-flowered; bracts lanceolate or ovate-lanceolate, acuminate, 3–6 lin. long; pedicels ½–¾ in. long; flowers large, blue, with
some darker stripes on the dorsal sepal, sometimes with the lip
creamy-white; dorsal sepal galeate, broadly ovate, acute, 7–8 lin.
long; spur broadly conical below, narrowed and subacute above,
3–4 lin. long; lateral sepals lanceolate-oblong, acute, 6–7 lin. long;
petals falcately bent, about 2½ lin. long, dilated at the base in front
into a roundish oblong lobe, apex obliquely dilated and acute or
sometimes 2-lobed; lip subsessile, ovate-oblong or elliptic, obtuse,
fringed or more deeply lacerate at the margin, about 5 lin. long;
column broad, about 1 lin. long; anther reflexed; rostellum erect,
rhomboid, 3-fid at the apex; stigma pulvinate. *Disa lacera,*
Sw. in Vet. Acad. Handl. Stockh. 1800, 212; *Thunb. Fl. Cap. ed.*
Schult. 12; *Lindl. Gen. & Sp. Orch.* 354; *Drège, Zwei Pfl. Docu-*
mente, 122; *Sond. in Linnæa,* xix. 97; *Journ. Hort.* 1888, ii. 220,
fig. 24; *N. E. Br. in Gard. Chron.* 1888, iv. 664; *Bolus in Journ.*
Linn. Soc. xxv. 202; *Durand & Schinz, Conspect. Fl. Afr.* v. 103;

Kränzl. Orch. Gen. et Sp. i. 797, *partly*; *Schlechter in Engl. Jahrb.*
xxxi. 287. *D. lacera, var. multifida, N. E. Br. in Gard. Chron.*
1888, iv. 664, *fig.* 93; *Bot. Mag. t.* 7066. *D. barbata, Lindl. Gen.*
& Sp. Orch. 354, *partly, not of Sw. D. venusta, Bolus in Journ.*
Linn. Soc. xx. 482, *Trans. S. Afr. Phil. Soc.* v. 170, *t.* 9, *and*
Orch. Cap. Penins. 170, *t.* 9; *N. E. Br. in Gard. Chron.* 1885,
xxiv. 232; *Durand & Schinz, Conspect. Fl. Afr.* v. 110.

SOUTH AFRICA : without locality, *Masson* ! *Brown, Waldegrave* ! *Harvey*, 140 !
Sieber, 200 !
COAST REGION : Cape Div. ; Cape Flats, *Burchell*, 151 ! *Wolley-Dod*, 391 !
Bolus, 4566 ! *Ecklon & Zeyher, Pappe, Wallich* ! *Schlechter* ; Table Mountain,
Harvey ! *Prior* ! Stellenbosch Div. ; near Eerste River, *Trimen*. Caledon Div. ;
Vogelgat, 150 ft., *Schlechter*, 9544 ! near Caledon, *Bowie*. Swellendam Div. ;
near Swellendam, *Mund, Zeyher*. Riversdale Div. ; between Garcias Pass and
Krombeks River, *Burchell*, 7182 ! George Div. ; near George, 600 ft., *Rehmann*,
529 ! Knysna Div. ; between Knysna and Plettenbergs Bay, *Pappe* ! near Knysna,
Bowie ! *Schlechter*, 5928, *Penther*, 50 ! Uniondale Div. ; between Welgelegen and
Onzer, in Lange Kloof, 2000 ft., *Drège*, 2211a ! Uitenhage Div. ; Van Stadens
Mountains and Flats near them, 1000 ft., *MacOwan*, 1045 ! *Zeyher*, 628 ! *Cooper*,
1464 ! Port Elizabeth Div. ; around Krakakamma, *Burchell*, 4572 ! between
Krakakamma and upper part of Maitland River, *Burchell*, 4592 ! near Port
Elizabeth, 200 ft., *Hallack*. Albany Div. ; near Grahamstown, 2200 ft., *Bowie* !
MacOwan.

Dr. Bolus has labelled one of Masson's drawings *Disa lugens*, Bolus, but from
the white, not green lip, I believe it belongs here, and his Herbarium contains a
specimen of this, but none of *D. lugens*.

7. **H. lugens** (Kränzl. Orch. Gen. et Sp. i. 806) ; plant 1¼–2½ ft.
high ; stem slender ; leaves radical, suberect, narrowly elongate-
linear, acute, rigid, keeled below, channelled above, ¾–1½ in. long ;
scapes 1¼–2½ ft. high, with several narrow acuminate sheaths ;
spikes 4–8 in. long, lax, many-flowered ; bracts ovate-lanceolate,
acuminate, 5–8 lin. long ; pedicels ¾–1 in. long ; flowers large,
metallic greenish-purple or lilac-purple, with a light green lip ;
dorsal sepal galeate, broadly ovate, acute, 6–8 lin. long ; spur
broadly conical below, narrowed and acute above, about ¼ in. long,
with the narrowed apex about 1 lin. long ; lateral sepals spreading,
oblong-lanceolate, acute, 5–6 lin. long ; petals strongly falcately
bent, about 3 lin. long, dilated at the base in front into an oblong
lobe, apex dilated and acute or 2-dentate ; lip sessile, broadly ovate-
oblong, deeply lacerate-multifid at the margin, 6–8 lin. long, longer
than the lateral sepals ; column broad, about 1 lin. long ; anther
reflexed ; connective exceeding the cells ; rostellum erect, 3-partite,
with linear lobes ; stigma pulvinate. *Disa lugens, Bolus in Trans.*
S. Afr. Phil. Soc. v. 171, *Orch. Cap. Penins.* 171, *Journ. Linn. Soc.*
xx. 483, xxv. 203, *and Ic. Orch. Austr.-Afr.* ii. *t.* 76 ; *N. E. Br. in*
Gard. Chron. 1885, xxiv. 232 ; *Durand & Schinz, Conspect. Fl. Afr.*
v. 105 ; *Schlechter in Engl. Jahrb.* xxxi. 288. *D. barbata, Lindl.*
Gen. & Sp. Orch. 354, *partly, not of Sw.*

SOUTH AFRICA : without locality, *Villet* ! *Harvey*, 116 ! *Hooker* ! *Yorke*, 34 !
COAST REGION : Cape Div. ; sandy heathy downs, eastward of Rondebosch and

Claremont, 60–80 ft., *Bolus*, 3810! *Bodkin in MacOwan & Bolus, Herb. Norm. Austr.-Afr.* 494! *Dümner*, 756! near Kuils River, *Pappe*, 39! 377! *Zeyher*, 1566! *Sturk*; Vygeskraal Farm, *Wolley-Dod*, 358! 1798! Stellenbosch Div.; near Eerste River, *Trimen*. Albany Div.; near Grahamstown, *MacOwan*, 700! *Atherstone*! *Bolton*! high hills of the New Years River, *Mrs. Barber*, 113! Zuurberg Range, *Mrs. Barber*! *Galpin*.

CENTRAL REGION: Somerset Div.; Somerset, *Mrs. Barber*!

8. **H. Baurii** (Kränzl. Orch. Gen. et Sp. i. 804); plant $\frac{3}{4}$–1$\frac{1}{2}$ ft. high; stem slender; leaves radical, 4–7, suberect with recurved apex, linear-filiform, acute, rigid, $\frac{1}{2}$–1 ft. long, shrivelled at flowering time; scapes $\frac{3}{4}$–1$\frac{3}{4}$ ft. high, with several narrow acuminate sheaths; spikes 4–6 in. long, lax, 8–12-flowered; bracts lanceolate or ovate-lanceolate, acuminate, 4–8 lin. long; pedicels 6–9 lin. long; flowers rather large, blue; dorsal sepal galeate, broadly ovate, acute or apiculate, 6–7 lin. long; spur broadly conical below, upper part linear or clavate-linear, 3$\frac{1}{2}$–4 lin. long; lateral sepals elliptic-oblong, apiculate, 5–6 lin. long; petals falcately bent, about 3 lin. long, dilated at the base in front into a rounded auricle, apex dilated and shortly and unequally 2-lobed; lip broadly elliptic or suborbicular, 4–5 lin. long, deeply lacerate, segments often dilated above; column broad, 2$\frac{1}{4}$ lin. long; anther reflexed, connective narrow; rostellum 3-fid, side lobes somewhat 2-lobed; stigma pulvinate. *Disa Baurii, Bolus in Journ. Linn. Soc.* xxv. 174, *fig.* 12, 203; *Durand & Schinz, Conspect. Fl. Afr.* v. 100; *Schlechter in Engl. Jahrb.* xxxi. 289.

KALAHARI REGION: Transvaal; grassy hills on Woodbush Mountains, *Barber*, 12! Saddleback, near Barberton, 4000–5000 ft., *Galpin*, 427! *Culver*, 20! *Thorncroft*, 2478! near Verrers Poort, Middelburg, 5000 ft., *Bolus*, 9788! between Pretoria and the Drakensberg Range, *Bolus*.

EASTERN REGION: Tembuland; Bazeia Mountains, 3000 ft., *Baur*, 814!

9. **H. tripartita** (Rolfe); plant about 10 in. high; stem rather slender; leaves about 7, somewhat arching, narrowly linear, acute, slightly broader in the upper half, 5–6 in. long, with a few membranous sheaths below; scapes about 10 in. high, with several lanceolate acuminate sheaths, 1-flowered; bracts lanceolate, aristate-acuminate, 9–10 lin. long; pedicels about 1 in. long; flowers large; dorsal sepal galeate, broadly ovate, subobtuse, 7–8 lin. long; spur conical, with obtuse or rounded apex, about 3 lin. long; lateral sepals ovate or ovate-oblong, subobtuse, spreading, 5–6 lin. long; petals strongly falcate, base obliquely dilated into a rounded crenulate lobe, about 2 lin. broad, apex 2-lobed, with acute front lobe and broader crenulate back lobe; lip unguiculate; claw 3–5 lin. long; limb abruptly dilated, 3-lobed, with linear acute crenulate lobes, side lobes 3 lin. long, front lobe twice as long; column broad, 2 lin. long; anther reflexed; rostellum 3-lobed; stigma pulvinate. *Disa tripartita, Lindl. Gen. & Sp. Orch.* 353; *Drège, Zwei Pfl. Documente,* 139; *Bolus in Journ. Linn. Soc.* xxv. 203; *Durand & Schinz, Conspect. Fl. Afr.* v. 109; *Kränzl. Orch. Gen. et Sp.* i. 797. *D. spathulata, Schlechter in Engl. Jahrb.* xxi. 283, *partly.*

COAST REGION : Albany Div. ; near Bushmans River, under 1000 ft., *Drège*, 3577a !

Only known from Drège's original specimen. Lindley remarks that he had not been able to ascertain the form of the petals, but he had not dissected the loose flower preserved in a capsule in his own Herbarium. Schlechter includes the species under *Disa spathulata,* Sw.

10. **H. atropurpurea** (Rolfe) ; plant about 6 in. high ; leaves about 9, suberect or somewhat curved, narrowly linear, acute, 3–4 in. long, 1–1½ lin. broad, narrowed at the base ; scape about 6 in. high, with 2 lanceolate acuminate sheaths, 1–2-flowered ; bracts elliptic-lanceolate, acute or acuminate, cucullate, 6–8 lin. long ; pedicels about 7 lin. long ; flowers large, dark purple ; dorsal sepal galeate, broadly ovate, acute or shortly acuminate, 6–7 lin. long, with incurved margin ; spur nearly globose, 1 lin. long ; lateral sepals ovate-oblong, acute, concave, 5–6 lin. long ; petals 2–2½ lin. long, somewhat curved, dilated at the base, constricted in the middle, dilated at the apex into a 2-lobed limb ; lip narrowly unguiculate ; claw 2½–4 lin. long ; limb dilated, broadly cordate-ovate, crenulate, 3–5-lobed in front, 3–6 lin. long and about as broad, front lobe acute or acuminate ; column broad, 1½–2 lin. long ; anther reflexed ; connective as long as the cells ; rostellum erect, 3-dentate, with acute teeth ; stigma ovate. *Disa atropurpurea, Sond. in Linnæa,* xix. 96 ; *N. E. Br. in Gard. Chron.* 1886, xxv. 532 ; *Bot. Mag. t.* 6891 ; *Bolus in Journ. Linn. Soc.* xxv. 203 ; *Durand & Schinz, Conspect. Fl. Afr.* v. 100 ; *Kränzl. Orch. Gen. et Sp.* i. 794. *D. spathulata, var. atropurpurea, Schlechter in Engl. Jahrb.* xxxi. 284.

COAST REGION : Tulbagh Div. ; near Tulbagh Waterfall, *Ecklon & Zeyher* !

11. **H. spathulata** (Rolfe) ; plant ½–1 ft. high ; stem rather slender ; leaves 8–12, suberect or somewhat arching, narrowly linear, slightly broader in the upper half, acute or acuminate, 3–5 in. long ; scapes ½–1 ft. high, with 2–4 lanceolate acuminate sheaths ; spikes 1–5-flowered ; bracts lanceolate, aristate-acuminate ; pedicels ¾–1 in. long ; flowers large, pink or purple ; dorsal sepal galeate, broadly ovate, acute, 6–8 lin. long ; spur broadly conical, with oblong obtuse apex, about 3 lin. long ; lateral sepals ovate, with shortly falcate obtuse apex, 4–5 lin. long ; petals falcate, about 4 lin. long, obliquely dilated at the base in front, constricted in the middle, dilated and unequally 2-lobed at the apex ; lip long-unguiculate ; claw ½–1¼ in. long ; limb broadly cordate, crenulate, often more or less 3-lobed above, 4–5 lin. broad, front lobe acute, side lobes varying from rounded to acute ; column broad, 1½–2 lin. long ; anther reflexed ; rostellum 3-lobed, with acute teeth, the lateral pair longer and complicate ; stigma pulvinate. *Orchis spathulata, Linn. f. Suppl.* 398. *Satyrium spathulatum, Thunb. Prodr.* 5. *Disa spathulata, Sw. in Vet. Acad. Handl. Stockh.* 1800, 213 ; *Thunb. Fl. Cap. ed. Schult.* 15 ; *Bauer, Illustr. Orch. Gen.*

t. 14 ; *Quart. Journ. Sci. & Arts*, iv. 206, *t.* 6, *fig.* 3 ; *Lindl. Gen. & Sp. Orch.* 353 ; *Drège, Zwei Pfl. Documente*, 98, 102 ; *Harv. Thes. Cap.* i. 54, *t.* 86 ; *Bolus in Journ. Linn. Soc.* xxv. 203 ; *Durand and Schinz, Conspect. Fl. Afr.* v. 108 ; *Kränzl. Orch. Gen. et Sp.* i. 794 ; *Schlechter in Engl. Jahrb.* xxxi. 283, *partly.* *D. propinqua, Sond. in Linnæa*, xix. 95. *D. propinqua, var. trifida, Sond. l.c.* 96.

SOUTH AFRICA : without locality, *Masson*! *Auge*! *Mund & Maire, Rogers.*

COAST REGION : Clanwilliam Div. ; near Brakfontein, *Ecklon & Zeyher*! near Olifants River, behind Modder Fontein, 500 ft., *Schlechter*, 4997 ; near Zwartbosch Kraal, 4000–5000 ft., *Schlechter*, 5165.   Piquetberg Div. ; Piquetberg Mountains, *Thunberg.*   Malmesbury Div.; near Malmesbury, *Kässner, Schlechter.*   Tulbagh Div. ; near Tulbagh, *Ecklon & Zeyher, Pappe*! *Bolus*! *Kässner.*   Paarl Div. ; between Paarl and Lady Gray Railway Bridge, under 1000 ft., *Drège*! Dassen Berg, near Mamre, under 500 ft., *Drège*, 1234!

12. **H. multifida** (Rolfe) ; plant about 1¼ ft. high ; stem rather slender ; leaves few, suberect, filiform, acute, about 5½ in. long, with a few narrow membranous sheaths at the base; scapes about 1½ ft. high, with several lanceolate acute sheaths ; spikes about 3 in. long, about 4-flowered ; bracts oblong-lanceolate, acuminate, 5–6 lin. long ; pedicels 6–7 lin. long ; flowers rather large ; dorsal sepal galeate, ovate, with acuminate reflexed apex, about ½ in. long ; spur narrowly conical, with acute apex, about 4 lin. long ; lateral sepals narrowly ovate or triangular-ovate, very acuminate, 4–5 lin. long ; petals 2-lobed ; lower lobe broadly roundish-oblong, obscurely crenulate, 2 lin. long ; upper lobe 3½ lin. long, broadly oblong at the base, dilated, truncate, crenulate and obliquely apiculate at the apex ; lip long-unguiculate ; claw ½ in. long ; limb somewhat dilated and broken up into several narrow, somewhat branched filaments ; column broad, 1 lin. long ; anther reflexed ; rostellum 3-lobed ; stigma pulvinate. *Disa multifida, Lindl. Gen. & Sp. Orch.* 353; *Drège, Zwei Pfl. Documente*, 73 ; *Bolus in Journ. Linn. Soc.* xxv. 203 ; *Durand & Schinz, Conspect. Fl. Afr.* v. 104 ; *Schlechter in Engl. Jahrb.* xxxi. 285. *D. lacera, Kränzl. Orch. Gen. et Sp.* i. 797, *partly.*

COAST REGION : Clanwilliam Div. ; Blue Berg, 4000–5000 ft., *Drège*, 3577b !

Only known from Drège's original specimen preserved in Lindley's Herbarium. The petals have not previously been described.   Kränzlin has confused the species with *Disa lacera*, Sw.

13. **H. charpentieriana** (Kränzl. Orch. Gen. et Sp. i. 807) ; plant 1¼–1½ ft. high ; stem rather slender ; leaves 5–9, suberect or somewhat arching, elongate-linear or subfiliform, acute, 8–15 in. long ; scapes 1¼–1½ ft. high, with several lanceolate or acuminate sheaths ; spikes 3–5 in. long, 2–6-flowered ; bracts elliptic or elliptic-lanceolate, acuminate, sheathing, 4–6 lin. long ; pedicels ½–¾ in. long ; flowers large, greenish suffused with dull violet-blue ; dorsal sepal galeate, ovate, acute or acuminate, 6–9 lin. long ; spur conical, with slender acuminate curved apex, 3–5 lin. long ; lateral sepals ovate, acuminate,

spreading or deflexed, 5–6 lin. long; petals ascending, falcate-oblong, dilated and rounded at the base, oblique and deeply fringed or lacerate at the apex, 4–5 lin. long; lip long-unguiculate; claw slender, 1¾–2½ in. long; limb somewhat dilated, about 4 lin. long, broken up into several narrow somewhat fleshy filaments; column broad, about 2 lin. long; anther reflexed; rostellum 3-lobed, with acuminate lobes; stigma pulvinate. *Disa charpentieriana, Reichb. f. in Linnæa,* xx. 688; *N. E. Br. in Gard. Chron.* 1885, xxiv. 231; *Hook. Ic. Pl. t.* 1841; *Bolus in Journ. Linn. Soc.* xxv. 203; *Durand & Schinz, Conspect. Fl. Afr.* v. 101; *Schlechter in Engl. Jahrb.* xxxi. 285; *Bolus, Ic. Orch. Austr.-Afr.* ii. *t.* 77. *D. macroglottis, Sond. ex Drège in Linnæa,* xx. 219, *name only*; *Reichb. f. in Reichb. Ic. Fl. Germ.* xiii. *p.* viii. *t.* 354, *fig.* 21–23.

SOUTH AFRICA : without locality, *Gueinzius, Yorke*!
COAST REGION : Tulbagh Div. ; Winterhoek Range, 3000 ft., *Bodkin in Herb. Bolus* ! Stellenbosch Div. ; Hottentots Holland Mountains, *Pappe,* 371 ! Caledon Div. ; Zwart Berg, near Caledon, 1000–2000 ft., *Zeyher* ! 3918 ! near Villiersdorp, 1300 ft., *Bolus,* 5278 ! near Caledon, *Miss Guthrie in Herb. Bolus* ; hills between Houw Hoek and Palmiet River, 1200 ft., *Bolus,* 5278.
CENTRAL REGION : Prince Albert Div. ; summit of the Zwartberg Pass, near Prince Albert, 5300 ft., *Bolus,* 11645.

Bolus remarks that the long lip is very remarkable, and as it sways about in the wind suggests a possible means of attraction for insects.

XXXIVA. FORFICARIA, Lindl.

Sepals free, odd one inferior, horizontally spreading, spathulate, concave, without a spur. *Petals* somewhat oblique, entire, lanceo-late, recurved above the middle, lying on the odd sepal. *Lip* superior, erect or somewhat recurved, reniform-orbicular, obscurely trilobed and ciliate at the apex, without a spur. *Column* short. *Anther* suberect, 2-celled ; cells distinct and parallel ; pollinia solitary in each cell, granular, attached by short caudicles to two distinct glands, seated at the apex of the rostellum. *Ovary* straight ; rostellum short, produced on each side of the column into adnate ciliate wings ; stigma flat.

An erect herb, with several narrow grass-like radical leaves, and a few basal sheaths ; flowers rather small, pale yellow with red-purple petals and lip, borne on a slender somewhat elongated raceme.

DISTRIB. A single species, limited to the south-west corner of Cape Colony.

This curious plant is much like *Herschelia* in habit, but differs in its untwisted ovary and superior lip, as well as in the concave, not galeate, odd sepal and the separate glands of the pollinia.

By an oversight this genus was omitted from the key.

1. F. graminifolia (Lindl. Gen. & Sp. Orch. 362) ; plant 1–1½ ft. high ; stem rather slender ; leaves radical, rather numerous, sub-erect, narrowly linear or grass-like, acuminate, 6–8 in. long ; scapes

1–1½ ft. high, with several narrow acuminate sheaths above ; spikes
3–6 in. long, rather lax, many-flowered ; bracts lanceolate or ovate-
lanceolate, very acuminate, 4–8 lin. long ; pedicels about ½ in. long ;
flowers medium-sized, light yellow with red-purple petals and lip ;
dorsal sepal galeate, broadly ovate, acute, 5–6 lin. long, saccate
behind ; lateral sepals obliquely ovate-lanceolate, acute or shortly
acuminate, 5–6 lin. long ; petals falcate-oblong, the upper half
narrowed, recurved, subobtuse and ciliate, about 3 lin. long ; lip
subreniform, somewhat recurved, obscurely 3-lobed and ciliate, about
3 lin. long ; column short ; anther suberect ; rostellum short, produced into broad ciliate wings, adnate to the sides of the column ;
stigma pulvinate ; capsule oblong, about 8 lin. long. *Drège, Zwei
Pfl. Documente,* 82 ; *Sond. in Linnæa,* xix. 105 ; *Kränzl. Orch. Gen.
et Sp.* i. 723. *Disa Forficaria, Bolus, Ic. Orch. Austr.-Afr.* i. *t.* 87 ;
Schlechter in Engl. Jahrb. xxxi. 297.

CoAST REGION : Worcester Div. ; Dutoits Kloof, 3000–4000 ft., *Drège,* 2211b !
Stellenbosch Div. ; Hottentots Holland Mountains, *Ecklon & Zeyher.* Knysna
Div. ; stony hill sides, Forest Hall, near Plattenbergs Bay, 650 ft., *Miss Newdigate
in Herb. Bolus.*

XXXV. PENTHEA, Lindl.

Sepals free, odd one superior, erect, spathulate, flat or concave,
without a spur. *Petals* erect, oblique, with a small basal auricle,
free from the column. *Lip* inferior, narrowly linear, without a spur.
Column short. *Anther* suberect, 2-celled ; cells distinct and parallel ;
pollinia solitary in each cell, granular, attached by short caudicles
to two distinct glands, situated on the arms of the rostellum. *Ovary*
twisted ; rostellum erect, 3-partite, with divaricate side lobes ;
stigma cushion-shaped. *Capsule* oblong.

Terrestrial herbs, with ovoid-oblong tubers : leaves radical or cauline, narrow,
reduced upwards into the bracts ; flowers medium-sized, red or yellow, produced
in short corymbose spikes ; bracts narrow.

DISTRIB. Species 2, limited to the south-west corner of Cape Colony.

This genus was primarily based upon *Disa patens,* Thunb., and *D. filicornis,*
Thunb. (*Lindl. Nat. Syst. Bot. ed.* ii. 446), to which Lindley added others with
inverted flowers, now referred to *Orthopenthea,* Rolfe. It is readily separated from
Disa by the spathulate spurless dorsal sepal.

Flowers yellow ; lateral sepals narrow and acuminate ... (1) **patens.**

Flowers crimson or rosy ; lateral sepals broadly oblong
      and apiculate ...   . (2) **filicornis.**

1. **P. patens** (Lindl. Gen. & Sp. Orch. 362) ; plant 3–10 in. high ;
stem somewhat slender ; leaves radical, numerous, the lower spreading, the rest suberect, linear or narrowly lanceolate-linear, acute,
½–1 in. long ; scapes 3–10 in. long, with numerous narrow acuminate
sheaths ; spikes short, loosely subcorymbose, 1–5- (rarely to 8-)
flowered ; bracts lanceolate, acute or acuminate, 4–8 lin. long ;

pedicels ½–1 in. long; flowers rather large, bright yellow; dorsal
sepal nearly flat, broadly ovate or cordate-ovate, shortly acuminate,
6–8 lin. long; spur obsolete; lateral sepals narrowly ovate from a
broad oblique base, with a narrow falcate acuminate apex, 6–8 lin.
long; petals falcate-oblong, acute, with a rounded basal auricle in
front, 3–4 lin. long; lip narrowly linear, subfiliform above, 4 lin.
long; column 4 lin. long; anther incurved, acuminate, furnished
with a narrow petaloid wing on each side; rostellum erect, 3-partite,
with oblong divaricate side lobes; front lobe triangular, acute,
concave; stigma pulvinate. *Drège, Zwei Pfl. Documente,* 74, 82,
89; *Krauss in Flora,* 1845, 306, *and Beitr. Fl. Cap- und Natal.* 158;
Sond. in Linnæa, xx. 220. *Ophrys patens, Linn. f. Suppl.* 404; *Murr.
Syst. Veg.* xiv. 814. *Serapias patens, Thunb. Prodr.* 3. *Disa patens,
Sw. in Vet. Acad. Handl. Stockh.* 1800, 214; *N. E. Br. in Gard. Chron.*
1885, xxiv. 232; *Bolus, Ic. Orch. Austr.-Afr.* ii. *t.* 69; *Durand &
Schinz, Conspect. Fl. Afr.* v. 105; *Kränzl. Orch. Gen. et Sp.* i. 742;
Schlechter in Engl. Jahrb. xxxi. 229. *D. tenuifolia, Sw. in Vet. Acad.
Handl. Stockh.* 1800, 214; *Bolus in Trans. S. Afr. Phil. Soc.* v. 157,
Orch. Cap. Penins. 157, *and Journ. Linn. Soc.* xxv. 199.

SOUTH AFRICA: without locality, *Masson*! *Roxburgh*! *Brown*! *Hesse,* 13!
Sieber! *Bunbury*! *Grey*! *Scully*! *Leibold.*
    COAST REGION: Clanwilliam Div.; Ezels Bank, 3000–4000 ft., *Drège,* 1232 c.
Piquetberg Div.; near Pikeniers Kloof, 1000–5000 ft., *Zeyher,* 1580! Tulbagh
Div.; near Tulbagh Waterfall, *Ecklon & Zeyher*! Worcester Div.; Dutoits Kloof,
3000–4000 ft., *Drège,* 1232 b! Cap. Div.; Table Mountain, 800–3500 ft., *Burchell,*
656! *Harvey*! *Prior*! *Drège,* 1232 a! *Ecklon,* 343! *Bolus*! *Wolley-Dod,* 882!
Rehmann, 570! *Kässner, Schlechter,* 145, 187; Muizen Berg, 1500 ft., *Bolus,* 3913!
Trimen! *Schlechter, Kässner*; Stein Berg, *Wolley-Dod,* 2112 b. Caledon Div.;
Zwart Berg, near Caledon, *Bowie*! Genadendal, *Prior*! Baviaans Kloof, *Krauss*!
Steenbrass River, *Mund.* Swellendam Div.; summit of a mountain peak near
Swellendam, *Burchell,* 7338! ridges near the Zondereinde River, *Zeyher,* 3932!
Langeberg Range, above Tradouw Pass, *Mund.* Riversdale Div.; Kampsche Berg,
Burchell, 7055! George Div.; near George, 650 ft., *Schlechter,* 5865! plains at
Woodville, 800 ft., *Galpin,* 4619!

2. **P. filicornis** (Lindl. Gen. & Sp. Orch. 361); plant 4–10 in.
high; stem rather stout; leaves radical and cauline, numerous, suberect, linear or narrowly lanceolate-linear, acute, narrowed upwards
into the bracts, ½–1½ in. long; scapes 4–10 in. high, with narrow
acute sheaths above; spikes short, loosely subcorymbose, 2–8-flowered;
bracts lanceolate, acute or acuminate, ½–1 in. long; pedicels ½–1 in.
long; flowers rather large, pink, spotted with purple, and the lateral
sepals suffused with purple; dorsal sepal galeate, obovate-spathulate,
apiculate, 6–8 lin. long; spur reduced to a nearly obsolete sac;
lateral sepals oblong, subobtuse or shortly apiculate, spreading,
6–8 lin. long; petals falcate-oblong, subobtuse, with a rounded basal
auricle in front, 3–4 lin. long; lip narrowly linear, 3–4 lin. long;
column 4 lin. long; anther suberect, furnished with a broad petaloid
wing-like appendage on each side; rostellum erect, 3-partite, with
subdivaricate side lobes; front lobe shorter and concave; stigma
pulvinate. *Drège, Zwei Pfl. Documente,* 82, 84, 114; *Sond. in Linnæa,*

xx. 220. *P. reflexa, Lindl. Gen. & Sp. Orch.* 361 ; *Drège, Zwei Pfl. Documente,* 125 ; *Krauss in Flora,* 1845, 306, *and Beitr. Fl. Cap- und Natal.* 158. *Orchis filicornis, Linn. f. Suppl.* 400 ; *Murr. Syst. Veg.* xiv. 811. *Disa filicornis, Thunb. Fl. Cap. ed. Schult.* 17 ; *N. E. Br. in Gard. Chron.* 1885, xxiv. 232 ; *Durand & Schinz, Conspect. Fl. Afr.* v. 103 ; *Kränzl. Orch. Gen. et Sp.* i. 741 ; *Schlechter in Engl. Jahrb.* xxxi. 228 ; *Bolus, Ic. Orch. Austr.-Afr.* ii. *t.* 68. *D. patens, Sw. in Vet. Acad. Handl. Stockh.* 1800, 214 ; *Bolus in Trans. S. Afr. Phil. Soc.* v. 157, *Orch. Cap. Penins.* 157, *and in Journ. Linn. Soc.* xxv. 200. *D. reflexa, Reichb. f. in Flora,* 1865, 182.

SOUTH AFRICA: without locality, *Auge* ! *Masson* ! *Mund* ! *Zeyher,* 1578 ! *Ecklon & Zeyher,* 105 ! *Hesse* ! *Harvey,* 145 ! *Rogers* !

COAST REGION : Tulbagh Div. ; near Tulbagh, *Zeyher.* Worcester Div. ; mountains near De Liefde, 2000–3000 ft., *Drège,* 1233 b ! Cape Div. ; Cape Flats, 50–100 ft., *Pappe,* 7 ! *Ecklon,* 245 ! *Bunbury* ! on mountains, *MacOwan & Bolus, Herb. Norm. Austr.-Afr.* 163 ! Muizenberg Mountain, 1100 ft., *Bolus,* 3365 ! Silvermine Valley, Cape Peninsula, *Wolley-Dod,* 2215 ! Table Mountain, 2500 ft., *Zeyher, Krauss, Schlechter,* 66. Stellenbosch Div. ; Lowrys Pass, 1000–2000 ft., *Drège,* 1233 a ! Caledon Div. ; Zwart Berg, near Caledon, *Zeyher,* 3931 ! near Zonder Einde River, *Pappe,* 4 ! Genadendal, *Prior* ! Houw Hoek, 1500 ft., *Bowie* ! *Schlechter,* 9413 ! near Palmiet River, *Leibold* ; Steenbrass River, *Ecklon & Zeyher.* Riversdale Div. ; Langeberg Range, near Riversdale, 1500–2500 ft., *Schlechter,* 1909. George Div. ; near George, 700 ft., *Bowie* ! *Schlechter,* 5865 ! *Penther, Krook.* Knysna Div. ; Bosch River, under 5000 ft., *Drège,* 8281 ! hills near Plettenbergs Bay, *Schlechter, Penther, Krook* ; near Knysna River, *Bowie* !

EASTERN REGION : Natal ; Oakford, *Rehmann,* 521 !

This species has been much confused with the preceding.

XXXVI. DISA, Berg.

Sepals free, odd one superior, concave, hood-shaped or helmet-shaped, with a more or less elongated spur or sac. *Petals* usually much smaller than the sepals, generally more or less adnate to the column at the base, often included within the dorsal sepal but quite free from it, very variable in shape. *Lip* anticous, sessile, usually small and narrow, without a spur. *Column* short. *Anther* suberect, horizontal, reclinate or reflexed, 2-celled, the cells distinct and parallel ; pollinia solitary in each cell, granular, attached by short or long caudicles to two distinct glands, seated at the apex or in the arms of the rostellum. *Ovary* twisted ; rostellum erect, sub-entire, bifid or trifid at the apex, sometimes with side processes, often more or less adnate to the base of the petals and sometimes forming a ridge upon them ; stigma cushion-like, seated in front of the rostellum. *Capsule* cylindric, clavate or narrowly ellipsoid.

Terrestrial herbs, with simple sessile tubers ; leaves appearing with or before the flowers, radical or cauline, sometimes produced upon separate growths, and the leaves of the flowering stem reduced to sheaths ; flowers large, medium-sized or small, corymbose, racemose or in lax or dense spikes, rarely solitary ; bracts usually narrow.

DISTRIB.—Species about 100, the majority South African, the remainder inhabiting the uplands of Tropical Africa, with a few in Madagascar.

The limits of the genus were extended by Bolus to include Lindley's genera *Monadenia, Herschelia, Forficaria, Penthea* and *Schizodium,* which are here restored.

*Flowers numerous, usually under ½ in. (often under
 ⅓ in.) long, but over ¼ in. in *D. Hallackii,* and
 arranged in a more or less dense elongated spike ;
 spur saccate or oblong, rarely slender, usually
 obtuse and shorter than the limb :

†Plant 2 to 10 in. high, occasionally taller, usually
 with dense and relatively long spikes and small
 flowers :

Leaves oblong or oblong-lanceolate   ...   ... (1) **cylindrica.**

Leaves linear or narrow :
  Spur reduced to a shallow sac or only a quarter
    as long as the limb :
    Dorsal sepal 2–2½ lin. long ; spur subobsolete (2) **neglecta.**

    Dorsal sepal 3–3½ lin. long ; spur distinct ... (3) **Pappei.**

  Spur oblong or narrow, simple, over a third as
    long as the limb :
    Leaves very narrow and more or less spirally
      twisted :
      Spur broad and about a third as long as the
        limb   ...   . (4) **brachyceras.**

      Spur slender and over half as long as the
        limb ...   . (5) **tenella.**

    Leaves linear, not twisted :
      Sepals about 1½ lin. long   ...   . (6) **micropetala.**

      Sepals 2 lin. or more long :
        Spur conspicuously sulcate behind   ... (7) **picta.**

        Spur not sulcate behind :
          Lateral sepals 2–2½ lin. long ; rostellum
            bifid   ...   . (8) **obtusa.**

          Lateral sepals 3–3½ lin. long ; rostellum
            trifid   ...   . (9) **tabularis.**

  Spur straight and filiform, with a short oblong
    sac at each side   ...   . (10) **tenuicornis.**

††Plant usually 10 in. or more high, if shorter with
 relatively short or somewhat lax spikes or larger
 flowers :

‡Spikes lax or ultimately somewhat lax ; leaves
 mostly cauline, rarely subradical :
Spur clavate or subclavate :
  Leaves narrowly elongate-linear   ...   ... (11) **longifolia.**

  Leaves lanceolate or oblong-lanceolate, nar-
    rowed at the base :
    Dorsal sepal acute or apiculate ; leaves
      narrow ...   . (12) **ocellata.**

    Dorsal sepal obtuse ; leaves usually broader (13) **uncinata.**

  Leaves oblong, not distinctly narrowed at the
    base   ...   . (14) **Tysoni.**

Spur oblong or conical :
  Spur broad and obtuse :
    Spur conical, about as long as the limb ... (15) **stachyoides.**

    Spur broadly conical, longer than the
      limb   ...   . (16) **aconitoides.**

Spur narrow and subacute　...　...　... (17) **stricta.**

‡‡Spike very dense; leaves radical or subradical
　　and elongate, unknown in *D. Hallackii* :
　Flowers under ¼ in. long :
　　Spur narrow or not inflated ; spike usually
　　　under 6 in. long :
　　　Lip rhomboid-oblong, gland-tipped　... (18) **sanguinea.**

　　　Lip linear or subspathulate, not gland-
　　　　tipped :
　　　　Spur short, not descending below the
　　　　　base of the limb　...　...　... (19) **Sankeyi.**

　　　　Spur long, descending much below the
　　　　　base of the limb :
　　　　　Leaves shortly oblong ; lip spathulate-
　　　　　　linear　...　...　...　... (20) **fragrans.**

　　　　　Leaves elongate-oblong; lip linear　... (21) **polygonoides.**

　　　　Spur inflated ; spike ½ to over 1 foot long　... (22) **chrysostachya.**

　Flowers over ½ in. long ...　...　...　... (23) **Hallackii.**

**Flowers numerous, usually over ½ in. long (often much
　larger), and arranged in a more or less dense
　elongated spike ; spur slender and acute, elongated
　from a conical base and usually longer than the
　limb, or clavate and much longer than the limb ;
　leaves usually broad ; lateral sepals usually short
　or broad :
　Flowers small : dorsal sepal 2-4 lin. long :
　　†Spur more or less recurved from about the
　　　middle :
　　　Spur under ¼ in. long or shorter than the
　　　　limb :
　　　　Cauline leaves oblong - lanceolate ; spikes
　　　　　3-5½ in. long, narrow　...　...　... (24) **extinctoria.**

　　　　Cauline leaves ovate-oblong; spikes 4-9 in.
　　　　　long, rather broader　...　...　... (25) **Macowani.**

　　　Spur over ¼ in. long or not shorter than the
　　　　limb :
　　　　Dorsal sepal about 3 lin. long ; spur about as
　　　　　long as the limb ...　...　...　... (26) **læta.**

　　　　Dorsal sepal about 4 lin. long ; spur longer
　　　　　than the limb　...　...　...　... (27) **rhodantha.**

　　††Spur ascending and nearly straight :
　　　Radical leaves elongate-oblong :
　　　　Spur subcylindrical, about twice as long as
　　　　　the limb　...　...　...　...　... (28) **caffra.**

　　　　Spur clavate, three times as long as the limb (29) **Galpinii.**

　　　Radical leaves broadly elliptic or orbicular　... (30) **ovalifolia.**

　Flowers large ; dorsal sepal ½ in. or more long :
　　Spur ¾ in. or less long :
　　　Leaves broadly oblong or ovate-oblong and
　　　　crowded :
　　　　Dorsal sepal obtuse ; lip broadly elliptic or
　　　　　suborbicular　...　...　...　... (31) **cornuta.**

　　　　Dorsal sepal acute ; lip oblong or narrowly
　　　　　elliptic　...　...　...　...　... (32) **macrantha.**

Leaves narrowly oblong and not crowded   ... (33) **Thodei**.

Spur 1 in. or more long:
  Lateral sepals 5–6 in. long:
    Lip narrowly ovate-oblong or lanceolate:
      Petals falcate-oblong   ...   . (34) **zuluensis**.
      Petals obliquely ovate   ...   . (35) **Scullyi**.
    Lip broadly rhomboid-ovate ...   . (36) **Cooperi**.
  Lateral sepals about 1 in. long ...   . (37) **crassicornis**.

***Flowers usually arranged in a short subcorymbose or
  rather elongated lax spike or raceme, very various
  in size, rarely solitary; spur short and broad, or
  if narrow not as long as the limb; lateral sepals
  usually short or broad:

†Dorsal sepal under ½ in. long:
  ‡Dorsal sepal entire:
    Flowers subcorymbose or in a short oblong
    compact spike:
      Spur broad and obtuse, a quarter to a third
      as long as the limb:
        Petals falcate - oblong, obliquely acute,
        crenulate   ...   . (38) **Vasselotii**.

        Petals falcate or sharply bent, with two
        diverging apical lobes ...   . (39) **falcata**.

      Spur narrow or subacute or rarely as long as
      the limb:

        Leaves spathulate with broad limb:
          Leaves and sheaths hairy   ...   ... (40) **glandulosa**.
          Leaves and sheaths smooth and densely
          lepidote   ...   . (41) **vaginata**.
        Leaves linear or lanceolate-linear ...   ... (42) **frigida**.

      Flowers in a somewhat elongated more or less
      lax raceme:
        Leaves cauline ...   . (43) **caulescens**.

        Leaves radical or subradical ...   . (44) **tripetaloides**.

  ‡‡Dorsal sepal bilobed or subtrilobed:
    Dorsal sepal subtrilobed ...   . (45) **triloba**.
    Dorsal sepal with two broadly oblong spreading
    lobes ...   . (46) **sagittalis**.

††Dorsal sepal ¾ in. or more long:
  Flowers rose-coloured, about 1½ in. across:
    Dorsal sepal obovate-oblong obtuse   ...   ... (47) **racemosa**.

    Dorsal sepal rhomboid-cuneate, acute ...   ... (48) **venosa**.
  Flowers scarlet, 3 in. or more across   ...   ... (49) **uniflora**.

****Flowers usually numerous and arranged in a lax or
  sometimes dense spike or raceme, rarely solitary;
  spur usually long and slender, subsaccate in
  D. maculata; leaves usually narrow or elongated;
  lateral sepals usually narrow or elongated:
†Flowers in short or elongated spikes or racemes,
  solitary in weak examples of *D. Marlothii*:

‡Spur not or scarcely longer than the limb or under
  5 lin. long:
  Flowers in a dense subcapitate or shortly oblong
  spike:

Dorsal sepal broadly ovate-orbicular ; lateral sepals broad and flat ... ... (50) **cephalotes.**

Dorsal sepal ovate-oblong; lateral sepals narrow and very concave ... ... (51) **Gerrardii.**

Flowers in a more or less lax elongated raceme : Lateral sepals under 4 lin. long and rather broad :
   Leaves narrowly linear or grass-like ... (52) **oreophila.**
   Leaves lanceolate-linear ... ... (53) **saxicola.**

Lateral sepals over 4 lin. long and more or less elongate :
   Leaves ensiform or broader :
      Spur shorter than or rarely slightly longer than the limb :
         Dorsal sepal 4–6 lin. long :
            Sepals ovate-oblong or rather broad (54) **montana.**
            Sepals oblong or rather narrow ... (55) **patula.**
         Dorsal sepal 8 lin. or more long :
            Petals linear - oblong or falcate-oblong, rather narrow :
               Spur about as long as the limb ... (56) **nervosa.**
               Spur distinctly shorter than the limb :
                  Petals about 4 lin. long ; lip oblong-lanceolate ... (57) **Kraussii.**
                  Petals about 1 in. long ; lip filiform .. .. (58) **Fanniniæ.**
               Petals lanceolate or ovate-lanceolate (59) **pulchra.**
      Spur twice as long as the limb ... ... (60) **Marlothii.**
   Leaves narrowly linear or grass-like :
      Sepals 6–7 lin. long ; lip lanceolate or narrowly lanceolate-oblong ... ... (61) **gladioliflora.**
      Sepals 4–5 lin. long ; lip spathulate-obovate ... ... (62) **capricornis.**
‡‡ Spur much longer than the limb or over 1 in. long :
   Sepals ¾ in. or more long :
      Leaves oblong-linear or elongate-linear :
         Flowers straw-yellow ; petals oblong ... (63) **Draconis.**
         Flowers lilac with some purple marks ; petals linear ... ... (64) **harveiana.**
      Leaves very narrow and grass-like ... ... (65) **schlechteriana.**
   Sepals ¼–⅓ in. long :
      Spur subhorizontal, about three times as long as the limb ... ... (66) **ferruginea.**
      Spur suberect, over four times as long as the limb ... ... (67) **porrecta.**
††Flowers solitary, blue :
   Flowers large with much elongated spur ... ... (68) **longicornu.**
   Flowers medium-sized with short subsaccate spur (69) **maculata.**

1. D. cylindrica (Sw. in Vet. Acad. Handl. Stockh. 1800, 213); plant ½–1 ft. high; stem usually rather stout; leaves radical and

cauline, suberect, oblong or oblong-lanceolate, subacute, often
numerous, 1½–4 in. long, reduced upwards into the bracts; scapes
½–1 ft. high, with a few leaf-like sheaths above; spikes oblong or
cylindrical, 1–4½ in. long, very dense and many-flowered; bracts
ovate-lanceolate, subacute, 5–7 lin. long; pedicels about 3 lin. long;
flowers small, dull yellow; dorsal sepal galeate, ovate-oblong,
obtuse, 2½–3 lin. long; spur saccate-oblong, obtuse, about one-sixth
as long as the limb; lateral sepals ovate-oblong, obtuse, recurved,
2½ lin. long; petals obliquely ovate, with broad obtuse fleshy apex,
and a short rounded basal lobe in front, 2 lin. long; lip linear-
oblong, obtuse, recurved, 1¾ lin. long; column short; anther
reflexed; rostellum erect, 3-lobed, with small rounded side lobes
and large cucullate obtuse front lobe; stigma pulvinate. *Thunb.
Fl. Cap. ed. Schult.* 13; *Lindl. Gen. & Sp. Orch.* 356; *Bolus in
Trans. S. Afr. Phil. Soc.* v. 153, *Orch. Cap. Penins.* 153, *Journ.
Linn. Soc.* xxv. 197, *and Ic. Orch. Austr.-Afr.* ii. *t.* 73; *Durand &
Schinz, Conspect. Fl. Afr.* v. 102; *Kränzl. Orch. Gen. et Sp.* i. 746,
and in Ann. Naturhist. Hofmus. Wien, xx. 7; *Schlechter in Engl.
Jahrb.* xxxi. 268. *D. bracteata, Lindl. Bot. Reg. t.* 324; *Lindl.
Gen. & Sp. Orch.* 354, *probably not of Sw.*; *Krauss in Flora,* 1845,
306, *and Beitr. Fl. Cap- und Natal.* 158; *Sond. in Linnæa,* xix. 97;
xx. 219. *Satyrium cylindricum, Thunb. Prodr.* 5. *Monadenia bracteata,
Durand & Schinz, Conspect. Fl. Afr.* v. 111; *Kränzl. Orch. Gen. et
Sp.* i. 810.

SOUTH AFRICA: without locality, *Masson, Thunberg! Lehmann! Grey! Harvey,*
139! 240! *Rogers! Prior!*
COAST REGION: Cape Div.; Devils Peak, *Bergius!* Table Mountain, 2500–
3500 ft., *Bolus,* 4537! and in *MacOwan & Bolus, Herb. Norm. Austr.-Afr.* 359!
Mrs. Jameson! Prior! Schlechter, 135! *Rehmann,* 578! *Kässner, Wolley-Dod,*
2117! Steenberg Rocks, *Wolley-Dod,* 2136! Summit of the Twelve Apostles,
Wolley-Dod, 3603! Caledon Div.; by the Zondereinde River, near Appels Kraal,
Zeyher, 3926! near Knoflook, *Pappe,* 51! Swellendam Div.; Baviaans Kloof,
Krauss, 1314. Riversdale Div.; Garcias Pass, 1200 ft., *Galpin,* 4617! Langeberg
Range, 1500 ft., *Schlechter.* George Div.; above Montagu Pass, 4000 ft.,
Schlechter, Penther, 180, 335, *Krook*; near George, *Penther,* 100. Knysna Div.;
Elands River, *Penther,* 294; near Knysna, *Penther,* 284.

Disa bracteata, Sw. in Vet. Acad. Handl. Stockh. 1800, 211, is a mystery.
It was quoted as "C. b. sp. Sparrman," and characterised as "galea obtusa,
calcare oblongo; labello lineari apice latiore; spica cylindrica, bracteis erectis
floribus longioribus." A sheet kindly lent by the authorities at Stockholm
contains three specimens, two of which belong to a ticket from an unknown
collector, inscribed "43. Im Sande zwischen den gefaulagen [?] auf der Platte
des Tafelberges. Jan. 24–24." This presumably means 1824, in which case it
cannot be Swartz's type. The other specimen is *Zeyher,* 3926, which is also at
Kew. All three are *Disa cylindrica,* Sw., as here understood. The plant
figured as *D. bracteata,* Sw., in the Botanical Register, t. 324, is apparently a
lax-flowered form of the same; but Robert Brown's wild specimens there alluded
to belong to *Monadenia micrantha,* Lindl., and are preserved at the British
Museum, labelled by Brown himself as "*Disa bracteata,* Sw." The description
alluded to in Bot. Reg. t. 324 is also preserved (but not the actual specimen),
and it shows that Brown at first described this plant as a new species, but
afterwards thought that it agreed "too closely to justify a separation," an opinion
in which I cannot concur. Bolus also thought *D. bracteata,* Sw., might be a
Monadenia, which is quite borne out by the long bracts. I believe it was

M. multiflora, Lindl., with which the phrase "calcare oblongo, labello lineari apice latiore" is fully in agreement. It does not describe the cylindrical spur and linear lip of *M. micrantha*, Lindl.

2. **D. neglecta** (Sond. in Linnæa, xix. 100); plant 4–7 in. high; stem moderately stout; leaves radical and cauline, few, suberect, the former linear-lanceolate, acute, about 3½ in. long, the latter linear, acuminate, 1½–3 in. long; scapes 4–7 in. long, with a few short acuminate sheaths above; spikes oblong, 1½–2½ in. long, somewhat dense, many-flowered; bracts ovate-lanceolate, acuminate, 5–8 lin. long; pedicels about 4 lin. long; flowers rather small, dull yellow, with purple spots and lines on the sepals, and a purple stain on the middle of the lip; dorsal sepal galeate, very broadly ovate, shortly apiculate, 3½–4 lin. long, with a narrow mouth, slightly gibbous at the base, but neither saccate nor spurred; lateral sepals oblong, subobtuse, slightly concave, suberect or adpressed to the dorsal sepal, 3½–4 lin. long; petals oblique, semiovate-oblong, obtuse, obscurely crenulate, 1½ lin. long; lip ovate-lanceolate, subobtuse, concave, incurved or suberect so as to nearly close the opening to the flower, about 1½ lin. long; column short; anther reclinate; rostellum erect, 3-dentate, middle lobe shortly 2-dentate; stigma pulvinate. *Bolus in Journ. Linn. Soc.* xxv. 198; *Durand & Schinz, Conspect. Fl. Afr.* v. 104; *Kränzl. Orch. Gen. et Sp.* i. 799; *Schlechter in Engl. Jahrb.* xxxi. 249. *D. lineata, Bolus in Journ. Linn. Soc.* xxii. 74, xxv. 199, *Trans. S. Afr. Phil. Soc.* v. 154, *t.* 18, *and Orch. Cap. Penins.* 154, *t.* 18; *Kränzl. Orch. Gen. et Sp.* i. 784.

SOUTH AFRICA: without locality, *Bergius, Brown*!
COAST REGION: Tulbagh Div.; mountains near Tulbagh, 2500–3000 ft., *Ecklon & Zeyher.* Cape Div.; moist slopes of mountains near Constantia, 2700 ft., *Bodkin in Herb. Bolus*, 4966! *MacOwan & Bolus, Herb. Norm. Austr.-Afr.* 405! top of Skeleton Ravine, *Wolley-Dod*, 3073! George Div.; Outeniqua Mountains, above Montagu Pass, 4000 ft., *Schlechter.*

3. **D. Pappei** (Rolfe); plant about 9 in. high; stem slender; leaves radical or subradical, few, somewhat spreading, linear-lanceolate, acute, about 2 in. long, with a few linear acute leaf-like sheaths above; scape about 9 in. high; spike about 2½ in. long, rather lax, many-flowered; bracts ovate-lanceolate with a long acuminate apex, 4–5 lin. long; pedicels 4 lin. long; flowers very small; dorsal sepal galeate, broadly ovate, obtuse, 2–2½ lin. long; spur saccate-oblong, obtuse, about a quarter as long as the limb; lateral sepals ovate-oblong, obtuse or subapiculate, 2–2½ lin. long; petals subulate-oblong, subobtuse, ¾ lin. long; lip linear, obtuse, about 1½ lin. long; column short; anther reclinate; rostellum tridentate; stigma pulvinate.

COAST REGION: Caledon Div.; Zondereinde River, near Knoflook, *Pappe*, 46!

4. **D. brachyceras** (Lindl. Gen. & Sp. Orch. 355); plant 2–4 in. high; stem rather slender; leaves radical and cauline, numerous,

suberect, linear, flexuous, acute, $\frac{1}{2}$–1$\frac{1}{4}$ in. long ; scapes 2–4 in. high ; spikes $\frac{1}{3}$–1$\frac{1}{2}$ in. long, dense, many-flowered ; bracts broadly ovate below, long-acuminate above, 3–6 lin. long ; pedicels about 2 lin. long ; flowers very small ; dorsal sepal galeate, broadly ovate, subacute or apiculate, 1$\frac{3}{4}$–2 lin. long ; spur broadly saccate, obtuse, about one-third as long as the limb ; lateral sepals ovate-oblong, subacute, 1$\frac{1}{2}$–1$\frac{3}{4}$ lin. long ; petals rhomboid-ovate, narrowed upwards and acute, about 1$\frac{1}{4}$ lin. long ; lip linear, obtuse, about 1$\frac{1}{4}$ lin. long ; column short ; anther reclinate ; rostellum very short, broadly rhomboid, obtuse ; stigma pulvinate ; capsule elliptic-oblong, about 2$\frac{1}{2}$ lin. long. *Drège, Zwei Pfl. Documente,* 119 ; *Sond. in Linnæa,* xix. 98, xx. 219 ; *Bolus in Journ. Linn. Soc.* xxv. 198 ; *Kränzl. Orch. Gen. et Sp.* i. 791. *D. tenella, var. brachyceras, Schlechter in Engl. Jahrb.* xxxi. 245.

SOUTH AFRICA : without locality, *Auge* !
COAST REGION : Worcester Div. ; between Slangenheuvel, French Hoek, and Donker Hoek, under 1000 ft., *Drège,* 1246 ! Caledon Div. ; Klein River Mountains, 1000–3000 ft., *Zeyher,* 54 ! near Caledon, *Templeman in Herb. Bolus* !
CENTRAL REGION : Ceres Div. ; Cold Bokkeveld, 3500 ft., *Schlechter,* 8713 !

Similar in habit to *D. tenella,* Sw., but very distinct in the shape of the spur and petals.

5. **D. tenella** (Sw. in Vet. Acad. Handl. Stockh. 1800, 212) ; plant 3–5 in. high ; stem rather slender ; leaves radical and cauline, numerous, suberect, linear, flexuous, acute, 1–2$\frac{1}{4}$ in. long ; scapes 3–5 in. high ; spikes $\frac{1}{2}$–2 in. long, rather dense, many-flowered ; bracts ovate-oblong below, linear-acuminate above, $\frac{1}{4}$–$\frac{3}{4}$ in. long ; pedicels about 2$\frac{1}{2}$ lin. long ; flowers very small, pinkish-purple, rarely yellowish or white ; dorsal sepal galeate, broadly ovate, obtuse, somewhat inflexed at the margin, 1$\frac{3}{4}$–2 lin. long ; spur narrowly conical, subobtuse, over half as long as the limb ; lateral sepals broadly oblong, obtuse or subapiculate, 1$\frac{3}{4}$–2 lin. long ; petals broadly ovate or rhomboid-ovate, subacute, 1–1$\frac{1}{4}$ lin. long ; lip subspathulate, obtuse, about 1$\frac{1}{4}$ lin. long ; column short ; anther reflexed ; rostellum short, subrhomboid, emarginate ; stigma pulvinate. *Thunb. Fl. Cap. ed. Schult.* 11 ; *Lindl. Gen. & Sp. Orch.* 355 ; *Drège, Zwei Pfl. Documente,* 119 ; *Sond. in Linnæa,* xix. 98, xx. 219 ; *N. E. Br. in Gard. Chron.* 1885, xxiv. 232 ; *Bolus in Trans. S. Afr. Phil. Soc.* v. 151, *Orch. Cap. Penins.* 151, *Journ. Linn. Soc.* xxv. 197, *and Ic. Orch. Austr.-Afr.* ii. *t.* 72 ; *Durand & Schinz, Conspect. Fl. Afr.* v. 108 ; *Kränzl. Orch. Gen. et Sp.* i. 791 ; *Schlechter in Engl. Jahrb.* xxxi. 245. *Orchis tenella, Linn. f. Suppl.* 400. *Satyrium tenellum, Thunb. Prodr.* 5.

SOUTH AFRICA : without locality, *Masson* ! *Brown, Mund & Maire, Trimen* ! *Rogers* ! *Zeyher* ! *Hesse,* 12 ! *Oldenburg,* 404 ! *Sieber,* 390 !
COAST REGION : Clanwilliam Div. ; Cederberg Range, near Clanwilliam, *Leipoldt,* 573 ; near Sneeuwkop, *Bodkin.* Malmesbury Div. ; Zwartland, *Mund.* Tulbagh Div. ; near Tulbagh, under 1000 ft., *Drège,* 487 ! Tulbagh Kloof, *Bolus.* Paarl Div. ; mountains near Wellington, *Miss Cummings in Herb. Bolus,* 6094 ; Cape Div. ; sand dunes near Wynberg, *Alexander* ! *Prior* ! sandy flat, Kenilworth

Racecourse, *Schlechter*, 1552, *Ecklon & Zeyher* ; eastern side of Table Mountain, near Constantia, *Ecklon & Zeyher.* Stellenbosch Div. ; Hottentots Holland, *Thunberg* ! near Stellenbosch, *Lloyd in Herb. Sanderson*, 928 ! *Miss Farnham in MacOwan & Bolus, Herb. Norm. Austr.-Afr.* 309 ! *Prior* ! Uitenhage Div. ; Uitenhage, *Ecklon* !

CENTRAL REGION : Ceres Div. ; Cold Bokkeveld, 3500 ft., *Marloth*, 573, *Schlechter*, 8913 !

The Uitenhage station rests on the authority of *Ecklon*, and requires confirmation.

6. **D. micropetala** (Schlechter in Engl. Jahrb. xx. Beibl. 50, 7) ; plant 2½–8 in. high ; stem stout ; leaves radical and cauline, sub-erect or somewhat spreading, linear, acute or acuminate, ¾–3 in. long ; scapes 2½–8 in. high ; spikes cylindrical, 1¼–5 in. long ; bracts lanceolate, acuminate, 4–5 lin. long ; pedicels 1½–2 lin. long ; flowers minute, dorsal sepal galeate, broadly ovate, obtuse or very shortly apiculate, 1½ lin. long ; spur oblong, obtuse, curved, about one-third as long as the limb ; lateral sepals ovate or ovate-oblong, obtuse, 1½ lin. long ; petals linear or falcate-linear, obtuse, ½ lin. long ; lip linear-oblong, obtuse, about ¾ lin. long ; column short ; anther reflexed ; rostellum minute ; stigma pulvinate ; capsule ovoid-oblong, about 2 lin. long. *Schlechter in Engl. Jahrb.* xxxi. 246, *t. 3, fig. A–D.*

COAST REGION : Caledon Div. ; mountains near Genadendal, 4800 ft., *Schlechter*, 9881 ! Swellendam Div. ; moist grassy slopes above Voormansbosch, *Ecklon & Zeyher* !

The two specimens cited here are very distinct in appearance, and there are also slight floral differences. Ecklon and Zeyher's specimen is the type of the species, but the specimen illustrated in *Engler's Jahrbücher* is *Schlechter*, 9881, which, curiously enough, is not cited in the text. Further materials are necessary to decide if both are forms of the same species.

7. **D. picta** (Sond. in Linnæa, xix. 99) ; plant 6–9 in. high ; stem rather stout ; leaves radical and cauline, numerous, suberect, linear, acuminate, 2–5 in. long ; scapes 6–9 in. high ; spikes 2½–5 in. long, dense, many-flowered ; bracts ovate or ovate-lanceolate, very acuminate, ½–¾ in. long ; pedicels 3–4 lin. long ; flowers small ; dorsal sepal galeate, broadly ovate, obtuse or apiculate, 2–2½ lin. long ; spur broadly oblong, obtuse, about a quarter as long as the limb, conspicuously sulcate behind ; lateral sepals oblong, obtuse, spreading, 2 lin. long ; petals linear-oblong, subacute, with a small rounded lobe in front, about ¾ lin. long ; lip linear-oblong, obtuse, 1½–1¾ lin. long ; column short ; anther reflexed ; rostellum suberect, short, 3-denticulate ; stigma pulvinate. *Sond. in Linnæa*, xx. 219 ; *Bolus in Journ. Linn. Soc.* xxv. 198 ; *Durand & Schinz, Conspect. Fl. Afr.* v. 106 ; *Kränzl. Orch. Gen. et Sp.* i. 745, *and in Ann. Naturhist. Hofmus. Wien*, xx. 7 ; *Schlechter in Engl. Jahrb.* xxxi. 247.

COAST REGION : Swellendam Div. ; by the Zondereinde River near Appels Kraal, *Zeyher*, 3923, *Pappe*, 60 ! mountains near Puspas Valley, 1500–2500 ft., *Ecklon & Zeyher.* Riversdale Div. ; grassy slopes on the Langeberg Range, near Riversdale, 1500–3000 ft., *Schlechter* !

8. D. obtusa (Lindl. Gen. & Sp. Orch. 355); plant ⅓–1¼ ft. high ;
stem moderately stout or rarely slender ; leaves radical and cauline,
numerous, suberect, linear, acute, 2–7 in. long ; scapes ⅓–1¼ ft. high,
with reduced leaf-like sheaths above ; spikes 1–8 in. long, dense,
many-flowered ; bracts lanceolate or linear-lanceolate, long-acuminate,
⅓–¾ in. long ; pedicels 3–4 lin. long ; flowers small, whitish or lilac,
with darker spots and stripes ; dorsal sepal galeate, broadly ovate or
ovate-orbicular, obtuse, 2½–3 lin. long ; spur broadly conical, obtuse,
about a quarter as long as the limb, not sulcate behind ; lateral
sepals oblong, obtuse, spreading, 2–2½ lin. long ; petals oblong or
falcate-oblong, obtuse, about 1 lin. long, with a short rounded basal
limb in front ; lip linear or linear-oblong, obtuse, about 1¼ lin. long ;
column short ; anther reflexed ; rostellum erect, shortly bifid ;
stigma pulvinate. *Krauss in Flora,* 1845, 307, *and Fl. Cap- und
Natal.* 158 ; *Bolus in Trans. S. Afr. Phil. Soc.* v. 153, *t.* 34, *Orch.
Cap. Penins.* 153, *t.* 34, *and Journ. Linn. Soc.* xxv. 198 ; *Durand &
Schinz, Conspect. Fl. Afr.* v. 105 ; *Kränzl. Orch. Gen. et Sp.* i. 792,
and in Ann. Naturhist. Hofmus. Wien, xx. 8 ; *Schlechter in Engl.
Jahrb.* xxxi. 247.

SOUTH AFRICA: without locality, *Masson* ! *Brown* ! *Bergius, Mrs. Jameson* !
Sieber !
COAST REGION : Cape Div. ; Table Mountain, 1000–3500 ft., *Mund & Maire,
Ecklon & Zeyher, Harvey* ! *Wilson, Rehmann,* 582 ! *Schlechter,* 162, *Bolus,* 4549 !
and in *MacOwan & Bolus, Herb. Norm. Austr.-Afr.* 336 ! Muizenberg, 1400 ft.,
Bolus, 4549 ! *Schlechter,* 162 ; Simons Bay, *Wright,* 133 ! 149 ! Simons Town,
Trimen ! marsh at source of Slangkop River, *Wolley-Dod,* 3212 ! Waai Vley,
Wolley-Dod, 2126 ! Knysna Div. ; Elands River, *Penther,* 295 ; Zitzikamma
River, *Penther,* 48.
EASTERN REGION : Natal ; Umgeni River, *Krauss,* 23.

I have not seen a Natal specimen, and Krauss' specimen may not belong here.

9. D. tabularis (Sond. in Linnæa, xix. 99) ; plant ⅓–1 ft. high ;
stem stout ; leaves radical and cauline, numerous, suberect, linear,
acuminate, 2–5 lin. long ; scapes ⅓–1 ft. high, with short leaf-like
sheaths above ; spikes 1–4 in. long, cylindrical, dense, many-flowered ;
bracts ovate-lanceolate, acute, 3–8 lin. long ; pedicels 3–5 lin. long ;
flowers rather small, tawny-yellow, with brown spots on the back of
the dorsal sepal and margins of the lateral sepals ; dorsal sepal
galeate, broadly ovate, subobtuse or apiculate, 3½–4 lin. long ; spur
broadly oblong or subsaccate, obtuse, about a quarter as long as the
limb ; lateral sepals oblong, obtuse, 3–3½ lin. long ; petals falcate-
oblong, obtuse, 1¼–1½ lin. long, with a round basal lobe in front ;
lip linear, obtuse, nearly as long as the lateral sepals ; column short ;
anther reclinate ; rostellum erect, short, emarginate, with a pair of
divaricate linear arms ; stigma pulvinate. *Sond. in Linnæa,* xx.
219 ; *N. E. Br. in Gard. Chron.* 1885, xxiv. 232 ; *Bolus in Trans.
S. Afr. Phil. Soc.* v. 152, *t.* 15 ; *Orch. Cap. Penins.* 152, *t.* 15, *and
Journ. Linn. Soc.* xxv. 198 ; *Durand & Schinz, Conspect. Fl. Afr.* v.
108 ; *Kränzl. Orch. Gen. et Sp.* i. 745 ; *Schlechter in Engl. Jahrb.*
xxxi. 248.

COAST REGION : Cape Div. ; among grasses on Table Mountain, 2500–3000 ft., *Ecklon & Zeyher* ! *Harvey* ! *Bolus,* 4819 ! and in *MacOwan & Bolus, Herb. Norm. Austr.-Afr.* 406 ! *Mrs. Jameson* ! near Maclears Beacon, 3300 ft., *Bolus,* 4819 ! slopes of Constantia Mountains, 2300 ft., *Bolus* !

10. **D. tenuicornis** (Bolus in Journ. Linn. Soc. xxii. 68); plant ½–1¼ ft. high; stem rather stout ; leaves radical and cauline, numerous, suberect, narrowly linear, acuminate, 3–5 in. long ; scapes ½–1¼ ft. high, with a few leaf-like sheaths above ; spikes oblong or somewhat elongate, 1½–5 in. long, dense, many-flowered ; bracts ovate-lanceolate, acuminate, ½–¾ in. long ; pedicels about ⅓ in. long ; flowers rather small, white, spotted and somewhat suffused with purple ; dorsal sepal galeate, broadly obovate, subobtuse or apiculate, 5–6 lin. long, with a short oblong sac on each side of the spur ; spur descending, cylindrical, subfiliform, obtuse, 1¼–2 lin. long ; lateral sepals falcate-oblong or ovate-oblong, subobtuse, 4–5 lin. long ; petals minute, lanceolate-oblong, subobtuse, with a broad rounded basal lobe, about 1 lin. long ; lip narrowly triangular-oblong, subobtuse, with incurved margin, 3–4 lin. long ; column short ; anther reclinate ; rostellum erect, subentire, broadly triangular, obtuse ; stigma pulvinate. *Bolus in Trans. S. Afr. Phil. Soc.* v. 151, *t.* 14, *Orch. Cap. Penins.* 151, *t.* 14, *and Journ. Linn. Soc.* xxv. 199 ; *Durand & Schinz, Conspect. Fl. Afr.* v. 108 ; *Kränzl. Orch. Gen. et Sp.* i. 792 ; *Schlechter in Engl. Jahrb.* xxxi. 244.

SOUTH AFRICA : without locality, *Masson* !
COAST REGION : Cape Div. ; lower slopes of Table Mountain, near Breakfast Camp on the Hout Bay Stream, 2500 ft., *Bolus,* 4967 ! and in *MacOwan & Bolus, Herb. Norm. Austr.-Afr.* 407 ! rocky kloof near Wynberg Reservoir, *Wolley-Dod,* 3175 ! George Div. ; near George, *Prior* !

11. **D. longifolia** (Lindl. Gen. & Sp. Orch. 349) ; plant ½–1 ft. high ; stem usually stout ; leaves radical and cauline, often numerous, erect, linear, acuminate, 3–5 in. long, somewhat reduced upwards ; scapes ½–1 ft. high, with leaf-like sheaths above ; spikes oblong, 1–3 in. long, dense, many-flowered ; bracts narrowly lanceolate or linear with a broader base, acuminate, ¾–1 (rarely 2) in. long ; pedicels ⅓–½ in. long ; flowers rather small, white, with a greenish-yellow spur, some similar markings at the base of the lateral sepals, and a purple stripe on the lip ; dorsal sepal galeate, broadly ovate, acute, 4–5 lin. long ; spur oblong, from a conical base, obtuse, scarcely half as long as the limb ; lateral sepals oblong, obtuse or apiculate, 4–5 lin. long ; petals falcate-linear from a broader base, acuminate, with a rounded basal lobe in front, about 1½ lin. long ; lip linear-oblong, subacute, papillose, about 2½ lin. long ; column short ; anther reflexed or almost pendulous ; rostellum ascending, 3-lobed, with short tooth-like side lobes, and a much larger rounded front lobe ; stigma pulvinate. *Drège, Zwei Pfl. Documente,* 76 ; *Bolus in Journ. Linn. Soc.* xxv. 198, *and Ic. Orch. Austr.-Afr.* i. *t.* 83 ; *Durand & Schinz, Conspect. Fl. Afr.* v. 104 ;

Kränzl. Orch. Gen. et Sp. i. 787 ; *Schlechter in Engl. Jahrb.* xxxi. 266.

COAST REGION : Piquetberg Div. ; Piquet Berg, 2000–3000 ft., *Drège*, 564 ! Tulbagh Div. ; near Tulbagh Waterfall, *Bolus.* Worcester Div. ; near Hex River, *Tyson in Herb. Bolus,* 6043 !
CENTRAL REGION : Ceres Div. ; among stones on the Skurfdeberg Range, near Ceres, 1800 ft., *Bodkin in Herb. Bolus,* 7325 !

12. **D. ocellata** (Bolus in Journ. Linn. Soc. xx. 477) ; plant 4–9 in. high ; stem rather slender ; leaves subradical, 3–5, erect or suberect, lanceolate or spathulate-lanceolate, acute or shortly acuminate, 1–2½ in. long ; scapes 4–9 in. long, with a few lanceolate acuminate sheaths above ; spikes cylindrical, 1½–4 in. long, usually lax ; bracts linear-lanceolate, acute, ¼–¾ in. long ; pedicels 4–5 lin. long ; flowers small, pale yellow, with a pair of brown eye-like spots on the side of the dorsal sepal ; dorsal sepal galeate, broadly ovate, acute or apiculate, subcrenulate, about 3 lin. long ; spur broadly conical at the base, much constricted above the middle, obovate and truncate at the apex, rather shorter than the limb ; lateral sepals oblique, narrowly ovate, acute, about 3 lin. long ; petals falcate-linear from a broad base, acuminate, with a rounded sub-crenulate basal lobe in front ; lip linear, acute, 2½ lin. long ; column short ; anther horizontal ; rostellum erect, emarginate ; stigma pulvinate. *Bolus in Trans. S. Afr. Phil. Soc.* v. 148, *t.* 5, *Orch. Cap. Penins.* 148, *t.* 5, *and Journ. Linn. Soc.* xxv. 199 ; *Durand & Schinz, Conspect. Fl. Afr.* v. 105 ; *Kränzl. Orch. Gen. et Sp.* i. 787 ; *Schlechter in Engl. Jahrb.* xxxi. 265, *excl. syn. Bolus. D. maculata, Harv. in Hook. Lond. Journ. Bot.* i. 15, *not of Linn. f.*

COAST REGION : Cape Div. ; Table Mountain, 3400–3500 ft., *Harvey* ! *Dümmer,* 945 ! *Bolus,* 4849 ! *MacOwan, Schlechter,* 86 ; Devils Mountain, *Brown* !

13. **D. uncinata** (Bolus in Journ. Linn. Soc. xx. 478) ; plant ⅓ to over 1 ft. high ; stem rather slender, flexuous ; leaves radical and cauline, 3–5, suberect or spreading, lanceolate or narrowly oblong-lanceolate, acute or acuminate, 2–6 in. long, membranous ; scapes ⅓ to over 1 ft. long, with a few linear-lanceolate acute sheaths above ; spikes 1–5 in. long, usually lax, few- to many-flowered ; bracts lanceolate or narrowly ovate-lanceolate, acuminate, ⅓–¾ in. long ; pedicels ⅓–½ in. long ; flowers rather small, whitish, with the lip and base of the petals pale yellow, and the spur and apex of the petals dusky-brown ; dorsal sepal galeate, broadly suborbicular-ovate, obtuse, subcrenulate, about 3 lin. long ; spur subglobose, obtuse, from a broadly conical base, constricted above the middle, about as long as the limb ; lateral sepals oblong, obtuse, about 3 lin. long ; petals uncinate-falcate, acuminate, with a broad basal rounded denticulate lobe in front, over 1½ lin. long ; lip lanceolate-linear, obtuse, subundulate, over 1½ lin. long ; column short ; anther reclinate ; rostellum erect, emarginate ; stigma pulvinate. *Bolus,*

Ic. Orch. Austr.-Afr. i. *t.* 82 ; *Durand & Schinz, Conspect. Fl. Afr.*
v. 109 ; *Kränzl. Orch. Gen. et Sp.* i. 786, *and in Ann. Naturhist.
Hofmus. Wien*, xx. 8.   *D. ocellata, Schlechter in Engl. Jahrb.* xxxi.
265, *partly.*

COAST REGION : Tulbagh Div. ; Mitchells Pass, 1100 ft., *Bolus,* 5279 !
Worcester Div. ; Bains Kloof, 2600 ft., *Hutton* ! *Cooper*, 3598 ! *Schlechter*, 9190 !
near Hex River, *Bolus*, 6095.   Cape Div. ; Table Mountain, *Prior* ! *Harvey* !
Chalvin, Schlechter, Dümmer, 804 ! Orange Kloof River, *Wolley-Dod*, 2228 !
Devils Mountain, *Brown* ! Caledon Div. ; Zwart Berg, near Caledon, *Bolus* ;
near Palmiet River, 900 ft., *Schlechter*, 5424. Swellendam Div. ; Langeberg
Range near Swellendam, *Bodkin*. George Div. ; moist places at Montagu Pass,
2700 ft., *Schlechter*, 5804 (distributed by error as 8804) ! *Penther*, 227 ! 337 (*ex
Kränzlin*) ; near George, *Penther*, 164, *Krook*. Albany Div. ; without precise
locality, *Cooper*, 1878 !

Quite distinct from *D. ocellata*, Bolus, to which it has been reduced by
Schlechter.

14. **D. Tysoni** (Bolus in Journ. Linn. Soc. xxv. 172, fig. 10) ;
plant 10–16 in. high ; stem rather stout ; leaves cauline, numerous,
suberect, oblong or ovate-lanceolate, subacute, sheathing at the base,
somewhat fleshy, 2–4½ in. long, decreasing upwards into the bracts ;
scapes 10–16 in. high ; spikes oblong, 3–4 in. long, dense, many-
flowered ; bracts ovate or ovate-lanceolate, acute or acuminate, ½–¾
in. long ; pedicels 5–6 lin. long ; flowers medium-sized, light green,
with white petals and a yellow lip ; dorsal sepal galeate, very
broadly ovate, obtuse, about 4 lin. long ; spur ascending, oblong,
obtuse, about half as long as the limb ; lateral sepals broadly ovate-
elliptic, obtuse, about 3½ lin. long ; petals falcate-oblong, obtuse,
about 2 lin. long, with an ample rounded somewhat concave basal
lobe in front ; lip ovate-oblong, obtuse, convex, nearly 3 lin. long ;
column short ; anther horizontal ; rostellum suberect, very short,
with a pair of broad diverging arms ; stigma pulvinate. *Bolus in
Journ. Linn. Soc.* xxv. 199, *and Ic. Orch. Austr.-Afr.* ii. *t.* 80 ;
Durand & Schinz, Conspect. Fl. Afr. v. 109 ; *Kränzl. Orch. Gen. et
Sp.* i. 786 ; *Schlechter in Engl. Jahrb.* xxxi. 254.

COAST REGION : Stutterheim Div. ; Mount Dohne, 4500 ft., *Bolus*, 8708 !
Sim, 960.
EASTERN REGION : Griqualand East ; grassy slopes above Beeste Kraal, near
Kokstad, 4800 ft., *Tyson*, 1609 ; summit of the Insizwa Range, 6800 ft., *Schlechter*,
6509.

15. **D. stachyoides** (Reichb. f. in Flora, 1881, 328) ; plant ½–1 ft.
high ; stem rather stout ; leaves cauline, 4–6, suberect, lanceolate
or oblong-lanceolate, acute or acuminate, somewhat fleshy, 1–2 in.
long ; scapes ½–1 ft. high, with a few leaf-like sheaths above ;
spikes cylindrical, 1½–4½ in. long, dense, many-flowered ; bracts
lanceolate or ovate-lanceolate, acute or acuminate, ⅓–¾ in. long ;
pedicels about ⅓ in. long ; flowers small, purple or pink, with paler
petals and lip ; dorsal sepal galeate, oblong, obtuse, with involute
margins, about 2½ lin. long ; spur stout, nearly straight, obtuse and

slightly compressed at the apex, as long as the limb; lateral sepals oblong, obtuse, concave, with a conical dorsal apiculus, 2½–3 lin. long; petals oblique, broadly ovate, obtuse, with a rounded basal lobe in front, 1½ lin. long; lip oblong, obtuse or apiculate, slightly thickened near the apex, 2–2¼ lin. long; column short; anther reflexed; rostellum shortly 3-lobed; stigma pulvinate. *Bolus in Journ. Linn. Soc.* xxv. 198; *Durand & Schinz, Conspect. Fl. Afr.* v. 108; *Kränzl. Orch. Gen. et Sp.* i. 755, *and in Ann. Naturhist. Hofmus. Wien,* xx. 7; *Schlechter in Engl. Jahrb.* xxxi. 262. *D. gracilis, Krauss in Flora,* 1845, 306, *and Beitr. Fl. Cap- und Natal.* 158, *not of Lindl. D. hemisphærophora, Reichb. f. Otia Bot. Hamb.* ii. 106.

KALAHARI REGION: Orange River Colony; Harrismith, *Sankey,* 262! and without precise locality, *Cooper,* 975! Transvaal; near Lydenburg, *Atherstone!* Fall Creek, *Mudd!* summit of Saddleback Range, near Barberton, 4000–5000 ft., *Galpin,* 715! *Culver,* 7! Elandspruit Mountains, 6600 ft., *Schlechter,* 3988! Pilgrims Rust, *Greenstock.* Houtbosch Mountains, 5000–7000 ft., *Schlechter.*

EASTERN REGION: Tembuland; Bazeia Mountains, 2500–3000 ft., *Baur,* 591! Griqualand East; near Fort Donald, 5000 ft., *Tyson,* 1595! and in *MacOwan & Bolus, Herb. Norm. Austr.-Afr.* 549! Insizwa Range, 6500 ft., *Schlechter!* *Krook, Penther,* 150; near Newmarket, *Krook, Penther,* 116. Natal; near Durban Bay, *Krauss,* 22; Inanda, 2000 ft., *Wood,* 164! 770! near Liddlesdale, 4000–5000 ft., *Wood,* 842; near Weenen, 4000 ft., *Wood*; Drakensberg Range, 5000–6000 ft., *Wood,* 5146! Dargle Farm, *Mrs. Fannin,* 92! Ixopo River, *Mrs. Clarke!* and without precise locality, *Buchanan!* *Gerrard,* 2182! *Mrs. Saunders.* Zululand, *Haygarth in Herb. Wood,* 7422!

16. **D. aconitoides** (Sond. in Linnæa, xix. 91); plant ¾–2 ft. high; stem moderately stout; leaves radical and cauline, suberect, the former linear or linear-oblong, acute, 3–5 in. long, the latter oblong or lanceolate-oblong, acute, 1–3 in. long, reduced upwards into the bracts; scapes ¾–2 ft. high, with oblong-lanceolate acute sheaths above; spikes 3–8 in. long, usually lax, many-flowered; bracts lanceolate or ovate-lanceolate, acute or acuminate, 4–7 lin. long; pedicels 3–5 lin. long; flowers small, white or lilac, with light purple spots; dorsal sepal galeate, erect, broadly ovate, obtuse, limb about 2½ lin. long; spur conical, obtuse, longer than the limb; lateral sepals oblique, ovate or ovate-oblong, obtuse, with an obtuse dorsal apiculus, about 2½ lin. long; petals falcate, somewhat constricted below, dilated above into an elliptic obtuse limb, with a small rounded basal lobe in front, 1½ lin. long; lip elliptic-spathulate, obtuse, convex, about 1½ lin. long; column short; anther reclinate; rostellum erect, subquadrate, shortly and obtusely 3-lobed; stigma pulvinate. *Sond. in Linnæa,* xx. 219; *Harv. Thes. Cap.* i. 26, *t.* 41; *Bolus in Journ. Linn. Soc.* xxv. 198, *and Ic. Orch. Austr.-Afr.* i. *t.* 79; *Durand & Schinz, Conspect. Fl. Afr.* v. 99; *Kränzl. Orch. Gen. et Sp.* i. 780, *and in Ann. Naturhist. Hofmus. Wien,* xx. 8; *Schlechter in Engl. Jahrb.* xxxi. 255.

SOUTH AFRICA: without locality, *Hallack!* *Miss Bowker!* *Mrs. Barber,* 523, partly!

COAST REGION: George Div.; near George, *Penther,* 102. Knysna Div.;

Zitzikamma River, *Penther*, 212. Humansdorp Div. ; near Storms River, 250 ft.,
Schlechter, 5985 ; Coldstream River, *Penther*, 166. Alexandria Div. ; Olifants
Hoek, near Bushmans River, 200 ft., *Ecklon & Zeyher*! Zuurberg Range, *Mrs.*
Barber, 446 ! Albany Div. ; near Grahamstown, 2000 ft., *MacOwan*, 699 !
Bolton ! *Miss Bowker* ! *Schönland*, *Galpin*, 306 ! Komgha Div. ; near Komgha,
2000 ft., *Flanagan*, 1035. Eastern Frontier, *Hutton* !
    KALAHARI REGION : Orange River Colony ; without precise locality, *Cooper*, 981 !
Transvaal; Elands Spruit Mountains, 7000 ft., *Schlechter*, 3851.
    EASTERN REGION : Natal ; Umtshanga, near Richmond, 2000–2500 ft.,
Sanderson, 109 F! 498 ! 735 ; Grey Town, *Mrs. Saunders*! Dargle Farm,
Sanderson, 745, *Mrs. Fannin*, 22 ! Inanda, *Wood*, 1418 ! near Howick, 3000–
4000 ft., *Wood*, 5136, 11820 !

17. D. stricta (Sond. in Linnæa, xix. 91); plant ¾–1½ ft. high ;

stem rather slender ; leaves cauline, numerous, suberect, linear,
acute or acuminate, 2–6 in. long ; scapes ¾–1½ ft. high, with leaf-
like sheaths above ; spikes cylindrical, 1–3 in. long, dense or somewhat lax ; bracts lanceolate, very acuminate, ½–1¼ in. long ;
pedicels ⅓–½ in. long ; flowers small, purple ; dorsal sepal galeate,
broadly ovate, subcompressed, with obtuse somewhat fleshy apex,
2½ lin. long ; spur broadly conical, obtuse, 2½ lin. long ; lateral
sepals ovate-oblong or broadly elliptic-oblong, obtuse, with an obtuse
dorsal apiculus, 2½ lin. long ; petals falcate-linear, subobtuse, 1¼ lin.
long, with a short rounded basal angle in front ; lip broadly elliptic-
oblong, obtuse, reflexed at the sides, 1½ lin. long ; column short ;
anther reclinate ; rostellum short, 3-lobed, with short side lobes,
and a longer fleshy front lobe ; stigma pulvinate. *Sond. in Linnæa,*
xx. 219 ; *Bolus in Journ. Linn. Soc.* xxv. 198, *and Ic. Orch. Austr.-*
Afr. i. *t.* 78 ; *Durand & Schinz, Conspect. Fl. Afr.* v. 108 ; *Kränzl.*
Orch. Gen. et Sp. i. 783 ; *Schlechter in Engl. Jahrb.* xxxi. 262.

    COAST REGION : Stockenstrom Div. ; slopes of Kat Berg, *Scully in Herb. Bolus,*
5911 ! Queenstown Div. ; Winterberg Range, 3500–6100 ft., *Ecklon & Zeyher*, 55 !
Zeyher ! Eastern Frontier, 2500 ft., *Hutton* ! King Williamstown Div. ; Mount
Kemp, near King Williams Town, 4000 ft., *Sim.* British Kaffraria ; on moun-
tains, *Mrs. Barber*, 27 !
    EASTERN REGION : Tembuland : Bazeia Mountains, 3000–4000 ft., *Baur*, 544 !
Griqualand East ; among stones at the summit of the Insizwa Range, 6800 ft.,
Schlechter.

18. D. sanguinea (Sond. in Linnæa, xix. 97) ; plant ¾–1¼ ft.

high ; stem stout ; leaves cauline, erect, oblong-lanceolate, acuminate, somewhat fleshy, imbricate, 1–3 in. long, decreasing upwards
into the bracts ; scapes ¾–1¼ ft. high, with lanceolate-acuminate
imbricate sheaths ; spikes ¾–1¼ in. long, very dense, many-flowered ;
bracts ovate-lanceolate, acuminate, ½–¾ in. long ; pedicels about ½ in.
long ; flowers small, crimson outside, pale rose within ; dorsal sepal
galeate, broadly ovate or subhemispherical, obtuse or obscurely
emarginate, about 2 lin. long ; spur oblong, obtuse, about half as
long as the limb ; lateral sepals broadly oblong, carinate, obtuse,
with a dorsal apiculus, 2 lin. long ; petals incurved within the
galea, falcate-linear from a broader base, subobtuse, 1½ lin. long,
with an ample rounded crenulate lobe in front ; lip rhomboid-

ovate, about 1 lin. long, with a small pellucid gland at the apex,
"resembling a dew-drop" (*Bolus*); column short; anther reflexed;
rostellum erect, shortly 3-lobed; stigma pulvinate. *Bolus in Journ.
Linn. Soc.* xxv. 198, *and Ic. Orch. Austr.-Afr.* i. *t.* 80; *Durand &
Schinz, Conspect. Fl. Afr.* v. 107; *Kränzl. Orch. Gen. et Sp.* i. 784;
Schlechter in Engl. Jahrb. xxxi. 252, *and in Ann. Naturhist.
Hofmus. Wien,* xx. 8. *D. Huttonii, Reichb. f. Otia Bot. Hamb.*
ii. 105.

COAST REGION: Swellendam Div.; Zuurbraak, *Penther,* 338. Fort Beaufort
Div.; damp stony places on the Winterberg Range, 3000–4000 ft., *Zeyher!*
Stockenstrom Div.; Kat Berg, *Hutton,* 53! King Williamstown Div.; Mount
Kemp, 2660–3000 ft., *Sim,* 1497.

19. D. Sankeyi (Rolfe); plant 5–7 in. high; stem rather stout;
leaves cauline or subradical, suberect, oblong or linear-oblong,
acute, 2–3½ in. long, rather fleshy, reduced upwards into the
bracts; scapes 5–7 in. high, with several lanceolate acuminate
sheaths above; spikes oblong, 1½–3 in. long, dense, many-flowered;
bracts ovate or ovate-oblong, 6–8 lin. long; pedicels about 4 lin.
long; flowers rather small, green and purple, fragrant (*Sankey*);
dorsal sepal subgaleate, obovate-oblong, obtuse, 3–3½ lin. long;
spur linear or subclavate, about half as long as the limb and not
descending below its base; lateral sepals oblong, obtuse, spreading,
3–3½ lin. long; petals suboblique, obovate-oblong, obtuse, 2½–3 lin.
long; lip linear-oblong, obtuse, 3–3½ lin. long; column nearly 2 lin.
long; anther reclinate; rostellum minute; stigma pulvinate.

KALAHARI REGION: Orange River Colony; Harrismith, *Sankey,* 264!

20. D. fragrans (Schlechter in Engl. Jahrb. xx. Beibl. 50, 40);
plant ½–¾ ft. high; stem stout; leaves cauline, 4–6, suberect or
somewhat spreading, oblong, subobtuse, spotted with purple,
2–3½ in. long, gradually reduced upwards into the sheaths; scapes
½–¾ ft. long, with several imbricating sheaths below the bracts;
spikes oblong, 2–4 in. long, very dense; bracts ovate or elliptic-
ovate, subobtuse, about 5 lin. long; pedicels 4 lin. long; flowers
small, pale pink or whitish, spotted with purple; dorsal sepal
galeate, elliptic-oblong, obtuse, 2½ lin. long; spur somewhat clavate,
as long or nearly as long as the limb; lateral sepals oblong, obtuse,
2½ lin. long; petals subspathulate-oblong, obtuse, 2¼ lin. long; lip
subspathulate-linear, obtuse, as long as the dorsal sepal; column
broad, 1½ lin. long; anther suberect, obtuse; rostellum erect,
3-lobed, with short side lobes and a triangular acute front lobe;
stigma pulvinate. *Kränzl. Orch. Gen. et Sp.* i. 748, *and in Ann.
Naturhist. Hofmus. Wien,* xx. 7; *Schlechter in Engl. Jahrb.*
xxxi. 223.

KALAHARI REGION: Orange River Colony; slopes of Quaqua Mountains and
Mopedis Peak, Witzies Hoek, 6800–8100 ft., *Thode,* 55! Mont aux Sources,
8000–9000 ft., *Thode;* Harrismith, *Krook, Penther,* 124. Transvaal; stony

places at the summit of Houtbosch Berg, 7000 ft., *Schlechter*, 4445, and without precise locality, *Molyneux*!
    EASTERN REGION : Griqualand East ; Isitsa Footpath, Maclear District, 7550 ft., *Galpin*, 6837 ! summit of Insiswa Range, 6500 ft., *Schlechter*. Natal ; Drakensberg Range, near Van Reenen, 6500 ft., *Schlechter, Krook, Penther*, 86 ; summit of Amawahqua Mountain, 6800 ft., *Wood*, 4565 !

21. **D. polygonoides** (Lindl. Gen. & Sp. Orch. 349) ; plant 1–1½ ft. high, with stout stem ; leaves usually numerous, suberect, radical and cauline, the lower linear-oblong to lanceolate, subacute, ½–1 ft. long, upper acuminate and gradually reduced upwards into the bracts ; scapes 1–1½ ft. high, with numerous acuminate sheaths above ; spikes oblong, 2–6 in. long, very dense ; bracts ovate or elliptic-ovate, acute or subobtuse, ⅓–½ in. long ; pedicels 4–5 lin. long ; flowers small, light or dark red, with some yellow in the centre ; dorsal sepal galeate, elliptic, obtuse, 3–3½ lin. long ; spur slenderly clavate, rather shorter than the limb ; lateral sepals elliptic-oblong, obtuse, 3 lin. long ; petals broadly obovate-elliptic, obtuse, concave, 2 lin. long ; lip linear, obtuse, as long as the dorsal sepal ; column broad, 1½ lin. long ; anther suberect, obtuse ; rostellum erect, 3 lobed ; side lobes with inflexed margin, and obscurely 3-lobed, front lobe concave, small ; stigma pulvinate. *Lindl. Gen. & Sp. Orch.* 349 ; *Drège, Zwei Pfl. Documente*, 149 ; *Krauss in Flora*, 1845, 306, *and Fl. Cap- und Natal.* 158 ; *N. E. Br. in Gard. Chron.* 1885, xxiv. 232 ; *Bolus in Journ. Linn. Soc.* xxv. 198, *and Ic. Orch. Austr.-Afr.* ii. *t.* 84 ; *Durand & Schinz, Conspect. Fl. Afr.* v. 106 ; *Kränzl. Orch. Gen. et Sp.* i. 747 ; *Schlechter in Engl. Jahrb.* xxxi. 222. *D. natalensis, Lindl. in Hook. Lond. Journ. Bot.* i. 16 ; *Krauss in Flora*, 1845, 307, *and Fl. Cap- und Natal.* 159.

    COAST REGION: George Div.; near George, *Prior*! Albany Div. ; near Grahamstown, 2000–2500 ft., *MacOwan*, 357! 472! *Bolton*! *Hutton*, 53! *Read*! Brookhuizens Poort, *Atherstone*, 3! Howisons Poort, *Hutton*! and without precise locality, *Mrs. Barber*, 523! Eastern Frontier of Cape Colony, *Prior*!
    CENTRAL REGION : Somerset Div. ; *Bowker*!
    KALAHARI REGION : Basutoland ; Mametsana, *Dieterlen*, 489!
    EASTERN REGION : Pondoland ; between Umtata River and St. Johns River, *Drège*, 4572! marshy ground, summit of Western Gate, Port St. John, 1250 ft., *Galpin*, 2871! Natal ; near Umlaas, *Krauss*, 334! near Durban, *Sanderson*, 24! 110! *Gueinzius*! *Wylie in Herb. Wood*, 18! *Gerrard*, 315! *McKen*, 749! *Plant*, 51! 66! *Peddie*! *Harvey*! *Mrs. Saunders*! Inanda, *Wood*, 278! Clairmont, *Wood*, 1094! Richmond, *Sanderson*, 480, partly! Alexandra District, *Rudatis*, 224! and without precise locality, *Sanderson*, 501! Delagoa Bay, in marshy ground, *Speke*, 8!

22. **D. chrysostachya** (Sw. in Vet. Acad. Handl. Stockh. 1800, 211) ; plant 1¼–3½ ft. high ; stem stout, often very stout ; leaves usually numerous, suberect or somewhat spreading, radical, elongate-linear or linear-oblong, acute, sheathing at the base, ⅓–1 ft. long ; cauline oblong or broad, acute, gradually passing upwards into the bracts ; scapes 1¼–3½ ft. high, with numerous imbricate sheaths above ; spikes ½–1½ ft. long, cylindrical, very dense ; bracts ovate-

lanceolate or elliptic-ovate, acute or subobtuse, $\frac{1}{3}-\frac{1}{2}$ in. long; pedicels 4–6 lin. long; flowers rather small, orange-coloured with the lip and petals yellow; dorsal sepal galeate, elliptic-obovate, subobtuse, $2\frac{1}{2}$–3 lin. long; spur inflated, elliptic-oblong, about as long as the limb; lateral sepals elliptic, obtuse, 2–$2\frac{1}{2}$ lin. long; petals obovate, obtuse, about 2 lin. long; lip linear, subacute, as long as the lateral sepals; column 1 lin. long; anther suberect, cells cucullate-connate; rostellum erect, 3-lobed, side lobes reflexed, much smaller than the front lobe; capsule ovoid-oblong, somewhat attenuate above, 8–9 lin. long. *Lindl. Gen. & Sp. Orch.* 349; *Drège, Zwei Pfl. Documente*, 122; *Bolus in Journ. Linn. Soc.* xxv. 197; *Durand & Schinz, Conspect. Fl. Afr.* v. 101; *Kränzl. Orch. Gen. et Sp.* i. 748, *and in Ann. Naturhist. Hofmus. Wien*, xx. 7; *Schlechter in Engl. Jahrb.* xxxi. 221. *D. gracilis, Lindl. Gen. & Sp. Orch.* 348; *Krauss in Flora*, 1845, 306, *and in Fl. Cap- und Natal.* 158.

SOUTH AFRICA: without locality, *Thunberg, Masson*! *Krebs, Hutton, Large*! *Mrs. Barber*! *Mrs. Saunders.*

COAST REGION: Cape Div.; Table Mountain, *Penther*, 138. Swellendam Div.; near Swellendam, *Bowie*! Riversdale Div.; between Welgelegen and Onzer, in Lange Kloof, 1500–2000 ft., *Drège*, 2212! George Div.; near George, *Bowie*! *Prior*! *Schlechter, Penther*, 104. Knysna Div.; near Knysna, 150 ft., *Schlechter*, 5917! *Penther*, 211; near Plettenbergs Bay, *Pappe*; near Keurbooms River, *Penther*, 84; Zout River, *Penther*, 52; Blauuw Krantz River, *Penther*, 43. Uitenhage Div.; *Zeyher.* Port Elizabeth Div.; Walmer, near Port Elizabeth, 100 ft., *Hallack in MacOwan & Bolus, Herb. Norm. Austr.-Afr.* 949! *Mrs. Holland*, 17! Bathurst Div.; between Riet Fontein and the sources of the Kasuga River, *Burchell*, 4129! Albany Div.; "common throughout Albany in marshy places," *Hutton*! Komgha Div.; near Komgha, *Krook, Penther*, 143. Eastern Frontier of Cape Colony, *Hutton*!

KALAHARI REGION: Orange River Colony, *Cooper*, 979! 1095! Basutoland; Mohopung, *Dieterlen*, 131! Transvaal; Little Lomati Valley, near Barberton, 3600 ft., *Culver*, 66; marsh near Botsabelo, 4800 ft., *Schlechter*, 3778; Belfast, *Miss Leendertz*, 2860! Swaziland; marshes at Forbes Concession, 4500 ft., *Galpin*, 717!

EASTERN REGION: Griqualand East; mountains near Clydesdale, 3500 ft., *Tyson*, 2904! Natal; near Fields Hill, Camperdown, 1000–2000 ft., *Sanderson*, 601! near Nottingham Road, 5000 ft., *Wood*, 1020, partly! Entakama, near Ixopo River, *Mrs. Clarke*! Oliviers Hoek, sources of Tugela River, 5000 ft., *Allison*, 31! Alexandra District, 2000 ft., *Rudatis*, 544! Richmond, *Sanderson*, 480, partly! and without precise locality, *Krauss*, 22, *Buchanan*! *Mrs. Fannin*, 14!

23. **D. Hallackii** (Rolfe); a stout plant over 8 in. high, base not seen; leaves cauline, oblong or lanceolate-oblong from a broader base, acute, about 4 in. long, rather fleshy; scape over 8 in. high, leafy to the base of the spike; spike oblong, 4 in. or more long, dense, many-flowered; bracts ovate or lanceolate-ovate, acute, $\frac{3}{4}-1\frac{1}{4}$ in. long; pedicels about $\frac{1}{2}$ in. long; flowers large, purple; dorsal sepal galeate, ovate, obtuse, 7 lin. long; spur oblong, obtuse or nearly truncate, somewhat flattened, nearly half as long as the limb; lateral sepals oblong or ovate-oblong, obtuse, with a short stout dorsal apiculus, about 7 lin. long; petals strongly 2-lobed;

upper lobe strongly curved, linear-oblong, somewhat dilated near the apex, and subacute, about 2½ lin. long; lower lobe obovate-oblong, subacute, 2¼ lin. long; lip lanceolate-oblong, subobtuse, about 3 lin. long; column short; anther reflexed; rostellum dilated, tridenticulate; stigma pulvinate.

COAST REGION: Port Elizabeth Div. ; wet places near Port Elizabeth, *Hallack*, 6213 !

Most like *D. cornuta*, Sw., in its robust habit and general appearance, but having smaller flowers and a very differently shaped spur.

24. D. extinctoria (Reichb. f. in Flora, 1881, 328); plant 1–1¾ ft. high; stem moderately stout; leaves radical and cauline, suberect, submembranous, the former subspathulate-linear, subacute, 3–4 in. long, the latter oblong-lanceolate, acute or acuminate, 1–2 in. long; scapes 1–1¾ ft. high, with many lanceolate very acuminate sheaths above, ½–1¼ in. long; spikes cylindrical, 3–5½ in. long, dense, many-flowered; bracts lanceolate or ovate-lanceolate, very acuminate, 4–8 lin. long; pedicels 3–4 lin. long; flowers small, scarlet or crimson; dorsal sepal galeate, very broadly ovate, obtuse, with incurved margin, about 2 lin. long; spur cylindrical from a broadly conical base, obtuse, sharply reflexed below the middle, about as long as the limb; lateral sepals ovate-oblong, obtuse, with an acuminate dorsal apiculus; petals broadly rhomboid-ovate, subobtuse, 1¾–2 lin. long; lip linear or subspathulate-linear, obtuse; column short; anther reflexed; rostellum 3-lobed, with obtuse inflexed apex, and narrow acute side lobes; stigma pulvinate; capsule elliptic-oblong, about 4 lin. long. *Bolus in Journ. Linn. Soc.* xxv. 198; *Durand & Schinz, Conspect. Fl. Afr.* v. 102; *Kränzl. Orch. Gen. et Sp.* i. 759; *Schlechter in Engl. Jahrb.* xxxi. 241.

KALAHARI REGION: Transvaal; marshy places, Lomati Valley, near Barberton, 3500 ft., *Galpin*, 716! *Culver*, 10! near Lydenberg, 4000 ft., *Atherstone*! near Botsabelo, 4000 ft., *Schlechter*. Mac Mac, *Mudd* ! hills near Barberton, *Dyce* !
EASTERN REGION: Natal ; grassy slopes near Inanda, *Wood*. Zululand, *Gerrard*.

25. D. Macowani (Reichb. f. Otia Bot. Hamb. ii. 106); plant 1–1¾ ft. high; stem stout; leaves radical and cauline, suberect, somewhat fleshy, the former elongate-linear, acute or acuminate, 6–11 in. long, the latter ovate-oblong, acute, 1½–3 in. long; scapes 1–1¾ ft. high, with lanceolate acute sheaths above; spikes cylindrical, 4–9 in. long, very dense and many-flowered; bracts ovate-lanceolate, acuminate, 4–8 lin. long; pedicels about 4 lin. long; flowers small, pink; dorsal sepal galeate, broadly ovate, obtuse or minutely apiculate, 2½–3 lin. long; spur descending, cylindrical, obtuse, nearly straight, rather shorter than the limb; lateral sepals obliquely ovate, apiculate, spreading, 2¼–3 lin. long; petals obliquely ovate, subobtuse, 2–2¼ lin. long; lip subspathulate-oblong, 2–2¼ lin. long; column very short; anther reflexed;

rostellum broadly triangular-oblong, truncate; stigma pulvinate.
Bolus in Journ. Linn. Soc. xxv. 198; *Durand & Schinz, Conspect.*
Fl. Afr. v. 104; *Kränzl. Orch. Gen. et Sp.* i. 754, *and in Ann.*
Naturhist. Hofmus. Wien, xx. 7. *D. versicolor, Schlechter in Engl.*
Jahrb. xxxi. 240, *partly, not of Reichb. f.*

COAST REGION: Victoria East Div.; Hogsback Range, 5500 ft., *Galpin*, 7993!
Cathcart Div. ; near Cathcart, 2600 ft., *Flanagan*, 1686.

CENTRAL REGION: wet places at the summit of Bosch Berg, 4800 ft., *MacOwan*,
1123 ! *Scully.*

KALAHARI REGION : Orange River Colony; without precise locality, *Cooper*,
1095 ! Transvaal ; Lomati Valley, near Barberton, 3900 ft., *Galpin*, 717 ! 1152 ;
marshes near Botsabelo, 5000 ft., *Schlechter*, 4060 ! Witbank, Middelburg,
5300 ft., *Galpin*, 7247 !

EASTERN REGION : Transkei ; summit of Mount Bazeia, 2500 ft., *Baur*, 592 !
Pondoland ; foot of Mount Eulenzi, near Fort William, 2500 ft., *Tyson*, 2697 !
Griqualand East ; plain at summit of the Insizwa Range, 6500 ft., *Schlechter*,
Krook, Penther, 143. Natal; near Lambonjwa River, *Wood*, 3421 ! Oliviers
Hoek, 4000 ft., *Allison*, 35 ! Mooi River, 4000–5000 ft., *Wood*, 5361 ; near
Highlands, 5000 ft., *Schlechter*, 6849. Swaziland, *Henderson* and *Forbes* (*ex
Galpin*).

Allied to the Angolan *D. versicolor*, Reichb. f., with which it has been united,
but is not identical.

26. **D. læta** (Reichb. f. Otia Bot. Hamb. ii. 106); plant 1¼–1½ ft.
high; stem stout; leaves not seen; scapes 1¼–1½ ft. high, with
several narrow acute sheaths above; spikes oblong, 3–4 in. long,
dense, many-flowered; bracts ovate-lanceolate, acuminate, about
¾ in. long; pedicels about ½ in. long; flowers medium-sized; dorsal
sepal galeate, broadly triangular-ovate, obtuse, about 3½ lin. long;
spur cylindrical, curved from near the narrowly conical base, about
as long as the limb ; lateral sepals ovate-oblong, acute or subacute,
about 4 lin. long; petals narrowly ovate-oblong, subobtuse, 2¼ lin.
long; lip subspathulate-linear, subobtuse, about 3 lin. long; column
short; anther suberect; rostellum triangular-oblong, subobtuse,
short; stigma pulvinate. *Bolus in Journ. Linn. Soc.* xxv. 198;
Durand & Schinz, Conspect. Fl. Afr. v. 104; *Kränzl. Orch. Gen. et
Sp.* i. 753. *D. hircicornis, Schlechter in Engl. Jahrb.* xxxi. 242,
partly, not of Reichb. f. D. Culveri, Schlechter in Engl. Jahrb. xx.
Beibl. 50, 17.

KALAHARI REGION : Transvaal ; Little Lomati Valley, near Barberton, 3500 ft.,
Culver, 75 ; marshes at Botsabelo, near Middelburg, 4800 ft., *Schlechter*, 4063.

EASTERN REGION : Natal; without precise locality, *Mrs. Fannin*, 53 ! *Sanderson.*

D. læta, Reichb. f., was based upon a Kew specimen, but I cannot find what
has become of it. I follow Reichenbach in considering it distinct from the
Tropical African *D. hircicornis*, Reichb. f., to which Schlechter reduces it, as
well as his own *D. Culveri*. The latter I have not seen, and have left here.

27. **D. rhodantha** (Schlechter in Engl. Jahrb. xx. Beibl. 50, 40,
by error *rodantha*); plant 1–1¾ ft. high; stem stout; leaves radical
and cauline, the former in short lateral tufts, suberect or somewhat
spreading, linear or oblong-linear, acute or acuminate, 4–8 in. long;

scapes 1–1¾ ft. high, with several short acuminate sheaths above ; spikes oblong, 2–6 in. long, dense, many-flowered ; bracts ovate-lanceolate, acute or acuminate, 6–10 lin. long ; pedicels about 5 lin. long ; flowers medium-sized, rose-coloured ; dorsal sepal galeate, broadly triangular-ovate, subobtuse or apiculate, about 4 lin. long ; spur cylindrical from a conical base, strongly curved from above the middle, much longer than the limb ; lateral sepals obliquely ovate-oblong, obtuse, with a short stout dorsal apiculus, about 4 lin. long ; petals oblong-lanceolate, subacute, rather fleshy, about 3 lin. long ; lip subspathulate-oblong, obtuse, about 3 lin. long ; column short ; anther suberect ; rostellum emarginate or shortly bilobed, with very short side lobes ; stigma pulvinate. *Schlechter in Engl. Jahrb.* xxxi. 242, *t.* 5, *fig. F–L.*

KALAHARI REGION : Transvaal ; marshes near Brug Spruit, between Middelburg and Pretoria, 4600 ft., *Schlechter*, 3756 ! and without precise locality, *Hallack* !

EASTERN REGION : Griqualand East ; marshes on the Zuurberg Range, 4500 ft., *Schlechter*, 6583 ! Natal ; swamp near Nottingham Road, 4000–5000 ft., *Wood*, 1020, partly ! swamps at Van Reenen, *Wood*, 4527 !

28. **D. caffra** (Bolus in Journ. Linn. Soc. xxv. 171, 172, fig. 9) ; plant 1–1¼ ft. high ; stem moderately stout ; leaves cauline, sub-erect, few, lanceolate or linear-lanceolate, subacute, reduced upwards into the bracts, 2–4½ in. long, rather fleshy ; scapes 1–1¼ ft. high, with narrow acute sheaths above ; spikes 1–2½ in. long, somewhat dense, 8–12-flowered ; bracts ovate-lanceolate, acuminate, 6–9 lin. long ; pedicels 5–6 lin. long ; flowers medium-sized, deep vinous-purple ; dorsal sepal galeate, very broadly ovate, obtuse or truncate, about 4 lin. long ; spur cylindrical from a conical base, obtuse, curved, longer than the limb ; lateral sepals broadly elliptic, obtuse, with a short stout dorsal apiculus ; petals falcate-lanceolate, sub-acute, 2½ lin. long ; lip linear from a somewhat broader base, subobtuse ; column short ; anther reflexed ; rostellum somewhat triangular, subacute, with acute side lobes ; stigma pulvinate. *Bolus in Journ. Linn. Soc.* xxv. 199 ; *Durand & Schinz, Conspect. Fl. Afr.* v. 101 ; *Kränzl. Orch. Gen. et Sp.* i. 820 ; *Schlechter in Engl. Jahrb.* xxxi. 258.

EASTERN REGION : Pondoland ; among grasses near Umkwani River, 200 ft., *Tyson in Herb. Bolus*, 2611 ; in swampy ground at the summit of West Gate, Port St. John, 1200 ft., *Galpin*, 3417 !

29. **D. Galpinii** (Rolfe) ; plant 1½ ft. high ; stem stout ; leaves rather fleshy, cauline only seen, and those oblong and imperfect at the apex, reduced upwards into the bracts ; scape 1½ ft. high, with several acute imbricate sheaths above ; spike subcylindrical, 5 in. long, dense, many-flowered ; bracts ovate-lanceolate, acuminate, ¾–1 in. long ; pedicels 4–6 lin. long ; flowers rather small, fleshy, "red" (*Galpin*) ; dorsal sepal galeate, broadly ovate, shortly and broadly apiculate, 2½ lin. long ; spur clavate, 7 lin. long ; lateral sepals oblong, obtuse, with a stout obtuse dorsal apiculus ; petals

oblique, broadly rounded at the base, broadly oblong and obtuse above, crenulate, 1¼ in. long; lip oblong, obtuse, 2 lin. long; column short; anther reflexed; rostellum broadly triangular, subobtuse; stigma pulvinate.

EASTERN REGION : Griqualand East; near a streamlet at Midlothian Farm, Maclear District, 5600 ft., *Galpin,* 6838 !

30. **D. ovalifolia** (Sond. in Linnæa, xix. 93); plant 6–8 in. high; stem very stout; leaves radical, 2 or 3, horizontally spreading, broadly elliptic or suborbicular, obtuse or subapiculate, fleshy, 1–1¾ in. long; scapes 6–8 in. high, with several broadly oblong obtuse or subacute imbricate sheaths ; spikes oblong, 1½–3 in. long, rather dense and many-flowered ; bracts ovate-oblong, acute or acuminate, ½–1¼ in. long; pedicels 7–8 lin. long; flowers medium-sized, white ; dorsal sepal galeate, funnel-shaped, with a short broadly ovate obtuse limb, about 3 lin. broad, continued behind in a slender cylindrical spur, the whole 6–7 lin. long; lateral sepals oblong, obtuse, spreading, somewhat concave, 4½–5 lin. long; petals falcate-oblong, obtuse, incurved, with a rounded basal lobe in front, about 2½ lin. long; lip linear-oblong, obtuse, fleshy, about 5 lin. long; column short; anther reclinate; rostellum suberect, emarginate, with diverging arms; stigma pulvinate. *Bolus in Journ. Linn. Soc.* xxv. 200, *and Ic. Orch. Austr.-Afr.* i. *t.* 29 ; *Durand & Schinz, Conspect. Fl. Afr.* v. 105 ; *Kränzl. Orch. Gen. et Sp.* i. 773 ; *Schlechter in Engl. Jahrb.* xxxi. 251. *D. pallidiflora, Bolus ex Schlechter in Engl. Jahrb.* xxxi. 252.

COAST REGION : Clanwilliam Div.; near Brack Fontein, *Zeyher* ! foot of Olifants River Mountains, 400 ft., *Schlechter.* Malmesbury Div. ; near Berg River, *Mund.*
CENTRAL REGION : Ceres Div. ; Cold Bokkeveld, near Gydouw, 3500 ft., *Bolus,* 1097 ! 7326 ! and in *MacOwan & Bolus, Herb. Norm. Austr.-Afr.* 1097 ! *Schlechter,* 8892 !

31. **D. cornuta** (Sw. in Vet. Acad. Handl. Stockh. 1800, 210); plant ½–1½ ft. high; stem very stout; leaves cauline, numerous, suberect, oblong to ovate-oblong, subacute, very fleshy, 3–6 in. long, gradually reduced upwards into the bracts; scapes ½–1½ ft. high; spikes oblong, 3–9 in. long, very dense, many-flowered ; bracts ovate or ovate-oblong, acute or acuminate, ¾–1½ in. long; pedicels about ¾ in. long; flowers large, the dorsal sepal lead-purple, the apex of the lip blackish-purple, and the rest of the flower whitish or light yellow; dorsal sepal galeate, broadly ovate, obtuse, with an orbicular mouth, somewhat depressed at the apex, 8–9 lin. long; spur cylindrical from a broadly conical base, obtuse, more or less curved and deflexed, about as long as the limb; lateral sepals broadly elliptic-oblong, obtuse, with a short dorsal apiculus; petals falcate-lanceolate, with an acute incurved apex, and a nearly orbicular basal lobe in front, 4–4½ lin. long; lip broadly elliptic or suborbicular, obtuse, deflexed at the sides, about 5 lin. long;

232 ORCHIDEÆ (Rolfe). [*Disa.*

column short; anther reflexed; rostellum erect, bifid, with a pair
of diverging arms; stigma pulvinate. *Thunb. Fl. Cap. ed. Schult.* 7;
Lindl. Gen. & Sp. Orch. 349; *Drège, Zwei Pfl. Documente,* 76, 81,
89, 124; *Krauss in Flora,* 1845, 307, *and Fl. Cap- und Natal.* 159;
Sond. in Linnæa, xix. 90, xx. 218; *N. E. Br. in Gard. Chron.* 1885,
xxiv. 231; *Bolus in Trans. S. Afr. Phil. Soc.* v. 149, *Orch. Cap.
Penins.* 149, *Journ. Linn. Soc.* xxv. 197, *and Ic. Orch. Austr.-Afr.* ii.
t. 81; *Durand & Schinz, Conspect. Fl. Afr.* v. 101; *Kränzl. Orch.
Gen. et Sp.* i. 767, *excl. var., and in Ann. Naturhist. Hofmus. Wien,*
xx. 8; *Schlechter in Engl. Jahrb.* xxxi. 256. *Orchis cornuta, Linn.
Sp. Pl. ed.* ii. 1330; *Houtt. Handl.* xii. 456, *t.* 86, *fig.* 2. *Satyrium
cornutum, Thunb. Prodr.* 5.

32. **D. macrantha** (Sw. in Vet. Acad. Handl. Stockh. 1800, 210);
plant 1-2¼ ft. high; stem very stout; leaves cauline, somewhat
numerous, suberect, oblong or ovate-oblong, subacute, very fleshy,
4-9 in. long, reduced upwards into the bracts; scapes 1-2¼
ft. high; spikes ¾-1 ft. long, dense, many-flowered; bracts
ovate or ovate-lanceolate, acute or acuminate, ¾-1½ in. long;
pedicels about ¾ in. long; flowers large, the dorsal sepal dull lead-
purple or rather pale, the rest of the flower yellowish, with some
brown markings on the lip; dorsal sepal galeate, broadly ovate,
subacute, with an orbicular mouth, 7-8 lin. long; spur cylindrical
from a broadly conical base, obtuse, usually somewhat curved but
chiefly near the apex, distinctly longer than the limb; lateral sepals
broadly ovate-oblong, subobtuse, with a short dorsal apiculus; petals
falcate-lanceolate, with an acute incurved apex, and an ovate-
orbicular basal lobe in front, 4-4½ lin. long; lip oblong or elliptic,
subobtuse, somewhat deflexed at the sides, about 5 lin. long; column
short; anther reflexed; rostellum erect, bifid, with a pair of
diverging arms; stigma pulvinate. *Thunb. Fl. Cap. ed. Schult.* 8;

Lindl. Gen. & Sp. Orch. 349 ; *N. E. Br. in Gard. Chron.* 1886, xxv.
499, *partly. D. cornuta, Drège, Zwei Pfl. Documente,* 99, *not of
Sw. D. cornuta, Hook. in Bot. Mag. t.* 4091, *not of Sw. D. æmula,
Bolus in Journ. Linn. Soc.* xxii. 69, xxv. 199, *Trans. S. Afr. Phil.
Soc.* v. 150, *Orch. Cap. Penins.* 150, *and Ic. Orch. Austr.-Afr.* ii.
t. 82, 83 ; *Durand & Schinz, Conspect. Fl. Afr.* v. 99 ; *Schlechter in
Engl. Jahrb.* xxxi. 257. *D. cornuta, var. æmula, Kränzl. Orch. Gen.
et Sp.* i. 768.

COAST REGION : Malmesbury Div. ; Groene Kloof, Mamre, 300 ft., *Bolus,* 4330 !
Paarl Div. ; by the Berg River, near Paarl, under 500 ft., *Drège,* 1262 d ; Cape
Div. ; sands about Salt River, near Cape Town, *Harvey* ! near Tygerberg,
MacOwan in Herb. Bolus, 4330 ! Rapenburg, below 100 ft., *Goatcher in Herb.
Bolus,* 6862 ! Vyges Kraal Farm, *Wolley-Dod,* 397 !

For notes on the confusion which has existed between this species and *D. crassicornis,* Lindl., see J. Medley Wood in *Gard. Chron.* 1886, xxv. 43, and N. E.
Brown, *l.c.* 499.

33. **D. Thodei** (Schlechter ex Kränzl. Orch. Gen. et Sp. i. 796) ;
plant about 1 ft. high ; stem moderately stout ; leaves cauline,
erect, oblong, acute, 2–3 in. long, submembranous ; scapes about
1 ft. high, with a few broad acuminate sheaths above ; spikes about
2 in. long, somewhat lax, about 6-flowered ; bracts oblong-
lanceolate, acuminate, $\frac{3}{4}$–1 in. long ; pedicels about $\frac{3}{4}$ in. long ;
flowers rather large, submembranous, purple or pale pink ; dorsal
sepal galeate, broadly ovate, apiculate or subobtuse, 6–7 lin. long ;
spur cylindrical, straight, subobtuse, 8–9 lin. long ; lateral sepals
narrowly ovate, apiculate, 6–7 lin. long ; petals ovate, acute or
shortly acuminate, 5–6 lin. long ; lip oblong-lanceolate, acute,
5–6 lin. long ; column short ; anther reclinate ; rostellum 3-lobed ;
stigma pulvinate.

KALAHARI REGION : Orange River Colony ; grassy banks of a streamlet on the
slopes of the Caledon Range, 7900–8300 ft., *Thode,* 53 !

Schlechter has included this in his *D. Cooperi,* var. *Scullyi,* but both it and
D. Scullyi, Bolus, are distinct from *D. Cooperi,* Reichb. f.

34. **D. zuluensis** (Rolfe) ; base of plant not seen ; scape stout,
with narrow acuminate sheaths above ; spike 5 in. long, somewhat
lax, 15–20-flowered ; bracts ovate-lanceolate, acuminate, 1–1$\frac{1}{2}$ in.
long ; pedicels 6–8 lin. long ; flowers large ; dorsal sepal galeate,
obovate-oblong, shortly acuminate, $\frac{1}{2}$ in. long ; spur cylindrical
from a narrowly conical base, very slightly thickened at the apex,
curved, 1–1$\frac{1}{4}$ in. long ; lateral sepals obliquely ovate-oblong, shortly
acuminate, about $\frac{1}{2}$ in. long ; petals falcate-oblong, subacute, with a
short rounded auricle at the base in front, 5 lin. long ; lip lanceolate,
subacute, about $\frac{1}{2}$ in. long ; column short ; anther reflexed ; rostellum
broadly triangular ; stigma pulvinate.

EASTERN REGION : Zululand, *Gerrard,* 2170 !

35. **D. Scullyi** (Bolus in Journ. Linn. Soc. xxii. 70) ; plant
1$\frac{1}{2}$–1$\frac{3}{4}$ ft. high ; stem rather stout ; leaves cauline, few, suberect,

oblong, subacute, submembranous, 2–4 in. long; scapes 1½–1¾ ft.
high, with numerous broad acuminate sheaths above; spikes
3–6 in. long, rather dense, many-flowered; bracts oblong-lanceolate,
acuminate, 1–1¼ in. long; pedicels 8–10 lin. long; flowers rather
large, somewhat fleshy, rose-coloured; dorsal sepal galeate, very
broadly ovate, subobtuse or apiculate, 4–5 lin. long; spur ascending,
strongly curved, cylindrical from a narrowly conical base, slightly
thickened and acute above, 14–15 lin. long; lateral sepals ovate,
obtuse, 5 lin. long, with a verrucose dorsal apiculus; petals
broadly cordate-ovate, obtuse, 3½–4 lin. long; lip narrowly ovate-
oblong, very obtuse, minutely papillose, 3½–4 lin. long; column
short; anther reflexed; rostellum 3-lobed; stigma pulvinate.
Bolus in Journ. Linn. Soc. xxv. 199; *Durand & Schinz, Conspect. Fl.*
Afr. v. 108; *Kränzl. Orch. Gen. et Sp.* i. 796. *D. Cooperi, var.*
Scullyi, Schlechter in Engl. Jahrb. xxxi. 236, *excl. syn. D. Thodei.*

COAST REGION: Stockenstrom Div.; Menzies Berg, near Stockenstrom, *Scully,*
5915! Chumie Berg and Gaikas Kop, *Mrs. Barber,* 23!

Quite distinct from *D. Cooperi,* Reichb. f., to which Schlechter has referred it
as a variety.

36. **D. Cooperi** (Reichb. f. in Flora, 1881, 328); plant 1½–2 ft.
high; stem very stout; leaves radical and cauline, the former
springing from a distinct bud, suberect, lanceolate, 4–16 in. long,
the latter erect, ovate-oblong, acute, 1–3 in. long, somewhat fleshy;
scapes 1¼–2 ft. high, with very broad acuminate imbricate sheaths
above; spikes 6–10 in. long, usually dense, many-flowered; bracts
ovate-oblong, acute or acuminate, 1–1¾ in. long; pedicels ¾–1 in.
long; flowers rather large, fleshy, rose-pink with a yellow lip;
dorsal sepal galeate, very broadly ovate, apiculate, 5–6 lin. long;
spur ascending, strongly curved, cylindrical from a narrowly
conical base, subobtuse, 1¼–1¾ in. long; lateral sepals ovate,
obtuse, 5–6 lin. long, with an obtuse dorsal apiculus, 1–2 lin.
long; petals broadly oblong or elliptic-oblong, obtuse, obscurely
auricled about the middle in front, about 3½ lin. long; lip broadly
rhomboid-ovate, obtuse, crenulate, about 3½ lin. long; column
short; anther reflexed; rostellum 3-lobed; stigma pulvinate.
Bolus in Journ. Linn. Soc. xxv. 198, *and Ic. Orch. Austr.-Afr.*
ii. *t.* 65, *excl. syn.*; *Bot. Mag. t.* 7256; *N. E. Br. in Gard. Chron.*
1892, xii. 268, 269, *fig.* 45; *Durand & Schinz, Conspect. Fl. Afr.*
v. 101; *Kränzl. Orch. Gen. et Sp.* i. 795; *Schlechter in Engl. Jahrb.*
xxxi. 236, *excl. syn. and var. Scullyi.*

KALAHARI REGION: Orange River Colony; near Besters Vley, 5000 ft., *Miss*
Jacobsz in Herb. Bolus, Flanagan; near Zaai Hoek, *Thode*; near Harrismith,
Sankey, 266! *Cooper,* 1098! 1871! Basutoland; Mametsana, *Dieterlen,* 132!
Transvaal; near Lydenburg, *Atherstone!* margins of marshes near Middelburg,
4900 ft., *Schlechter,* 4104! Ermelo, *Miss Leendertz,* 3000!
   EASTERN REGION: Griqualand East; Matatiele, 5000 ft., *Tyson,* 1606! Natal;
near Mooi River, 4000–5000 ft., *Wood,* 4493, 5359; Van Reenen, 5500 ft.,
Wood, 8751; Oliviers Hoek, 5000 ft., *Allison,* 27! Dargle Farm, *Mrs. Fannin,* 2!
Swaziland; between Swaziland and Carolina, 6900 ft. (*fide Bolus*).

37. D. crassicornis (Lindl. Gen. & Sp. Orch. 348); plant 1–3 ft. high; stem very stout; leaves radical and cauline, suberect, the former oblong-lanceolate or oblong-ligulate, acute, attenuated at the base, ½–1 ft. long, the latter oblong-lanceolate, acute, 3–5 in. long, somewhat fleshy; scapes 1–3 ft. long, with oblong-lanceolate acuminate sheaths above; spikes 4–10 in. long, dense or somewhat lax, usually many-flowered; bracts oblong-lanceolate, very acuminate, 1½–3 in. long; pedicels about 1½ in. long; flowers very large, sub-membranous, pale yellow or cream-coloured, spotted with pale purple; dorsal sepal galeate, broadly ovate, subobtuse, 1 in. or more long; spur curved, cylindrical from a broadly conical base, obtuse, 1½–1¾ in. long; lateral sepals oblique, narrowly ovate, subobtuse, 1 in. long; petals oblique, broadly ovate, subobtuse, about ¾ in. long; lip oblong-lanceolate, obtuse, about 1 in. long; column ⅓ in. long; anther reflexed; rostellum short, obscurely 3-lobed; stigma pulvinate. *Drège, Zwei Pfl. Documente,* 53; *Bolus in Journ. Linn. Soc.* xxv. 198; *Durand & Schinz, Conspect. Fl. Afr.* v. 101; *Kränzl. Orch. Gen. et Sp.* i. 766, *and in Ann. Naturhist. Hofmus. Wien,* xx. 7; *Schlechter in Engl. Jahrb.* xxxi. 237. *D. megaceras, Hook. f. in Bot. Mag. t.* 6529. *D. macrantha, Hemsl. in Garden,* 1880, xvii. 494, *t.* 235, *not of Sw.*

SOUTH AFRICA : without precise locality, *Large*!

COAST REGION : Knysna Div.; Zitzikamma River, *Penther,* 73. Uitenhage Div. ; Uitenhage, *Zeyher,* 628 ! Bedford Div.; Kaga Berg, *MacOwan.* Fort Beaufort Div. ; Winter Berg, *Zeyher, Mrs. Barber.* Stockenstrom Div. ; Kat Berg, in moist ravines, 8000 ft., *Hutton*! *Scully,* 181; 3000 ft., *Galpin,* 1742 ! King Williamstown Div. ; Keiskamma Hoek, *Cooper,* 1291 ! Komgha Div. ; near Komgha, 2000 ft., *Flanagan,* 543.

CENTRAL REGION : Somerset Div. ; summit of Bosch Berg, 4500 ft., and hills near Somerset, *Tuck in Herb. MacOwan,* 529 ! *Cooper,* 529 ! Aliwal North Div. ; Witteberg Range, 7000–8000 ft., *Drège,* 3577 !

EASTERN REGION : Transkei ; Krielis Country, *Bowker,* 212 ! Kentani District, 1200 ft., *Miss Pegler,* 315 ! Pondoland ; Mount Puguwan, 3000 ft., *Tyson.* Griqualand East ; mountain sides near Clydesdale, 3500 ft., *Tyson,* 2900 ! near Matatiele, 8000 ft., *Mrs. Cowen*! Natal ; near Richmond, 2000 ft., *Weir in Herb. Sanderson,* 498 ! between Richmond and Maritzburg, 2000 ft., *Sanderson,* 498 ! Dargle Farm, *Mrs. Fannin,* 6 ! Highlands, *Gerrard,* 1561 ! and without precise locality, *Hallack* !

38. D. Vasselotii (Bolus ex Schlechter in Verhandl. Bot. Ver. Brandenb. xxxv. 47, name only); plant 3–7 in. high; stem moderately stout ; leaves mostly radical, tufted, spreading, linear or lanceolate-oblong, acute or subobtuse, ½–1¼ in. long; scapes 3–7 in. high, with a few reduced acuminate sheaths above ; spikes subcorymbose, 2–10-flowered ; bracts ovate or ovate-lanceo-late, acuminate, 4–6 lin. long; pedicels 4–5 lin. long; flowers rather small, white ; dorsal sepal galeate, broadly ovate, obtuse, about 4½ lin. long ; spur conical-oblong, obtuse, a quarter to a third as long as the limb ; lateral sepals elliptic-ovate, obtuse, with a short blunt dorsal apiculus ; petals falcate-oblong, obliquely acute, crenulate, about 2 lin. long ; lip subspathulate-linear, obtuse, about

2 lin. long ; column short ; anther reflexed ; rostellum suberect,
shortly bilobed; stigma pulvinate.   *Schlechter in Engl. Jahrb.*
xxxi. 273.

CoAST REGION : Riversdale Div. ; mountains near Riversdale, 2000–3000 ft.,
Schlèchter, 2219.   Knysna Div. ; mountains near Knysna, 3000–4000 ft., *Vasselot
de Regnier in Herb. Bolus*, 10485 ! *Foucard.*

39. D. falcata (Schlechter in Verhandl. Bot. Ver. Brandenb.
xxxv. 46) ; plant ½–1 ft. high ; stem moderately stout ; leaves
radical and cauline, suberect, linear or the lower subspathulate-
linear, acute or acuminate, 1–4½ in. long ; scapes ½–1 ft. high, with -
numerous acuminate sheaths above ; spikes subcorymbose or oblong,
up to 2 in. long, somewhat dense, few- to many-flowered ; bracts
ovate-lanceolate, acute or acuminate, ⅓–½ in. long ; pedicels 3–5 lin.
long ; flowers rather small, white ; dorsal sepal galeate, broadly
ovate, subobtuse, about 4 lin. long ; spur oblong, obtuse, curved,
about a third as long as the limb ; lateral sepals obliquely ovate-
oblong, subacute, about 4 lin. long ; petals falcate and subobtuse or
sharply bent above the middle, 2–2½ lin. long, with two narrow
diverging apical lobes ; lip lanceolate, acute, recurved, about 2 lin.
long, with a pair of small rounded auricles at the base ; column
short ; anther reflexed ; rostellum erect, emarginate, with a pair of
divaricate linear side lobes ; stigma pulvinate. *Schlechter in Engl.
Jahrb.* xxxi. 274.

CoAST REGION : Swellendam Div. ; moist slopes at the summit of the Langeberg
Range, near Zuurbraak, 3100–3500 ft., *Schlechter*, 2178 !

40. D. glandulosa (Burch. ex Lindl. Gen. & Sp. Orch. 351) ;
plant 3–7 in. high ; stems rather stout ; leaves radical, somewhat
spreading, spathulate or ovate-lanceolate, acute or subacute,
glandular-pubescent, ½–1 in. long ; scapes 3–7 in. high, with
several oblong-lanceolate acute somewhat pubescent sheaths ; spikes
½–1 in. long, compact, 2–10-flowered ; bracts oblong-lanceolate,
acute, somewhat pubescent, 4–8 lin. long ; pedicels 6–8 lin. long ;
flowers small, rose-pink ; dorsal sepal galeate, ovate, subobtuse,
3–3½ lin. long ; spur rather stout from a conical base, rather shorter
than the limb ; lateral sepals elliptic-oblong, obtuse, 3–3½ lin. long ;
petals broadly ovate-oblong, obtuse, somewhat oblique, about 2 lin.
long ; lip obovate-oblong, obtuse, about 2 lin. long ; column short ;
anther reflexed ; rostellum erect, short, scarcely exceeding the
pulvinate stigma. *Bolus in Trans. S. Afr. Phil. Soc.* v. 159, *t.* 35 ;
Orch. Cap. Penins. 159, *t.* 35, *and Journ. Linn. Soc.* xxv. 200 ; *Durand
& Schinz, Conspect. Fl. Afr.* v. 103 ; *Kränzl. Orch. Gen. et Sp.* i. 773 ;
Schlechter in Engl. Jahrb. xxxi. 234.

SOUTH AFRICA : without locality, *Mund & Maire.*
CoAST REGION : Cape Div. ; between rocks on Muizen Berg, 1600 ft., *Bolus*,
4540 ! and in *MacOwan & Bolus, Herb. Norm. Austr.-Afr.* 169 ! Table Mountain,
3000 ft., *Bolus*, 4540 ! Stellenbosch Div. ; Lowrys Pass, 4200 ft., *Schlechter*,
7218 ! Swellendam Div. ; summit of a mountain peak near Swellendam, *Burchell*,
7337 ! Langeberg Range, near Zuurbraak, 2900 ft., *Schlechter*, 2107 !

41. D. vaginata (Harv. ex Lindl. in Hook. Lond. Journ. Bot.
i. 15); plant 3–7 in. high; stem rather stout; leaves on the lower
part of the stem, 3–5, erect or somewhat spreading, subsessile,
lanceolate to ovate-lanceolate, acute, glabrous, somewhat fleshy,
½–1¼ in. long; scapes 3–7 in. high, with several oblong-lanceolate
acute resin-dotted sheaths above; spikes 1–1½ in. long, compact,
2–10-flowered; bracts oblong-lanceolate, acute, glabrous or slightly
scabrous, 4–8 lin. long; pedicels 6–8 lin. long; flowers small, rose-
pink, with darker rose dots on the base of the dorsal sepal and
petals; dorsal sepal galeate, ovate, subobtuse, 2½–3 lin. long; spur
slender from a conical base, about as long as the limb; lateral
sepals broadly oblong, obtuse, 2½–3 lin. long; petals falcate-oblong,
obtuse, somewhat concave at the apex, 2–2½ lin. long; lip
obovate-oblong, truncate, 2–2½ lin. long; column short; anther
horizontal; rostellum erect, scarcely exceeding the pulvinate
stigma. *Sond. in Linnæa,* xix. 93; *Bolus in Trans. S. Afr.
Phil. Soc.* v. 159, *Orch. Cap. Penins.* 159, *Journ. Linn. Soc.*
xxv. 200, *and Ic. Orch. Austr.-Afr.* ii. *t.* 71; *Durand & Schinz,
Conspect. Fl. Afr.* v. 109; *Kränzl. Orch. Gen. et Sp.* i. 772;
Schlechter in Engl. Jahrb. xxxi. 233. *D. modesta, Reichb. f. in
Linnæa,* xx. 690, *and in Flora,* 1883, 463.

SOUTH AFRICA : without locality, *Bergius, Mund, Kitching* ! *Prior* !
COAST REGION : Paarl Div. ; Drakenstein Mountains, near Dutoits Kloof (*ex
Bolus*). Cape Div. ; Table Mountain, 2500–3000 ft., *Bolus,* 3898 ! and in
MacOwan & Bolus, Herb. Norm. Austr.-Afr. 164 ! *Harvey* ! *Rehmann.* 576 !
Kässner, Schlechter, 205. Waai Vley and Steenberg, *Wolley-Dod,* 2113 ! Muizen
Berg, 1400–1800 ft., *Bolus,* 3898 ! *Schlechter.* Devils Mountain, 3000 ft.,
Schlechter, 68 ! Caledon Div. ; cliffs at Bee River, near Villiersdorp and
French Hoek, 2500 ft., *Bolus,* 3898 B !

42. D. frigida (Schlechter in Engl. Jahrb. xx. Beibl. 50, 18);
plant 5–10 in. long; stem rather slender; leaves subcauline, 3–4,
suberect, linear, acuminate, 2½–4 in. long; scapes 5–10 in. high,
with 2 or 3 narrow acuminate sheaths above : spikes 1–2 in. long,
lax, 6–10-flowered; bracts linear-lanceolate, acuminate, ¼–½ in.
long; pedicels 4–5 lin. long; flowers small, rose or rarely white;
dorsal sepal galeate, broadly ovate, obtuse, about 2½ lin. long;
spur slender, curved, two-thirds as long as the dorsal sepal;
lateral sepals oblong, shortly and bluntly apiculate, about 2½ lin.
long; petals oblique, ovate, subobtuse, about 1½ lin. long; lip
subspathulate-linear, obtuse, 2 lin. long; column short; anther
reflexed ; rostellum erect, short, minutely tridentate ; stigma
pulvinate. *Schlechter in Engl. Jahrb.* xxxi. 269, *t.* 4, *fig. A–D.
D. cephalotes, Kränzl. Orch. Gen. et Sp.* i. 758, *partly.*

KALAHARI REGION : Orange River Colony; summit of Mont aux Sources,
Drakensberg Range, 9500–10000 ft., *Thode,* 54 !

43. D. caulescens (Lindl. Gen. & Sp. Orch. 351); plant ½–1¼ ft.
high; stem rather slender ; leaves cauline, 5–9, rather lax, lanceolate, acute, 1–2½ in. long, reduced upwards into the bracts; scapes

½–1¼ ft. high, with a few linear or linear-lanceolate sheaths above ; spikes 1–3 in. long, lax, usually rather few-flowered ; bracts lanceolate, acute, ⅓–¾ in. long ; pedicels ⅓–½ in. long ; flowers small, rather membranous, white with light purple petals ; dorsal sepal galeate, broadly ovate, 4–5 lin. long ; spur rather slender, from a conical base, about half as long as the limb ; lateral sepals elliptic-oblong, obtuse, with a short conical dorsal apiculus, 4–5 lin. long ; petals broadly obovate-elliptic, emarginate or shortly bilobed, about 2½ lin. long ; lip linear, subobtuse, about 3 lin. long ; column short ; anther reclinate ; rostellum short, emarginate, with a pair of divaricate arms ; stigma pulvinate. *Drège, Zwei Pfl. Documente,* 81 ; *Sond. in Linnæa,* xix. 93, xx. 219 ; *Bolus in Journ. Linn. Soc.* xxv. 198, *and Ic. Orch. Austr.-Afr.* i. t. 31 ; *Durand & Schinz, Conspect. Fl. Afr.* v. 101 ; *Kränzl. Orch. Gen. et Sp.* i. 789 ; *Schlechter in Engl. Jahrb.* xxxi. 264.

SOUTH AFRICA : without locality, *Verreaux*! *Zeyher,* 1554!

COAST REGION : Tulbagh Div. ; between New Kloof and Pikiniers Kloof, *Zeyher,* 1568 ! stony places above Tulbagh Waterfall, 2000–3000 ft., *Ecklon & Zeyher*! *Pappe*! Worcester Div. ; Dutoits Kloof, 2000–3000 ft., *Drège,* 1246 ! *Tyson* ; Malbroks Kloof, *Drège,* 1248 b ! Worcester, *Zeyher*! Baines Kloof, 2000 ft., *Schlechter,* 9152 !

CENTRAL REGION : Ceres Div. ; rivulets at the foot of the Skurfdeberg Range, near Ceres, 1800 ft., *Bolus,* 7449 !

44. **D. tripetaloides** (N. E. Br. in Gard. Chron. 1889, v. 360) ; plant ¾–1½ ft. high ; stem moderately stout ; leaves radical, usually numerous, arranged in a rosulate tuft, lanceolate or oblong-lanceolate, acute or shortly acuminate, 1¼–4 in. long ; scapes ¾–1½ ft. high, with several narrow acute sheaths ; spikes 2–5 in. long, lax, usually many-flowered ; bracts elliptic-lanceolate or ovate-lanceolate, acute, ½–¾ in. long ; pedicels ¾–1 in. long ; flowers medium-sized, pale pink or white, with carmine spots on the sepals (yellow in var. *aurata*) ; dorsal sepal galeate, broadly ovate or orbicular-ovate, obtuse, 4–5 lin. long ; spur broadly conical or oblong from a broadly conical base, obtuse, one-third to half as long as the limb ; lateral sepals ovate-oblong or ovate-elliptic, obtuse, 4–5 lin. long ; petals falcate-oblong, obtuse, about 2 lin. long ; lip linear or oblong-linear, obtuse, about 2 lin. long ; column short ; anther reclinate ; rostellum erect, tall, emarginate at the apex, with short side lobes ; stigma pulvinate. *Bolus in Journ. Linn. Soc.* xxv. 199 ; *Bot. Mag. t.* 7206 ; *Durand & Schinz, Conspect. Fl. Afr.* v. 109 ; *Kränzl. Orch. Gen. et Sp.* i. 788, *and in Ann. Naturhist. Hofmus. Wien,* xx. 8 ; *Schlechter in Engl. Jahrb.* xxxi. 263. *D. venosa, Lindl. Gen. & Sp. Orch.* 351, *not of Sw. D. excelsa, Sw. in Vet. Acad. Handl. Stockh.* 1800, 213 ; *Thunb. Fl. Cap. ed. Schult.* 14, *partly. Orchis tripetaloides, Linn. f. Suppl.* 398 ; *Murr. Syst. Veg.* xiv. 807. *Satyrium excelsum, Thunb. Prodr.* 5, *partly.*

VAR. β, **aurata** (Bolus, Ic. Orch. Austr.-Afr. i. t. 30) ; flowers bright yellow, without spots ; spurs rather shorter and stouter than in the type. *Schlechter in Engl. Jahrb.* xxxi. 264.

SOUTH AFRICA : without locality, *Sparrman* ! *Brown* ! *Masson* ! *Carmichael* !
COAST REGION : Stellenbosch Div. ; Hottentots Holland Mountains, *Bowie* !
Zeyher. Caledon Div. ; Knoflooks Kraal and Little Houw Hoek, 1000–3000 ft.,
Zeyher, 3916 ! Houw Hoek, 2500 ft., *Schlechter,* 9410 ! *Bowie* ! trunks of fallen
trees in Grietjesgat River, near Palmiet River, 800 ft., *Bolus,* 4209 ! Riversdale
Div. ; by mountain rivulets near Riversdale, *Schlechter.* George Div. ; Wolf
Drift, Malgaten River, *Burchell,* 6123 ! George, *Prior* ! *Bowie* ! *Penther,* 101 ;
near rivulets above Montagu Pass, 4000 ft., *Schlechter,* 5843, *Krook.* Knysna
Div. ; mountains near Knysna, 200 ft., *Newdigate in MacOwan, Herb. Austr.-Afr.*
1650 ! *Penther,* 2. Uniondale Div. ; Lange Kloof, *Thunberg* ! Humansdorp Div. ;
Kromme River, *Thunberg* ! Humansdorp, *Harvey,* 896 ! Uitenhage Div. ; Water-
fall on Van Stadens Berg, *MacOwan,* 1095 ! at the sources of Bulk River and
Van Stadens River, *MacOwan,* 1095 ! Var. β, Swellendam Div. ; Zuurbraak,
Schlechter, 2148 ! on a mountain peak near Swellendam, *Burchell,* 7409 ! wet
mountain slopes near Swellendam, 2800 ft., *Bolus,* 7339 ! and in *MacOwan &*
Bolus, Herb. Norm. Austr.-Afr. 1098 !
EASTERN REGION : Natal ; on wet banks, Murchison, Alfred Co., 2000 ft.,
Wood, 1981 !

45. D. triloba (Lindl. Gen. & Sp. Orch. 351) ; plant 4–6 in. high ;
stem stout ; leaves radical, 3–6, oblong or lanceolate-oblong, subacute, 1–2 in. long ; scapes 4–6 in. high, with numerous oblong
subacute sheaths ; spikes 1–1½ in. long, compact, 4–9-flowered ;
bracts oblong or lanceolate-oblong, acute, 6–9 lin. long ; pedicels
6–9 lin. long ; flowers medium-sized ; dorsal sepal subgaleate,
obovate-oblong, subtrilobed above, subacute, somewhat attenuate
below, about ½ in. long ; spur very slender, from a narrowly
conical base, 2 lin. long ; lateral sepals oblong, subobtuse, somewhat spreading, 4–5 lin. long ; petals linear-oblong, from a broader
base, subobtuse, 3–4 lin. long ; lip linear, 4–5 lin. long ; column
scarcely 1 lin. long ; anther reclinate ; rostellum obtusely 3-lobed ;
stigma pulvinate. *Drège, Zwei Pfl. Documente,* 114 ; *Bolus in*
Journ. Linn. Soc. xxv. 200 ; *Durand & Schinz, Conspect. Fl. Afr.* v.
109 ; *Kränzl. Orch. Gen. et Sp.* i. 772. *D. sagittalis, var. triloba,*
Schlechter in Engl. Jahrb. xxxi. 232.

COAST REGION : Worcester Div. ; on mountains near De Liefde, 1000–2000 ft.,
Drège, 1236 !

46. D. sagittalis (Sw. in Vet. Acad. Handl. Stockh. 1800, 212) ;
plant ¼–1 ft. high ; stem moderately stout ; leaves radical, 4–9,
oblong or lanceolate-oblong, acute, submembranous, 1–5 in. long ;
scapes ¾–1 ft. high, with numerous oblong acute subimbricate
sheaths ; spikes 1–6 in. long, usually lax, few- to many-flowered ;
bracts oblong or ovate-oblong, acute or subacute, membranous, 6–9
lin. long ; pedicels slender, ¾–1 in. long ; flowers medium-sized,
white ; dorsal sepal subgaleate, 4–5 lin. long, oblong below, extended
into two broadly oblong spreading lobes above, each nearly 3 lin.
long ; spur slenderly conical, 2½ lin. long ; lateral sepals oblong,
subobtuse, about 4 lin. long ; petals linear, obtuse, about 4 lin. long,
with a broad auriculate basal lobe in front ; lip linear, obtuse, about
4 lin. long ; column 1 lin. long ; anther reclinate ; rostellum obtusely
3-lobed ; stigma pulvinate. *Thunb. Fl. Cap. ed. Schult.* 9 ; *Lindl.*

Gen. & Sp. Orch. 350 ; *Sond. in Linnæa*, xix. 93. *Bolus in Journ. Linn.*
Soc. xxv. 200, *and Ic. Orch. Austr.-Afr.* i. *t.* 32 ; *Bot. Mag. t.* 7403 ;
Durand & Schinz, Conspect. Fl. Afr. v. 107 ; *Kränzl. Orch. Gen. et*
Sp. i. 770, *and in Ann. Naturhist. Hofmus. Wien*, xx. 8 ; *Schlechter*
in Engl. Jahrb. xxxi. 232. *D. attenuata, Lindl. Gen. & Sp. Orch.*
351 ; *Drège, Zwei Pfl. Documente*, 121, 130 ; *Sond. in Linnæa*, xix.
93, xx. 219 ; *Bolus in Journ. Linn. Soc.* xxv. 200. *Orchis sagittalis,*
Linn. f. Suppl. 399. *Satyrium sagittale, Thunb. Prodr.* 5.

SOUTH AFRICA : without locality, *Masson* ! *Verreaux* ! *Hesse*, 16 ! *Zeyher*, 3920 !
COAST REGION : Swellendam Div. ; mountains near Swellendam, *Bowie* ! *Mund* !
Langeberg Range, near Zuurbraak, 1200 ft., *Schlechter* ; Buffeljagts River, *Penther*,
310. Mossel Bay Div. ; Gauritz River Bridge, 400 ft., *Schlechter*. George Div. ;
near George, *Alexander* ! rocky hill near the East end of Lange Vallei, *Burchell*,
5708 ! Gwyang River, 450 ft., *Bolus*, 5535 ! Montagu Pass, 1500 – 3000 ft.,
Schlechter, 5794 ! *Penther*, 336 ! Klip River, near Keurbooms River, 2000–
3000 ft., *Drège*, 8277 a ! Outeniqua Mountains, 1500–3000 ft., *Thunberg* !
Rehmann, 28 ! *Penther, Krook, Young.* Knysna Div. ; Baak Hill and hills near
Plettenbergs Bay, *Burchell*, 5344 ! *Mund & Maire, Bowie* ! Humansdorp Div. ;
Humansdorp, *Tyson*, 2974, *Galpin*, 4620 ! Uitenhage Div. ; hill near the mouth
of the Zwartkops River, under 500 ft., *Drège*, 8277 b ! Uitenhage, *Tredgold*, 35 !
mountains near Koega River, *Pappe*, 59 ! *Zeyher*, 1006 ! Port Elizabeth Div. ;
near Port Elizabeth, *Hewitson* ! *Hall* ! Alexandria Div. ; Zuurberg Range, *Mrs.*
Barber, 76 ! 435 ! Albany Div. ; Bothas Berg, 2200 ft., *MacOwan*, 396 ! *Galpin*,
307 ; Howisons Poort, at the edge of the Krantz, *MacOwan*, 396 ! near
Grahamstown, *Bolton* ! and without precise locality, *Mrs. Hutton* ! Stockenstrom
Div. ; Kat Berg, *Scully*, 262 ! King Williamstown Div. ; Perie Forest, *Sim.*
Eastern Frontier of Cape Colony, *Hutton* !

47. **D. racemosa** (Linn. f. Suppl. 406) ; plant 1–2¼ ft. high ;
stem rather stout, somewhat flexuous ; leaves radical or subradical,
spreading or suberect, numerous, lanceolate, acute, membranous,
3–5 in. long ; scapes 1–2¼ ft. high, with several narrow
acuminate sheaths ; spikes short, lax, 1–4-flowered ; bracts ovate
or ovate-lanceolate, acute or acuminate, ¾–1 in. long ; pedicels
about 1 in. long ; flowers large, bright rose ; dorsal sepal subgaleate, broadly ovate or suborbicular, subobtuse, 10–12 lin. long ;
spur broadly sac-shaped, obtuse, very short ; lateral sepals broadly
elliptic, obtuse, 10–12 lin. long ; petals broadly falcate-oblong,
incurved over the anther and subacute at the apex, 5–6 lin. long ;
lip linear, acute, about 4 lin. long ; column about 5 lin. long ;
anther ascending ; rostellum erect, 3-lobed ; side lobes divaricate
and furnished with a petaloid appendage behind ; front lobe concave, obtuse ; stigma pulvinate. *N. E. Br. in Gard. Chron.* 1885,
xxiv. 232 ; *Bolus in Trans. S. Afr. Phil. Soc.* v. 155, *Orch. Cap.*
Penins. 155, *Journ. Linn. Soc.* xxv. 199, *and Ic. Orch. Austr.-Afr.* ii.
t. 85 ; *Hook. f. in Bot. Mag. t.* 7021 ; *Reichb. f. in Gard. Chron.*
1887, ii. 809, 1888, iii. 592, 593, *fig.* 81 ; *Journ. Hort.* 1888, ii. 22,
fig. 25 ; *Warn. Orch. Alb.* viii. *t.* 356 ; *The Garden*, 1891, xxxix.
10, *t.* 786 ; *Durand & Schinz, Conspect. Fl. Afr.* v. 107 ; *Kränzl.*
Orch. Gen. et Sp. i. 740 ; *Schlechter in Engl. Jahrb.* xxxi. 226.
D. secunda, Sw. in Vet. Acad. Handl. Stockh. 1800, 213 ; *Thunb. Fl.*
Cap. ed. Schult. 14 ; *Lindl. Gen. & Sp. Orch.* 348 ; *Drège, Zwei Pfl.*

Documente, 81. *D. venosa, Sond. in Linnæa,* xx. 219, *not of Sw.*
Satyrium secundum, Thunb. Prodr. 4.

VAR. β, **isopetala** (Bolus in Trans. S. Afr. Phil. Soc. v. 156); petals nearly as
large as the spurless dorsal sepal; lip nearly as large as the lateral sepals. *Bolus,*
Orch. Cap. Penins. 156.
A peloriate form of the species.

SOUTH AFRICA : without locality, *Harvey! Yorke!*
COAST REGION : Tulbagh Div. ; near Tulbagh Waterfall, *Zeyher ;* Witzenberg
Range, *Mund.* Worcester Div. ; Dutoits Kloof, 2000–3000 ft., *Drège,* 1239 a !
and without precise locality, *Cooper,* 1612 ! 1684 ! 1685 ! Paarl Div. ; French
Hoek, *Thunberg!* Cape Div. ; Table Mountain, 2500 ft., *Bolus,* 4888 ! *Harvey!*
Barkly! Schlechter, 94 ! Constantia Berg, 1500 ft., *Schlechter,* 186 ! head of
Silvermine Valley, *Wolley-Dod,* 2138 ! Red Hill, *Mrs. Jameson!* Caledon Div.;
Genadendal, *Roser!* Houw Hoek, 2550 ft., *Pappe,* 42 ! *Schlechter,* 9410 ! *Bowie!*
Knoflooks Kraal and Klein Houw Hoek, 3000 ft., *Zeyher,* 3915 ! Swellendam
Div. ; on a mountain peak near Swellendam, *Burchell,* 7408 ! Humansdorp
Div. ; near Zitzikamma, 200 ft., *Schlechter.* Port Elizabeth Div. ; near Port
Elizabeth, *Hallack.* Albany Div. ; Howisons Poort, near Grahamstown, 2200–
2500 ft., *MacOwan,* 387 ! *Galpin,* 272, 292 ! *Schönland, Hutton!*

48. D. venosa (Sw. in Vet. Acad. Handl. Stockh. 1800, 213) ;
plant ¾–1½ ft. high ; stem somewhat flexuous, moderately slender ;
leaves radical or subradical, 4–8, suberect or somewhat spreading, lanceolate or linear-lanceolate, acute, 1–3 in. long ; scapes
¾–1½ ft. high, with several narrow acuminate sheaths ; spikes
⅓–¾ ft. high, lax, 5–9-flowered ; bracts ovate-lanceolate, acuminate,
8–10 lin. long ; pedicels about ¾ in. long ; flowers large, bright
rose ; dorsal sepal subgaleate, rhomboid-cuneate in outline, subobtuse, about ¾ in. long ; spur broadly sac-shaped, obtuse, very
short ; lateral sepals elliptic-oblong, obtuse, about ¾ in. long ; petals
broadly oblong, incurved over the anther and subobtuse at the
apex, 4–5 lin. long ; lip linear-subulate, acute, 5–6 lin. long ; column.
about 4 lin. long ; anther ascending ; rostellum erect, 3-lobed ; side
lobes incurved, furnished with a petaloid appendage behind ; front
lobe concave, obtuse ; stigma pulvinate. *Thunb. Fl. Cap. ed.*
Schult. 15 ; *Krauss in Flora,* 1845, 306, *and Fl. Cap- und Natal.* 158 ;
Sond. in Linnæa, xix. 93, xx. 219 ; *N. E. Br. Gard. Chron.* 1885,
xxiv. 232 ; *Bolus in Trans. S. Afr. Phil. Soc.* v. 156, *Orch. Cap.*
Penins. 156, *Journ. Linn. Soc.* xxv. 199, *and Ic. Orch. Austr.-Afr.* ii.
t. 86 ; *Kränzl. Orch. Gen. et Sp.* i. 741, *and in Ann. Naturhist. Hofmus.*
Wien, xx. 6. *D. excelsa, Thunb. Fl. Cap. ed.* 1, i. 78, *partly, not of Sw.* ;
and ed. Schult. 14. *D. secunda, Lindl. Gen. & Sp. Orch.* 348, *partly,*
not of Sw. ; *Drège, Zwei Pfl. Documente,* 83. *D. racemosa, var.*
venosa, Schlechter in Engl. Jahrb. xxxi. 227.

SOUTH AFRICA : without locality, *Thunberg! Masson, Villet!*
COAST REGION : Clanwilliam Div. ; Krakadouw Pass, Cederberg Range,
3000 ft., *Leipold,* 603. Paarl Div. ; Drakenstein Mountains, 2000–3000 ft.,
Drège, 1239 b ! French Hoek, *Schlechter,* 9311 ! Cape Div. ; Table Mountain,
2400–3000 ft., *Mund,* 54 ! *Harvey!* above Kirstenbosch, *Bolus,* 4845 ! Constantia
Mountains, 3000 ft., *Schlechter,* 451. Stellenbosch Div. ; Hottentots Holland,
Krauss, 1307. Knysna Div. ; Coldstream River, *Penther,* 167.
CENTRAL REGION : Ceres Div. ; Skurfdeberg Range, near Ceres, *Bolus,* 4845.

The Dutoits Kloof specimen of Drège, cited by Bolus and others, may be erroneous, for the original specimen in Lindley's Herbarium is labelled by Drège himself 1239 b, and the ticket is still gummed to the specimen as originally issued.

49. **D. uniflora** (Berg. Descr. Pl. Cap. 348); plant 1–2½ ft. high; stem stout, leafy; leaves lanceolate to elongate-linear, acute, usually somewhat spreading, 3–8 in. long, gradually passing upwards into acuminate sheaths; scapes 1–2½ ft. high, with numerous acuminate sheaths above; spikes 1–2-flowered, rarely 3–5-flowered; bracts lanceolate, acuminate, 1¼–2 in. long; pedicels 1–1¾ in. long; flowers very large, brilliant scarlet, with darker veins on the dorsal sepal, and some yellow or orange on the petals; dorsal sepal galeate, broadly ovate; acute or apiculate, 1½–2¼ in. long; spur narrowly conical, about ½ in. long; lateral sepals broadly ovate, abruptly acuminate, 1½–2¼ in. long; petals obovate or obovate-oblong, with incurved subobtuse apex, ¾–1 in. long; lip linear-lanceolate, acuminate, ½–¾ in. long, recurved at the apex; column ¾–1 in long; anther reclinate, narrow and acute; rostellum erect, somewhat elongate, with short divaricate side lobes; stigma pulvinate; capsule oblong, 1½–2 in. long. *N. E. Br. in Gard. Chron.* 1885, xxiv. 232; *Bolus in Trans. S. Afr. Phil. Soc.* v. 147, *Orch. Cap. Penins.* 147, *Journ. Linn Soc.* xxv. 197, *and Ic. Orch. Austr.-Afr.* ii. *t.* 63; *Durand & Schinz, Conspect. Fl. Afr.* v. 109; *Kränzl. Orch. Gen. et Sp.* i. 765, *and in Ann. Naturhist. Hofmus. Wien*, xx. 7; *Schlechter in Engl. Jahrb.* xxxi. 225. *D. grandiflora, Linn. f. Suppl.* 406; *Sw. in Vet. Acad. Handl. Stockh.* 1800, 210; *Ker in Quart. Journ. Sci. & Arts*, iv. 205, *t.* 6, *fig.* 1; *Thunb. Fl. Cap. ed. Schult.* 7; *Lindl. Bot. Reg. t.* 926, *Sert. Orch. t.* 49, *and Gen. & Sp. Orch.* 347; *Drège, Zwei Pfl. Documente*, 77, 82; *Bot. Mag. t.* 4073; *Krauss in Flora*, 1845, 307, *and Fl. Cap- und Natal.* 158; *Sond. in Linnæa*, xx. 218; *Fl. des Serres, t.* 160; *R. Trimen in Journ. Linn. Soc.* xix. 233, *fig.* 1; *Gard. Chron.* 1882, xvii. 402, *fig.* 62, 1888, iv. 665, *fig.* 94; *Lindenia*, vii. *t.* 308; *Reichenbachia, ser.* 2, i. 33, *t.* 15; *Fl. Mag.* 1862, *t.* 69; *Gartenfl.* xl. 176, 177, *fig.* 48, 49; *Rev. Hort.* 1903, 84, *with plate. D. Barelli, Puydt, Orch.* 275, *t.* 18. *Satyrium grandiflorum,* `Thunb. Prodr.* 4.

SOUTH AFRICA: without locality, *Masson*! *Oldenburg*, 1011! *Villet, Sieber*, 246! *Yorke*! *Robertson*! *Mrs. Allport*!
COAST REGION: Clanwilliam Div.; Cederberg Range, 1600 ft., *Mader*! *MacOwan*, 2031! *Leipold*, 602. Tulbagh Div.; Great Winterhoek Mountain, near Tulbagh, 2000–3000 ft., *Drège.* Worcester Div.; Dutoits Kloof, 3000–4000 ft., *Drège*, 1240 a! Paarl Div.; French Hoek Mountains, *McGibbon*! Cape Div.; Table Mountain, 1100–3500 ft.. at margins of streams. *Bergius, Thunberg*! *Masson, Burchell*, 858! *Mund, Ecklon & Zeyher, Wallich*! *Krauss, Bunbury*! *Prior*! *Alexander*! *Harvey*! *Bolus*, 4662! *MacOwan & Bolus, Herb. Norm. Austr.-Afr.* 559! *Kässner, Schlechter*, 299; *Penther*, 153, 327; Constantia Mountains, 1550 ft., *Schlechter*, 185; Waai Vley, *Wolley-Dod*, 879!
CENTRAL REGION: Ceres Div.; Gydouw Mountains (*fide Bolus*).

50. **D. Cephalotes** (Reichb. f. Otia Bot. Hamb. ii. 106); plant ½–1½ ft. high; stem rather slender; leaves cauline, 3–5, suberect,

linear-lanceolate, acuminate, 3–8 in. long ; scapes $\frac{1}{2}$–$1\frac{1}{2}$ ft. high, with several lanceolate acuminate sheaths above ; spikes subcapitate or shortly oblong, $\frac{1}{2}$–1 in. long, dense, many-flowered ; bracts lanceolate, very acuminate, 4–8 lin. long ; pedicels 4–7 lin. long ; flowers small, white, with a few purple spots on the sepals ; dorsal sepal galeate, broadly ovate or suborbicular, minutely apiculate, with much incurved margin, about $2\frac{1}{2}$ lin. long, papillose ; spur cylindrical from a short conical base, obtuse, somewhat curved, nearly twice as long as the limb ; lateral sepals broadly ovate-oblong, obtuse, with a short dorsal apiculus ; petals oblique, broadly semi-ovate-orbicular, subobtuse, $1\frac{1}{4}$ lin. long ; lip linear-oblong, obtuse, about 2 lin. long ; column short ; anther reflexed ; rostellum ascending, shortly bifid, with short rounded side lobes ; stigma pulvinate. *Bolus in Journ. Linn. Soc.* xxv. 198, *and Ic. Orch. Austr.-Afr.* i. *t.* 81 ; *Durand & Schinz, Conspect. Fl. Afr.* v. 101 ; *Kränzl. Orch. Gen. et Sp.* i. 758, *excl. syn. Schlechter ; Schlechter in Engl. Jahrb.* xxxi. 268.

COAST REGION : Stockenstrom Div. ; west slope of Elands Berg, 5000 ft., *Scully*, 408 ! Queenstown Div. ; Hanglip Mountain, near Queenstown, 6100 ft., *Galpin*, 1777 ! Stutterheim Div. ; Dohne Mountain, *Sim.*

CENTRAL REGION : Somerset Div. ; grassy slopes of Bosch Berg beyond Besters Hoek, 4000 ft., *MacOwan*, 1533 !

KALAHARI REGION : Orange River Colony ; Besters Vallei, near Harrismith, *Miss Jacobsz.*

EASTERN REGION : Tembuland ; Mount Engcobo, 4200 ft., *Bolus*, 8707 ! Natal ; slopes of the Drakensberg Range, near Tugela River, 5000 ft., *Buchanan in Herb. Sanderson* ! stony hills near Van Reenen, 7000 ft., *Schlechter*, 6933 ; Oliviers Hoek, 5000 ft., *Allison*, 2 !

51. **D. Gerrardii** (Rolfe) ; plant about 8 in. high ; stem rather stout ; leaves radical and cauline, few, narrowly lanceolate, acuminate, 2–5 in. long, reduced upwards into the bracts ; scapes about 8 in. long, with several narrow acute sheaths above ; spikes subcapitate or shortly oblong, about 1 in. long, dense, many-flowered ; bracts narrowly lanceolate, acuminate, 6–8 lin. long ; pedicels 4–6 lin. long ; flowers small ; dorsal sepal galeate, oblong or ovate-oblong, obtuse, about $2\frac{1}{2}$ lin. long ; spur cylindrical from a narrowly conical base, twice as long as the limb ; lateral sepals narrowly oblong, obtuse, very concave, with a short broad dorsal apiculus, 3 lin. long ; petals obliquely ovate-oblong, obtuse or obscurely tridenticulate, 2 lin. long ; lip linear, obtuse, 3 lin. long ; column short ; anther reflexed ; rostellum short, obtuse ; stigma pulvinate.

EASTERN REGION : Natal ; *Gerrard*, 2178 !

52. **D. oreophila** (Bolus in Journ. Linn. Soc. xxv. 170, fig. 8) ; plant $\frac{1}{2}$–$\frac{3}{4}$ ft. high ; stem slender ; leaves radical and cauline, numerous, suberect, narrowly linear and grass-like, acute, 3–8 in. long ; scapes $\frac{1}{2}$–$\frac{3}{4}$ ft. high, with a few very narrow sheaths above ;

spikes 2–3 in. long, rather lax, many-flowered; bracts lanceolate or narrowly ovate-lanceolate, long-acuminate, 3–6 lin. long; pedicels slender, 5–6 lin. long; flowers small, pale pink with darker spots on the sepals; dorsal sepal galeate, ovate-oblong, obtuse, often with a dorsal apiculus, 3–3½ lin. long; spur cylindrical from a conical base, subobtuse longer than the limb; lateral sepals ovate-oblong, obtuse, with a long slender dorsal apiculus, 3–3½ lin. long; petals broadly oblong, obtuse, with a large rounded basal lobe in front, about 1½ lin. long; lip spathulate from a slender base, obtuse, 3–3½ lin. long; column short; anther reflexed; rostellum subrhomboid, with an emarginate apex; stigma pulvinate. *Bolus in Journ. Linn. Soc.* xxv. 199, *and Ic. Orch. Austr.-Afr.* ii. *t.* 78; *Durand & Schinz, Conspect. Fl. Afr.* v. 105; *Kränzl. Orch. Gen. et Sp.* i. 781, *and in Ann. Naturhist. Hofmus. Wien,* xx. 8; *Schlechter in Engl. Jahrb.* xxxi. 270.

KALAHARI REGION : Orange River Colony; Harrismith, *Sankey,* 255 !
EASTERN REGION : Transkei ; Intwanazana, near Gat Berg, 4000 ft., *Baur,* 738 ! Tembuland ; Engcobo Mountain, 4500 ft., *Bolus,* 8737. Griqualand East; Mount Currie, 7500 ft., *Tyson,* 1073 ! Insizwa Range, *Schlechter, Krook, Penther,* 149. Natal ; Drakensberg Range, near Oliviers Hoek, 5000 ft., *Wood,* 3413.

53. **D. saxicola** (Schlechter in Engl. Jahrb. xx. Beibl. 50, 41) ; plant 6–10 in. long; stem rather slender ; leaves cauline, 3–6, sub-erect or somewhat arcuate, lanceolate or linear-lanceolate, acute or acuminate, 2–5 in. long, 2–8 lin. broad ; scapes 6–10 in. long, with one or two narrow acuminate sheaths above ; spikes 1½–3 in. long, somewhat lax, many-flowered ; bracts lanceolate or ovate-lanceolate, acuminate, ¼–¾ in. long ; pedicels 5–7 lin. long; flowers rather small, white with purple spots ; dorsal sepal galeate, broadly ovate, obtuse, about 3 lin. long; spur subcylindrical from a broadly conical base, subacute, much longer than the limb; lateral sepals broadly oblong, obtuse or truncate, about 3 lin. long; petals broadly falcate-oblong, obtuse, somewhat dilated at the base in front, 1½ lin. long; lip spathulate from a narrow base, obtuse, 2 lin. long; column short; anther reflexed; rostellum erect, shortly tridentate, with obtuse lobes; stigma pulvinate. *N. E. Br. in Dyer, Fl. Trop. Afr.* vii. 281 ; *Kränzl. Orch. Gen. et Sp.* i. 781 ; *Schlechter in Engl. Jahrb.* xxxi. 271, *t.* 4, *fig. E–G.*

KALAHARI REGION : Transvaal ; rock fissures near Botsabelo, 4900 ft., *Schlechter,* 4091 ; Berg Plateau, *Mudd* !
EASTERN REGION : Swaziland ; among rocks and on trees at the summit of Devils Bridge Mountain, 5000 ft., on the Swaziland border, *Galpin,* 714.

Also in Nyasaland, Tropical Africa.

54. **D. montana** (Sond. in Linnæa, xix. 90) ; plant 1½–2 ft. high ; stem leafy ; leaves rigid, ensiform, veined ; scapes 1½–2 ft. high ; spikes lax, many-flowered ; bracts very acuminate or setaceous, the lower 1 in. long, longer than the flower ; pedicels slender, about 7 lin. long; flowers erect, medium-sized, rosy-red ; dorsal sepal

galeate, broadly ovate, subobtuse, about 5 lin. long; spur very narrowly conical, quite slender towards the apex, subobtuse, rather shorter than the limb ; lateral sepals oblique, ovate-oblong, obtuse, with a short dorsal apiculus, about 5 lin. long; petals narrowly ovate-triangular, acute, 2 lin. long ; lip oblong, obtuse, about 5 lin. long ; column short ; anther reflexed ; rostellum erect, shortly bifid, with a pair of diverging arms ; stigma pulvinate. *Sond. in Linnæa,* xix. 90 ; *Reichb. f. in Linnæa,* xx. 692 ; *Bolus in Journ. Linn. Soc.* xxv. 198 ; *Kränzl. Orch. Gen. et Sp.* i. 768. *D. pulchra, var. montana, Schlechter in Engl. Jahrb.* xxxi. 259.

Coast Region: Queenstown Div. ; Winterberg Range, between Tarka and Katberg, 5000–6000 ft., *Zeyher* !

Eastern Region : Griqualand East ; Vaal Bank, near Kokstad, 4000–5000 ft., *Haygarth,* 856.

I have only seen a single flower of the type specimen, preserved in the Lindley Herbarium.

55. **D. patula** (Sond. in Linnæa, xix. 94) ; plant $\frac{3}{4}$–$1\frac{1}{4}$ ft. high ; stem stout ; leaves cauline, numerous, suberect, lanceolate to ovate-lanceolate, acute or acuminate, 2–5 in. long, reduced upwards into the bracts ; scapes $\frac{3}{4}$–$1\frac{1}{4}$ ft. high ; spikes 2–6 in. long, dense or rarely lax, many-flowered ; bracts lanceolate or oblong-lanceolate, acuminate, $\frac{3}{4}$–1 in. long ; pedicels $\frac{1}{2}$–$\frac{3}{4}$ in. long ; flowers medium-sized, pink or rose, sometimes spotted and lined with purple ; dorsal sepal galeate, broadly oblong, obtuse, with a somewhat involute margin, 4–6 lin. long ; spur cylindrical from a narrowly conical base, rather slender, suboutuse, somewhat curved, rather longer than the limb ; lateral sepals oblong, obtuse, with a broad obtuse dorsal apiculus, 4–6 lin. long ; petals obliquely ovate-oblong, obtuse, $3\frac{1}{2}$–4 lin. long ; lip linear, slightly thickened above, 5–6 lin. long ; column short ; anther reflexed ; rostellum short, 3-dentate ; stigma pulvinate. *Bolus in Journ. Linn. Soc.* xxv. 201 ; *Kränzl. Orch. Gen. et Sp.* i. 773. *D. stenoglossa, Bolus in Journ. Linn. Soc.* xxv. 173, 199, *fig.* 11. *Monadenia leydenbergensis, Kränzl. Orch. Gen. et Sp.* i. 811.

Coast Region: Albany Div. ; on a flat hill top above Featherstone Kloof, near Grahamstown, 2300 ft., *MacOwan,* 678 ! Howisons Poort Hills, 2200 ft., *Galpin,* 3089 ! Stockenstrom Div. ; Kat Berg, 2000 ft., *Hutton* ! Queenstown Div. ; slopes of the Winterberg Range, *Zeyher* ! *Mrs. Barber,* 522 !

Kalahari Region: Transvaal ; Mac Mac, *Mudd* ! near Lydenberg, *Atherstone* ! *Wilms,* 1364.

Eastern Region : Transkei ; Tsomo, *Mrs. Barber,* 682 ! Krielis Country, *Bowker,* 353 ! and without precise locality, *Hallack* ! Natal ; hill near Mooi River, 4200 ft., *Wood,* 4077 ! Oliviers Hoek, 5000 ft., *Allison,* N ! and without precise locality, *Mrs. Saunders.*

This has smaller flowers and relatively much longer leaves than *D. nervosa,* Lindl., with which it has been confused by Schlechter. I cannot, however, separate *D. stenoglossa,* Bolus, of which Mrs. L. Bolus has kindly sent a flower from the type specimen.

56. **D. nervosa** (Lindl. Gen. & Sp. Orch. 352) ; plant $1\frac{1}{2}$–2 ft. high ; stem stout ; leaves cauline, suberect, 5–8, linear or narrowly

oblong-linear, subacute, somewhat fleshy, 4–8 in. long; scapes 1½–2 ft. high, with narrow leaf-like sheaths above; spikes 4–6 in. long, oblong, somewhat dense, many-flowered; bracts lanceolate or oblong-lanceolate, very acuminate, 1–1½ in. long; pedicels slender, ¾–1 in. long; flowers large, deep pink (*Wood*) or rosy purple (*Allison*); dorsal sepal galeate, oblong or elliptic-oblong, subobtuse, with involute margins, 8–12 lin. long; spur cylindrical from a short conical base, curved, subobtuse, about as long as the limb; lateral sepals oblong, obtuse, with a short dorsal apiculus, 8–10 lin. long; petals falcate-oblong, obtuse, 7–8 lin. long; lip filiform, slightly thickened near the apex, 9–12 lin. long; column short; anther reflexed; rostellum broad, obtuse; stigma pulvinate. *Drège, Zwei Pfl. Documente,* 149, 152; *Gard. Chron.* 1894, xvi. 308, 309, *fig.* 41; *Bolus in Journ. Linn. Soc.* xxv. 200, *and Ic. Orch. Austr.-Afr.* i. *t.* 84; *Durand & Schinz, Conspect. Fl. Afr.* v. 104; *Kränzl. Orch. Gen. et Sp.* i. 774; *Schlechter in Engl. Jahrb.* xxxi. 260, *excl. all syn.*

SOUTH AFRICA : without precise locality, *Large* !

EASTERN REGION : Pondoland; between Umtata and St. Johns River, *Drège*, 4561 a ! between St. Johns River and Umtsikaba River, *Drège*, 4561 b ! Natal; Kirkmans Cutting, *Sanderson*, 1047 ! rocky hill near Bothas, 2000 ft., *Wood*, 4821 ! Van Reenen, 5000–6000 ft., *Wood*, 6564 ! Sevenfontein, 3000–4000 ft., *Wylie in Herb. Wood*, 8099 ! Polela, Ixopo River, *Mrs. Clarke*, 66 ! and without precise locality, 4000 ft., *Allison* !

57. **D. Kraussii** (Rolfe); plant 2¼ ft. high; stem very stout; leaves cauline, rather numerous, suberect, linear, acuminate, rigid, strongly nerved, 5–8 in. long, diminishing upwards into the bracts; scapes 2¼ ft. high; spikes 8 in. long, cylindrical, somewhat dense, many-flowered; bracts linear-lanceolate from a broader base, very acuminate, ¾–1¼ in. long; pedicels ¾–1 in. long; flowers large; dorsal sepal galeate, oblong or elliptic-oblong, subobtuse, 1 in. long; spur cylindrical from a broadly conical base, somewhat curved, subacute, about 8 lin. long; lateral sepals oblong or elliptic-oblong, subacute, 1 in. long; petals linear-oblong, acute or acuminate, about 4 lin. long; lip oblong-lanceolate, subacute, about 1 in. long; column short; anther reflexed; rostellum short, obtuse; stigma pulvinate. *D. nervosa, Krauss in Flora,* 1845, 306, *and Beitr. Fl. Cap- und Natal.* 158, *not of Lindl.*

EASTERN REGION : Natal, on mountains between the Umgeni River and Mooi River, 4000 ft., *Krauss*, 15 !

Quite distinct from *D. nervosa*, Lindl., to which it was referred by Krauss.

58. **D. Fanniniæ** (Harv. MSS.); plant about 2¼ ft. high; stem very stout; leaves cauline, numerous, suberect, rather rigid, oblong-linear, acute or acuminate, 3–5 in. long, narrowed upwards into the bracts; scapes 2¼ ft. high, with short leaf-like sheaths above; spikes about 6 in. long, cylindrical, rather dense, many-flowered; bracts

lanceolate, acuminate, 1–1¼ in. long ; pedicels ¾–1 in. long ; flowers large, rose-coloured ; dorsal sepal galeate, oblong or elliptic-oblong, subobtuse, with involute margin, 1–1¼ in. long ; spur cylindrical from a conical base, subobtuse, about half as long as the limb ; lateral sepals oblong, obtuse, with a short dorsal apiculus, 1–1¼ in. long ; petals oblong or slightly falcate-oblong, obtuse, about 1 in. long ; lip filiform, slightly thickened above, about 1 in. long ; column short ; anther reflexed ; rostellum broad, obtuse ; stigma pulvinate.

EASTERN REGION : Natal ; Dargle Farm, *Mrs. Fannin,* 1 ! Ingoma, *Gerrard,* 1545 ! and without precise locality, *Sanderson in Herb. Wood* ! *Mrs. Saunders in Herb. Wood,* 3899 ! *Gerrard,* 2177 ! 2179 !

59. **D. pulchra** (Sond. in Linnæa, xix. 94) ; plant 1½–2¼ ft. high ; stem rather stout ; leaves cauline, numerous, suberect, linear or oblong-linear, acuminate, 3–8 in. long, reduced upwards into the bracts ; scapes 1½–2¼ ft. high ; spikes 5–10 in. long, usually somewhat lax, many-flowered ; bracts linear or lanceolate-linear, very acuminate or subsetaceous, 1–2 in. long ; pedicels slender, ¾–1¼ in. long ; flowers large, bright pink or rose-coloured ; dorsal sepal galeate, elliptic or elliptic-lanceolate, acute, 1–1½ in. long ; spur cylindrical from a narrowly conical base, slender above, subobtuse, 7–9 lin. long ; lateral sepals elliptic-lanceolate, acuminate, 1–1½ in. long ; petals lanceolate or narrowly ovate-lanceolate, acute or acuminate, 5–6 lin. long ; lip narrowly ovate or elliptic-ovate, subobtuse, 10–13 lin. long ; column short ; anther reflexed ; rostellum erect, bifid, with a pair of diverging arms ; stigma pulvinate. *Sond. in Linnæa,* xx. 219 ; *N. E. Br. in Gard. Chron.* 1885, xxiv. 232, *and* 1896, xx. 778, 785, *fig.* 141 ; *Bolus in Journ. Linn. Soc.* xxv. 201 ; *Durand & Schinz, Conspect. Fl. Afr.* v. 106 ; *Kranzl. Orch. Gen. et Sp.* i. 771 ; *Schlechter in Engl. Jahrb.* xxxi. 259, *excl. var.*

COAST REGION : Fort Beaufort Div. ; mountains near Kat River, *Pappe,* 73 ! Victoria East Div. ; Hogsback Mountains, 6300 ft., *Galpin,* 7992 ! Stockenstrom Div. ; Kat Berg, 5000–5800 ft., *Scully in Herb. Bolus,* 5910 ! *Galpin,* 1680 ! Queenstown Div. ; Winterberg Range, *Ecklon & Zeyher* ! *Zeyher* ! *Mrs. Barber,* 528 !
KALAHARI REGION : Orange River Colony ; Nelsons Kop, *Cooper,* 978 !
EASTERN REGION : Griqualand East ; grassy slopes near Fort Donald, 5000 ft., *Tyson,* 1597 ; Insizwa Range, 6800 ft., *Schlechter,* 6463 ! Natal ; near Polela, *Clarke* ; Imyassuti River, near Emangweni, 6000–7000 ft., *Thode* ; near Mooi River, *Sanderson,* 1055 ! Swaziland ; Illalikulu, 4100 ft., *Miss Stewart* !

60. **D. Marlothii** (Bolus in Trans. S. Afr. Phil. Soc. xvi. 148) ; plant 5–10 in. high ; stem rather slender ; leaves radical, 5–7, somewhat spreading, lanceolate or oblong-lanceolate, acute or shortly acuminate, 1–2½ in. long ; scapes 5–10 in. high, with several short acuminate sheaths ; spikes short, few-flowered, occasionally 1-flowered ; bracts oblong-lanceolate, acute, ¼–½ in. long ; pedicels slender, ½–1 in. long ; flowers medium-sized, rose-

pink, with carmine spots on the petals and dorsal sepal; dorsal
sepal subgaleate, obovate-orbicular, subtruncate, about 4 lin. long;
spur subcylindrical from a narrowly conical base, somewhat narrowed
upwards, subobtuse, twice as long as the limb; lateral sepals broadly
oblong, obtuse, with a short dorsal apiculus, 4–4½ lin. long; petals
falcate-oblong, obtuse, about 2½ lin. long; lip oblong, obtuse,
2½–3 lin. long; column short; anther reflexed; rostellum erect,
oblong, narrowed at the apex; stigma pulvinate. *Bolus, Ic. Orch.
Austr.-Afr.* ii. *t.* 67.

Coast Region: Worcester Div.; Sand-drift Kloof, Hex River Mountains,
3200 ft., *Marloth*, 2378; Matroos Berg, 4600 ft., *Bolus*, 6371!
Central Region: Ceres Div.; Cold Bokkeveld, on the Skurfdeberg Range,
near Klein Vley, 6000 ft., *Schlechter*, 10204!

61. D. gladioliflora (Burch. ex Lindl. Gen. & Sp. Orch. 352);
plant 1¼–2 ft. high; stem rather slender; leaves radical, rather
numerous, suberect, narrowly linear or grass-like, acute, 6–8 in.
long; scapes 1¼–2 ft. high, with several narrow acuminate sheaths
above; spikes 3–5 in. long, lax, many-flowered; bracts narrowly
lanceolate, very acuminate, ¾–1¼ in. long; pedicels slender, 1–1¼ in.
long; flowers medium-sized, rather membranous; dorsal sepal
galeate, ovate-oblong, obtuse or apiculate, 6–7 lin. long; spur
slender, curved from a short conical base, subobtuse, about half as
long as the limb; lateral sepals oblong, subobtuse, 6–7 lin. long;
petals falcate-lanceolate, acute, nearly 2 lin. long, with a short
rounded basal lobe in front; lip lanceolate or narrowly lanceolate-
oblong, subobtuse, 5–6 lin. long; column short; anther reflexed;
rostellum erect, short, bidentate; stigma pulvinate. *Bolus in Journ.
Linn. Soc.* xxv. 200; *Durand & Schinz, Conspect. Fl. Afr.* v. 103;
Kränzl. Orch. Gen. et Sp. i. 775; *Schlechter in Engl. Jahrb.* xxxi. 296,
partly, excl. syn.

Coast Region: Knysna Div.; on the Parde Berg, *Burchell*, 5184!

62. D. capricornis (Reichb. f. in Linnæa, xx. 689); plant 10–15 in.
high; stem rather slender; leaves radical, rather numerous, suberect, narrowly linear or grass-like, acute, 3–8 in. long; scapes
10–15 in. high, with several narrow acuminate sheaths above;
spikes 2½–4 in. long, lax, many-flowered; bracts narrowly lanceolate, very acuminate, ½–¾ in. long; pedicels slender, ¾–1 in. long;
flowers rather small, membranous, rose-coloured; dorsal sepal
galeate, ovate-oblong, obtuse, 4–5 lin. long; spur slender, curved
from a short conical base, nearly half as long as the limb; lateral
sepals oblong, obtuse, 4–5 lin. long; petals falcate-oblong, subobtuse,
with a short rounded basal lobe in front, about 1½ lin. long; lip
spathulate-obovate, obtuse, about 4 lin. long: column short; anther
reflexed; rostellum erect, short, bidentate; stigma pulvinate. *D.
gladioliflora, Bolus, Ic. Orch. Austr.-Afr.* ii. *t.* 79, *not of Reichb. f.*

Coast Region: Swellendam Div.; mountains near Tradouw, *Mund & Maire*!
Langeberg Range, near Swellendam, 3000 ft., *Borcherds in MacOwan & Bolus*,

Herb. Norm. Austr.-Afr. 1049! Riversdale Div. ; slopes of the Langeberg Range, near Riversdale, 1000 ft., *Schlechter*, 2218. Stutterheim Div. ; Dohne Mountain, 4000 ft., *Sim*, 960. *Bolus*, 8708.

EASTERN REGION : Griqualand East ; Beeste Kraal, near Kokstad, 5300 ft., *Tyson*, 1609 ; Insizwa Mountains, 6900 ft., *Schlechter*, 6509.

63. **D. Draconis** (Sw. in Vet. Acad. Handl. Stockh. 1800, 210) ; plant 1–1¾ ft. high ; stem rather stout ; leaves subradical, 2–4, linear or oblong-linear, acute or subacute, 3–9 in. long ; scapes 1–1¾ ft. high, with numerous broad acute or acuminate sheaths ; spikes 2–5 in. long, 6- to many-flowered ; bracts lanceolate, acuminate, 1–1½ in. long ; pedicels 1–1¼ in. long ; flowers large, straw-yellow ; dorsal sepal subgaleate, obovate or obovate-oblong, obtuse, somewhat reflexed, 8–11 lin. long ; spur filiform, somewhat curved, 1½–2¼ in. long ; lateral sepals spreading, oblong, subobtuse, 10–12 lin. long ; petals oblong or subfalcate-oblong, obtuse, recurved, with a small rounded basal auricle in front, 7–8 lin. long ; lip linear, obtuse, 7–8 lin. long ; column short ; anther pendulous ; rostellum erect, subquadrate, obtuse ; stigma pulvinate. *Thunb. Fl. Cap. ed. Schult.* 10 ; *Lindl. Gen. & Sp. Orch.* 352 ; *Drège, Zwei Pfl. Documente*, 68, 78, 87 ; *Sond. in Linnæa*, xix. 94, *and* xx. 219 ; *N. E. Br. in Gard. Chron.* 1885, xxiv. 231 ; *Bolus in Journ. Linn. Soc.* xxv. 200, *and Ic. Orch. Austr. Afr.* ii. *t.* 74 ; *Durand & Schinz, Conspect. Fl. Afr.* v. 102 ; *Kränzl. Orch. Gen. et Sp.* i. 776 ; *Schlechter in Engl. Jahrb.* xxxi. 231. *Orchis Draconis, Linn. f. Suppl.* 400. *Satyrium Draconis, Thunb. Prodr.* 5.

SOUTH AFRICA : without locality, *Oldenburg*, 1849 ! *Masson* ! *Brown* ! *Leibold*.
COAST REGION : Clanwilliam Div. ; Brakfontein, *Mrs. Vanschou* ! Malmesbury Div. ; Groene Kloof (Mamre), *Harvey*, 243 ! *Zeyher*, 1565 ! Tulbagh Div. ; New Kloof, 2000–3000 ft., *Drège*, 1237 b ! *Ecklon & Zeyher*, 52 ! mountains above Tulbagh Waterfall, *Pappe* ! *Bolus*, 5280 ! Worcester Div. ; Bains Kloof, 1500 ft., *Schlechter*, 9123 ! and without precise locality. *Cooper*, 1624 ! 1681 ! Paarl Div. ; Paarl Mountain, 1000–2000 ft., *Drège*, 1237 a ! French Hoek, *Thunberg* ! Cape Div. ; Kuils River, *Pappe*, 47 ! Table Mountain, above Orange Kloof, 1000 ft., *Schlechter*. Stellenbosch Div. ; near Stellenbosch, 300 ft., *Farnham in MacOwan & Bolus, Herb. Norm. Austr.-Afr.* 334 ! Caledon Div. ; Genadendal, *Roser* !
CENTRAL REGION : Calvinia Div. ; Roggeveld, *Thunberg* !
WESTERN REGION : Little Namaqualand ; Lily Fontein, 4000–5000 ft., *Drège*, 1237 c !

64. **D. harveiana** (Lindl. in Hook. Lond. Journ. Bot. i. 15) ; plant ¾–1½ ft. high ; stem moderately stout ; leaves subradical, 2–5, linear or oblong-linear, acute or acuminate, spreading or sub-erect, 3–7 in. long ; scapes ¾–1½ ft. high, with numerous narrow acute sheaths ; spikes 2–3 in. long, lax, 3–10-flowered ; bracts lanceolate or linear-lanceolate, acuminate, ¾–1¼ in. long ; pedicels 1–1¼ in. long ; flowers large, lilac or rosy lilac, with some purple marks at the base of the dorsal sepal and petals ; dorsal sepal subgaleate, obovate or obovate-oblong, obtuse, somewhat reflexed, 8–10 lin. long ; spur filiform, somewhat curved, 1–1½ in. long ; lateral sepals spreading, oblong, obtuse, 9–11 lin. long ; petals

linear or subfalcate-linear, obtuse, recurved, with a rounded basal
auricle in front, 7–9 lin. long ; lip linear, obtuse, about $\frac{3}{4}$ in. long ;
column short ; anther reflexed ; rostellum erect, 3-dentate, front
lobe subretuse, side lobes acute ; stigma pulvinate. *Bolus in
Trans. S. Afr. Phil. Soc.* v. 158 ; *Orch. Cap. Penins.* 158, *and in
Journ. Linn. Soc.* xxv. 200 ; *Durand & Schinz, Conspect. Fl. Afr.*
v. 103 ; *Kränzl. Orch. Gen. et Sp.* i. 775. *D. Draconis, var.
harveyana, Schlechter in Engl. Jahrb.* xxxi. 231.

COAST REGION : Cape Div. ; Table Mountain, 1000–3000 ft., *Harvey* ! *Bolus,*
3304 ! *Prior* ! *Schlechter,* 90 ! *Kässner* ; Lower Plateau, *Wolley-Dod,* 2324 !
Devils Peak, *Bolus in MacOwan & Bolus, Herb. Norm. Austr.-Afr.* 162 !
between Constantia Mountains and Hout Bay, *Wolley-Dod,* 801 !

65. **D. schlechteriana** (Bolus in Trans. S. Afr. Phil. Soc. xvi. 149) ;
plant $1\frac{1}{4}$–$1\frac{3}{4}$ ft. high ; stem moderately stout ; leaves radical, 7–11,
erect, narrowly linear, acuminate, rigid, prominently veined, about
1 ft. long ; scapes $1\frac{1}{4}$–$1\frac{3}{4}$ ft. high, with several narrow acuminate
adpressed sheaths above ; spikes 3–$5\frac{1}{2}$ in. long, lax, 6–12-flowered ;
bracts lanceolate, acuminate, 8–10 in. long ; pedicels about 1 in.
long ; flowers large, rather membranous, drab or cream-coloured,
with narrow purple lines on the dorsal sepal ; dorsal sepal galeate,
broadly ovate, acute or shortly acuminate, 9–11 lin. long ; spur
filiform from a narrowly conical base, somewhat curved, about
twice as long as the limb ; lateral sepals oblong-lanceolate, acute
or acuminate, 9–11 lin. long ; petals reflexed, very acuminate,
oblong below the bend, with a broad rounded auricle at the
base in front, nearly $\frac{1}{2}$ in. long ; lip lanceolate-oblong or narrowly
elliptic, subobtuse, pandurately constricted near the base, 6–8 lin.
long ; column short ; anther reflexed : rostellum erect, oblong,
tridentate at the apex, slightly incurved ; stigma pulvinate. *Bolus,
Ic. Orch. Austr.-Afr.* ii. *t.* 75.

COAST REGION : Riversdale Div. ; rocky mountain sides at Kleenberg, between
Garcias Pass and Muis Kraal, 1800 ft., *Mrs. Luyt in Herb. Bolus,* 10571 !

66. **D. ferruginea** (Sw. in Vet. Acad. Handl. Stockh. 1800, 210) ;
plant $\frac{3}{4}$–$1\frac{1}{2}$ (rarely 2) ft. high ; stem moderately stout ; leaves
radical, 4–8, arching, linear, acute, 5–10 in. long, generally withered
at flowering time ; scapes $\frac{3}{4}$–$1\frac{1}{2}$ (rarely 2) ft. high, with numerous
narrow acuminate sheaths above ; spikes oblong, $1\frac{1}{2}$–3 in. long,
rather broad, dense, many-flowered ; bracts ovate-lanceolate, long-
acuminate and setaceous, 5–8 lin. long ; pedicels $\frac{1}{2}$–$\frac{3}{4}$ in. long ;
flowers medium-sized, submembranous, orange-vermilion or rarely
cream-coloured ; dorsal sepal galeate, broadly ovate, subobtuse,
about 4 lin. long ; spur horizontal, broadly conical below, attenuate,
somewhat curved and subacute above, 10–12 lin. long ; lateral sepals
ovate, obtuse, with a long slender bristle behind the apex, about
$4\frac{1}{2}$ lin. long ; petals falcate-ovate, acuminate, with a rounded auriculate basal lobe in front, about $2\frac{1}{2}$ lin. long ; lip lanceolate or oblong-

lanceolate, acuminate, about 4½ lin. long ; column very short, apiculate ; anther reflexed ; rostellum ascending, very short ; stigma pulvinate. *Thunb. Fl. Cap. ed. Schult.* 11 ; *Harv. in Hook. Ic. Pl. t.* 214 ; *Krauss in Flora,* 1845, 307, *and Fl. Cap- und Natal.* 158, *excl. syn.* ; *N. E. Br. in Gard. Chron.* 1885, xxiv. 232 ; *Bolus in Trans. S. Afr. Phil. Soc.* v. 172, *Orch. Cap. Penins.* 172, *Journ. Linn. Soc.* xxv. 203, *and Ic. Orch. Austr.-Afr.* i. *t.* 38 ; *Durand & Schinz, Conspect. Fl. Afr.* v. 102 ; *Kränzl. Orch. Gen. et Sp.* i. 760, *and in Ann. Naturhist. Hofmus. Wien,* xx. 7 ; *Schlechter in Engl. Jahrb.* xxxi. 295. *Satyrium ferrugineum, Thunb. Prodr.* 5. *D. porrecta, Ker in Quart. Journ. Sci. & Arts,* v. 104, *t.* 1, *fig.* 1 ; *Hook. Ic. Pl. t.* 214 ; *Sond. in Linnæa,* xix. 95, *excl. syn.*

SOUTH AFRICA: without locality, *Bergius, Masson* ! *Grey* ! *Trimen* ! *Villet* ! *Bunbury* !

COAST REGION: Cape Div. ; Table Mountain, common from 1800–3500 ft., *Thunberg* ! *Ecklon & Zeyher, Pappe, Krauss,* 1329 ; *Harvey* ! *Prior* ! *MacOwan,* 2419 ! *Bolus,* 4764 ! and in *MacOwan & Bolus, Herb. Norm. Austr.-Afr.* 165 ! *Mrs. Jameson* ! *Kässner, Schlechter,* 574 ; False Bay, *Robertson* ! *Wolley-Dod,* 1024 ! *Penther,* 152. Stellenbosch Div. ; mountains of Lowrys Pass, Hottentots Holland, *Burchell,* 8199 ! Swellendam Div. ; near Swellendam, *Kennedy,* 31 ! mountain slopes near Tradouw, *Mund.*

67. D. porrecta (Sw. in Vet. Acad. Handl. Stockh. 1800, 211) ; plant 1–1½ ft. high ; stem moderately stout ; leaves radical, numerous, suberect or somewhat spreading, elongate-linear, acute, 5–7 in. long ; scapes 1–1½ ft. high, with numerous narrow acuminate sheaths above ; spikes oblong, 2–3½ in. long, rather dense, many-flowered ; bracts ovate or lanceolate-ovate, long-acuminate, 5–8 lin. long ; pedicels ½–¾ in. long ; flowers medium-sized, rather fleshy, deep vermilion, with the petals, lip, and the margin of the sepals yellow ; dorsal sepal galeate, shortly and very broadly ovate, obtuse or emarginate, undulate, limb 2½–3 lin. long, broader than long ; spur ascending, narrowly conical, long-attenuate, 1–1¼ in. long ; lateral sepals elliptic-suborbicular, obtuse, concave, with a slender dorsal bristle 3–4 lin. long ; petals nearly quadrate, obtuse and obliquely apiculate, with an auriculate basal lobe in front, about 1¼ lin. long ; lip oblong or lanceolate-oblong, subobtuse, about 3 lin. long ; column very short, apiculate ; anther reflexed ; rostellum erect, minutely tridentate ; stigma pulvinate. *Lindl. Gen. & Sp. Orch.* 352, *partly* ; *Bolus in Journ. Linn. Soc.* xxv. 175, 204, *and Ic. Orch. Austr.-Afr.* i. *t.* 39 ; *Durand & Schinz, Conspect. Fl. Afr.* v. 106 ; *Schlechter in Engl. Jahrb.* xxxi. 294. *D. Zeyheri, Sond. in Linnæa,* xix. 95, *and* xx. 219 ; *Reichb. f. in Flora,* 1883, 461 ; *N. E. Br. in Gard. Chron.* 1885, xxiv. 232.

COAST REGION: Uniondale Div. ; on a rocky hill, near Haarlem, *Burchell,* 5014 ! Uitenhage Div. ; on Van Stadens Berg, nearest to Galgebosch, *Burchell,* 4693 ! Elands River Mountains, *Ecklon & Zeyher,* 1532 ! Albany Div. ; Brookhuizens Poort, near Grahamstown, 2400 ft., *Glass, South,* 505, and without precise locality, *Mrs. Barber,* 146 ! Eastern Frontier, *Hutton* !

CENTRAL REGION : Somerset Div. ; Bosch Berg, near Somerset East, 4000 ft.,

MacOwan, 1478 ! 1532! *Bowker*! and without precise locality, *Mrs. Barber*, 146 !
Graaff Reinet Div. ; Koudeveld Mountains, 4450 ft., *Bolus*, 1298 !
  KALAHARI REGION : Basutoland ; Lekholela, *Dieterlen*, 479 !
  EASTERN REGION : Transkei, Krielis Country, *Bowker*!

  68. D. longicornu (Linn. f. Suppl. 406) ; plant 4–10 in. high ;
stem somewhat slender ; leaves radical or subradical, 4–6, spreading, lanceolate from a narrow base, acute or acuminate, 1½–5 in.
long ; scapes 4–10 in. high, with 2 or 3 narrow acute sheaths
above, 1-flowered ; bracts lanceolate, acuminate, ¾–1 in. long ;
pedicels about ¾ in. long ; flowers large, lilac-blue, with a carmine
anther ; dorsal sepal galeate, broadly ovate, subobtuse or apiculate,
with a very wide mouth, limb ¾–1 in. long ; spur elongate from a
broadly conical base, curved, subobtuse, 1–1½ in. long ; lateral
sepals elliptic-oblong, subobtuse or apiculate, ¾–1 in. long ; petals
linear, from a broader base, long-attenuate above, with a subacute
ovate-triangular lobe in front below the middle, turned backwards
into the spur and extending nearly to its apex ; lip lanceolate-
elliptic, subobtuse, 8–10 lin. long ; column short ; anther reflexed ;
rostellum erect, short, with short suberect side lobes ; stigma pulvinate.   *Thunb. Prodr. 4, and Fl. Cap. ed. Schult.* 8 ; *Sw. in Vet. Acad.
Handl. Stockh.* 1800, 210 ; *Lam. Encycl. t.* 727, *fig.* 2 ; *Lindl. Gen.
& Sp. Orch.* 350 ; *Sond. in Linnæa,* xix. 92 ; *N. E. Br. in Gard.
Chron.* 1885, xxiv. 232 ; *Bolus in Trans. S. Afr. Phil. Soc.* v. 145,
t. 6, *Orch. Cap. Penins.* 145, *t.* 6, and *Journ. Linn. Soc.* xxv. 197 ;
Durand & Schinz, Conspect. Fl. Afr. v. 104 ; *Kränzl. Orch. Gen. et
Sp.* i. 766 ; *Schlechter in Engl. Jahrb.* xxxi. 274 ; *Marloth in Wiss.
Ergebn. Deutsch. Tiefsee-Exped.* ii. iii. 396.

  SOUTH AFRICA : without locality, *Villet* ! *Barkly* !
  COAST REGION : Cape Div. ; Table Mountain, in damp shaded fissures of rocks,
2100–3000 ft., *Thunberg* ! *Harvey* ! *Ecklon & Zeyher, Bolus,* 4818 ! and in
MacOwan & Bolus, Herb. Norm. Austr.-Afr. 161 ! *Prior* ! *Barkly* ! *Kitching* !
Bonny ! *Kässner, Schlechter,* 83, rocks near Waai Vley, *Wolley-Dod,* 2331 !
Stellenbosch Div. ; mountains near Stellenbosch, 3000 ft., *Dyke.*

  69. D. maculata (Linn. f. Suppl. 407) ; plant 4–9 in. long ; stem
rather slender ; leaves radical, rosulate, 5–8, spreading, lanceolate,
acute or acuminate, narrowed at the base, 1–2 in. long ; scapes
4–9 in. long, with several narrow acute sheaths above, 1-flowered ;
bracts lanceolate or ovate-lanceolate, acute or acuminate, ⅓–½ in.
long ; pedicels slender, ¾ in. long ; flowers rather large, bright blue,
faintly striped with green, petals yellow with purple apex ; dorsal
sepal galeate, broadly ovate, apiculate, about 8 lin. long, obtusely
saccate at the back ; lateral sepals elliptic-oblong, apiculate or
subacute, about 8 lin. long ; petals turned backwards into the sac
and inflexed at its apex, with a long slender base, sharply inflexed
above the middle, then dilated into an obovate-oblong limb, with
dentate apex, 4 lin. long to the bend, and 2 lin. long above it ;
lip lanceolate-oblong, subobtuse, 5 lin. long ; column short ;
anther reflexed ; rostellum erect, shortly bifid, with short suberect side lobes ; stigma prominent, somewhat 3-lobed.   *Thunb.*

Prodr. 4, *and Fl. Cap. ed. Schult.* 14 ; *Bolus in Trans. S. Afr. Phil. Soc.* v. 146, *t.* 7, *Orch. Cap. Penins.* 146, *t.* 7, *and Journ. Linn. Soc.* xx. 478, xxv. 197 ; *Kränzl. Orch. Gen. et Sp.* i. 782 ; *Schlechter in Engl. Jahrb.* xxxi. 275. *Schizodium maculatum, Lindl. Gen. & Sp. Orch.* 360.

SOUTH AFRICA : without locality, *Masson* !
COAST REGION : Tulbagh Div. ; near Winter Hoek, *Thunberg* ! Cape Div. ; moist clefts of rocks, Muizen Berg, 1200–1600 ft., *Bodkin, Bolus,* 4843 ! and in *MacOwan & Bolus, Herb. Norm. Austr.-Afr.* 160 ! lower slopes of Constantia Mountains, 2800 ft., *Bodkin, Wolley-Dod,* 2150 ; Caledon Div. ; moist rock on Houw Hoek Mountains, *Bolus.*

XXXVII. SCHIZODIUM, Lindl.

Odd *sepal* superior, with an oblong or narrow spur behind, the limb often narrow or acuminate and reflexed ; lateral sepals free, more or less spreading, often narrow or acuminate. *Petals* free, ascending, often narrow and unequally bilobed at the apex, more or less adnate to the rostellum at the base, the outer lobe sometimes much elongated and filiform. *Lip* pandurate or subpandurate, with a broad or concave base, often narrowed or acuminate in front, sometimes with a filiform appendage, spurless. *Column* short. *Anther* reflexed, with two parallel cells ; pollinia attached by short caudicles to separate glands. *Rostellum* erect, emarginate or retuse, with short side lobes. *Stigma* cushion-shaped. *Capsule* oblong or obovate-oblong, ribbed.

Terrestrial herbs, with simple thickened tubers ; stems erect, very slender and rigid, often strongly flexuous at the base ; leaves spathulate or elliptic-ovate, in a spreading radical tuft ; flowers in short or lax racemes, medium-sized ; bracts usually ovate-lanceolate.

DISTRIB. Species 8, almost limited to the south-western corner of South Africa.

Schizodium was reduced by Bolus to a section of *Disa*, but afterwards restored to generic rank. It differs markedly in the structure of the lip and in habit. Bolus remarks that the structure of the scape is quite unique among South African Orchids, and appears to afford a maximum of strength with a minimum of resistance to the wind on the wind-swept plains where these plants usually grow.

Sepals broadly elliptic-oblong, obtuse ... ... (1) **flexuosum.**
Sepals linear-lanceolate, subobtuse or acute :
  Spur not constricted at the base :
    Spur broad and straight ... ... (2) **inflexum.**
    Spur narrowed upwards and distinctly curved ... (3) **rigidum.**
  Spur clavate, or distinctly constricted at the base :
    Dorsal sepal 3½–4 lin. long ... ... (4) **obliquum.**
    Dorsal sepal 2½–3 lin. long :
      Spur narrowly clavate ; petals obliquely apiculate (5) **obtusatum.**
      Spur broadly clavate ; petals shortly 2-fid ... (6) **clavigerum.**
Sepals very narrow, long-acuminate :
  Leaves with broad entire limb ; front lobe of petals
    short ... ... (7) **arcuatum.**
  Leaves with narrow crisped-undulate limb ; front lobe
    of petals twice as long as the other .. ... (8) **longipetalum.**

1. **S. flexuosum** (Lindl. Gen. & Sp. Orch. 359); plant ½–1 ft.
high; stem slender, flexuous; leaves radical, spathulate; limb
ovate or elliptic-oblong, obtuse, 5–10 lin. long; petiole 3–5 lin. long;
scapes ½–1 ft. high, with several oblong obtuse suberect sheaths;
spikes 1–3 in. long, lax, 2–6-flowered, rarely 1-flowered; bracts
ovate or ovate-oblong, obtuse, rather fleshy, 3–4 lin. long; pedicels
4–6 lin. long; flowers rather large, white with yellow petals and
lip, the latter usually having a few brown spots; dorsal sepal erect,
suborbicular, obtuse, sharply reflexed at the sides, 3½–4 lin. long;
spur oblong, obtuse, curved, rather more than half as long as the
limb; lateral sepals obovate or suborbicular from a narrowed base,
obtuse, spreading, 3½–4 lin. long; petals erect, linear from an
obliquely ovate base, subobtuse, 3–3½ lin. long; lip subpandurate,
porrect, 5–6 lin. long, base dilated and concave, limb broad and
very undulate, with a sigmoidly deflexed rostrate or acuminate apex;
column short; anther very reflexed; rostellum erect, minute;
stigma pulvinate; capsule obovate, shortly stalked, 4–5 lin. long.
Drège. Zwei Pfl. Documente, 100, 101; *Sond. in Linnæa*, xix. 103;
Drège in Linnæa, xx. 219; *N. E. Br. in Gard. Chron.* 1885, xxiv.
331; *Durand & Schinz, Conspect. Fl. Afr.* v. 113; *Kränzl. Orch.
Gen. et Sp.* i. 726, *and in Ann. Naturhist. Hofmus. Wien*, xx. 5;
Schlechter in Engl. Jahrb. xxxi. 300; *Bolus, Ic. Orch. Austr.-Afr.*
ii. *t.* 92. *Orchis flexuosa, Linn. Pl. Afr. Rar.* 26, *Amœn. Acad.* vi.
108, *and Sp. Pl. ed.* ii. 1331. *Satyrium flexuosum, Thunb. Prodr.* 5.
Disa flexuosa, Sw. in Vet. Acad. Handl. Stockh. 1800, 212; *Thunb.
Fl. Cap. ed. Schult.* 9; *Bolus in Trans. S. Afr. Phil. Soc.* v. 160,
Orch. Cap. Penins. 160, *and Journ. Linn. Soc.* xxv. 201.

SOUTH AFRICA: without locality, *Bergius, Masson*! *Roxburgh*! *Leibold, Ludwig,
Rogers*! *Prior*! *Trimen*! *Faure*! *Pappe*! *Lehmann*! *Sieber*, 449!
COAST REGION: Malmesbury Div.; between Malmesbury and Groene Kloof, 300
ft., *Thunberg*! *Bolus*, 4332! *Schlechter*, 1661, *Kässner*; near Hopefield, *Schlechter*,
5309! Tulbagh Div.; mountains near Tulbagh Waterfall, 600–1500 ft., *Kässner,
Schlechter*! Worcester Div.; near Hex River, 1600 ft., *Tyson*, 642. Paarl Div.;
Klein Drakenstein Mountains, under 1000 ft., *Drège*, 1238! Flats between
Paarl and French Hoek, under 500 ft., *Drège*! French Hoek, *Grey*! near Paarl,
MacOwan, 2475! and in *MacOwan & Bolus, Herb. Norm. Austr.-Afr.* 172!
Klapmuts, *Wolley-Dod*, 496! Cape Div.; Cape Flats, about Doorn Hoogte,
Zeyher, 3927! *Ecklon & Zeyher*, 63! *Pappe*, 17! near Rondebosch, *Bolus*;
between Cape Town and Wynberg, *Ecklon & Zeyher*! *Bolus.* Caledon Div.;
Caledon, *Prior.* Div.? Joosten Berg, *Pappe*!

2. **S. inflexum** (Lindl. Gen. & Sp. Orch. 360); plant 4–13 in.
high; stem slender, slightly flexuous; leaves radical, spathulate,
usually dull red at flowering time, ½–1 in. long, with elliptic or
elliptic-lanceolate obtuse limb and short petiole; scapes 4–13 in.
high, with a few narrowly lanceolate acute sheaths; spikes lax,
1–3-flowered; bracts oblong-lanceolate, acute, 3–5 lin. long; pedicels
4–6 lin. long; flowers medium-sized, delicate flesh-pink, with
carmine spots and lines on the lip and carmine tips to the petals;
dorsal sepal erect, somewhat recurved, oblong, obtuse, concave,

4 lin. long; spur oblong, obtuse, three-fourths as long as the limb; lateral sepals lanceolate-oblong, obtuse, $4\frac{1}{2}$ lin. long; petals erect, oblong, broader at the base and with a broad tooth in front, bidentate with a short tooth at the apex, $2\frac{1}{2}$ lin. long; lip subpandurate, $5\frac{1}{2}$ lin. long, with a dilated and concave base, limb oblong, somewhat crenulate, with a long subulate or filiform apex; column short; anther reflexed; rostellum short, erect; stigma pulvinate; capsule clavate-oblong, stalked, 4–5 lin. long.   *Sond. in Linnæa*, xix. 103; *Durand & Schinz, Conspect. Fl. Afr.* v. 113; *Kränzl. Orch. Gen. et Sp.* i. 727, *and in Ann. Naturhist. Hofmus. Wien*, xx. 6; *Schlechter in Engl. Jahrb.* xxxi. 304; *Bolus, Ic. Orch. Austr.-Afr.* ii. *t.* 93. *Disa torta, Sw. in Vet. Acad. Handl. Stockh.* 1800, 211, *partly.   D. inflexa, Mund ex Lindl. Gen. et Sp. Orch.* 360; *Bolus in Trans. S. Afr. Phil. Soc.* v. 162, *t.* 22, *fig.* 12–14, *Orch. Cap. Penins.* 162, *t.* 22, *fig.* 12–14, *and Journ. Linn. Soc.* xxv. 201.

SOUTH AFRICA : without locality, *Hesse*, 15! *Scully* !
COAST REGION : Cape Div. ; Table Mountain, 2000–3500 ft., *Burchell*, 655 ! *Zeyher, Harvey* ! *Pappe* ! *Bolus*, 3882 ! *Schlechter*, 123 ! and in *MacOwan & Bolus, Herb. Norm. Austr.-Afr.* 1375 ! *Kässner, Wolley-Dod*, 2125 ! *Dümmer*, 553 a ! Cape Flats, *Pappe* ! Tulbagh Div. ; Steendaal Mountains, *Pappe* ! Winterhoek Mountain, *Ecklon & Zeyher.* Stellenbosch Div. ; Hottentots Holland, *Prior* ! Caledon Div. ; Houw Hoek Mountains, 2200 ft., *Schlechter*, 5488 ! Palmiet River, 700 ft., *Schlechter*, 5415. Swellendam Div. ; mountain peak near Swellendam, *Burchell*, 7307 ! Riversdale Div. ; summit of Kampsche Berg, *Burchell*, 7107 ! George Div. ; mountains of Montagu Pass, *Rehmann*, 29 ! Eastern District of Cape Colony, *Hallack*, 2 !
CENTRAL REGION : Ceres Div. ; mountains near Ceres, *Bolus.*

3. **S. rigidum** (Lindl. Gen. & Sp. Orch. 360) ; plant 5–9 in. high ; stem slender, flexuous ; leaves radical, spathulate, $\frac{1}{2}$–1 in. long, with broadly elliptic-oblong obtuse or apiculate limb ; petiole 2–3 lin. long ; scapes 5–9 in. high, with 2 or 3 oblong-lanceolate acute suberect sheaths ; spikes 1–4-flowered, lax ; bracts ovate-lanceolate, acute or subacute, 3–4 lin. long ; pedicels 4–6 lin. long ; flowers rather large, pink, with yellow tips to the petals and four rows of violet dots on the lip ; dorsal sepal erect, oblong or obovate-oblong, obtuse, concave, 5–6 lin. long ; spur elongate-oblong from a narrowly conical base, subobtuse, somewhat curved, nearly as long as the limb ; lateral sepals falcate-oblong, subacute, 5–6 lin. long ; petals narrowly oblong from a broader base, unequally and shortly bifid at the apex, 4 lin. long ; lip narrowly subpandurate behind, somewhat concave at the base, 8 lin. long, abruptly narrowed from about the middle into a long linear subobtuse apical lobe ; column short ; anther reflexed ; rostellum short, erect ; stigma pulvinate. *Drège, Zwei Pfl. Documente*, 120 ; *Sond. in Linnæa*, xix. 103 ; *Drège in Linnæa*, xx. 220 ; *N. E. Br. in Gard. Chron.* 1885, xxiv. 331 ; *Schlechter in Engl. Jahrb.* xxxi. 303 ; *Bolus, Ic. Orch. Austr.-Afr.* ii. *tt.* 94, 95. *S. bifidum, Reichb. f. in Flora*, 1883, 460 ; *Durand & Schinz, Conspect. Fl. Afr.* v. 113 ; *Kränzl. Orch. Gen. et Sp.* i. 729, *partly, and in Ann. Naturhist. Hofmus. Wien*, xx. 6 ; *Schlechter in*

Engl. Jahrb. xxxi. 302, *partly, excl. var., not of Reichb. f. Satyrium bifidum, Thunb. Prodr.* 5. *Disa bifida, Sw. in Vet. Acad. Handl. Stockh.* 1800, 212; *Thunb. Fl. Cap. ed. Schult.* 9; *Bolus in Trans. S. Afr. Phil. Soc.* v. 163, *Orch. Cap. Penins.* 163, *and Journ. Linn. Soc.* xxv. 201.

SOUTH AFRICA: without locality, *Oldenburg*! *Thunberg*! *Masson*! *Brown, Gueinzius, Harvey, Prior, Trimen*! *Rogers*!
COAST REGION: Cape Div.; sandy places near Cape Town, *Bolus*, 3742! 7019 β! Table Mountain, *Bolus*! Stellenbosch Div.; Stellenbosch, *Miss Lloyd in Herb. Sanderson*, 930! *Fielder*! Lowrys Pass, 300–400 ft., *Schlechter*, 1175! Caledon Div.; Zwart Berg, 1000–2000 ft., *Zeyher*, 101! Swellendam Div.; Sparrbosch, under 1000 ft., *Drège*, 1231e! near Swellendam, *Zeyher*, 3929! *Borcherds.* Port Elizabeth Div.; near Port Elizabeth, *Mrs. Holland*, 7! *Kemsley.*

4. **S. obliquum** (Lindl. Gen. & Sp. Orch. 359); plant 3–11 in. high; stem slender, flexuous; leaves radical, spathulate, ½–1 in. long, with ovate-oblong or rounded subobtuse limb; petiole 3–5 lin. long; scapes 3–11 in. high, with 2 or 3 oblong-lanceolate suberect sheaths; spikes 1–6-flowered, lax; bracts ovate-lanceolate, acute, 3–4½ lin. long; pedicels 5–8 lin. long; flowers medium-sized, flesh-coloured with purple blotches on the lip, a darker apex and some purple lines at the apex of the petals; dorsal sepal erect, elliptic-oblong or lanceolate-oblong, subacute, 3½–4 lin. long; spur clavate-oblong, nearly as long as the limb; lateral sepals oblong or lanceolate-oblong, subobtuse, 4½–5½ lin. long; petals oblong, unequally bilobed at the apex, with an ample rounded or angular basal limb in front, 2¼–2½ lin. long; lip subpandurate-oblong, 4–5 lin. long, concave at the base, terminating in a long linear obtuse fleshy apex; column short; anther reflexed; rostellum short, erect; stigma pulvinate. *Sond. in Linnæa,* xix. 103; *Drège in Linnæa,* xx. 220; *Durand & Schinz, Conspect. Fl. Afr.* v. 114, *excl. syn. Lindl.*; *Kränzl. Orch. Gen. et Sp.* i. 728, *and in Ann. Naturhist. Hofmus. Wien,* xx. 6. *S. Gueinzii, Reichb. f. in Linnæa,* xx. 694, *and Walp. Ann.* i. 804; *Kränzl. Orch. Gen. et Sp.* i. 730. *S. bifidum, Schlechter in Engl. Jahrb.* xxxi. 302, *partly, excl. syn., not of Reichb. f.*; *Bolus, Ic. Orch. Austr.-Afr.* ii. *t.* 96, *excl. all syn. Disa obliqua, Bolus in Trans. Phil. Soc.* v. 162, *partly, Orch. Cap. Penins.* 162, *partly, and Journ. Linn. Soc.* xxv. 201. *D. Gueinzii, Bolus in Journ. Linn. Soc.* xxv. 201.

SOUTH AFRICA: without locality, *Brown*! *Bunbury*! *Trimen, Menzies*! *Gueinzius, Rogers*! *Alexander*! *Zeyher*, 3930!
COAST REGION: Tulbagh Div.; Tulbagh, *Schlechter*! Cape Div.; Table Mountain, 1800–2500 ft., *Harvey*, 118! *Bolus*! *Schlechter*, 1314! Simons Bay, *Wright*, 132! Muizen Berg, 1200 ft., *MacOwan & Bolus, Herb. Norm. Austr.-Afr.* 173! Cape Flats, at Doorn Hoogte, *Zeyher*, 1571! *Pappe*, 18! Wynberg, *Prior*! Devils Peak, *Wilms*, 3662! Flat near Claremont, *Wolley-Dod*, 606! Farmer Pecks Valley, *Wolley-Dod*, 1390! Mountains near Nord Hoek, *Wolley-Dod*, 2746! Caledon Div.; Genadendal, *Pappe*! behind Steenberg Crag, *Wolley-Dod*, 1291!

5. **S. obtusatum** (Lindl. Gen. & Sp. Orch. 359); plant 3–5 in. high; stem slender, flexuous; leaves radical, spathulate, 4–7 lin.

long, with an ovate subacute limb and very short broad petiole; scapes 3–5 in. high, with 2 or 3 lanceolate-oblong subacute sheaths; spikes 1–4-flowered, lax; bracts ovate-lanceolate, acute, 2½–3 lin. long; pedicels 3–4 lin. long; flowers small; dorsal sepal erect, elliptic-oblong, subobtuse, concave, 3 lin. long; spur clavate, obtuse, rather shorter than the limb; lateral sepals oblong, subobtuse, 3 lin. long; petals narrowly oblong, very obliquely apiculate, with a prominent rounded angle in front, 2 lin. long; lip subpandurate, 3 lin. long, somewhat concave at the base, abruptly terminating in a short linear fleshy obtuse apex; column short; anther reflexed; rostellum short, erect; stigma pulvinate. *Drège, Zwei Pfl. Documente,* 81, 83; *Sond. in Linnæa,* xix. 103; *Drège in Linnæa,* xx. 219; *Kränzl. Orch. Gen. et Sp.* i. 729. *S. bifidum, Schlechter in Engl. Jahrb.* xxxi. 302, *partly, excl. var., not of Reichb. f.*

SOUTH AFRICA: without locality, *Hesse!*
COAST REGION: Worcester Div.; Dutoits Kloof, 2000–3000 ft., *Drège,* 1231 c! Paarl Div.; Drakenstein Mountains, 3000–4000 ft., *Drège.* Caledon Div.; Klein River Mountains, 1000–2000 ft., *Zeyher,* 3930! Genadendal, *Roser!*

6. **S. clavigerum** (Lindl. Gen. & Sp. Orch. 360); plant 4–5 in. high; stem slender, flexuous; leaves radical, spathulate, 4–6 lin. long, with ovate subacute limb and rather short petiole; scapes 4–5 in. high, with 2 or 3 lanceolate-oblong subacute sheaths; spikes 1–4-flowered, lax; bracts ovate-lanceolate, acute, 2½–3 lin. long; pedicels 4–5 lin. long; flowers small, flesh-pink, with some darker spots on the lip, the apex of the lip and petals red; dorsal sepal erect, elliptic-oblong, subacute, 3 lin. long; spur clavate or sub-clavate, obtuse, two-thirds as long as the limb; lateral sepals oblong, subacute, 3½ lin. long; petals narrowly oblong, unequally and shortly bifid at the apex, with an ample rounded basal lobe in front, 2 lin. long; lip pandurate-oblong, 3 lin. long, concave at the base, terminating in a short linear obtuse fleshy apex in front; column short; anther reflexed; rostellum short, erect; stigma pulvinate. *Drège, Zwei Pfl. Documente,* 102; *Sond. in Linnæa,* xix. 103; *Durand & Schinz, Conspect. Fl. Afr.* v. 113; *Kränzl. Orch. Gen. et Sp.* i. 727, *and in Ann. Naturhist. Hofmus. Wien,* xx. 6. *S. bifidum, var. clavigerum, Schlechter in Engl. Jahrb.* xxxi. 303. *Disa clavigera, Bolus in Trans. S. Afr. Phil. Soc.* v. 163, *Orch. Cap. Penins.* 163, *Journ. Linn. Soc.* xxv. 201, *and Ic. Orch. Austr.-Afr.* i. t. 34.

SOUTH AFRICA: without locality, *Bowie,* 6!
COAST REGION: Paarl Div.; between Paarl Mountain and Paarde Berg, *Drège,* 1231 d! Cape Div.; rocky places on Table Mountain, *Ecklon & Zeyher.* Stellenbosch Div.; near Stellenbosch, 300 ft., *Miss Farnham in Herb. Bolus,* 5923! Steenbrass River, 1500 ft., *Schlechter,* 5386! near Lowrys Pass, 6000 ft., *Schlechter,* 1144! Caledon Div.; slopes of Zwart Berg, 1700 ft., *Bolus.*

7. **S. arcuatum** (Lindl. Gen. & Sp. Orch. 359); plant 4–9 in. high; stem very slender, rigid, flexuous; leaves radical, spathulate, 4–9 lin. long, with elliptic-lanceolate or ovate-elliptic subobtuse

limb and short petiole ; scapes 4–9 in. high, with a few short
lanceolate-oblong suberect sheaths ; spikes short, lax, 1–4-flowered :
bracts oblong-lanceolate, acute, 3–4 lin. long ; pedicels 4–8 lin.
long ;
flowers medium-sized, pale lilac to rose-red with some greenish
spots on the base of the lip, and purple spots on the petals ; dorsal
sepal recurved from an erect base, linear-lanceolate, subobtuse, with
incurved margins, 5 lin. long ; spur oblong, obtuse, more than half
as long as the limb ; lateral sepals linear-lanceolate from a broader
base, subobtuse, obliquely recurved and with incurved margins,
5–5½ lin. long ; petals erect, with an auriculate lobe in front, then
sharply bent and angled behind, shortly and unequally bifid at the
apex, 2½ lin. long ; lip pandurate, 4½ lin. long, base dilated and
concave, limb orbicular, crenulate, concave, 1¼ lin. broad, terminating
in a long subulate or filiform beak at the apex ; column short ;
anther reflexed ; rostellum short and broad ; stigma pulvinate ;
capsule obovate-oblong, stalked, 4–5 lin. long. *Drège, Zwei Pfl.
Documente*, 120 ; *Sond. in Linnæa*, xix. 103 ; *Drège in Linnæa*, xx.
219 ; *Krauss in Flora*, 1845, 306, *and Fl. Cap- und Natal.* 157.
S. biflorum, Durand & Schinz, Conspect. Fl. Afr. v. 113 ; *Kränzl.
Orch. Gen. et Sp.* i. 725, *and in Ann. Naturhist. Hofmus. Wien*, xx.
5 ; *Schlechter in Engl. Jahrb.* xxxi. 301. *Satyrium cornutum, Linn.
Pl. Afr. Rar.* 27. *Orchis satyroides, Linn. Amœn. Acad.* vi. 109.
Orchis biflora, Linn. Sp. Pl. ed. ii. 1330. *Satyrium tortum, Thunb.
Prodr.* 5. *Disa torta, Sw. in Vet. Acad. Handl. Stockh.* 1800, 211 ;
Thunb. Fl. Cap. ed. Schult. 10, *partly* ; *Bolus in Trans. S. Afr. Phil.
Soc.* v. 161, *Orch. Cap. Penins.* 161, *Journ. Linn. Soc.* xxv. 201, *and
Ic. Orch. Austr.-Afr.* i. *t.* 33.

SOUTH AFRICA : without locality, *Oldenburg* ! *Mund et Maire.*
COAST REGION : Clanwilliam Div. ; Elandsfontein, *Penther*, 256 ! Piquetberg Div. ;
Piquet Berg, 400 ft., *Schlechter*, 7894 ; Malmesbury Div. ; Groene Kloof, *Thun-
berg* ! *Drège.* Tulbagh Div. ; near Artois, 600 ft., *Bolus*, 5445 ! flat near Tulbagh
Road, 400 ft., *Schlechter*, 1433 ! Cape Div. ; mountains near Constantia, 2000 ft.,
Krauss, 1326. Stellenbosch Div. ; near Stellenbosch, 300 ft., *Bolus*, 5922 !
Miss Lloyd in Herb. Sanderson, 929 ! *Drège, Miss Farnham in Herb. Bolus*, 5922 !
and in *MacOwan & Bolus, Herb. Norm. Austr.-Afr.* 1376 ! *Marloth, Prior* !
Caledon Div. ; Caledon, under 1000 ft., *Drège* ; Klein River Mountains, 1000–
3000 ft., *Zeyher*, 3928 ! hills near Bot River, 400 ft., *Bolus*, 5444 ! Zwart Berg,
near the Hot Springs, 1000–2000 ft., *Ecklon & Zeyher*, 64 ! Genadendal, *Roser* !
WESTERN REGION : Vanrhynsdorp Div. ; near Olifants River, 400–1000 ft.,
Schlechter, 5012, 5103.

8. **S. longipetalum** (Lindl. Gen. & Sp. Orch. 359) ; plant 3–6 in.
high ; stem slender, flexuous ; leaves radical, spathulate, 4–8 lin.
long, with lanceolate-oblong or elliptic obtuse or subacute, very
crisped-undulate limb and short petiole ; scapes 3–6 in. high, with
2 or 3 oblong-lanceolate acute sheaths ; spikes 1–5-flowered, lax ;
bracts ovate-lanceolate, acuminate, 3–3½ lin. long ; pedicels 3–4 lin.
long ; flowers medium-sized, greenish-yellow, with purple spots on
the limb of the lip and base of the lateral sepals, the filiform
processes of the lip and petals dull purple ; dorsal sepal erect,
oblong-lanceolate, acute, recurved, 3½–4 lin. long ; spur conical-

oblong, subobtuse, more than half as long as the limb; lateral
sepals oblong-lanceolate, very acuminate, 5 lin. long; petals broadly
oblong at the base, very unequally bilobed, limb 1½ lin. long, front
lobe filiform, more than twice as long as the limb, back lobe short
and obtuse; lip pandurate, 5–5½ lin. long, concave at the base;
limb obcordate, undulate, 2 lin. long; terminal appendage filiform,
4 lin. long; column short; anther reflexed; rostellum erect,
acutely tridentate; stigma pulvinate. *Drège, Zwei Pfl. Documente,*
86; *Durand & Schinz, Conspect. Fl. Afr.* v. 114; *Kränzl. Orch. Gen.
et Sp.* i. 725. *S. antenniferum, Schlechter in Engl. Jahrb.* xxiv. 426,
and xxxi. 305; *Bolus, Ic. Orch. Austr.-Afr.* i. *t.* 88. *S. bifidum,
Schlechter in Engl. Jahrb.* xxxi. 302, *partly, not of Reichb. f.   Disa
longipetala, Bolus in Journ. Linn. Soc.* xxv. 201.

COAST REGION: Piquetberg Div.; plain at summit of Piquetberg Range,
1600 ft., *Schlechter,* 5248! Paarl Div.; Paarl Mountain, 1000–2000 ft., *Drège,*
8273!

XXXVIII. BROWNLEEA, Harv.

Odd *sepal* superior, galeate, extending behind in an extinguisher-
shaped or elongated spur; lateral sepals free, more or less spreading.
Petals erect, more or less adhering to the inner sides of the mouth
of the dorsal sepal and adnate to the column at the base. *Lip*
usually narrow or minute, entire, sometimes dilated and clasping
the base of the column, spurless. *Column* short. *Anther* horizontal
or reflexed, with two parallel cells, produced in front into channels,
which continue up the back of the rostellum; pollinia solitary in
each cell, granular, attached by short caudicles to separate glands.
Rostellum erect, more or less distinctly bilobed; lobes notched at
the apex, tuberculate on the sides at the base. *Stigma* seated at
the base of the rostellum, vertical or oblique, cushion-shaped.
Capsule cylindrical or oblong, ribbed.

Terrestrial herbs, with simple sessile tubers; stem rather slender, erect;
leaves cauline, few to several, usually more or less spreading, broad or narrow;
flowers in dense or lax spikes or racemes, rarely solitary, small, medium-sized or
occasionally large; bracts usually narrow.

DISTRIB. Species about 13, the majority South African, with two in Tropical
Africa and one in Madagascar.

The genus chiefly differs from *Disa* in having the petals more or less adnate to
the dorsal sepal.

Spur about 2 lin. long:
  Flowers in a dense cylindrical spike; spur much
    curved ... ... (1) **parviflora**.
  Flowers in a subcapitate or shortly oblong spike; spur
    nearly straight ... ... (2) **Galpini**.
Spur ½ to over 1 in. long:
  Spikes several- to many-flowered:
    Lip minute, much shorter than the lateral sepals:
      Leaves linear-lanceolate to oblong-lanceolate,
        usually suberect:

Flowers pink or rose - coloured, unspotted;
  petals not crenate at the apex:
    Spur about as long as the dorsal sepal; petals
      falcate-oblong   ...   . (3) **recurvata.**
    Spur much longer than the dorsal sepal;
      petals ovate-lanceolate   ...   . (4) **Nelsoni.**
Flowers markedly spotted (in the dried state);
  petals crenate at the apex ...   . (5) **natalensis.**
Leaves elliptic-lanceolate or ovate, spreading, very
  membranous ; spur ⅔-1 in. long:
    Leaves ⅓-1 in. broad ; flowers pink and white   (6) **Woodii.**
    Leaves ¾-2 in. broad ; flowers lilac-blue with
      violet spots ...   . (7) **cœrulea.**
Lip linear, as long as the lateral sepals   ...   ... (8) **pentheriana.**
Spikes 1- to 2-flowered:
    Leaves suberect ; spur 1-1¼ in. long   ...   ... (9) **macroceras.**
    Leaves recurved ; spur over 1½ in. long   ...   ... (10) **monophylla.**

1. B. parviflora (Harv. ex Lindl. in Hook. Lond. Journ. Bot.
i. 16) ; plant 1-1½ ft. high ; stem rather slender ; leaves cauline,
1-2, suberect, sessile, lanceolate or oblong-lanceolate, acute or
acuminate, 3-5 in. long, ⅓-¾ in. broad ; scapes 1-1½ ft. high, with
2 or 3 narrow acuminate sheaths above ; spikes cylindrical, 2-4 in.
long, dense, many-flowered ; bracts ovate-lanceolate, very acuminate,
⅓-½ in. long ; pedicels about 2 lin. long ; flowers very small, white,
with the spur and nerves of the petals greenish, and some purple
on the anther ; dorsal sepal galeate, ovate, subacute, somewhat
incurved, 2 lin. long ; spur much curved, stout or subclavate, obtuse,
rather longer than the limb ; lateral sepals ovate-elliptic, acute,
united to the middle, 2 lin. long ; petals obliquely ovate, subobtuse,
subauriculate at the base in front, cohering to the dorsal sepal,
2 lin. long ; lip minute, linear from a rather broader base, not
½ lin. long ; column short ; anther reflexed ; rostellum ovate, erect,
emarginate, about ½ lin. long, bituberculate at the base ; stigma
pulvinate ; capsule oblong, about 3 lin. long. *Sond. in Linnæa,*
xix. 107 ; *Bolus in Journ. Linn. Soc.* xxv. 204, *and Ic. Orch. Austr.-*
Afr. i. *t.* 43 ; *Durand & Schinz, Conspect. Fl. Afr.* v. 114 ; *Kränzl.*
Orch. Gen. et Sp. i. 733, *and in Ann. Naturhist. Hofmus. Wien,* xx. 6 ;
Schlechter in Engl. Jahrb. xxxi. 312. *Disa parviflora, Reichb. f.*
Otia Bot. Hamb. ii. 119.

COAST REGION : Albany Div. ; near Grahamstown, 2400 ft., *Glass.* Bedford
Div. ; Kaga Berg, 4500 ft., *Weale,* 913 ! Victoria East Div. ; Hogs Back,
Victoria East, *Scully,* 234 ! Stockenstrom Div. ; Kat Berg, 2000-3000 ft., *Zeyher,*
Hutton ! Queenstown Div. ; Hanglip, near Queenstown, 4500 ft., *Galpin,* 1506 !
Woodvale, 4600 ft., *Galpin,* 8182 ! Stutterheim Div. ; Dohne Hill, *Sim,* 23.
King Williamstown Div. ; near King Williamstown, *Brownlee.* British Kaffraria,
Brownlee !
  CENTRAL REGION : Somerset Div. ; Bosch Berg, 4300 ft., *MacOwan,* 1530 !
  KALAHARI REGION : Transvaal ; Saddleback Range, near Barberton, 4500-
5000 ft., *Galpin,* 1229, *Culver,* 49 ; Houtbosch Mountains, above Mamavolo,
6800 ft., *Schlechter,* 4711 ! Belfast, *Miss Doidge,* 4797 !

EASTERN REGION : Transkei ; Tsomo, *Mrs. Barber*, 848 ! Griqualand East, Ingeli Range, 4500–6000 ft., *Tyson*, 1080 ! Enyembi Bush, in the Umzinkulu District, 4500–5000 ft., *Tyson*, 2066 ! Natal ; dry rocks near Great Noodsberg, *Buchanan* ! stony places near Van Reenan, 7000 ft., *Schlechter*, 6942 ; grassy hills at Umkomaas, 4000–5000 ft., *Wood*, 971 !

2. **B. Galpini** (Bolus, Ic. Orch. Austr.-Afr. i. t. 42) ; plant 1½–2¼ ft. high ; stem rather slender ; leaves cauline, 2–3, suberect, sessile, lanceolate or linear-lanceolate, acute or acuminate, 2–4 in. long, 2–3 lin. broad, the upper usually small and narrow ; scapes 1½–2¼ ft. high ; spikes subcapitate or shortly oblong, 1–1½ in. long, dense, many-flowered ; bracts ovate-lanceolate, acute or acuminate, 5–8 lin. long ; pedicels 4–5 lin. long ; flowers small, white, with a few purple spots on the petals ; dorsal sepal galeate, lanceolate, acuminate, 3½ lin. long, somewhat recurved near the apex ; spur cylindrical from a conical base, obtuse, about 2 lin. long, slightly curved ; lateral sepals oblong-lanceolate, subobtuse, 3½ lin. long ; petals oblique, oblong, obtuse, very undulate on the outer margin, somewhat broader near the base, cohering with the margin of the dorsal sepal to near the apex, 3½ lin. long ; lip erect, linear, minute, about ½ lin. long ; column short ; anther reflexed ; rostellum erect, 1 lin. long, with parallel oblong lobes, bitubercled at the base ; stigma pulvinate ; capsule fusiform-oblong, 5–6 lin. long. *Kränzl. Orch. Gen. et Sp.* i. 733, *and in Ann. Naturhist. Hofmus. Wien*, xx. 6 ; *Schlechter in Engl. Jahrb.* xxxi. 310.

VAR. **major** (Bolus, Ic. Orch. Austr.-Afr. i. under t. 42) ; flowers a fourth larger than in the type, with broader more lobed and concave petals, and the lip twice as large. *Kränzl. Orch. Gen. et Sp.* i. 733 ; *Schlechter in Engl. Jahrb.* xxxi. 310.

KALAHARI REGION : Orange River Colony ; Mont aux Sources, 7000–8000 ft., *Thode.* Transvaal ; damp slopes near Barberton, 4500 ft., *Galpin*, 1255 ! marshes on Houtbosch Mountains, 7200 ft., *Schlechter*, 4391 ! Bamboo Mountain, 6500 ft., *Miss Doidge*, 5575 !

EASTERN REGION : Griqualand East ; mountains near Kokstad, 5000 ft., *Tyson*, 1084 ! Natal ; hills near Van Reenans Pass, 4000–6000 ft., *Wood*, 5660 ! *Krook*, 88 ! Var. β : Griqualand East ; southern slopes of Mount Currie, 6000 ft., *Tyson*, 1074.

3. **B. recurvata** (Sond. in Linnæa, xix. 107) ; plant 1–1½ ft. high ; stem rather slender ; leaves cauline, 1–3, suberect, sessile, lanceolate or linear-lanceolate, attenuate or acuminate, 2½–4 in. long, 3–5 lin. broad, the upper much smaller ; scapes 1–1¼ ft. high ; spikes 2–3½ in. long, rather lax, usually many-flowered ; bracts ovate-lanceolate, acute or acuminate, ½–1 in. long ; pedicels 5–6 lin. long ; flowers medium-sized, pale pink, sometimes streaked with purple, and the anther red-brown ; dorsal sepal galeate, ovate-lanceolate, acuminate, 4½–5 lin. long, somewhat recurved near the apex ; spur cylindrical from a conical base, obtuse, 4 lin. long, somewhat curved ; lateral sepals ovate-lanceolate, subacute, 5 lin. long ; petals oblique, falcate-oblong, obtuse, somewhat undulate on the outer margin above, subauriculate below, cohering to the margin

of the dorsal sepal to near the apex, $4\frac{1}{2}$ lin. long; lip erect, linear,
minute, about $\frac{1}{2}$ lin. long; column short; anther reflexed, clavate;
rostellum erect, $1\frac{1}{4}$ lin. long, with parallel oblong arms, bituberculate at the base; stigma pulvinate; capsule fusiform-oblong,
5–6 lin. long. *Reichb. f. in Walp. Ann.* i. 800; *Harv. Thes. Cap.* ii.
3, *t.* 104; *Bolus in Journ. Linn. Soc.* xxv. 204, *and Ic. Orch. Austr.-
Afr.* i. *t.* 41; *Durand & Schinz, Conspect. Fl. Afr.* v. 114; *Kränzl.
Orch. Gen. et Sp.* i. 732; *Schlechter in Engl. Jahrb.* xxxi. 309. *Disa
recurvata, Reichb. f. Otia Bot. Hamb.* ii. 119.

COAST REGION: Uitenhage Div.; Elands River Mountains, *Ecklon & Zeyher*!
Albany Div.; above Howison's Poort, near Grahamstown, 2400 ft., *Glass*, and in
MacOwan & Bolus, Herb. Norm. Austr.-Afr. 1377! Stockenstrom Div.; Kat
Berg, 3000–4000 ft., *Hutton*! Cathcart Div.; Toise River Station, 4000 ft.,
Flanagan, 2256. Eastern Frontier of Cape Colony, *MacOwan*, 526!
CENTRAL REGION: Somerset Div.; Bosch Berg, 4500 ft., *MacOwan*, 526!
Scott-Elliot, 483!
KALAHARI REGION: Basutoland; near Satsannas Peak, Drakensberg Range,
8750 ft., *Galpin*, 6840!
EASTERN REGION: Tembuland; Bazeia Mountains, 3500 ft., *Baur*, 625!

4. B. Nelsoni (Rolfe); plant about 8 in. high; stem rather
slender; leaves cauline, 2 or 3, suberect, sessile, lanceolate or oblong-
lanceolate, subacute, $2-3\frac{1}{2}$ in. long, 3–4 lin. broad, the upper much
smaller; scapes about 8 in. high; spikes oblong, rather lax, about
10-flowered; bracts ovate-lanceolate, very acuminate, 5–8 lin. long;
pedicels about 5 lin. long; flowers medium-sized, rose-purple; dorsal
sepal galeate, ovate-lanceolate, acuminate, about 4 lin. long; spur
cylindrical from a conical base, obtuse and somewhat clavate at
the apex, somewhat curved, 6–8 lin. long; lateral sepals elliptic-
ovate, shortly acuminate, about 4 lin. long; petals oblique, ovate-
lanceolate, subacute, cohering with the margin of the dorsal sepal
to near the apex, with an acute angle below the middle behind,
$3\frac{1}{2}$ lin. long; lip erect, $\frac{1}{2}$ lin. long, 3-lobed, with linear obtuse front
lobe and short rounded side lobes; column short; anther reflexed;
rostellum erect, rather broad, 1 lin. long, 3-dentate; stigma
pulvinate.

KALAHARI REGION: Transvaal; without precise locality, *Nelson*, 16!

5. B. natalensis (Rolfe); plant $\frac{3}{4}$–1 ft. high; stem rather slender;
leaves cauline, 2, suberect, sessile, linear-lanceolate, acuminate, 3–4
in. long, 2–3 lin. broad; scapes $\frac{3}{4}$–1 ft. high; spikes $1\frac{1}{2}$–$2\frac{1}{2}$ in. long,
lax, 6–8-flowered; bracts ovate-lanceolate, very acuminate, 7–10 lin.
long; pedicels 5–6 lin. long; flowers medium-sized, irregularly
spotted; dorsal sepal galeate, ovate-lanceolate, obtuse, somewhat
recurved near the apex, about 5 lin. long; spur cylindrical from a
conical base, obtuse, slightly clavate at the apex, 5–7 lin. long;
lateral sepals oblong-lanceolate, subacute, about 5 lin. long; petals
falcate-oblong, obtuse, crenate at the apex, auriculate at the base
in front, cohering with the margin of the dorsal sepal to near the

apex, 5 lin. long ; lip obovate, obtuse, ½ lin. long ; column short ; anther reflexed ; rostellum erect, 2-lobed, 1¼ lin. long, with bidentate lobes ; stigma pulvinate.

EASTERN REGION : Natal ; near Byrne, *Wood,* 3177 !

6. **B. Woodii** (Rolfe) ; plant ¾–1¼ ft. high ; stem slender ; leaves cauline, 3, more or less spreading, subsessile or shortly petiolate, lanceolate or elliptic-lanceolate, acuminate, 2–3¾ in. long, ⅓–1 in. broad, decreasing upwards into the bracts, membranous ; scapes ¾–1¼ ft. high ; spikes 1–2 in. long, lax, few- to 8-flowered ; bracts lanceolate or ovate-lanceolate, acuminate, 5–9 lin. long ; pedicels about ½ in. long ; flowers medium-sized, pink and white (*Wood*) ; dorsal sepal galeate, ovate-lanceolate, very acuminate, somewhat recurved, 4 lin. long ; spur cylindrical from a conical base, obtuse and shortly clavate, curved, 9–11 lin. long ; lateral sepals elliptic-lanceolate, subacute, somewhat curved, 4½ lin. long ; petals elliptic-oblong, obtuse, undulate on the upper margin, with an acute angle near the base behind, cohering to the margin of the dorsal sepal, 4 lin. long ; lip linear, minute ; column short ; anther reflexed ; rostellum erect, 3-lobed, with shortly bidentate side lobes ; stigma pulvinate ; capsule fusiform-oblong, 6–7 lin. long.

EASTERN REGION : Natal ; in stony bush near Murchison, 2000 ft., *Wood,* 1982 ! 3179 !

Distinguished from *B. cœrulea,* Harv., by its narrower leaves and smaller, differently-coloured flowers. The lip and appendages of the column are very membranous and difficult to make out from dried specimens.

7. **B. cœrulea** (Harv. ex Lindl. in Hook. Lond. Journ. Bot. i. 16) ; plant ½–1½ ft. high ; stem usually rather slender ; leaves cauline, 3 or 4, more or less spreading, ovate or elliptic-ovate, acute or acuminate, shortly petiolate, 2–6 in. long, ¾–2 in. broad, decreasing upwards into the bracts, membranous ; scapes ½–1½ ft. high ; spikes 1–3 in. long, lax, usually many-flowered ; bracts ovate-lanceolate, very acuminate, ½–1 in. long ; pedicels ½–¾ in. long ; flowers rather large, lilac-blue with darker spots ; dorsal sepal galeate, ovate-lanceolate, very acuminate, somewhat recurved, 4–5 lin. long ; spur cylindrical from a conical base, obtuse and shortly clavate at the apex, somewhat curved, ¾–1 in. long ; lateral sepals elliptic-lanceolate, acute, about 5 lin. long ; petals oblique, oblong-lanceolate, subobtuse, undulate on the upper margin, with an acute angle near the base behind, cohering with the margin of the dorsal sepal, 5 lin. long ; lip very short, linear, adnate to the column ; column short ; anther reflexed, somewhat curved at the apex ; rostellum erect, rather tall, with an acute tooth at the apex, bituberculate at the base ; stigma pulvinate ; capsule fusiform-oblong, 7–8 lin. long. *Harv. Thes. Cap.* ii. 2, *t.* 103, *excl. syn.* ; *Bolus in Journ. Linn. Soc.* xxv. 204, *and Ic. Orch. Austr.-Afr.* i. *t.* 40 ; *Bot. Mag. t.* 7309, *excl. syn.* ; *Durand & Schinz, Conspect. Fl. Afr.* v. 114, *excl. syn. Sond.* ;

Kränzl. Orch. Gen. et Sp. i. 731 ; *Schlechter in Engl. Jahrb.* xxxi.
308. *Disa cœrulea, Reichb. f. Otia Bot. Hamb.* ii. 119.

COAST REGION : Bathurst Div. ; kloof near Bathurst, *Atherstone*, 30 ! Albany
Div. ; Howisons Poort, 2000 ft., *Hutton* ! *Pappe* ! *MacOwan*, 232 ! *Glass* ; hills
near Grahamstown, *Mrs. Barber*, 504 ! Stockenstrom Div. ; Kat Berg, 2000 ft.,
Hutton ! King Williamstown Div. ; Perie Forest, 3000–3600 ft., *Scott-Elliot*, 913 !
Sim, 14 ; near King Williamstown, *Brownlee* ! British Kaffraria ; Hangmans Hill,
D'Urban, 107 ! and without precise locality, *Brownlee* ! Eastern Frontier of
Cape Colony, *Hutton* !
KALAHARI REGION : mountains near Barberton, 4000–5000 ft., *Galpin*, 1257,
Culver, 88 ; Houtbosch Berg, *Schlechter*.
EASTERN REGION : Tembuland ; near Bazeia, 2500–3000 ft., *Baur*, 638.
Natal ; near Krans Kop, *McKen*, 22 ! Ingoma, *Gerrard*, 1543 ! Great Noodsberg,
3000–4000 ft., *Wood*, 5379 ! Zululand ; Ngoya, 1000–2000 ft., *Wylie in Herb.
Wood*, 7501 !

B. macroceras, Sond., and *B. madagascarica*, Ridl., have been cited as
synonyms of *B. cœrulea*, Harv., but I believe that both are distinct.

8. **B. pentheriana** (Kränzl. in Ann. Naturhist. Hofmus. Wien,
xx. 6) ; plant 1 ft. high, slender ; leaves subradical and cauline, the
2 lower linear-lanceolate, acute, narrowed into a petiole at the base,
about 4 in. long, 5 lin. broad, with 2 or 3 smaller sheath-like leaves
above, decreasing into the bracts ; scapes 1 ft. high ; spikes with
few remote flowers ; bracts oblong-lanceolate, acute, convolute, $\frac{3}{4}$ in.
long ; pedicels longer than the bracts ; flowers medium-sized ;
dorsal sepal broadly oblong, rounded at the apex, flat or scarcely
concave ; spur conical, attenuate, obtuse and somewhat inflated at
the apex, sepal and spur together about 14 lin. long ; lateral sepals
oblong, acute, scarcely concave, $3\frac{1}{2}$ lin. long ; petals oblique, narrowed
upwards from a broad rounded base, obtuse at the apex, $2\frac{1}{2}$ lin.
long ; lip linear, obtuse, $3\frac{1}{2}$ lin. long ; column short ; rostellum
rectilinear in front, with involute margins.

COAST REGION : George Div. ; Montague Pass, *Penther*, 189.

9. **B. macroceras** (Sond. in Linnæa, xix. 106) ; plant $\frac{1}{2}$–1 ft. high ;
stem slender, somewhat flexuous ; leaves cauline, 2 or 3, suberect,
oblong-lanceolate, acute or acuminate, sheathing at the base, the
lower $2\frac{1}{2}$–3 lin. long, 4–6 lin. broad, the upper much smaller ; scapes
$\frac{1}{2}$–1 ft. high, 1- or 2-flowered ; bracts ovate-lanceolate, acuminate,
$\frac{3}{4}$ in. long ; pedicels $\frac{3}{4}$–1$\frac{1}{4}$ in. long ; flowers rather large ; dorsal
sepal galeate, lanceolate, subobtuse, somewhat recurved, about $\frac{1}{2}$ in.
long ; spur cylindrical from a narrowly conical base, obtuse and
shortly subclavate at the apex, curved, 1–1$\frac{1}{4}$ in. long ; lateral sepals
oblong-lanceolate, subobtuse, not diverging, $\frac{1}{2}$ in. long ; petals
obliquely oblong, subobtuse, cohering with the margin of the dorsal
sepal, about $\frac{1}{2}$ in. long ; lip very narrow, 1 lin. long ; column 1$\frac{1}{2}$ lin.
long ; anther reflexed ; rostellum erect, oblong, shortly 2-lobed ;
stigma pulvinate. *Reichb. f. in Walp. Ann.* i. 800. *Disa macroceras, Reichb. f. Otia Bot. Hamb.* ii. 119.

COAST REGION: Stockenstrom Div. ; Kat River Mountains, *Zeyher* !

Quite distinct from *B. cærulea*, Harv., under which Harvey and all subsequent authors have included it, perhaps without seeing an original specimen.

10. **B. monophylla** (Schlechter in Engl. Jahrb. xx. Beibl. 50, 18) ; plant 6–7 in. high ; stem slender ; leaves cauline, 1 or 2, spreading or recurved, narrowly lanceolate or linear, acuminate, the lower 2 in. long, the upper smaller ; scapes 6–7 in. high, 1- or 2-flowered ; bracts spreading or recurved, ovate-lanceolate, very acuminate, ¾–1 in. long ; pedicels about 1 in. long ; flowers medium-sized, white ; dorsal sepal galeate, linear-lanceolate from a broader base, acute, recurved, 5 lin. long ; spur cylindrical from a narrowly conical base, obtuse and shortly clavate at the apex, nearly straight, over 1½ in. long ; lateral sepals oblong-lanceolate, subacute, 7 lin. long ; petals obliquely lanceolate-oblong, acute, upper margin undulate, cohering with the margin of the dorsal sepal, 5 lin. long ; lip erect, linear, minute ; column very short ; anther reflexed, shorter than the rostellum ; rostellum large, erect, 2-partite at the apex ; stigma pulvinate. *Schlechter in Engl. Jahrb.* xxxi. 307.

KALAHARI REGION : Orange River Colony ; rocky grassy ledges, near summit of Mopedis Peak, Witzies Hoek, *Thode*, 52!

XXXIX. CERATANDRA, Eckl.

Odd *sepal* inferior, deflexed, rather narrow, concave ; lateral sepals approximate or connivent at the margin into a broad limb and erect behind the lip. *Petals* narrow, concave, united with the dorsal sepal into a concave deflexed limb. *Lip* superior, broadly unguiculate and adnate to the column between the arms of the rostellum ; limb free, ascending, broadly cordate or auriculate below, with an obovate-oblong truncate fleshy appendage about the middle. *Column* suberect or spreading, short, produced upwards into a pair of suberect oblong rostellary arms holding at their apex the glands of the pollinia. *Anther-cells* lying on the arms of the rostellum and opening by a narrow slit ; pollinia coarsely granular, didymous, attached by long slender stipites to broad glands situated at the apex of the rostellary arms. *Stigmas* two, cushion-shaped, situated on the face of the column near the base. *Capsule* oblong, ribbed.

A stout terrestrial herb ; leaves numerous, cauline, narrow, suberect or some-what spreading ; flowers medium-sized, yellow, borne in a dense oblong spike or raceme ; bracts linear.

DISTRIB. A single species, limited to the south-western corner of Cape Colony.

The genus is here limited to the original *C. chloroleuca*, Eckl., which differs from those subsequently added by its posticous lip, and by the peculiar structure of the column.

1. **C. chloroleuca** (Eckl. ex Bauer, Ill. Orch. Gen. t. 16); plant
⅓–1 ft. high; stem rather stout, somewhat flexuous; leaves radical
and cauline, numerous, suberect or the lower spreading, linear,
acuminate, 1–3½ in. long; scapes ⅓–1 ft. high; spikes 1–6 in. long,
dense, many-flowered; bracts linear, acuminate, ½–1 in. long;
pedicels 5–6 lin. long; flowers medium-sized, sepals green, rest of
flower bright yellow; odd sepal oblong-lanceolate, subobtuse, 5½ lin.
long; lateral sepals spreading, ovate-oblong, obtuse, concave or the
upper margin conduplicately folded, 5 lin. long; petals elliptic-
lanceolate, subobtuse, subauriculate above the base, conduplicately
folded and concave, 5½ lin. long; lip broadly unguiculate; limb
broadly cordate, subobtuse or apiculate, 2 lin. long, with oblong
obtuse basal auricles, appendage obovate-oblong, truncate, about
half as long as the limb; column short; arms of the rostellum
incurved and erect, oblong, obtuse, 3 lin. long, with recurved outer
margin; capsule oblong, 5–7 lin. long. *Lindl. Gen. & Sp. Orch.*
364; *Drège, Zwei Pfl. Documente,* 76, 82, 85; *Sond. in Linnæa,* xix.
108, xx. 220; *Bolus in Trans. S. Afr. Phil. Soc.* v. 193, *Orch. Cap.
Penins.* 193, *and Journ. Linn. Soc.* xxv. 209. *C. atrata, Durand &
Schinz, Conspect. Fl. Afr.* v. 123; *Schlechter in Bull. Herb. Boiss.*
1ʳᵉ *sér.* vi. 907; *Kränzl. Orch. Gen. et Sp.* i. 866. *C. auriculata,
Lindl. Gen. & Sp. Orch.* 364; *Kränzl. Orch. Gen. et Sp.* i. 867, *and
in Ann. Naturhist. Hofmus. Wien,* xx. 11. *Ophrys atrata, Linn.
Mant.* 121; *Thunb. Prodr.* 2. *Pterygodium atratum, Sw. in Vet.
Acad. Handl. Stockh.* 1800, 218; *Thunb. Fl. Cap. ed. Schult.* 24.
Hippopodium atratum, Harv. ex Lindl. Gen. & Sp. Orch. 364.

SOUTH AFRICA: without locality, *Masson*! *Brown, Mund, Bergius, Leibold,
Harvey*! *Sieber,* 29! *Thom,* 277! *Ecklon*! *Ecklon & Zeyher,* 67.
COAST REGION: Piquetberg Div.; Piquet Berg, 2000–3000 ft., *Drège.* Malmesbury Div.; Zwartland, *Thunberg*! between Klip Fontein and Drei Fontein, *Zeyher,*
1577! between Groene Kloof and Saldanha Bay, under 500 ft., *Drège,* ex *Schlechter.*
Tulbagh Div.; Winterhoek Mountains, *Pappe*! near Tulbagh Waterfall, 1000–
2000 ft., *Ecklon & Zeyher*! Worcester Div.; Dutoits Kloof, 3000–4000 ft., *Drège,*
1241 b! near Worcester, 2000 ft., *Ecklon & Zeyher.* Paarl Div.; Paarl Mountain,
1000–2000 ft., *Drège,* 1241 a! Cape Div.; hills and flats around Cape Town,
50–3500 ft., *Bolus,* 4546! *Ecklon & Zeyher, Rehmann,* 572! *Bunbury,* 261!
Prior! *Kässner, Schlechter,* 125, *Dümmer,* 555! *Ecklon,* 165! *Pappe*! *Wolley-
Dod,* 665! 2127! 2249! Red Hill, *Mrs. Jameson*! Simonstown, *Brown*! Stellen-
bosch Div.; Hottentots Holland, *Mund.* Caledon Div.; Genadendal, *Prior*!
Swellendam Div.; Langeberg Range, near Zuurbraak, 1500 ft., *Schlechter,* 5707.
Riversdale Div.; between Little Vet River and Garcias Pass, *Burchell,* 6900!
near Garcias Pass, *Burchell,* 7151! near Riversdale, 900 ft., *Schlechter,* 1727.
George Div.; Montagu Pass, *Schlechter,* 5811! *Penther,* 208; plains of George,
Bowie! *Penther,* 97. Knysna Div.; Knysna, 250 ft., *Newdigate in MacOwan,
Herb. Austr.-Afr.* 1651!

XL. CERATANDROPSIS, Rolfe.

Odd *sepal* superior, ovate or oblong, concave; lateral sepals broad,
free, somewhat spreading. *Petals* somewhat oblique, united with
the dorsal sepal into a concave spreading limb, more or less
auriculate at the outer base. *Lip* inferior, broadly unguiculate

and adnate to the column between the arms of the rostellum; limb free, broadly cordate-ovate, without an appendage. *Column* short, suberect or ascending, produced upwards into a pair of broad diverging arms. *Anther-cells* adnate to the arms of the rostellum; pollinia solitary in each cell, coarsely granular, attached by a slender stipes to a broad rounded gland situated at the apex of the rostellary arms. *Stigma* cushion-shaped, somewhat bilobed. *Capsule* oblong, ribbed.

Stout erect herbs; leaves numerous, cauline, narrow, suberect or somewhat spreading; flowers medium-sized or small, yellow or dull red and white, arranged in a shortly oblong or subcapitate spike; bracts linear.

DISTRIB. Species 2, limited to the south-western corner of Cape Colony.

Differs from *Ceratandra*, Eckl., in having an inferior unappendaged lip.

Flowers deep yellow; sepals about ½ in. long ...   ... (1) **grandiflora.**
Flowers dull red and white; sepals about ¼ in. long   ... (2) **globosa.**

1. **C. grandiflora** (Rolfe); plant 6–10 in. high; stem moderately stout; leaves radical and cauline, suberect or the lower spreading, very numerous and crowded, linear, acuminate, ¾–1¾ in. long; scapes 6–10 in. high; spikes 1–2½ in. long, dense, many-flowered; bracts linear from a broader base, long-acuminate, 6–9 lin. long; pedicels 5–6 lin. long; flowers rather large, deep yellow; dorsal sepal oblong-lanceolate, subobtuse, 6 lin. long; lateral sepals somewhat spreading, ovate-oblong, obtuse, concave and somewhat incurved at the base, 5½ lin. long; petals oblique, semiovate-oblong, subobtuse, prominently auriculate above the base, condupli-cately folded and concave, 6 lin. long; lip broadly unguiculate; limb broadly cordate-ovate, subobtuse or apiculate, convex but without appendage, 3 lin. long, broader than long; column short; arms of the rostellum incurved and erect, broadly oblong, obtuse, 2 lin. long, with recurved outer margin. *Ceratandra grandiflora, Lindl. Gen. & Sp. Orch.* 364; *Drège, Zwei Pfl. Documente*, 136; *Sond. in Linnæa*, xix. 108, xx. 220; *Bolus in Journ. Linn. Soc.* xxv. 209; *Durand & Schinz, Conspect. Fl. Afr.* v. 123; *Schlechter in Bull. Herb. Boiss.* 1ʳᵉ *sér.* vi. 909; *Kränzl. Orch. Gen. et Sp.* i. 869.

SOUTH AFRICA : without locality, *Auge* ! *Nelson* ! *Rogers* !
COAST REGION : George Div. ; hills near the Krakadakow, *Prior* ! Langeberg Range, above Montagu Pass, 1500 ft., *Schlechter*, 5790, *Penther, Krook*. Knysna Div. ; Hartbeeste Vlagte, *Mund*, 1820, Belvedere, *Rehmann*, 414 ! Uniondale Div. ; Lange Kloof, 2000–4000 ft., *Ecklon & Zeyher*, 68 ! Humansdorp Div. ; near Storms River, 200 ft., *Schlechter*. Uitenhage Div. ; Van Stadensberg, *Zeyher*, 284 ! Port Elizabeth Div. ; sandy plains at Algoa Bay, 50 ft., *MacOwan*, 1219 ! Cape Recife, *MacOwan* ! Walmer Plains, 50–100 ft., *Hallack* ! between Port Elizabeth and Van Stadens River, *Hall* ! Alexandria Div. ; Zuurberg Range, 2000–3000 ft., *Drège*, 8282 ! Albany Div. ; near Grahamstown, 2200–2400 ft., *MacOwan, Galpin*, 309, *Glass, Tuck* !

2. **C. globosa** (Rolfe); plant 4–14 in. high; stem rather stout, slightly flexuous; leaves radical and cauline, suberect or the lower spreading, numerous, linear, acuminate, 2–3 in.

long; scapes 4–14 in. high; spikes $\frac{1}{2}$–2 in. long, dense, many-flowered; bracts linear, acuminate, 5–9 lin. long; pedicels 5–6 lin. long; flowers rather small, sepals dull red, petals and lip white; dorsal sepal narrowly ovate, subacute, $2\frac{1}{2}$–3 lin. long; lateral sepals spreading, ovate-oblong, obtuse, subconduplicately folded and obtuse, $2\frac{1}{2}$–3 lin. long; petals broadly unguiculate, with broadly and obliquely ovate obtuse limb, $3\frac{1}{2}$ lin. long, outer basal margin subauriculate; lip broadly unguiculate; limb broadly cordate-ovate, obtuse or subapiculate, convex but without append-age, $1\frac{3}{4}$ lin. long, broader than long; column short; arms of the rostellum incurved and erect, broadly oblong, obtuse, about 1 lin. long, with recurved outer margin; capsule oblong, 5–7 lin. long. *Ceratandra globosa, Lindl. Gen. & Sp. Orch.* 364; *Drège, Zwei Pfl. Documente,* 73, 74, 82; *Bolus in Trans. S. Afr. Phil. Soc.* v. 192; *Orch. Cap. Penins.* 192, *and Journ. Linn. Soc.* xxv. 209; *Durand & Schinz, Conspect. Fl. Afr.* v. 123; *Schlechter in Bull. Herb. Boiss.* 1re *sér.* vi. 910, *excl. syn.*; *Kränzl. Orch. Gen. et Sp.* i. 867. *C. parviflora, Lindl. Gen. & Sp. Orch.* 364; *Krauss in Flora,* 1845, 306, *and Fl. Cap- und Natal.* 158; *Bolus in Trans. S. Afr. Phil. Soc.* v. 192, *Orch. Cap. Penins.* 192, *and Journ. Linn. Soc.* xxv. 209; *Durand & Schinz, Conspect. Fl. Afr.* v. 123; *Kränzl. Orch. Gen. et Sp.* i. 868.

SOUTH AFRICA: without locality, *Kitching*!
COAST REGION: Clanwilliam Div.; near Honig Vallei and Koude Berg, 3000–4000 ft., *Drège,* 1243 b! Ezelsbank, 3000–4000 ft., *Drège,* Worcester Div.; Dutoits Kloof, 3000–4000 ft., *Drège.* Cape Div.; east side of Table Mountain, about 3200 ft., *Bolus,* 4565! *Burchell,* 560! 652! *Prior*! *Schlechter,* 102. Caledon Div.; sides of Baviaans Kloof, *Krauss,* 1313. Swellendam Div.; summit of mountain peak near Swellendam, *Burchell,* 7336; Langeberg Range, near Zuurbraak, 3000–5000 ft., *Schlechter, Bodkin.* Knysna Div.; plains of Plettenbergs Bay, *Bowie,* 10! *Schlechter*!

XLI. EVOTA, Rolfe.

Odd *sepal* superior, oblong, concave; lateral sepals broad, free and spreading. *Petals* broadly unguiculate, somewhat oblique, united with the dorsal sepal into a broad spreading limb. *Lip* inferior, broadly unguiculate and adnate to the face of the column between the arms of the rostellum; limb lunate or divergently bilobed, spreading or ascending, with a large subquadrate or bilobed fleshy appendage. *Column* short; arms of the rostellum laterally dilated, not elongated. *Anther-cells* lateral, curved; pollinia solitary in each cell, coarsely granular, attached by a slender stipes to a small gland. *Stigma* cushion-shaped or bilobed, situated at the base of the column. *Capsule* oblong, ribbed.

Slender erect herbs with fleshy roots; leaves radical and cauline, lax above, suberect or somewhat spreading; flowers rather large, brownish-green and yellow, borne in a lax or somewhat dense raceme; bracts ovate.

DISTRIB. Species 3, limited to the south-western corner of Cape Colony.

This genus is based upon *Ceratandra* section *Evota*, Lindl., and is characterised by the laterally dilated, not elongated arms of the rostellum and the large appendage of the lip.

Appendage of lip with a pair of elongated decurved apical
 horns ... ... (1) **bicolor.**
Appendage of lip without apical horns :
 Appendage of lip very broad, laterally dilated ... (2) **harveyana.**
 Appendage of lip subquadrate ... ... (3) **affinis.**

1. **E. bicolor** (Rolfe); plant 4–9 in. high; stem rather slender, flexuous; leaves radical and cauline, suberect, lax, linear, acuminate, broader and sheathing at the base, ½–1½ in. long; scapes 4–9 in. high; spikes 1–2¼ in. long, lax, 1–8-flowered; bracts ovate or ovate-lanceolate, acute or shortly acuminate, concave, 3–5 lin. long; pedicels 5–8 lin. long; flowers rather large, sepals light green, petals and lip light yellow, the latter with some red stripes on the disc; dorsal sepal elliptic-lanceolate, obtuse, 5–6 lin. long; lateral sepals spreading, obliquely and broadly ovate, apiculate, concave, 5–6 lin. long; petals obliquely cuneate-obovate, crenulate, ½ in. long; lip broadly unguiculate; limb oblong and conduplicate at the base, divergently bilobed above or lunate, 4 lin. long, 5 lin. broad, with obtuse crenulate somewhat incurved lobes and a broad truncate tubercle about the centre; appendage bipartite, falcately decurved, oblong, obtuse, 3 lin. long, with a pair of elongated decurved apical horns; column short; arms of the rostellum rounded, ascending, very short; anther-cells 4 lin. long. *Ceratandra bicolor, Sond. ex Drège in Linnæa,* xx. 220 ; *Bolus in Trans. S. Afr. Phil. Soc.* v. 190, *t.* 21, *Orch. Cap. Penins.* 190, *t.* 21, *and in Journ. Linn. Soc.* xx. 487, xxv. 209 ; *Schlechter in Bull. Herb. Boiss.* 1re *sér.* vi. 859 ; *Kränzl. Orch. Gen. et Sp.* i. 870. *C. harveyana, Sond. in Linnæa,* xix. 108, *not of Lindl.* xx. 220.

SOUTH AFRICA : without locality, *Verreaux* !
COAST REGION : Piquetberg Div. ; Twentyfour Rivers, *Zeyher* ! Tulbagh Div. ; mountains near Tulbagh, *Zeyher* ! *Pappe,* 30 ! *Ecklon & Zeyher,* 66 ! Paarl Div. ; French Hoek, 2500 ft., *Schlechter,* 9296 ! Cape Div. ; among Restiaceæ, especially after bush fires, upper slopes and top of Muizenberg Range, 1300 ft., *Bolus, Bodkin in MacOwan & Bolus, Herb. Norm. Austr.-Afr.* 340 ! Table Mountain, 2200–2400 ft., in Orchid Valley, and above Klassenbosch, *Bolus,* 4564 ! *Rehmann,* 566 ! *Dümmer,* 967 ! *Wolley-Dod,* 2208 ! *Schlechter,* 95 ; Cape Flats, *Harvey* !
CENTRAL REGION : Ceres Div. ; between Witsen Berg and Skurfde Berg, 2000–5000 ft., *Zeyher,* 1574 !

2. **E. harveyana** (Rolfe); plant 4–9 in. high; stem moderately stout, flexuous; leaves radical and cauline, suberect, lax, linear, acuminate, broader and sheathing at the base, ½–1 in. long; scapes 4–9 in. high; spikes 1–2 in. long, somewhat dense, 4–10- (rarely 1-) flowered; bracts ovate, acuminate, concave, 4–6 lin. long; pedicels 4–5 lin. long; flowers rather large, sepals brownish-green, petals and lip yellow; dorsal sepal elliptic-lanceolate, subobtuse, 4½–5 lin. long; lateral sepals spreading, elliptic-ovate, obtuse, concave, 5 lin. long; petals obliquely cuneate-obovate, crenulate, with the front

margin rounded, 5 lin. long; lip broadly unguiculate; limb broadly
oblong, 3½ lin. long, 4 lin. broad, divergently bilobed above, with
oblong somewhat recurved lobes, and a broadly dilated transverse
suborbicular crest below the insertion of the lobes; appendage
obcordate-reniform, 3 lin. long, 5 lin. broad, with a pair of incurved
auricles, but without long decurved apical horns; column broad;
arms of the rostellum auriculate, 1½ lin. long. *Ceratandra harveyana,
Lindl. Gen. & Sp. Orch.* 365; *Bolus in Trans. S. Afr. Phil. Soc.* v. 191,
Orch. Cap. Penins. 191, *Journ. Linn. Soc.* xxv. 209, *and in Hook. Ic.
Pl. t.* 1840; *Durand & Schinz, Conspect. Fl. Afr.* v. 123; *Schlechter
in Bull. Herb. Boiss.* 1ʳᵉ sér. vi. 905; *Kränzl. Orch. Gen. et Sp.* i. 869.

COAST REGION : Cape Div. ; Table Mountain, 2200 ft., *Bolus,* 4548 ! *Wolley-Dod,*
2185 ! *Zeyher, Harvey* ; sand dunes near Wynberg, *Becker.*

3. **E. affinis** (Rolfe); plant 4–5 in. high; leaves radical, linear,
involute, cauline, broader at the base; scapes 4–5 in. high; spikes
oblong, somewhat dense, 10–12-flowered; bracts ovate; sepals ovate-
oblong, obtuse; petals unguiculate, cuneate, with crisped margin,
rounded at the outer basal margin; lip hastate, acute; appendage
subquadrate, acute, fleshy; arms of the rostellum ovate. *Ceratandra
affinis, Sond. in Linnæa,* xix. 108; *Reichb. f. in Walp. Ann.* i. 804;
Bolus in Journ. Linn. Soc. xxv. 209; *Durand & Schinz, Conspect.
Fl. Afr.* v. 122; *Schlechter in Bull. Herb. Boiss.* 1ʳᵉ sér. vi. 911;
Kränzl. Orch. Gen. et Sp. i. 870.

COAST REGION : Worcester Div. ; Hex River, *Ecklon & Zeyher* !
Only known to me from the description.

XLII. OMMATODIUM, Lindl.

Odd *sepal* superior, lanceolate-oblong, nearly flat; lateral sepals
free, spreading, ovate. *Petals* oblique, obcordately bilobed, united
with the dorsal sepal into a broad spreading hood, which is strongly
inflexed at the oblique apex, not concave or saccate below. *Lip*
ascending and adnate to the face of the column below; limb sharply
reflexed, broadly triangular-hastate, with a spreading tooth-like fold
on either side near the base; appendage erect, shortly unguiculate,
obovate and denticulate at the apex. *Column* very short, broadly
dilated, produced in front into two oblong spreading arms. *Anther*
suberect; cells distant; pollinia solitary in each cell, coarsely granular, attached basally to a short caudicle and rounded gland.
Stigmas 2, cushion-shaped, situated at the base of the appendage.
Capsule oblong, ribbed.

A terrestrial herb with small undivided sessile tubers ; leaves few, radical or
subradical, ovate or broad, spreading, few ; flowers in a somewhat dense erect
spike, yellow ; bracts ovate, spreading or reflexed.

DISTRIB.—A single species, limited to the south-western corner of Cape Colony.

Differs from *Pterygodium* in the inverted anther, with the glands situated below.
The plant is much like *Satyrium* in habit.

1. O. Volucris (Lindl. Gen. & Sp. Orch. 365); plant $\frac{1}{2}$–$1\frac{1}{4}$ ft. high; stem rather stout, nearly straight; leaves 2 or 3, sessile and amplexicaul, ovate-oblong or cordate-oblong, subobtuse, the lower 3–4 in. long and subradical, upper much smaller and more erect; scapes $\frac{1}{2}$–$1\frac{1}{4}$ ft. high; spikes 2–4 in. long, somewhat dense, many-flowered; bracts ovate, acute, spreading or reflexed, 4–5 lin. long; pedicels 3–4 lin. long; flowers rather small, pale or sulphur-yellow; dorsal sepal lanceolate-oblong, obtuse, 3 lin. long; lateral sepals obliquely and broadly ovate, very obtuse, 3 lin. long; petals cohering with the dorsal sepal into a somewhat spreading limb, oblique, very broadly obovate, $3\frac{1}{2}$ lin. long; unequally bilobed at the apex, inner lobe oblong, obtuse, outer broadly rounded, undulate, with inflexed upper margin; lip broadly triangular-hastate or subtrilobed, very acute, $2\frac{1}{2}$ lin. long, with a spreading tooth-like fold on either side near the base; appendage erect, shortly unguiculate, obovate, denticulate or obtuse, concave, over $1\frac{1}{2}$ lin. long; column adnate to the lip; arms of the rostellum falcate-oblong, acute, adnate to the connective of the anther; stigmas small, placed on each side of the appendage of the lip. *Ophrys Volucris, Linn. f. Suppl.* 403. *O. triphylla, Thunb. Prodr.* 2. *Pterygodium Volucris, Sw. in Vet. Acad. Handl. Stockh.* 1800, 218; *Thunb. Fl. Cap. ed. Schult.* 22; *Ker in Quart. Journ. Sci. & Arts,* ix. 310, *t.* 4, *fig.* 2; *Bolus in Trans. S. Afr. Phil. Soc.* v. 188, *Orch. Cap. Penins.* 188, *in Journ. Linn. Soc.* xxv. 209, *and Ic. Orch. Austr.-Afr.* i. *t.* 100; *Durand & Schinz, Conspect. Fl. Afr.* v. 118; *Schlechter in Bull. Herb. Boiss.* 1re *sér.* vi. 805; *Kränzl. Orch. Gen. et Sp.* i. 854, *and in Ann. Naturhist. Hofmus. Wien,* xx. 10.

SOUTH AFRICA: without locality, *Masson*! *Auge*! *Nelson*! *Harvey,* 134! *Grey*! *Bergius, Leibold*.

COAST REGION: Clanwilliam Div.; near Brakfontein, *Ecklon & Zeyher,* 34! *Schlechter,* 5230. Malmesbury Div.; near Groene Kloof, 400 ft., *Bolus,* 4333! Tulbagh Div.; near Tulbagh, *Zeyher* (?). Paarl Div.; French Hoek, *Bolus*. Cape Div.; Eastern side of Table Mountain, *Bolus,* 4886, *Kässner*; Kloof between Table and Lion Mountains, *Brown*! Devils Peak, 1300 ft., *Schlechter*; near Muizenberg, *Schlechter,* 1481! near Little Lions Head, *Wolley-Dod,* 1662! near Chapmans Bay, *Wolley-Dod,* 1745! beyond Sandhills on Cape Flats, *Wolley-Dod,* 1791! Stellenbosch Div.; Lowrys Pass, *Schlechter*. Caledon Div.; near Caledon, 1000 ft., *fide Bolus*. Swellendam Div.; near Swellendam, *fide Bolus*.

WESTERN REGION: Little Namaqualand; near Klipfontein, 3000 ft., *fide Bolus*.

XLIII. **PTERYGODIUM**, Sw.

Odd *sepal* superior, more or less concave; lateral sepals free, spreading, ascending or reflexed. *Petals* oblique and united to the margins of the dorsal sepal into a more or less flattened hood, not contracted in front, not saccate at the base. *Lip* united to the face of the column and adnate to it between the stigmatic lobes and arms of the rostellum, narrowed into a claw below, dilated into a

reflexed undivided variously shaped limb above, usually smaller
than the petals, produced above the junction with the column into
a large erect variously shaped fleshy appendage. *Column* short,
dilated, produced in front into two horizontal arms, holding at their
extremities the glands of the pollinia. *Anther-cells* more or less
distant; pollinia solitary in each cell; caudicle and gland uppermost. *Stigma* situated between the arms of the rostellum, horseshoe-
shaped or sometimes divided into two more or less distant cushion-
like tuberculate lobes. *Capsule* cylindrical or obovate, with
prominent ribs.

Terrestrial herbs, with small undivided sessile tubers; leaves cauline, lanceolate
or oblong, usually suberect, sometimes few; flowers in dense or sometimes lax
spikes, generally yellow or white; bracts lanceolate or ovate-lanceolate.

DISTRIB. Species 11, all extra-tropical South African.

Schlechter has extended this genus so as to include the whole of *Corycium,*
Sw., which he then divides into three sections, *Eupterygodium, Corycium* and
Eleuterocorycium, the latter containing four species, one of which was included in
Corycium by Bolus, where the three others are better placed. With this alteration
Pterygodium becomes a natural and well-defined genus.

Leaves in a radical or subradical more or less spreading
   tuft:
  Flowers under $\frac{1}{2}$ in. broad, pale sulphur-yellow;
    appendage of lip long and narrow ... ... (1) **alatum.**
  Flowers over $\frac{1}{2}$ in. long, bright yellow; appendage of
    lip short and broad ... ... (2) **caffrum.**
Leaves cauline or few and suberect:
  Stem 6 to 15 in. high, moderately stout, usually
    several- to many-flowered:
    Petals usually 6 lin. or more long:
      Appendage of lip triangular-lanceolate or ovate-
        oblong, 4 lin. long:
        Leaves 3; appendage of lip usually entire ... (3) **acutifolium.**
        Leaves 2; appendage of lip usually serrate ... (4) **catholicum.**
      Appendage of lip cruciform, about 5 lin. long ... (5) **cruciferum.**
    Petals 3$\frac{1}{2}$ to 5 lin. long;
      Appendage of lip 3-lobed, 2$\frac{1}{4}$ lin. long ... ... (6) **hastatum.**
    Appendage of lip oblong:
      Appendage of lip 3$\frac{1}{2}$ to 5 lin. long:
        Flowers pale yellow; lip oblong, denticulate
          at the apex ... ... (7) **Newdigatæ.**
        Flowers white and green; lip very broad,
          emarginate or bilobed at the apex ... (8) **leucanthum.**
      Appendage of lip 1$\frac{1}{4}$ lin. long ... ... (9) **Cooperi.**
  Stem 5 to 6 in. high, very slender, one- to few-
    flowered:
    Stem 3- to 4-leaved, 2$\frac{1}{2}$ in. long; sepals subequal ... (10) **pentherianum.**
    Stem 1- to 2-leaved; lower leaves $\frac{3}{4}$ to 1 in. long;
      dorsal sepal shorter and narrower than the
      lateral ... ... (11) **platypetalum.**

1. **P. alatum** (Sw. in Vet. Acad. Handl. Stockh. 1800, 218),
plant 3–6 in. high, turning black in drying; stem moderately stout;

leaves radical and cauline, 7–12, the lower spreading, oblong-lanceolate, acute, ¾–1¾ in. long, the upper more erect and gradually reduced upwards into the bracts; scapes 3–6 in. high; spikes 1–2½ in. long, rather dense, usually many-flowered; bracts ovate, acute, 4–5 lin. long; pedicels 3–4 lin. long; flowers rather small, pale sulphur-yellow; dorsal sepal lanceolate-oblong, apiculate, 3¼ lin. long; lateral sepals spreading, lanceolate-oblong, subobtuse or obscurely apiculate, 3½ lin. long; petals cohering with the dorsal sepal into a large obovate-orbicular limb, 3½ lin. long, outer margin crisped and irregularly crenulate; lip 3-lobed, about 1½ lin. long, by thrice as broad; front lobe broadly triangular, acute; side lobes rounded obovate, crenulate; appendage erect, clavate, some-what 3-lobed, nearly 3 lin. long, with short broad side lobes, an obtuse front lobe and two keels on the lower two-thirds; arms of the rostellum diverging and ascending; anther-cells very short. *Thunb. Fl. Cap. ed. Schult.* 24; *Ker in Quart. Journ. Sci. & Arts,* viii. 221, *t.* 3, *fig.* 2; *Lindl. Gen. & Sp. Orch.* 366; *Drège, Zwei Pfl. Documente,* 100, 114; *Sond. in Linnæa,* xx. 220; *Bolus in Trans. S. Afr. Phil. Soc.* v. 187, *Orch. Cap. Penins.* 187, *and in Journ. Linn. Soc.* xxv. 208; *Durand & Schinz, Conspect. Fl. Afr.* v. 116; *Kränzl. Orch. Gen. et Sp.* i. 855, *and in Ann. Naturhist. Hofmus. Wien,* xx. 10; *Schlechter in Bull. Herb. Boiss.* 1ʳᵉ sér. vi. 806. *Ophrys alata, Thunb. Prodr.* 2.

SOUTH AFRICA: without locality, *Masson*! *Brown*! *Mund*! *Thom*! *Harvey,* 132!

COAST REGION: Clanwilliam Div.; by the Olifants River and near Brakfontein, *Ecklon & Zeyher,* 69! Malmesbury Div.; hills near Malmesbury, *Bachmann.* Worcester Div.; mountains near De Liefde, 1000–2000 ft., *Drège,* 1255 b! Paarl Div.; Drakenstein Mountains, *Rehmann,* 2233! Klein Drakenstein Mountains, 5000–6000 ft., *Drège,* 1255 a! near Wellington, *Miss Cummings.* Cape Div.; mountains near Cape Town, 50–1860 ft., *Thunberg*! *Ecklon,* 673! *Bolus,* 3930! *Pillans, Wilms,* 3652! *Marloth, Schlechter,* 1355! 1356! 1455, *Rogers*! *Wolley-Dod,* 500! Stellenbosch Div.; near Stellenbosch, *Lloyd in Herb. Sanderson,* 925! *Marloth, Prior*! Lowrys Pass, 250 ft., *Schlechter,* 1153, *Miss Farnham in MacOwan & Bolus, Herb. Norm. Austr.-Afr.* 337! Caledon Div.; Zwart Berg, near Caledon, 400–2000 ft., *Zeyher,* 3944, *Bolus.* Swellendam Div.; near Swellendam, *Zeyher*! 3948!

2. **P. caffrum** (Sw. in Vet. Acad. Handl. Stockh. 1800, 218); plant ⅓–1 ft. high; stem stout, often very stout; leaves radical and cauline, the lower 3 or 4 ovate or ovate-oblong, subobtuse, more or less spreading, 1–3 in. long, the upper much smaller, suberect, often sheath-like and reduced upwards into the bracts; scapes ⅓–1 ft. high; spikes 1½–4 in. long, often dense and many-flowered; bracts ovate or ovate-lanceolate, subacute, ½–1 in. long; pedicels 5–6 lin. long; flowers large, bright yellow; dorsal sepal oblong-lanceolate, subobtuse, 4 lin. long; lateral sepals widely spreading or ascending, obliquely ovate, shortly acuminate, deeply concave or subsaccate, 4 lin. long; petals cohering with the dorsal sepal into an expanded limb, obliquely obovate, subobtuse at the inner angle, with a rounded minutely crenulate margin, somewhat

concave, 4 lin. long; lip reniformly obovate, cuneate at the base, bilobed above, 4 lin. long, 6–9 lin. broad, with a crenulate margin; appendage ovate-oblong, obtuse, arched at the apex, with two diverging keels beneath, 3½ lin. long; column short; arms of the rostellum oblong, obtuse, concave, about 1 lin. long; capsule fusiform-oblong, about 5 lin. long. *Thunb. Fl. Cap. ed. Schult.* 23; *Lindl. Gen. & Sp. Orch.* 367; *Drège, Zwei Pfl. Documente,* 98; *Krauss in Flora,* 1845, 306, *and Fl. Cap- und Natal.* 158; *Sond. in Linnæa,* xx. 220; *Bolus in Trans. S. Afr. Phil. Soc.* v. 186, *t.* 22, *fig.* 22, *Orch. Cap. Penins.* 186, *t.* 22, *fig.* 22, *and in Journ. Linn. Soc.* xxv. 208; *Durand & Schinz, Conspect. Fl. Afr.* v. 117; *Schlechter in Bull. Herb. Boiss.* 1ʳᵉ *sér.* vi. 812; *Kränzl. Orch. Gen. et Sp.* i. 859, *and in Ann. Naturhist. Hofmus. Wien,* xx. 10. *Ophrys caffra, Thunb. Prodr.* 2.

SOUTH AFRICA : without locality, *Masson* ! *Brown* ! *Forster* ! *Harvey,* 146 ! *Shepherd,* 20 ! *Mund, Zeyher,* 3940 !

COAST REGION : Malmesbury Div. ; near Groene Kloof, *Bolus,* 3936 ; near Darling, 100 ft., *Schlechter,* 5341 ! *Penther, Krook.* Paarl Div. ; near Paarl, *Thunberg* ! between Paarl and Lady Grey Railway Bridge, under 1000 ft., *Drège,* 1254 a ! 1254 d ! Cape Div. ; Cape Flats, near Wynberg, 50–100 ft., *Ecklon & Zeyher, Webb, Krauss,* 1322, *Bolus,* 3936 ! *Schlechter,* 1556 ! Table Mountain, below 500 ft., *Ecklon & Zeyher,* 72 ! *Bolus,* 3936 ! *Bergius* ; Muizen Berg, 800 ft., *MacOwan & Bolus, Herb. Norm. Austr.-Afr.* 176 ! Red Hill, *Wolley-Dod,* 1799 ! Orange Kloof, near the Waterfall, *Wolley-Dod,* 2128 ! near Maitland, *Wolley-Dod,* 3123 ! near Tyger Berg, *Bolus,* 3936 ! Simons Bay, *Wright* ! Devils Mountain, *Ecklon, Pappe, Bunbury.* Stellenbosch Div. ; Hottentots Holland Mountains, *Pappe,* 23 ! *Prior* ! George Div. ; near George, *Bowie* ! Knysna Div. ; near Knysna, *Wallich.*

CENTRAL REGION : Ceres Div. ; Skurfdeberg Range, *Pappe* !

3. **P. acutifolium** (Lindl. Gen. & Sp. Orch. 366); plant ½–1¼ or rarely 2 ft. high; stem stout, somewhat flexuous; leaves 3–4, cauline, suberect, oblong-lanceolate, acute or apiculate, the lower 3–5 or rarely 7 in. long, the upper much smaller; scapes ½–1¼ or rarely 2 ft. high; spikes 1½–4 in. long, rather dense and often many-flowered ; bracts ovate or ovate-lanceolate, acute or shortly acuminate, 7–10 lin. long; pedicels 5–7 lin. long; flowers large, deep golden-yellow; dorsal sepal ovate-lanceolate, subobtuse, sub-concave, with a prominent rounded sac behind the sharply reflexed apex, 5–6 lin. long; lateral sepals widely spreading, obliquely ovate, shortly acuminate, subsaccate at the base, 5–6 lin. long; petals cohering with the dorsal sepal into a spreading limb, very oblique, broadly semiobovate, acute at the inner angle, rounded or subauriculate at the base, 6–7 lin. long; lip sharply reflexed, broadly triangular-ovate, acute, very undulate, 3 lin. long; appendage erect, triangular-oblong, with a recurved subobtuse apex, entire or somewhat crenulate at the margin, 4 lin. long; column short ; arms of the rostellum ascending, oblong, obtuse, 2 lin. long; capsule oblong, about 8 lin. long. *Drège, Zwei Pfl. Documente,* 82 ; *Sond. in Linnæa,* xx. 220 ; *Bolus in Trans. S. Afr. Phil. Soc.* v. 185, *Orch. Cap. Penins.* 185, *and in Journ. Linn. Soc.* xxv. 208 ; *Durand & Schinz, Conspect. Fl. Afr.* v. 116 ; *Schlechter in Bull. Herb. Boiss.*

1ʳᵉ *sér.* **vi.** 815 ; *Kränzl. Orch. Gen. et Sp.* i. 856, *and in Ann. Naturhist. Hofmus. Wien,* xx. 10.

SOUTH AFRICA : without locality, *Gronovius* ! *Brown* ! *Leibold, Villet* ! *Webb* ! *Zeyher,* 1573, partly ! *Harvey,* 137 !

COAST REGION : Malmesbury Div. ; near Groene Kloof, *Bolus.* Tulbagh Div. ; Tulbagh Waterfall, *Pappe* ! *Ecklon,* 74 ! Witsen Berg, *Zeyher,* 3941 ! New Kloof at the foot of Winterhoek Mountains, *Zeyher* ! 1572 ; Worcester Div. ; Dutoits Kloof, 3000 – 4000 ft., *Drège,* 1254 c ! Cape Div. ; Top of Table Mountain, *Burchell,* 654 ! 2000–3000 ft., *Bolus, Zeyher, Pappe,* 22 ! *Schlechter,* 93, *Milne,* 149 ! Lower Plateau, *Wolley-Dod,* 2114 ! Muizen Berg, 1500 ft., *Bolus,* 4334 ! and *MacOwan & Bolus, Herb. Norm. Austr.-Afr.* 175 ! *Schlechter,* 161. Simons Berg, *Mrs. Jameson* ! Devils Mountain, *Rehmann,* 957 ! Swellendam Div. ; near Swellendam, *Bowie* ! Riversdale Div. ; Langeberg Range, near Riversdale, 1500 ft., *Schlechter.* George Div. ; near George, *Bowie* ! Knysna Div. ; near Knysna, *Trimen* !

Very nearly allied to *P. catholicum,* but distinguished by its more robust habit, more numerous acute leaves, and (according to Bolus) large deep golden-yellow flowers. It also grows at a greater altitude and flowers later than *P. catholicum.* The two have been mixed under Zeyher, 1573.

4. P. catholicum (Sw. in Vet. Acad. Handl. Stockh. 1800, 218) ; plant ½–1 ft. high ; stem moderately stout, somewhat flexuous ; leaves 2 or rarely 3, distant, sessile, sheathing at the base, somewhat spreading, oblong or elliptic-oblong, subobtuse or apiculate, the lower 2–5 in. long, upper much smaller ; scapes ½–1 ft. high ; spikes 1–2 in. long, loosely 3–8-flowered ; bracts ovate-lanceolate, shortly acuminate, 5–7 lin. long ; pedicels 4–5 lin. long ; flowers large, pale sulphur-yellow, rarely with orange-red petals and lip ; dorsal sepal ovate-lanceolate, subobtuse, subconcave, subacute, with a distinct rounded sac near the apex, 5 lin. long ; lateral sepals widely spreading, obliquely ovate, shortly acuminate and curved at the apex, subsaccate at the base, 5 lin. long ; petals cohering with the dorsal sepal into an expanded limb, very oblique, broadly elliptic-oblong, subacute at the inner angle, broadly auriculate at the base, with a crenulate outer margin, 5–6 lin. long ; lip broadly rhomboid-orbicular, subobtuse, very undulate, reflexed, 2 lin. long ; appendage ovate-oblong, obscurely 3-lobed about the middle, obtuse, recurved at the apex, base broad, upper half narrow and crenulate, 4 lin. long ; column short ; arms of the rostellum auriculate, concave, 1½ lin. long. *Thunb. Fl. Cap. ed. Schult.* 22 ; *Ker in Quart. Journ. Sci. and Arts,* vi. 46, *t.* 1, *fig.* 3 ; *Lindl. Gen. & Sp. Orch.* 366 ; *Drège, Zwei Pfl. Documente,* 116 ; *Krauss in Flora,* 1845, 306, *and Fl. Cap- und Natal.* 158 ; *Sond. in Linnæa,* xx. 220 ; *Bolus in Trans. S. Afr. Phil. Soc.* v. 184, *Orch. Cap. Penins.* 184, *and in Journ. Linn. Soc.* xxv. 208 ; *Durand & Schinz, Conspect. Fl. Afr.* v. 117 ; *Schlechter in Bull. Herb. Boiss.* 1ʳᵉ *sér.* vi. 813 ; *Kränzl. Orch. Gen. et Sp.* i. 856, *excl. syn. Linn. f., and in Ann. Naturhist. Hofmus. Wien,* xx. 10. *Ophrys catholica, Linn. Amœn. Acad.* vi. 110, *and Sp. Pl. ed.* ii. 1344. *Ophrys alaris, Linn. f. Suppl.* 404. *Arethusa alaris, Thunb. Prodr.* 3.—*O. affinis, flore luteo, Buxb. Pl. Min. Cogn. Cent.* iii. 13, *t.* 21.

T 2

SOUTH AFRICA: without locality, *Sparrman*! *Brown*! *Masson*! *Oldenburg*, 85!
Carmichael! *Zeyher*, 1573, partly! 3942! 3943! *Harvey*, 147! *Thom*, 38! 1175!
Hesse, 4! *Mund*! *Wright*, 740!

COAST REGION: Tulbagh Div.; Tulbagh, *Pappe*, 25! *Schlechter*. Worcester Div.;
without precise locality, *Cooper*, 1672! 1698! Cape Div.; Hills and Flats near
Cape Town, *Thunberg*! *Wallich*! *Krauss*, 1324, *Zeyher*, 3941! *Prior*! *Milne*, 2!
Pappe, 24! *Ecklon & Zeyher*! *Bunbury*, 259! *Alexander*! *Bolus*, 3931! and in
MacOwan & Bolus, Herb. Norm. Austr.-Afr. 174! *Bowie*! *Bergius, Mund,
Tyson*, 2086, *Kässner, Ecklon*, 672! *Wilms*, 3658! *Wolley-Dod*, 501! *Schlechter*,
1339! Constantia Mountains, *Zeyher*! *Rogers*, 377! 448! Muizen Berg, *Wolley-Dod*,
1930! Waai Vley, *Wolley-Dod*, 2180! Red Hill, *Mrs. Jameson*! Simons Bay,
Wright! *MacGillivray*, 470! Stellenbosch Div.; near Stellenbosch, *Lloyd in
Herb. Sanderson*, 935! Caledon Div.; Genadendal, 3000–4000 ft., *Drège*, 1254 a!
Swellendam Div.; near Swellendam, *Bowie*! Mossel Bay Div.; Attaquas Kloof,
Gill! George Div.; near George, *Bowie*! *Prior*! Port Elizabeth Div.; Algoa
Bay, *Forbes*!

5. P. cruciferum (Sond. in Linnæa, xix. 109); plant 6–9 in.
high; stem moderately stout, nearly straight; leaves 2, cauline or
the lower subradical, suberect, oblong-lanceolate, acute or subacute,
3–4 in. long; scapes 6–9 in. high; spikes 2–3 in. high, somewhat
lax, 2–6-flowered; bracts elliptic-lanceolate or ovate-lanceolate,
acute, 6–8 lin. long; pedicels 5–8 lin. long; flowers large, sulphur-
yellow; dorsal sepal oblong-lanceolate, subacute, somewhat concave
at the base, 8–9 lin. long; lateral sepals somewhat spreading,
ovate-lanceolate, acuminate, 7–8 lin. long; petals completely
cohering to the dorsal sepal, forming a spreading limb, obliquely
obovate-orbicular, very concave, with subcrenulate outer margin,
7–8 lin. long; lip 3-lobed; side lobes auriculate, about 2 lin. long;
front lobe strongly recurved, narrowly subulate-linear, 2 lin. long;
appendage erect, cruciform, 5 lin. long, side lobes recurved, linear-
oblong, obtuse, 2 lin. long; front lobe rather longer and bilobed at
the apex; column short; arms of the rostellum diverging, oblong,
obtuse, 1½ lin. long; stigmas distant. *Bolus in Trans. S. Afr. Phil.
Soc.* v. 186, *t.* 22, *fig.* 18–21, *Orch. Cap. Penins.* 186, *t.* 22, *fig.* 18–21,
and in Journ. Linn. Soc. xxv. 208; *Durand & Schinz, Conspect.
Fl. Afr.* v. 117; *Kränzl. Orch. Gen. et Sp.* i. 855; *Schlechter in
Bull. Herb. Boiss.* 1re *sér.* vi. 808.

SOUTH AFRICA: without locality, *Burmann, Masson*! *Oldenburg, Marloth.*
COAST REGION: Cape Div.; among burnt shrubs on the slopes above Van
Camps Bay, 300 ft., *Bolus*, 4964! and in *MacOwan & Bolus, Herb. Norm. Austr.-
Afr.* 316! near Wynberg, *Bolus, Schlechter*, near Simons Bay, *Brown*! Caledon
Div.; Houw Hoek, 800 ft., *Schlechter*. Swellendam Div.; near Swellendam,
Mund! Uitenhage Div.; near Bethelsdorp, *Ecklon & Zeyher*.

6. P. hastatum (Bolus in Journ. Linn. Soc. xxv. 177, 178, fig. 14);
plant ½–1 ft. high; stem somewhat flexuous, moderately slender;
leaves 2, cauline, suberect or somewhat spreading, elliptic-oblong or
lanceolate-oblong, apiculate, the lower 2¼–4½ in. long, upper much
smaller; scapes ½–1 ft. high; spikes 2–3 in. long, lax, 7–10-flowered;
bracts ovate-lanceolate, acuminate, 4–7 lin. long; pedicels 5–6 lin.
long; flowers medium-sized, whitish, with green sepals; dorsal
sepal elliptic-oblong, subobtuse, very concave, 3½ lin. long; lateral

sepals spreading, elliptic-ovate, acute, concave, 3½ lin. long; petals cohering with the dorsal sepal into a somewhat expanded limb, obliquely quadrate-orbicular, apiculate at the inner angle, with an undulate crenulate outer margin, 3½–4 lin. long; lip 3-lobed; side lobes auriculate; front lobe reflexed, oblong, obtuse, crenulate, 1¼ lin. long; appendage erect, 3-lobed, 2¼ lin. long, with two prominent diverging keels behind, front lobe broadly rhomboid, subobtuse, side lobes much smaller; column short; arms of the rostellum obovate, 1 lin. long; stigma widely separated. *Bolus in Journ. Linn. Soc.* xxv. 208; *Durand & Schinz, Conspect. Fl. Afr.* v. 117; *Schlechter in Bull. Herb. Boiss.* 1ʳᵉ *sér.* vi. 809; *Kränzl. Orch. Gen. et Sp.* i. 857.

KALAHARI REGION : Orange River Colony ; rocky grassy spots on summit of Quaqua Mountains, Witzies Hoek, 7400 ft., *Thode,* 57 ! and without precise locality, *Cooper,* 1090 ! Transvaal ; near Barberton, 3500–5000 ft., *Galpin,* 1256, *Culver,* 87.

EASTERN REGION : Transkei ; near Tsomo, *Mrs. Barber,* 832 ! Natal ; Oliviers Hoek, 5000 ft., *Allison,* T ! near Van Reenen, 5000 ft., *Schlechter,* 6923.

7. **P. Newdigatæ** (Bolus, Ic. Orch. Austr.-Afr. i. t. 99, fig. A); plant 6–12 in. high ; stem moderately stout, flexuous ; leaves 2 or 3, cauline, oblong-lanceolate, acute, the lower 3–4 lin. long, somewhat spreading, upper much smaller and suberect ; scapes 6–12 in. high ; spikes 1–2 in. long, few- (occasionally 1-) flowered ; bracts ovate or ovate-lanceolate, acute, 6–9 lin. long ; pedicels 5–6 lin. long ; flowers small, pale yellow ; dorsal sepal oblong-lanceolate, acute, concave, 4½ lin. long ; lateral sepals spreading, obliquely ovate, acute, 4½–5 lin. long, very concave ; petals cohering with the dorsal sepal into a somewhat spreading limb, obliquely ovate, subobtuse and slightly lacerate at the apex, very concave, 4½–5 lin. long, the front margin sinuate near the base ; lip strongly reflexed, ovate-oblong, lacerate at the broad apex, 4 lin. long ; appendage erect, oblong, obtuse, incurved and denticulate at the apex, about 3 lin. long ; column short ; arms of the rostellum oblong, obtuse, about 1 lin. long ; capsule fusiform-oblong, about ½ lin. long. *Schlechter in Bull. Herb. Boiss.* 1ʳᵉ *sér.* vi. 814; *Kränzl. Orch. Gen. et Sp.* i. 860, *and in Ann. Naturhist. Hofmus. Wien,* xx. 11.

VAR. β, **cleistogamum** (Bolus, Ic. Orch. Austr.-Afr. i. t. 99, fig. B) ; differs from the type in having more numerous and self-fertilised flowers, the perianth remaining closed, a concave acute lip, free from the column, to which latter the staminodial appendage is attached. *Kränzl. Orch. Gen. et Sp.* i. 861, *and in Ann. Naturhist. Hofmus. Wien,* xx. 11. *P. cleistogamum, Schlechter in Bull. Herb. Boiss.* 1ʳᵉ *sér.* vi. 816.

COAST REGION : Knysna Div. ; Forest Hall, near Plettenbergs Bay, *Miss Newdigate in Herb. Bolus,* 164 ! in *Herb. MacOwan,* 1960 ! and *MacOwan, Herb. Austr.-Afr.* 1960 ! Var. β : Knysna Div. ; Forest Hall, near Plettenbergs Bay (with the type), *Miss Newdigate in Herb. Bolus,* 164 ! near Knysna, *Schlechter,* 5908 ! Albany Div. ; Coldstream Farm, near Grahamstown, 2000 ft., *Glass,* 479, *Penther,* 41 !

The column is very variable in the abnormal cleistogamic form, three conditions being figured by Bolus. Schlechter makes it a species, under the name of

P. cleistogamum, but it is clearly only a peloriate self-fertilising form of *P. Newdigatæ.*

8. **P. leucanthum** (Bolus in Trans. S. Afr. Phil. Soc. xvi. 150) ; plant 6–14 in. high; stem moderately slender, nearly straight ; leaves 3–5, suberect, lanceolate or oblong-lanceolate, acute or subacute, 1½–3 in. long, ½–1 in. broad, the upper and lower usually much smaller ; scapes 6–14 in. high ; spikes 1–2 in. long, somewhat lax, 7–10-flowered ; bracts lanceolate, acuminate, 5–7 lin. long ; pedicels about 4 lin. long ; flowers medium-sized, sepals pale green, petals and lip white, with the incurved tip of the appendage very dark green ; dorsal sepal elliptic-oblong, obtuse or apiculate, concave, 4 lin. long ; lateral sepals spreading, obliquely ovate-oblong, subacute or apiculate, subsaccate at the base, 3½ lin. long ; petals cohering with the dorsal sepal into a somewhat expanded limb, semiorbicular, subapiculate at the inner angle, conduplicately concave, with the front margin somewhat incurved, 4 lin. long ; lip sharply reflexed, transversely oblong, emarginate or somewhat bilobed, under 1 lin. long, much broader than long ; appendage erect, oblong from a rather broader base, 3½ lin. long, with bilobed incurved fleshy apex, each lobe saccate ; column short ; arms of the rostellum curved, oblong, obtuse, under 1 lin. long. *Bolus, Ic. Orch. Austr.-Afr.* ii. *t.* 99.

EASTERN REGION : Tembuland ; steep rocky places on the Engcobo Mountain, Fingoland, 4300 ft., *Bolus*, 8704 ! *Flanagan* ; slopes of Kwenkwe Mountain, near Gat Berg, Maclear District, 5500 ft., *Flanagan.*
KALAHARI REGION : Orange River Colony ; grassy slopes of Mopedis Peak, Witzies Hoek, 8100–8400 ft., *Thode*, 57 b !

9. **P. Cooperi** (Rolfe) ; plant ¾–1¼ ft. high ; stem moderately stout ; leaves cauline, sessile, 2 or 3, oblong or linear-oblong, acute or subacute, 2–4½ in. long ; scapes ¾–1¼ in. high ; spikes 2–3 in. long, dense, many-flowered ; bracts ovate or ovate-lanceolate, acute or acuminate, 6–8 lin. long ; pedicels about 5 lin. long ; flowers medium-sized ; dorsal sepal elliptic-oblong, subobtuse, concave, 2½–3 lin. long ; lateral sepals spreading, obliquely ovate, obtuse, 2½–3 lin. long, nearly as broad as long ; petals cohering with the dorsal sepal into a somewhat expanded limb, oblique, obovate-orbicular, 2½–3 lin. long, broader than long, unequally bilobed at the apex, upper lobe rather narrow, obtuse, lower lobe broadly rounded, subundulate ; lip broadly unguiculate ; limb broadly deltoid-triangular, apiculate or acute, concave, 1¼ lin. long, broader than long ; appendage obovate-oblong, truncate, 1¼ lin. long ; column short ; arms of the rostellum diverging, obovate-oblong, obtuse, curved, over 1 lin. long.

SOUTH AFRICA : without locality, *Cooper*, 1882 !
EASTERN REGION : Transkei ; under rocks near the Tsomo River, *Mrs. Barber*, 854 !

A very distinct species, which is unfortunately unlocalised in Cooper's original list.

10. P. pentherianum (Schlechter in Engl. Jahrb. xxiv. 432);
plant 4–6 in. high; stem somewhat flexuous, slender; leaves 3–4,
suberect or somewhat spreading, lanceolate-oblong, sheathing at the
base, the lower 2¼ in. long, ¾ in. broad, upper much smaller and
narrower; scapes 4–6 in. high, 2–4-flowered; bracts lanceolate,
acute, as long as the pedicels; flowers medium-sized, sulphur-yellow;
dorsal sepal oblong-lanceolate, acute, concave, 4–4½ lin. long;
lateral sepals spreading, obliquely ovate-lanceolate, acute, very concave, 4–4½ lin. long; petals cohering with the dorsal sepal into a
somewhat expanded limb, obliquely and broadly obovate, obtuse,
concave, 4–4½ lin. long, rather broader than long; lip deflexed,
bilobed, 2½ lin. long, with oblong or subquadrate minutely denticulate lobes; appendage erect, fleshy, nearly oblong, conspicuously
tridentate at the apex, 2–2½ lin. long. *Schlechter in Bull. Herb.
Boiss.* 1ʳᵉ *sér.* vi. 811; *Kränzl. Orch. Gen. et Sp.* i. 858.

COAST REGION: Piquetberg Div.; Berg River hills, near Piquetberg, 2000 ft.,
Schlechter, 3278.

Only known to me from the description.

11. P. platypetalum (Lindl. Gen. & Sp. Orch. 366); plant 3–6 in.
high; stem slender, flexuous; leaves cauline, 2, sessile, suberect,
lanceolate or oblong-lanceolate, subacute, the lower ¾–1 in. long, the
upper smaller and more acute; scapes 3–6 in. high, 1–2- (rarely 3-)
flowered; bracts ovate-spathaceous, acute, 4–5 lin. long; pedicels
2–3 lin. long; flowers rather large, pale yellow; dorsal sepal galeate,
ovate-lanceolate, acute or acuminate, deeply concave or subsaccate
at the base, 2–2½ lin. long; lateral sepals spreading, obliquely ovate,
acute, conduplicately folded at the base in front, 3½ lin. long;
petals united to the dorsal sepal at the base into a spreading limb,
broadly orbicular or reniform-orbicular, with sinuate or subcrenulate
outer margin, bilobed at the adjacent inner angles, very concave,
4½ lin. long, broader than long; lip 3-lobed, saddle-shaped, 3½ lin.
long, with rounded reflexed side lobes and a narrowly triangular
subobtuse front lobe; appendage elliptic-spathulate, obtuse, sharply
reflexed into the sac of the dorsal sepal, subconduplicate at the base,
1½ lin. long; column short; arms of the rostellum curved, oblong,
obtuse, about 1 lin. long. *Drège, Zwei Pfl. Documente,* 81, 98;
Krauss in Flora, 1845, 306, *and Fl. Cap.- und Natal.* 158; *Sond. in
Linnæa,* xix. 109; *Bolus in Trans. S. Afr. Phil. Soc.* v. 184; *Orch.
Cap. Penins.* 184, *in Journ. Linn. Soc.* xxv. 208, *and Ic. Orch. Austr.-
Afr.* i. *t.* 48; *Schlechter in Bull. Herb. Boiss.* 1ʳᵉ *sér.* vi. 807. *P.
alare, Durand & Schinz, Conspect. Fl. Afr.* v. 116; *Kränzl. Orch.
Gen. et Sp.* i. 858, *and in Ann. Naturhist. Hofmus. Wien,* xx. 10.
P. catholicum, var. *minor, Thunb. Fl. Cap. ed. Schult.* 22. Not
Arethusa alaris, Thunb. Prodr. 3.

SOUTH AFRICA: without locality, *Masson*!
COAST REGION: Malmesbury Div.; hills near Malmesbury, *Bachmann*. Paarl
Div.; marshes near the Berg River, below 100 ft., *Drège*; near Wellington,
Miss Cummings. Worcester Div.; Dutoits Kloof, 2000–3000 ft., *Drège*, 8280 a!

Cape Div. ; near Constantia, *Krauss*, 1325 ; mountains behind Houts Bay, 1800 ft., *Marloth in Herb. Bolus*, 4945 ! Devils Mountain, *Pillans* ; eastern slopes of Vlagge Berg, 1000 ft., *Schlechter*, 1455 ! Stellenbosch Div. ; near Stellenbosch, *Marloth, Miss Farnham in MacOwan & Bolus, Herb. Norm. Austr.-Afr.* 1379 ! Lowrys Pass, 250 ft., *Schlechter*, 1153 ! Caledon Div. ; Houw Hoek Mountains, *Ecklon & Zeyher*, 115 ! *Schlechter*, 5473, Zwart Berg, near Caledon, 400–2000 ft., *Zeyher*, 3944 ! *Bolus*. Swellendam Div. ; Puspas Valley, near Swellendam, *Zeyher, Bolus*.

XLIV. ANOCHILUS, Rolfe.

Odd *sepal* inferior, more or less concave ; lateral sepals free, spreading or reflexed. *Petals* oblique, more or less united to the dorsal sepal into a large concave limb, sometimes ultimately free, not saccate at the base. *Lip* superior, united to the face of the column between the stigmatic lobes and arms of the rostellum, narrowed into a claw below, dilated into a broad incurved or reflexed limb above, produced above the junction with the column into a large deflexed, somewhat bilobed appendage. *Column* short, dilated, produced in front into two short broadly rounded or oblong spreading arms carrying the glands of the pollinia. *Anther-cells* more or less distant ; pollinia solitary in each cell, coarsely granular, caudicle and gland uppermost. *Stigma* situated between the arms of the rostellum, cushion-shaped or somewhat bilobed. *Capsule* oblong, ribbed.

Terrestrial herbs, with small undivided sessile tubers ; leaves cauline, oblong or ovate-oblong, suberect or somewhat spreading ; flowers in dense erect spikes, greenish-yellow or smoky-grey ; bracts ovate or ovate-lanceolate.

DISTRIB.—Species 2, western or central.

Distinguished from *Pterygodium* by the superior lip and other correlated differences. It is based upon *Pterygodium* section *Anochilus*, Schlechter.

Flowers greenish-yellow ; lip obovate flabellate from a
    narrow base   ...   . (1) **inversum**.

Flowers smoky-grey ; lip oblong ...   . (2) **Flanagani**.

1. A. **inversum** (Rolfe) ; plant $\frac{3}{4}$–$1\frac{1}{2}$ ft. high ; stem stout, straight ; leaves cauline, 4–6, somewhat spreading, oblong from a broader sheathing base, acute or subacute, 3–6 in. long ; scapes $\frac{3}{4}$–$1\frac{1}{2}$ ft. high ; spikes oblong, 4–6 in. long, dense, many-flowered ; bracts ovate-lanceolate, acute, $\frac{3}{4}$–1 in. long ; pedicels 6–8 lin. long ; flowers rather large, greenish-yellow, with some brown stripes on the lip and sometimes on the petals ; odd sepal elliptic-oblong, with a short reflexed obtuse apex, 4 lin. long ; lateral sepals spreading, obliquely ovate-oblong, obtuse, rounded, conduplicately folded and concave at the base, 4 lin. long ; petals obovately suborbicular, obtuse, conduplicately folded and concave at the upper basal margin, 4 lin. long ; lip broadly unguiculate ; limb reflexed, obovate-flabellate from a cuneate oblong base, minutely crenulate, concave, $2\frac{1}{2}$ lin. long ; appendage oblong or obovate-oblong, emarginate or shortly

bilobed, about 3 lin. long; column short; arms of the rostellum broadly oblong, obtuse, 1 lin. long. *Ophrys inversa, Thunb. Prodr.* 2. *Pterygodium inversum, Sw. in Vet. Acad. Handl. Stockh.* 1800, 218; *Thunb. Fl. Cap. ed. Schult.* 23; *Ker in Quart. Journ. Sci. & Arts,* ix. 310, *t.* 4, *fig.* 1; *Lindl. Gen. & Sp. Orch.* 367; *Drège, Zwei Pfl. Documente,* 68, 113; *Bolus, Ic. Orch. Austr.-Afr.* i. *t.* 97, *and in Journ. Linn. Soc.* xxv. 208; *Durand & Schinz, Conspect. Fl. Afr.* v. 117; *Schlechter in Bull. Herb. Boiss.* 1ʳᵉ *sér.* vi. 855; *Kränzl. Orch. Gen. et Sp.* i. 861, *and in Ann. Naturhist. Hofmus. Wien,* xx. 11.

SOUTH AFRICA : without locality, *Masson* !

COAST REGION : Piquetberg Div. ; near Piquetberg, *Thunberg* ! near Piqueniers Kloof, 500 ft., *Schlechter,* 4959 ! Malmesbury Div. ; near Riebecks Casteel, *Thunberg* ! between Groene Kloof and Saldanha Bay, *Drège,* 4782 b ! near Malmesbury, 300 ft., *Bolus,* 4335 ! *Schlechter.* Worcester Div. ; Hex River, *Mallison* (?) *in Herb. Wolley-Dod* ! Caledon Div. ; mountains between Bredecamps Vley and Botberg, *Zeyher* ; near Caledon, *Prior* !

WESTERN REGION : Little Namaqualand ; between Pedros Kloof and Lily Fontein, 3000–4000 ft., *Drège,* 8284 !

2. **A. Flanagani** (Rolfe); plant ¾–1 ft. high; stem stout, erect; leaves cauline, numerous, suberect, ovate or ovate-oblong, acuminate, broadly sheathing and imbricate at the base, 2–4 in. long, smoky-grey; scapes ¾–1 ft. high; spikes oblong, 2–3 in. long, dense, many-flowered; bracts broadly ovate, acuminate, 6–9 lin. long, smoky-grey; pedicels about 4 lin. long; flowers medium-sized, brown or smoky-grey with a dark maroon lip; odd sepal elliptic-oblong, obtuse, 3 lin. long; lateral sepals diverging, the basal fourth connate, obliquely ovate, obtuse, concave, 3 lin. long; petals obliquely decurved, ovate, obtuse, conduplicately folded, subsaccate at the base, 3 lin. long; lip superior, broadly unguiculate, limb erect, recurved, oblong, obtuse, papillose and with 2 rounded auricles at the base, 3½ lin. long; appendage broadly elliptic or suborbicular, obtuse or emarginate, about half as long as the limb; column short; arms of rostellum rounded, short. *Pterygodium Flanagani, Bolus, Ic. Orch. Austr.-Afr.* i. *t.* 98; *Schlechter in Bull. Herb. Boiss.* 1ʳᵉ *sér.* vi. 854; *Kränzl. Orch. Gen. et Sp.* i. 862.

CENTRAL REGION : Molteno Div. ; Broughton, near Molteno, 6300 ft., *Flanagan,* 1639 !

XLV. CORYCIUM, Sw.

Odd *sepal* erect, narrow, concave; lateral sepals usually united, narrowed at the base and connate into a concave limb, sometimes free and spreading. *Petals* oblique, falcately curved and united to the margins of the dorsal sepal into a hood, usually somewhat contracted in front, concave or obliquely saccate at the base. *Lip* ascending along the face of the column and adnate to it between the stigmatic lobes and arms of the rostellum, narrowed into a claw below, dilated into a reflexed, transversely lunate or bilobed, more rarely oblong or lanceolate limb above; produced above the junction with the column into a large reflexed or erect variously shaped fleshy appendage. *Column* short, dilated, produced in front into two horizontal arms, holding at their extremities the glands of the pollinia. *Anther-cells* more or less distant, placed sometimes in front of, sometimes behind the arms of the rostellum, erect or ascending, glands of the pollinia uppermost. *Stigma* posticous, pulvinate, lunate or bilobed, or with two lateral and distant stigmas. *Capsule* elliptic or oblong, usually much narrowed above, ribbed.

Terrestrial herbs, with undivided, sometimes stalked tubers ; leaves radical and cauline, linear or oblong, suberect or sometimes spreading, flat or sometimes crisped ; flowers small or medium-sized, in erect often dense spikes ; bracts lanceolate.

DISTRIB. Species 14, all extratropical South African.

A natural and well-defined genus, readily separated from *Pterygodium* by the contracted flowers and concave or saccate petals.

Lateral sepals free or nearly so and more or less
    spreading :
  Leaves linear or lanceolate-linear, narrow :
    Limb of lip broadly hastate or flabellate :
      Appendage of lip bilobed or bipartite :
        Limb of lip broadly hastate, subacute, sub-
          entire ... ... (1) **rubiginosum.**
        Limb of lip flabellate, obtuse and crenulate ... (2) **venosum.**
      Appendage of lip galeate, with an incurved
        beak-like apex ... ... (3) **carnosum.**
    Limb of lip strongly three-lobed, with diverging
      subulate side lobes ... ... (4) **tricuspidatum.**
  Leaves oblong or linear-oblong, ¾–1¾ in. broad ... (5) **magnum.**
Lateral sepals more or less united into a concave limb
    situated behind the lip :
  Plant turning black in drying ; flowers dark purple ... (6) **nigrescens.**
  Plant not turning black in drying ; flowers yellow,
    greenish or dusky-brown :
    Leaves with strongly undulate or crisped margin ... (7) **crispum.**
    Leaves with flat or nearly uniform margin :
      Flowers yellow or greenish ; leaves not withered
        at flowering time :
        Flowers longer than broad, over ¼ in. long :
          Limb of lip divergently bilobed ... ... (8) **orobanchoides.**
          Limb of lip obovate-orbicular ... ... (9) **deflexum.**
        Flowers globose, under ¼ in. long :
          Limb of lip broadly dilated in front ; flowers
            yellow :
            Limb of lip shortly and obcordately bi-
              lobed ... ... (10) **bicolorum.**
            Limb of lip broadly dilated from a cuneate
              base, emarginate or shortly bilobed ... (11) **excisum.**
            Limb of lip ligulate or lanceolate ; flowers
              white ... ... (12) **bifidum.**
      Flowers smoky or ashy-brown or the leaves more
        or less shrivelled at flowering time :
        Limb of lip linear or oblong-linear ... ... (13) **microglossum.**
        Limb of lip obovate-oblong ... ... (14) **vestitum.**

1. C. rubiginosum (Rolfe); plant 7–10 in. high; stem rather slender, flexuous, leafy; leaves cauline, suberect, linear-lanceolate, acuminate, $\frac{3}{4}$–1$\frac{1}{4}$ in. long; scapes 7–10 in. high; spikes 1–2$\frac{1}{2}$ in. long, somewhat dense, many-flowered; bracts ovate-lanceolate or oblong-lanceolate, acute, 5–7 lin. long; pedicels 3–5 lin. long; flowers rather small, dull red with the appendage of the lip yellow; dorsal sepal narrowly ovate, subacute, concave, 2$\frac{1}{2}$–3 lin. long; lateral sepals spreading, ovate, very concave, about 3$\frac{1}{4}$ lin. long; petals erect, obversely triangular from a cuneate base, subacute at the inner angle, then subtruncate, with an undulate margin, 3 lin. long; lip broadly unguiculate; limb reflexed, broadly hastate, subacute, carinate, 2$\frac{1}{2}$ lin. long, and fully as broad; appendage erect, elliptic-obovate, bilobed at the apex, 1–1$\frac{1}{2}$ lin. long; column broad; arms of the rostellum rounded, about 1 lin. long. *Ptery-godium rubiginosum, Sond. in Linnæa,* xx. 220, *name only; Bolus in Journ. Linn. Soc.* xx. 486, xxv. 208, *and Ic. Orch. Austr.-Afr.* i. *t.* 50; *Durand & Schinz, Conspect. Fl. Afr.* v. 118; *Schlechter in Bull. Herb. Boiss.* 1re *sér.* vi. 820; *Kränzl. Orch. Gen. et Sp.* i. 863.

COAST REGION: Tulbagh Div.; mountain sides near Tulbagh Waterfall, 1300–1500 ft., *Guthrie.* Cape Div.; Constantia Mountains, 3000 ft., *Bodkin.* Stellenbosch Div.; Jonkerhoek, near Stellenbosch, 3700 ft., *Marloth.* Swellendam Div.; near Zondereinde River, 500–1000 ft., *Zeyher,* 3946! *Pappe,* 34!

2. C. venosum (Rolfe); plant $\frac{1}{2}$–$\frac{3}{4}$ ft. high; stem rather stout, leafy; leaves cauline, rather numerous, suberect, linear, acute, $\frac{3}{4}$–2 in. long; scapes $\frac{1}{2}$–$\frac{3}{4}$ ft. high; spikes oblong, 2$\frac{1}{2}$–3$\frac{1}{2}$ in. long, dense, many-flowered; bracts deltoid-lanceolate, long-acuminate, 4–8 lin. long; pedicels 3–4 lin. long; flowers rather small, sepals green and dull purple, petals and appendage of lip pink, lip with radiating pink stripes on a pale ground, tubercle of lip green, and anthers red-brown; dorsal sepal reflexed, elliptic-ovate, obtuse, concave, scarcely 2 lin. long; lateral sepals spreading, elliptic-ovate, concave, scarcely 2 lin. long; petals deflexed, broadly unguiculate, flabellate, crenulate and concave above, 2$\frac{1}{2}$ lin. long by nearly as broad; lip broadly unguiculate; limb broadly flabellate, deeply crenate, 2$\frac{1}{2}$ lin. broad, subconcave, with a prominent rhomboid fleshy crest; appendage erect, bipartite, with flattened horn like obtuse curved lobes, 1$\frac{1}{2}$ lin. long; column short; arms of the rostellum auriculate, fleshy, short. *Pterygodium venosum, Lindl. Gen. & Sp. Orch.* 367; *Harv. Thes. Cap.* i. 59, *t.* 94; *Bolus in Journ. Linn. Soc.* xxv. 208, *and Ic. Orch. Austr.-Afr.* i. *t.* 49; *Durand & Schinz, Conspect. Fl. Afr.* v. 118; *Kränzl. Orch. Gen. et Sp.* i. 863. *Ceratandra venosa, Schlechter in Engl. Jahrb.* xxiv. 433, *and in Bull. Herb. Boiss.* 1re *sér.* vi. 906.

SOUTH AFRICA: without locality, *Reeves! Masson!*
CAPE REGION: Tulbagh Div.; Great Winterhoek, 3000 ft., *Pappe!* mountain sides near Tulbagh Waterfall, 1300–1500 ft., *Guthrie.* Caledon Div.; near Palmiet River, *Bolus,* 5281; Villiersdorp, 2000 ft., *Bolus,* 5281! sands of Houw Hoek, *Bowie!*

3. **C. carnosum** (Rolfe); plant $\frac{1}{2}$–$1\frac{1}{4}$ ft. high, turning very black in drying; stem moderately stout, flexuous; leaves cauline, 4–7, sessile, linear, attenuate or acuminate above, 2–5 in. long, reduced upwards into the bracts; scapes $\frac{1}{2}$–$1\frac{1}{4}$ ft. high; spikes 2–5 in. long, very dense and many-flowered; bracts ovate or ovate-lanceolate, acute, 4–7 lin. long; pedicels about 4 lin. long; flowers rather small, contracted at the mouth, purple or light purple, with a whitish lip; dorsal sepal narrowly ovate, acute, concave, 3 lin. long; lateral sepals subconnivent, broadly and obliquely ovate, subacute or apiculate, $2\frac{1}{2}$–3 lin. long; petals broadly and obliquely obovate, subacute at the inner angle, truncate or sinuate and incurved at the apex, deeply concave, 4 lin. long; lip broadly unguiculate; limb reflexed, broadly flabellate from a cuneate base, emarginate, 2 lin. broad; appendage erect, galeate, bent forwards and rounded in the middle, with a beak-like incurved obtuse apex, $1\frac{1}{2}$ lin. long; column very short, with a transverse row of thick horizontal hairs at the base behind; arms of the rostellum rounded, about 1 lin. long. *Pterygodium carnosum, Lindl. Gen. & Sp. Orch.* 367; *Drège, Zwei Pfl. Documente,* 82; *Krauss in Flora,* 1845, 306, *and Fl. Cap- und Natal.* 158; *Sond. in Linnæa,* xx. 220; *Bolus in Trans. S. Afr. Phil. Soc.* v. 189, *t.* 12; *Orch. Cap. Penins.* 189, *t.* 12, *and in Journ. Linn. Soc.* xx. 486, xxv. 209; *Durand & Schinz, Conspect. Fl. Afr.* v. 117; *Schlechter in Bull. Herb. Boiss.* 1re sér. vi. 818; *Kränzl. Orch. Gen. et Sp.* i. 864.

SOUTH AFRICA : without locality, *Oldenburg,* 1096 !
COAST REGION : Worcester Div. ; Dutoits Kloof, 3000–4000 ft., *Drège,* 1244 ! Cape Div. ; Table Mountain, 2200–3500 ft., *Bolus,* 3879, 4547 ! *Mund* ! *Prior* ! *Kitching* ! *Harvey* ! Muizen Berg, 1300 ft., *Bolus in MacOwan & Bolus, Herb. Norm. Austr.-Afr.* 182 ! *Schlechter,* 163 ; near Constantia, *Wolley-Dod,* 2151 ! Steen Berg, *Wolley-Dod,* 2155 ! Waai Vley, *Wolley-Dod,* 2237. Stellenbosch Div. ; Jonkers Hoek, *Zeyher,* 3950 ! Caledon Div. ; near Genadendal, *Krauss.* Swellendam Div. ; near the Zondereinde River, *Zeyher,* 3950 ! *Pappe* ! 33 ! Riversdale Div. ; Langeberg Range, near Riversdale, 1400 ft., *Schlechter,* 2026. Humansdorp Div. ; near Storms River, 300 ft., *Schlechter,* 5959 !
EASTERN REGION : Natal ; Shafton, Howick, *Mrs. Hutton,* 152 !

4. **C. tricuspidatum** (Bolus in Journ. Linn. Soc. xxv. 176, fig. 13); plant $1\frac{1}{4}$ ft. high, all but the individual flowers turning black in drying; stems stout, straight, leafy; leaves cauline, somewhat spreading, lanceolate, acuminate, 2–$3\frac{1}{2}$ lin. long; scapes $1\frac{1}{4}$ ft. high; spikes 4–6 in. long, dense, many-flowered; bracts long-acuminate from a broad sheathing base, 7–10 lin. long; pedicels about 4 lin. long; flowers rather small; dorsal sepal broadly ovate, obtuse, concave, $2\frac{1}{2}$ lin. long; lateral sepals spreading, broadly elliptic-oblong, obtuse, concave, $2\frac{1}{2}$ lin. long; petals obliquely ovate or suborbicular, subobtuse, very concave, $2\frac{1}{2}$ lin. long, forming with the dorsal sepal a globose galea; lip broadly unguiculate; limb reflexed, cuneately trilobed, $2\frac{1}{2}$ lin. long; side lobes diverging, oblong-lanceolate, acute : front lobe narrowly subulate, about a third as large as the side lobes ; appendage very small and fleshy, transversely oblong and somewhat

bilobed, with crisped-undulate margin, about a sixth as long as the limb and broader than long; column short; arms of the rostellum strongly incurved, very concave, broadly oblong, subobtuse, 1½ lin. long. *Bolus in Journ. Linn. Soc.* xxv. 207; *Durand & Schinz, Conspect. Fl. Afr.* v. 122; *Känzl. Orch. Gen. et Sp.* i. 875, *and in Ann. Naturhist. Hofmus. Wien,* xx. 11. *Pterygodium tricuspidatum, Schlechter in Bull. Herb. Boiss.* 1ʳᵉ sér. vi. 821.

CENTRAL REGION : Cradock Div. ; near Cradock, *Cooper,* 1321, partly !
EASTERN REGION : Griqualand East ; foot of the Zuurberg Range, 4800 ft., *Schlechter,* 6594. Natal ; near Charlestown, 5000, *Wood.*

5. **C. magnum** (Rolfe); plant 2–2½ ft. high; stem stout, often very stout, leafy and with 2 or 3 short basal sheaths; leaves cauline, 5–7, sessile, suberect, oblong-lanceolate, acute or acuminate, sheathing at the base, 4–9 in. long, ¾–1¾ in. broad; scapes 2–2½ ft. high; spikes 6–18 in. long, cylindric, dense, many-flowered; bracts lanceolate, acuminate, ¾–1½ in. long; pedicels 6–8 lin. long; flowers medium-sized, green striped with light red; dorsal sepal oblong-lanceolate, shortly acuminate, concave, 5–5½ lin. long; lateral sepals spreading, ovate-lanceolate, acuminate, subconcave, 4½–5 lin. long; petals obliquely suborbicular-oblong, obtuse, with the outer margin strongly fringed, 5–5½ lin. long; lip sharply reflexed, cuneate-obovate or flabellate, obtuse, fringed at the apex, 2 lin. long; appendage erect, unguiculate, cucullate, 4½–5 lin. long, broadly deltoid or subtrilobed, obtuse, saccate behind, with a pair of obtuse basal angles projected in front; column short; arms of the rostellum oblong, obtuse, 1 lin. long; capsule sessile, oblong, somewhat attenuate above, 8 lin. long. *Pterygodium magnum, Reichb. f. in Flora,* 1867, 117; *Bolus in Journ. Linn. Soc.* xxii. 75, xxv. 208; *Durand & Schinz, Conspect. Fl. Afr.* v. 117; *Schlechter in Bull. Herb. Boiss.* 1ʳᵉ sér. vi. 817; *Känzl. Orch. Gen. et Sp.* i. 860, *and in Ann. Naturhist. Hofmus. Wien.* xx. 10.

COAST REGION : Bedford Div. ; summit of Kaga Berg, 3000 ft., *Hutton !* Albany Div. ; mountains near Grahamstown, 2000 ft., *Glass in MacOwan & Bolus, Herb. Norm. Austr.-Afr.* 1537 !
CENTRAL REGION : Somerset Div. ; slopes of Bosch Berg, *MacOwan.*
KALAHARI REGION : Orange River Colony ; Harrismith, *Sankey,* 273 ! Transvaal ; near Barberton, 4000–5000 ft., *Culver*; Houtbosch Mountains, 6500 ft., *Schlechter,* 4475.
EASTERN REGION : Transkei ; Tsomo, *Mrs. Barber,* 843 ! Griqualand East ; Pumungwana Mountains, near Clydesdale, 3500 ft., *Tyson,* 2707 ! *Glass.* Natal ; near the Drakensberg Range, *Wood,* 3424 ! Dargle Farm, *Mrs. Fannin,* 78 ! hillsides near Lynedoch, 4000–5000 ft., *Wood,* 1014 ! Inanda, *Wood,* 909 ! Shafton, Howick, *Mrs. Hutton,* 229 ! and without precise locality, *Cooper,* 3621 ! *Hallack,* 4 ! *Buchanan in Herb. Sanderson,* 1057 !

6. **C. nigrescens** (Sond. in Linnæa, xix. 110) ; plant ½–2 ft. high ; whole plant turning black in drying ; stem moderately stout, some-what flexuous ; leaves cauline, somewhat spreading, lanceolate, very acuminate, sheathing at the base, 2–6 in. long, reduced upwards

into the bracts; scapes $\frac{1}{2}$–2 ft. high; spikes cylindric, 2–6 in. long, dense, many-flowered; bracts ovate-lanceolate, acuminate, 5–9 lin. long; pedicels 4–5 lin. long; flowers small, globose, dark brown or dark purple; dorsal sepal suborbicular, obtuse, very concave, 2 lin. long; lateral sepals connate into a very broad suborbicular deflexed limb, obtuse or emarginate, very concave, $2\frac{1}{2}$–3 lin. long; petals obliquely orbicular, obtuse or obliquely apiculate, deeply saccate, with the outer margin broadly dilated and subconduplicate, $2\frac{1}{2}$ lin. long; lip broadly unguiculate; limb exserted and reflexed, pandurate-oblong, emarginate, concave, 2 lin. long; appendage forming a tall keel on the claw of the lip; column short; arms of the rostellum obliquely oblong, subacute, strongly incurved and twisted, about 2 lin. long. *Bolus in Journ. Linn. Soc.* xxv. 207; *Durand & Schinz, Conspect. Fl. Afr.* v. 122; *Kränzl. Orch. Gen. et Sp.* i. 876, *and in Ann. Naturhist. Hofmus. Wien,* xx. 11. *Pterygodium nigrescens, Schlechter in Bull. Herb. Boiss.* 1re *sér.* vi. 847.

COAST REGION: George Div.; near George, 600 ft., *Mund, Schlechter,* 2326. Alexandria Div.; near Quaggas Flats, *Zeyher.* Albany Div.; near Grahamstown, 2000–2200 ft., *Read*! *Glass, Zeyher.* Fort Beaufort Div.; Winterberg Range, *Mrs. Barber,* 437! 520! Stockenstrom Div.; Chumie Peak, near Stockenstrom, *Scully in Herb. Bolus,* 5919! Katberg, 2000 ft., *Hutton*!
CENTRAL REGION: Cradock Div.; near Cradock, *Cooper,* 1321, partly!
KALAHARI REGION: Orange River Colony; near Harrismith, *Sankey,* 267! Basutoland; Mametsana, *Dieterlen,* 484! Transvaal; Little Lomati Valley, near Barberton, 3500 ft., *Culver,* 59; near Bergendal, 6300 ft., *Schlechter,* 4013! Belfast, 6500 ft., *Miss Doidge,* 4807! Ermelo, *Miss Leendertz,* 2984!
EASTERN REGION: Transkei; Krielis Country, *Bowker,* 15! Tembuland; Bazeia Mountain, 2000 ft., *Baur,* 119! Griqualand East; Zuurberg Range, 4000–5000 ft., *Schlechter, Krook.* Natal; Pine Town, 1000 ft., *Sanderson,* 485! Inanda, 1800 ft., *Wood,* 530! 988! summit of the Drakensberg Range, on Satsannas Peak, 9200 ft., *Galpin,* 6844! near Bothas Hill, 2500 ft., *Wood*; near Van Reenen, 5500 ft., *Schlechter,* 6963; and without precise locality, *Gueinzius*! *Sanderson,* 482! *Buchanan*! *Mrs. Fannin*! 70! *Gerrard,* 748!

7. **C. crispum** (Sw. in Vet. Acad. Handl. Stockh. 1800, 222); plant 9–12 in. high; stem stout, nearly straight; leaves cauline, somewhat spreading, sessile from a sheathing base, linear-lanceolate or attenuate upwards and very acuminate, with strongly crisped-undulate margin, 3–5 lin. long; scapes 9–12 in. long; spikes oblong, 2–4 in. long, dense, many-flowered; bracts ovate, shortly acuminate, 6–8 lin. long; pedicels 5–6 lin. long; flowers medium-sized, pale yellow with the apex of the petals and limb of the lip deeper yellow; dorsal sepal elliptic-oblong, obtuse or slightly dilated and emarginate at the apex, $3\frac{1}{2}$ lin. long; lateral sepals connate to or beyond the middle into a suborbicular limb, 4 lin. long, slightly concave, lobes obtuse or apiculate; petals obliquely obovate-orbicular, conduplicately folded, obtuse at the inner angle, with very sinuate rounded margin, 4 lin. long, obtusely saccate at the base; lip broadly unguiculate; limb sharply reflexed, flabellate from a broadly cuneate base, emarginate, 2 lin. long, with rounded lobes; appendage bipartite, divaricately recurved, oblong, dilated, truncate

and crenulate at the apex, 2½ lin. long ; column short ; arms of the
rostellum rounded, short, recurved.　*Thunb. Fl. Cap. ed. Schult.* 20 ;
Lindl. Gen. & Sp. Orch. 368 ; *Bolus in Journ. Linn. Soc.* xxv. 207,
and Ic. Orch. Austr.-Afr. i. *t.* 45 ; *Durand & Schinz, Conspect. Fl.
Afr.* v. 121 ; *Kränzl. Orch. Gen. et Sp.* i. 872, *and in Ann. Naturhist.
Hofmus. Wien,* xx. 11.　*C. bicolor, Ker in Quart. Journ. Sci. & Arts,*
vi. 45, *t.* 1, *fig.* 1.　*Arethusa crispa, Thunb. Prodr.* 3.　*Pterygodium
crispum, Schlechter in Bull. Herb. Boiss.* 1ʳᵉ *sér.* vi. 849.—*Orchis
coccinea, foliis serratis, Buxb. Pl. Min. Cogn. Cent.* iii. 7, *t.* 11.

SOUTH AFRICA : without locality, *Masson* ! *Bergius, Leibold, Rogers, Grey* !
Hesse ! *Bachmann,* 35, 930, 931.
COAST REGION : Clanwilliam Div. ; sandy places near Clanwilliam, 300 ft.,
Schlechter, 5068, 8600 ! Piquetberg Div. ; foot of Piquetberg, 1000 ft., *Schlechter,
Penther.*　Malmesbury Div. ; near Groene Kloof, *Thunberg* ! *Bolus,* 3934 !
Lauuws Kloof, under 1000 ft., *Drège,* 587 ! Cape Div. ; Sandy Downs near Tiger
Berg, 100 ft., *Bolus,* 3934, *Schlechter* ; Rapen Berg, near Cape Town, *Guthrie* ;
between Cape Town and False Bay, *Mund* ; Sandhills near Duinefontein, *Wolley-
Dod,* 1627 ! Cape Flats, *Pappe,* 31 !
WESTERN REGION : Little Namaqualand ; hills near Vogel Klip, 2800 ft.,
Schlechter, 11302 !

8. **C. orobanchoides** (Sw. in Vet. Acad. Handl. Stockh. 1800,
222) ; plant 4–15 in. high ; stem stout, nearly straight ; leaves
cauline, somewhat spreading, sessile from a broad sheathing base,
linear-lanceolate or attenuate and long-acuminate, with entire
margin, 3–6 in. long, the upper rather shorter ; scapes 4–15 in.
high ; spikes oblong, 2–7 in. long, dense, many-flowered ; bracts
ovate or broadly ovate, acute or subacute, 4–6 lin. long ; pedicels
3–4 lin. long ; flowers medium-sized, greenish-yellow with crimson
tips to the petals, a white lip and crimson anther-cells ; dorsal
sepal obovate-oblong, obtuse, subconcave, 2½–3 lin. long ; lateral
sepals connate into an orbicular deflexed limb, obtuse or emarginate,
concave, 2¼–2½ lin. long ; petals very oblique, broadly oblong,
obscurely apiculate, with a rounded crenulate subconduplicate outer
margin, deeply saccate at the base, 3½ lin. long ; lip broadly unguiculate ; limb deflexed, divergently bilobed from a cuneate base,
1 lin. long ; lobes broadly oblong, obtuse, minutely crenulate ;
appendage rising shortly in front, then sharply reflexed and
extended into a pair of diverging subulate lobes 2 lin. long, which
curve round into the spur of the petals ; column short ; arms of the
rostellum broadly oblong, apiculate, short and spirally curved.
Ker in Quart. Journ. Sci. & Arts, viii. 222, *t.* 3, *fig.* 3 ; *Thunb. Fl.
Cap. ed. Schult.* 20 ; *Lindl. Gen. & Sp. Orch.* 369 ; *Bot. Reg.* 1838,
t. 45 ; *Drège, Zwei Pfl. Documente,* 98, 101 ; *Sond. in Linnæa,* xix.
111, xx. 221 ; *Bolus in Trans. S. Afr. Phil. Soc.* v. 181, *Orch. Cap.
Penins.* 181, *and Journ. Linn. Soc.* xxv. 207 ; *Durand & Schinz,
Conspect. Fl. Afr.* v. 122 ; *Kränzl. Orch. Gen. et Sp.* i. 873, *and in
Ann. Naturhist. Hofmus. Wien,* xx. 11.　*Satyrium orobanchoides, Linn.
f. Suppl.* 402 ; *Thunb. Prodr.* 6.　*Pterygodium orobanchoides,
Schlechter in Bull. Herb. Boiss.* 1ʳᵉ *sér.* vi. 848.

SOUTH AFRICA: without locality, *Masson*! *Thunberg*! *Brown*! *Lichtenstein,*
Mund, Bergius, Leibold, Roxburgh! *Niven*! *Rogers, Harvey,* 144! *Thom*! *Olden-*
burg, 683!
COAST REGION : Clanwilliam Div. ; by the Olifants River and near Brakfontein,
Zeyher, 3951! Olifants River Mountains near Clanwilliam, 300 ft., *Schlechter,*
Penther, Krook. Piquetberg Div. ; near Piquetberg Road, 500 ft., *Schlechter,*
2140. Paarl Div. ; by the Berg River, near Paarl, under 500 ft., *Drège*; Flats
between Paarl and French Hoek, *Drège,* 4782a! French Hoek Pass, *Harvey*!
Cape Div. ; sandy places near Cape Town, Rondebosch, etc., mostly under 100 ft.,
Bolus, 3935, and in *MacOwan & Bolus, Herb. Norm. Austr.-Afr.* 181 ! *Pappe,* 32!
Prior! *Rogers*! *Rehmann,* 1867! Devils Peak, *Wilms,* 3664! Stellenbosch Div. ;
near Stellenbosch, *Lloyd in Herb. Sanderson,* 936 ! *Brown*! Somerset West,
Ecklon & Zeyher, 76 ! *MacOwan, Tyson,* 2139, *Kässner.* Table Mountain, *Ecklon,*
217, *Dümmer,* 556! Lion Mountain, 200 ft., *Schlechter,* 1334! *Wolley-Dod,* 367 !
Green Point Common, *Wolley-Dod,* 1539! Nord Hoek, *Mrs. Jameson*! Caledon
Div.; Genadendal, *Prior*! Albany Div. : Grahamstown, *Atherstone*!

9. **C. deflexum** (Rolfe); plant 5–8 in. high; stem moderately
stout, somewhat flexuous ; leaves 3 or 4 spreading, linear, acuminate,
3–4 in. long, with somewhat broader sheathing base ; scapes 5–8 in.
high ; spikes 1–2½ in. long, somewhat dense, many-flowered; bracts
ovate, subacute, 5–6 lin. long; pedicels 3–4 lin. long; flowers
medium-sized, yellow with the appendage of the lip green ; dorsal
sepal obovate-oblong, obtuse, very concave at the apex, 3½ lin. long ;
lateral sepals connate nearly to or beyond the middle into an orbicular concave limb, 3 lin. long, lobes obtuse and conduplicately
folded at the outer margin ; petals obliquely ovate, obtuse with
broadly rounded outer margin, conduplicately folded, saccate at the
base, 3½ lin. long ; lip broadly unguiculate; limb sharply reflexed,
obovate-orbicular, obscurely apiculate, 1½ lin. long ; appendage
large and fleshy, hastate-reniform, very obtuse in front, 3 lin. broad ;
basal lobes extending behind in two falcately curved obtuse lobes,
about 1½ lin. long ; column short; arms of the rostellum twisted,
oblong, short. *Pterygodium deflexum, Bolus in Trans. S. Afr. Phil.*
Soc. xvi. 151, *and Ic. Orch. Austr.-Afr.* ii. *t.* 100.

COAST REGION : Clanwilliam Div. ; Koude Berg, near Wupperthal, 2500 ft.,
Bolus, 660 !

10. **C. bicolorum** (Sw. in Vet. Acad. Handl. Stockh. 1800, 222);
plant 4–15 in. high; stem stout, erect; leaves cauline, numerous,
somewhat spreading, sessile from a broad sheathing base, linear-
lanceolate, attenuate upwards and acuminate, 2–4 in. long, upper
somewhat smaller; scapes 4–15 in. high; spikes oblong, 2–7 in.
long, very dense, many-flowered ; bracts ovate, acute, 3–5 lin. long;
pedicels about 2 lin. long; flowers small, globose, clear sulphur-
yellow ; dorsal sepal elliptic-oblong, obtuse, 2 lin. long; lateral
sepals connate into an orbicular deflexed limb, emarginate, subconcave, 2 lin. long ; petals obliquely ovate, obtuse, conduplicately
folded, with a rounded basal auricle, broadly saccate at the base,
2 lin. long; lip broadly unguiculate; limb sharply reflexed, obcordately bilobed, ¾ lin. long, with broad obtuse lobes ; appendage rising

from the limb as a tall erect keel, then dilated into a broadly
hastate truncate limb, 1 lin. long; column short; arms of the
rostellum suborbicular, very short. *Bolus in Trans. S. Afr. Phil.
Soc.* v. 179, *Orch. Cap. Penins.* 179, *Journ. Linn. Soc.* xxv. 207, *and
Ic. Orch. Austr.-Afr.* i. *t.* 47; *Durand & Schinz, Conspect. Fl. Afr.* v.
121; *Kränzl. Orch. Gen. et Sp.* i. 874, *and in Ann. Naturhist. Hofmus.
Wien,* xx. 11. *C. bicolor, Thunb. Fl. Cap. ed. Schult.* 21; *Bauer, Ill.
Orch. Gen. t.* 15; *Lindl. Gen. & Sp. Orch.* 368; *Sond. in Linnæa,*
xix. 111, xx. 220. *Ophrys bicolor, Thunb. Prodr.* 2. *Pterygodium
bicolorum, Schlechter in Bull. Herb. Boiss.* 1re *sér.* vi. 851.

SOUTH AFRICA: without locality, *Forster*! *Thunberg*! *Masson, Niven*! *Nelson*!
COAST REGION: Tulbagh Div.; Tulbagh Waterfall, 1000 ft., *Zeyher, Ecklon &
Zeyher,* 75! *Bolus*; Great Winterhoek, *Pappe*! Worcester Div.; Bains Kloof,
1600 ft., *Schlechter,* 9140! Paarl Div.; hills near Wellington, *Miss Cummings.*
Cape Div.; mountains near Cape Town, *Ecklon & Zeyher*; slopes of Lion Head
Mountain, *Bolus,* 2856; Steen Berg, *Bergius.* Stellenbosch Div.; Hottentots
Holland, *Zeyher.* Caledon Div.; Genadendal, *Zeyher.* Swellendam Div.; hills
by the Buffeljagts River, 1000–2000 ft., *Zeyher,* 3952! Breede River, *Zeyher*!
Riversdale Div.; between Great Valsh River and Zoetmelks River, *Burchell,*
6605! near Vet River, 350 ft., *Schlechter,* 1978. George Div.; Outeniqua
Mountains, *Mund.* Knysna Div.; plains of Knysna, *Bowie*!
CENTRAL REGION: Ceres Div.; near Ceres, 1500 ft., *Bolus.*

11. C. excisum (Lindl. Gen. & Sp. Orch. 368); plant 4–7 in. high;
stem moderately stout, erect; leaves cauline, suberect or somewhat
spreading, linear from a broader sheathed base, attenuate and
narrowly acuminate above, 1–2½ lin. long; scapes 4–7 in. high;
spikes oblong, 1–2½ in. long, dense, many-flowered; bracts ovate or
elliptic-ovate, obtuse, 3–4 lin. long; pedicels 2–3 lin. long; flowers
rather small, light yellow, with the apex of the petals light yellow
and the lip with its appendage green; dorsal sepal lanceolate-oblong,
obtuse, slightly concave, 3 lin. long; lateral sepals connate into an
elliptic-orbicular limb, emarginate, very concave, 3 lin. long; petals
obliquely and broadly ovate, obtuse, conduplicately folded near the
inner margin, the outer margin broadly rounded or subauriculate,
very concave, 3 lin. long; lip broadly unguiculate; limb sharply
reflexed, broadly dilated from a cuneate base, emarginate or
shortly bilobed, crenulate in front, 1½ lin. long; appendage rising
from the limb as a tall erect keel, then dilated into a pair of almost
reniform convex lobes, about 1 lin. broad; column short; arms of
the rostellum rounded, rather short. *Drège, Zwei Pfl. Documente,*
109; *Sond. in Linnæa,* xix. 111, xx. 220; *Bolus in Trans. S. Afr.
Phil. Soc.* v. 182, *t.* 20, *Orch. Cap. Penins.* 182, *t.* 20, *and Journ.
Linn. Soc.* xxv. 207; *Kränzl. Orch. Gen. et Sp.* i. 874; *Durand &
Schinz, Conspect. Fl. Afr.* v. 122. *Pterygodium excisum, Schlechter
in Bull. Herb. Boiss.* 1re *sér.* vi. 852.

SOUTH AFRICA: without locality, *Harvey,* 143! 148!
COAST REGION: Clanwilliam Div.; between Lange Vallei and Olifants River,
1000–1500 ft., *Drège,* 8283! Tulbagh Div.; near Tulbagh Waterfall, *Zeyher,*
1576! Witsenberg Range, *Zeyher*; Great Winter Hoek, *Pappe*! Paarl Div.; sand
flats at Drakenstein, *Harvey*! Cape Div.; sandy places near Rondebosch, 50 ft.,

Bolus, 4832; and in *MacOwan & Bolus*, *Herb. Norm. Austr.-Afr.* 180 ! Tokay Flat, *Wolley-Dod*, 3650 ! Muizen Berg, 800 ft., *Bolus*. Bredasdorp Div. ; Koude River, 1000 ft., *Schlechter*, 9625 !

12. **C. bifidum** (Sond. in Linnæa, xix. 111); plant ½–1½ ft. high ; leaves distichous, narrow and long-acuminate from a broad sheathing base, 4 lin. long, the upper shorter and sheath-like; scapes ½–1½ ft. high ; spikes cylindric, 4 lin. long, very dense, many-flowered; bracts ovate, acute, shorter than the flower ; pedicels about 3 lin. long ; flowers rather small, white; dorsal sepal elliptic-oblong, with obtuse incurved apex, 2¼ lin. long ; lateral sepals united to the middle into a broadly suborbicular concave limb, 2½ lin. long, lobes obtuse ; petals obliquely ovate, very obtuse, conduplicately folded, saccate at the base, 2½ lin. long; lip broadly unguiculate ; limb reflexed, ovate-oblong, obtuse, 1½ lin. long ; appendage pandurate, with suborbicular minutely crenulate limb, 1¾ lin. long ; column short ; arms of the rostellum oblong, obtuse, incurved, 1¼ lin. long. *Reichb. f. in Walp. Ann.* i. 805 ; *Bolus in Trans. S. Afr. Phil. Soc.* v. 180, *Orch. Cap. Penins.* 180, *and in Journ. Linn. Soc.* xxv. 207 ; *Kränzl. Orch. Gen. et Sp.* i. 877, *and in Ann. Naturhist. Hofmus. Wien*, xx. 11. *C. ligulatum, Reichb. f. in Linnæa*, xix. 375. *Pterygodium bifidum, Schlechter in Bull. Herb. Boiss.* 1ʳᵉ sér. vi. 856 ; *Bolus in Trans. S. Afr. Phil. Soc.* xvi. 152.

COAST REGION : Cape Div.; mountains near Cape Town, *Ecklon & Zeyher* ! *Gueinzius*. Caledon Div. ; near Caledon, *Goatcher in Herb. Bolus*, 10568.

13. **C. microglossum** (Lindl. Gen. & Sp. Orch. 369) ; plant ½–1 ft. high ; stem very stout; leaves cauline, suberect, linear from a very broadly sheathing imbricate base, acuminate, 2–5 in. long ; scapes ½–1 ft. high; spikes oblong, 1½–5 in. long, very dense and many-flowered ; bracts ovate, acute, 6–9 lin. long ; pedicels 4–5 lin. long ; flowers medium-sized, smoky- or ashy-brown, with darker stripes on the petals, lip pale green, with emerald-green appendage, anther-cells rosy and rostellum purple ; dorsal sepal elliptic-oblong, obtuse, 3½ lin. long ; lateral sepals connate into an orbicular deflexed limb, emarginate or somewhat bilobed at the apex, very concave or sub-saccate at the base, 3 lin. long ; petals obliquely ovate, obtuse, with a broadly rounded or subauriculate basal angle, very concave at the base, somewhat conduplicately folded above, 3½ lin. long ; lip broadly unguiculate ; limb recurved, linear or oblong-linear, obtuse, 2½ lin. long ; appendage broadly unguiculate, broadly elliptic-ovate, obtuse or emarginate, somewhat convex and arching over the column, 2 lin. long ; column short; arms of the rostellum obovate-oblong, obtuse, curved, 1¼ lin. long. *Drège, Zwei Pfl. Documente*, 87 ; *Krauss in Flora*, 1845, 306, *and Fl. Cap- und Natal.* 158 ; *Sond. in Linnæa*, xix. 111, xx. 220 ; *Bolus in Journ. Linn. Soc.* xxv. 207, *and Ic. Orch. Austr.-Afr.* i. t. 46 ; *Kränzl. Orch. Gen. et Sp.* i. 876, *and in Ann. Naturhist. Hofmus. Wien*, xx. 11. *Pterygodium microglossum, Schlechter in Bull. Herb. Boiss.* 1ʳᵉ sér. vi. 853.

COAST REGION : Malmesbury Div. ; near Klipfontein, *Zeyher*, 1575 ! Tulbagh Div. ; near Tulbagh Waterfall, *Ecklon & Zeyher*, 77 ! Paarl Div. ; Paarl Mountain, 1000–2000 ft., *Drège*, 1242 ! Cape Div. ; Cape Flats, at Vyge Kraal, 60 ft., *Guthrie in Herb. Bolus*, 7098 ! *Schlechter*, 206 ; between Tokai and Steen Berg, 130 ft., *Fair.*

EASTERN REGION : Natal ; summit of Houtbosch Rand, 4000 ft., *Krauss.*

The Natal locality seems to require confirmation.

14. **C. vestitum** (Sw. in Vet. Acad. Handl. Stockh. 1800, 222) ; plant 10–11 in. high ; stem very stout, straight ; leaves numerous, the 2 or 3 lower oblong, obtuse, somewhat spreading, ¾–1½ in. long, upper reduced to broad imbricating sheaths ; scapes 10–11 in. high ; spikes 3–4 in. long, dense, many-flowered ; bracts broadly ovate, obtuse, 6–7 lin. long ; pedicels much shorter than the bracts ; flowers rather large ; dorsal sepal oblong, concave ; lateral sepals connate into a broad deflexed ovate limb, concave-ventricose at the base ; petals obliquely ovate-orbicular, obtuse with crenulate margin, subsaccate at the base, about 4 lin. long ; lip broadly unguiculate ; limb deflexed, obovate-oblong, crenulate, 1½ lin. long ; appendage strongly recurved, oblong, obtuse, conduplicate at the base, about 4 lin. long ; column short ; arms of the rostellum incurved, oblong, short. *Thunb. Fl. Cap. ed. Schult.* 21 ; *Lindl. Gen. & Sp. Orch.* 369 ; *Bolus in Journ. Linn. Soc.* xxv. 207 ; *Durand & Schinz, Conspect. Fl. Afr.* v. 122 ; *Kränzl. Orch. Gen. et Sp.* i. 873. *Ophrys volucris, Thunb. Prodr.* 2, not of *Linn. f. Pterygodium vestitum, Schlechter in Bull. Herb. Boiss.* 1ʳᵉ sér. vi. 857.

COAST REGION : Piquetberg Div. ; near Piquetberg and Verlooren Valley, *Thunberg* !

Only known from Thunberg's original specimens. The single flower at Kew is imperfect as to the apex of the sepals, and the whole flower is in bad condition.

XLVI. DISPERIS, Sw.

Dorsal *sepal* erect, galeate or calcarate ; lateral sepals oblique or horizontal, dorsally saccate or spurred about the middle. *Petals* oblique, falcately curved, united to the margins of the dorsal sepal, sometimes auriculate on the free margin near the base. *Lip* ascending along the face of the column and adnate to it between the stigmatic lobes and arms of the rostellum, narrowed into a claw below, dilated above into a variously shaped limb, bearing on its face a variously shaped appendage directed towards the apex of the dorsal sepal and sometimes incurved towards its mouth. *Column* erect, usually stout ; rostellum large, membranous, produced in front into two rigid cartilaginous diverging arms (which in the bud are enclosed within the sac of the lateral sepals), holding at their extremities the glands of the pollinia ; anther-bed horizontal or ascending. *Anther-cells* distinct, somewhat approximate ; pollinia with the granules usually large, and secund in a double row on the margin of the flattened caudicles, which curl up in a spiral on removal. *Stigma* bilobed, lobes situated on either side of the adnate claw of the lip, approximate or somewhat distant. *Capsule* cylindric or ovoid-oblong, ribbed.

Terrestrial herbs, usually small or slender, with ovoid tubers; leaves one to few, alternate or sometimes in a single opposite pair, or solitary; flowers in racemes or solitary; bracts medium-sized or large and leaf-like.

DISTRIB. Species about 55, most numerous in extra-tropical South Africa, extending to Tropical Africa, the Mascarene Islands, and South India, with a solitary representative in New Guinea.

*Dorsal sepal with a saccate or saccate-oblong galea, rarely
  twice as long as the limb:
  †Leaves alternate or rarely solitary:
    ‡Leaves narrow or more or less narrowed upwards,
      suberect:
      §Flowers small or medium-sized; sepals acute or
        shortly acuminate:
        Leaves linear; flowers pale green; dorsal sepal
          with a globose constricted galea ...   ... (1) **secunda.**

        Leaves lanceolate or oblong-lanceolate; flowers
          white, pink or marked with purple; dorsal
          sepal with oblong obtuse galea:
        Flowers 3 to many, rarely fewer in weak
          specimens:
          Lateral sepals 4–5 lin. long:
            Pedicels about half as long as the bracts:
              Leaves considerably larger than the
                bracts   ...   . (2) **stenoplectron.**

              Leaves rather small and sheath-like
                (unknown in 4, *D. ermelensis*):
                Bracts ovate, acute ...   . (3) **Cooperi.**

              Bracts ovate-lanceolate, attenuate
                or acuminate:
                Lateral sepals obovate-oblong,
                  saccate near the apex   ... (4) **ermelensis.**

                Lateral sepals ovate-lanceolate,
                  saccate about the middle   ... (5) **Buchananii.**

          Pedicels about three-quarters as long as
            the bracts:
            Appendage of lip triangular-lanceolate,
              cucullate, equalling or slightly
              exceeding the limb:
              Rostellum triangular-ovate, obtuse (6) **natalensis.**

              Rostellum shortly and broadly
                cordate ...   . (7) **Allisonii.**

            Appendage of lip linear, twice as long
              as the limb   ...   . (8) **kermesina.**

          Lateral sepals 2½–3 lin. long   ...   ... (9) **Tysoni.**
        Flowers 1–3:
          Lower leaves lanceolate or elliptic-oblong:
            Limb of lip peltate, bifid at the apex ... (10) **anomala.**

            Limb of lip not peltate, entire:
              Lateral sepals 3–4½ lin. long, acute or
                acuminate:
                Lateral sepals 3–3½ lin. long   ... (11) **concinna.**

Lateral sepals 4–4½ lin. long:
　　Appendage of lip ovate:
　　　Lateral sepals saccate below the
　　　　middle　...　...　... (12) **gracilis.**

　　　Lateral sepals saccate above the
　　　　middle　...　...　... (13) **Wealii.**

　　Appendage of lip linear ...　... (14) **bicolor.**

Lateral sepals 5–7 lin. long, very
　acuminate:
　　Appendage of lip broadly oblong,
　　　entire　...　...　...　... (15) **flava.**

　　Appendage of lip linear with a pair
　　　of triangular auricles at the
　　　base　...　...　...　... (16) **oxyglossa.**

Lower leaves oblong-linear ...　...　... (17) **paludosa.**

§§Flowers large, with caudate or caudate-acuminate
　　sepals　...　...　...　...　...　... (18) **capensis.**

‡‡Leaves usually broad and more or less spreading:
　Leaves narrowed or sheathing at the base or not
　　cordate-auriculate:
　　Flowers rather small and in distichous racemes (19) **cardiophora.**

　Flowers medium-sized, solitary or few:
　　Dorsal sepal 3 to 4 lin. long, with saccate
　　　galea　...　...　...　...　... (20) **villosa.**

　　Dorsal sepal 5 to 6 lin. long, with conical
　　　galea　...　...　...　...　... (21) **cucullata.**

Leaves sessile, broadly cordate-auriculate at the
　base:
　Bracts spathaceous or ovate-lanceolate and
　　suberect:
　　Galea of dorsal sepal short and broadly
　　　saccate, under 5 lin. long:
　　　Flowers yellow; dorsal sepal 3 lin. long ... (22) **bolusiana.**

　　　Flowers purple; dorsal sepal 4 lin. long ... (23) **purpurata.**

　　Galea of dorsal sepal conic-oblong, 5 to 7 lin.
　　　long　...　...　...　...　... (24) **macrocorys.**

　Bracts broadly ovate or cordate and more or less
　　spreading:
　　Dorsal sepal 2–3 lin. long:
　　　Sac of lateral sepals large, about as long as
　　　　the limb:
　　　　Appendage of lip obtuse ...　...　... (25) **Macowani.**

　　　　Appendage of lip shortly bifid at the
　　　　　apex ...　...　...　...　... (26) **Bodkini.**

　　　Sac of lateral sepals small, much shorter
　　　　than the limb:
　　　　Bracts 4–6 lin. long, generally longer
　　　　　than the pedicels　...　...　... (27) **micrantha.**

　　　　Bracts 3–4 lin. long, generally shorter
　　　　　than the pedicels　...　...　... (28) **disæformis.**

　　Dorsal sepal 4 lin. or more long:
　　　Leaves 1 or 2, broadly cordate, subacute:
　　　　Leaves 2, very unequal ...　...　... (29) **Thorncrofti.**

Leaves solitary ... ... (30) **lindleyana.**

Leaves 3, narrowly cordate, acute or
    acuminate ... ... (31) **Fanniniæ.**

††Leaves in an opposite pair :
  Lateral sepals about 3 lin. long, with a large conical
    sac ... ... (32) **virginalis.**

Lateral sepals about 5 lin. long, with a minute
  rounded sac below the middle ... ... (33) **Nelsonii.**

**Dorsal sepal narrowly elongate-oblong, galea about 3 times
  as long as the limb :
Leaves opposite or subopposite, usually ⅔–1 in. broad (34) **anthoceros.**

Leaves alternate or rarely subopposite, seldom over
  ½ in. broad :
Leaves subopposite and subequal ; rostellum rounded
  and very obtuse ... ... (35) **Woodii.**

Leaves alternate and very unequal ; rostellum
  obtusely apiculate ... ... (36) **stenoglossa.**

1. **D. secunda** (Sw. in Vet. Acad. Handl. Stockh. 1800, 220) ;
plant 3–8 in. high ; stem rather slender ; leaves cauline, alternate,
sessile, suberect, usually 2, linear or subulate, acute, sheathing at
the base, 1½–2½ in. long ; scapes 3–8 in. high ; spikes 1–2½ in. long,
subsecund, lax, 3–10-flowered ; bracts lanceolate or ovate-lanceolate,
acute or acuminate, ⅓–¾ in. long, the lower sometimes longer and
leaf-like ; pedicels 3–4 lin. long ; flowers medium-sized, pale sap-
green ; dorsal sepal galeate, broadly rhomboid-ovate, shortly
acuminate, deflexed in front, 2–2½ lin. long, with a broad transverse
mouth, and the sac nearly globose and somewhat constricted at the
base ; lateral sepals spreading, unguiculate, limb falcately ovate,
with an acuminate curved or hooked apex, 4–4½ lin. long, with a
broadly oblong spur about the middle ; petals cohering with the
dorsal sepal, broadly falcate-oblong, 3–3½ lin. long, acuminate and
incurved at the apex, auriculate at the base in front ; lip broadly
unguiculate ; limb broadly ovate, cymbiform, acuminate, about
1 lin. long ; appendage erect, linear, curved, about twice as long as
the limb ; column short ; rostellum broadly ovate, obtuse, convex ;
arms cartilaginous, much twisted, oblong from a broader base,
obtuse, 1¼ lin. long. *Ker in Quart. Journ. Sci. & Arts,* v. 105, *t.* 1,
fig. 3 ; *Thunb. Fl. Cap. ed. Schult.* 26 ; *Lindl. Gen. & Sp. Orch.* 370 ;
Drège, Zwei Pfl. Documente, 83 ; *Sond. in Linnæa,* xx. 221 ; *Krauss
in Flora,* 1845, 306, *and Fl. Cap- und Natal.* 158 ; *Bolus in Trans.
S. Afr. Phil. Soc.* v. 177, *t.* 11, *Orch. Cap. Penins.* 177, *t.* 11,
and in Journ. Linn. Soc. xxv. 206 ; *Schlechter in Bull. Herb. Boiss.*
1ʳᵉ *sér.* vi. 922. *D. circumflexa, Durand & Schinz, Conspect. Fl. Afr.*
v. 118 ; *Kränzl. Orch. Gen. et Sp.* i. 826. *Orchis circumflexa, Linn.
Sp. Pl. ed.* ii. 1344. *Arethusa secunda, Thunb. Prodr.* 3.

VAR. β, æmula (Schlechter in Bull. Herb. Boiss. 1ʳᵉ sér. vi. 923) ; flowers much
larger than in the type, with the dorsal sepal more strongly galeate, and the limb
larger. *D. circumflexa, var. æmula, Kränzl. Orch. Gen. et Sp.* i. 826.

SOUTH AFRICA : without locality, *Masson, Mund, Lalande, Leibold, Thom,* 700 !
Rogers ! *Harvey* ! 150 ! *Zeyher,* 3939 !

COAST REGION: Piquetburg Div.; near Piquiniers Kloof, 850 ft., *Schlechter*, 4945, *Penther*, *Krook*. Malmesbury Div. ; Saldanha Bay, *Bachmann*. Tulbagh Div. ; near Tulbagh Waterfall, 600 ft., *Schlechter*, 1413, *Kässner*. Paarl Div. ; Drakenstein Mountains, 2000–3000 ft., *Drège*, 8279 a ! Cape Div. ; mountains and Flats near Cape Town, 50–250 ft., *Thunberg* ! *Bergius*, *Ecklon* & *Zeyher*, 80 ! *Zeyher* ! *Pappe*, 8 ! *Krauss*, 1327, *Bolus*, 4817 ! *Prior* ! *Schlechter* ! *Wolley-Dod*, 1702 ! Albany Div. ; near Grahamstown, *Williamson* ! Var. β : Clanwilliam Div. ; Olifants River, 400 ft., *Schlechter*, 5013 ! *Penther*, *Krook* ; Olifants River Mountains, 1000 ft., *Schlechter* ; Koudeberg Range, near Wupperthal, 2400 ft., *Bolus*, 9091 !

2. **D. stenoplectron** (Reichb. f. Otia Bot. Hamb. ii. 102) ; plant ¾–1 ft. high ; stem moderately stout ; leaves cauline, alternate, sessile, 2–4, suberect, ovate to ovate-lanceolate, subacute, 1–2 in. long ; scapes ¾–1 ft. high ; spikes 2–4 in. long, lax, 6–10-flowered ; bracts ovate-lanceolate, cucullate, acute or acuminate, ½–1 in. long ; pedicels 4–6 lin. long ; flowers medium-sized, green and purple ; dorsal sepal galeate, broadly ovate, apiculate, about 5 lin. long ; lateral sepals spreading, 4½ lin. long, obliquely elliptic-oblanceolate from a narrow base, abruptly acuminate, with a conic-oblong sac below the middle ; petals cohering with the dorsal sepal, obliquely falcate-oblong, somewhat constricted in front about the middle, about 5 lin. long ; lip broadly unguiculate ; limb pandurate-oblong, subacute and incurved at the apex, with involute margin, and a conical tooth underneath above the middle ; disc papillose ; appendage erect, subulate-linear, obtuse, rather longer than the limb ; column short ; rostellum broadly and transversely oblong, obscurely tridentate at the apex ; arms cartilaginous, spreading, nearly straight, spathulate at the apex, 2 lin. long. *Bolus, Ic. Orch. Austr.-Afr.* i. *t.* 90, *and in Journ. Linn. Soc.* xxv. 206 ; *Durand & Schinz, Conspect. Fl. Afr.* v. 120 ; *Kränzl. Orch. Gen. et Sp.* i. 829 ; *Schlechter in Bull. Herb. Boiss.* 1ʳᵉ *sér.* vi. 921.

SOUTH AFRICA : without locality, *Zeyher.*
COAST REGION : Stutterheim Div. ; north side of Dohne Mountain, 4000 ft., *Sim*, 25 ! and in *Herb. Bolus*, 6267 ! *Flanagan*, 2309.
EASTERN REGION : Griqualand East ; south side of Mount Currie, 5000 ft., *Tyson*, 2525 !

3. **D. Cooperi** (Harv. Thes. Cap. ii. 47, t. 172) ; plant ¾–1¼ ft. high ; stem moderately stout, nearly straight ; leaves cauline, alternate, sessile, 2–4, suberect, ovate to ovate-lanceolate, acute, 1–1½ in. long ; scapes ¾–1¼ ft. high ; spikes 3–4 in. long, rather lax, many-flowered ; bracts ovate, cucullate at the base, acute or acuminate, ¾–1¼ in. long ; pedicels 6–8 lin. long ; flowers medium-sized, green and purple ; dorsal sepal galeate, suborbicular, nearly globose, obtuse, 4–5 lin. long, with an oblong mouth ; lateral sepals spreading, 4½ lin. long, broadly ovate-lanceolate, acute, with a short broadly conical obtuse sac about the middle ; petals cohering with the dorsal sepal, falcate-lanceolate, acute, spotted with purple on the outer margin, 4–4½ lin. long ; lip broadly unguiculate, limb cucullate, ovate, acute, 1½ lin. long ; appendage subulate-lanceolate, subacute,

cucullate, about twice as long as the limb ; column short ; rostellum large, broadly cordate-ovate, subobtuse, broader than long ; arms cartilaginous, twisted, oblong from a broader base, dilated at the apex, 3 lin. long. *Bolus in Journ. Linn. Soc.* xxv. 206 ; *Durand & Schinz, Conspect. Fl. Afr.* v. 119 ; *Schlechter in Bull. Herb. Boiss.* 1re sér. vi. 920 ; *Kränzl. Orch. Gen. et Sp.* i. 827.

KALAHARI REGION : Orange River Colony ; plain on the top of the Drakensberg Range, near Nelsons Kop, *Cooper,* 1100 !

Mrs. Fannin's specimen, included by Harvey in his original description, is different.

4. D. ermelensis (Rolfe) ; base of plant not seen ; stem moderately stout ; spikes about 4$\frac{1}{2}$ in. long, somewhat lax, 10–12-flowered ; bracts distichous, cucullate, cordate-lanceolate, acuminate, amplexi-caul, $\frac{3}{4}$–1$\frac{1}{4}$ in. long ; pedicels 8–9 lin. long ; flowers medium-sized, blue (*Todd*) ; dorsal sepal galeate, about 4$\frac{1}{2}$ lin. long ; limb broadly ovate-triangular, with a distinct narrowly acuminate reflexed apiculus ; galea broadly saccate ; lateral sepals spreading, obliquely obovate-oblong, narrowly acuminate at the apex, concave, 4 lin. long, with a short broadly conical sac near the apex ; petals falcate-oblong, subacute, about 4$\frac{1}{2}$ lin. long, with an undulate front margin ; lip broadly unguiculate, limb cucullate, triangular-lanceolate, acuminate, 2 lin. long ; appendage triangular-lanceolate, subobtuse, rather longer than the limb ; column short ; rostellum broadly cordate, 1$\frac{1}{2}$ lin. broad, base decurrent along the margin of the arms ; arms cartilaginous, triangular-oblong, obtuse, 2 lin. long ; capsule oblong, about 8 lin. long.

KALAHARI REGION : Transvaal ; in grass near Ermelo, *Todd in Herb. Wood.* 3176 !

5. D. Buchananii (Rolfe) ; plant over 1 ft. high ; stem moderately slender ; leaves sheath-like, suberect, oblong-linear, subobtuse, 1–1$\frac{1}{2}$ in. long ; scapes over 1 ft. high, with a few lanceolate sub-acute sheaths above ; spikes about 3 in. high, lax ; bracts ovate-lanceolate, acuminate ; pedicels 6–8 lin. long ; flowers medium-sized, lilac and greenish-white, $\frac{3}{4}$–1 in. long ; dorsal sepal galeate, 4 lin. long, with a triangular subacute deflexed limb and a rounded obtuse galea ; lateral sepals spreading, ovate-lanceolate, acuminate, 4 lin. long, with an oblong obtuse sac about the middle ; petals falcate-lanceolate, acute, 4 lin. long, outer margin undulate ; lip broadly unguiculate ; limb deflexed, triangular-lanceolate, subacute, cucul-late, about 3 lin. long ; appendage triangular-lanceolate, subacute, somewhat curved, cucullate, shorter than the limb ; column short ; rostellum ovate, subobtuse, reflexed at the margin, rather large ; arms cartilaginous, falcate-oblong, obtuse, 2 lin. long.

EASTERN REGION : Natal ; above Richmond, 2000–3000 ft., *Buchanan in Herb. Sanderson,* 1070 !

6. D. natalensis (Rolfe); plant 10–14 in. high; stem moderately stout; leaves 3 or 4, suberect and sheath-like, ovate or ovate-lanceolate, acute or shortly acuminate, sheathing at the base, 1¼–1¾ in. long; scapes 10–14 in. high; spikes 3–4 in. long, lax; bracts ovate-lanceolate, acuminate, cucullate, ½–1 in. long; pedicels 6–9 lin. long; flowers medium-sized; dorsal sepal galeate, 4–4½ lin. long, with rounded obtuse crenulate apex and a broadly saccate galea; lateral sepals spreading, ovate-lanceolate, acute, concave, 5 lin. long, with a short broadly rounded sac about the middle; petals falcate-lanceolate, acute, 4–4½ lin. long, outer margin undulate; lip broadly unguiculate; limb pandurate, acute, 2 lin. long; appendage triangular-oblong, subobtuse, cucullate, rather longer than the limb; column short; rostellum broadly triangular-ovate, obtuse, large; arms cartilaginous, twisted, broadly oblong, obtuse, 2½ lin. long; capsule oblong, about 7 lin. long. *D. Cooperi, Harv. Thes. Cap.* ii. 47, *partly.*

EASTERN REGION: Natal; Dargle Farm, *Mrs. Fannin*, 91!

Quite distinct from the plant figured as *D. Cooperi* by Harvey, though included under the same name by that author.

7. D. Allisonii (Rolfe); plant about 10 in. high; stem moderately stout; leaves 4, suberect, oblong or ovate-oblong, subobtuse, sheathing at the base, about 1 in. long; scapes about 10 in. high; spikes 4 in. long, rather lax; bracts distichous, ovate-lanceolate, acute, ¾–1 in. long, rather lax; pedicels 7–9 lin. long; flowers medium-sized, carmine-rose; dorsal sepal galeate, 4–4½ lin. long, with triangular obtuse deflexed limb and a broadly saccate galea; lateral sepals spreading, oblong-lanceolate, acuminate, 4 lin. long, with a short conic-oblong sac about the middle; petals falcate-oblong, subobtuse, 4–4½ lin. long, outer margin undulate; lip broadly unguiculate; limb pandurate, acute, 2 lin. long; appendage triangular-lanceolate, acute, cucullate, nearly as long as the limb; column short; rostellum broadly cordate, obtuse, deflexed at the apex, broader than long, with sinuate margin; arms cartilaginous, twisted, oblong, obtuse, about 2 lin. long.

EASTERN REGION: Natal; Oliviers Hoek, 5000 ft., *Allison*, 8!

8. D. kermesina (Rolfe); plant 8–12 in. high (base not seen); stem moderately slender; leaves 2, suberect, ovate-lanceolate, acute, sheathing at the base, 10–14 lin. long; scapes 8–12 in. high; spikes 2½–5 in. long, lax; bracts ovate or ovate-lanceolate, acute, 5–10 lin. long; pedicels 5–8 lin. long; flowers medium-sized, carmine; dorsal sepal galeate, 3½–4 lin. long, with short triangular acute deflexed limb and conic-oblong obtuse galea; lateral sepals spreading, elliptic-oblong, acuminate, concave, 3 lin. long, with a short conic-oblong sac about the middle; petals falcate-obovate, acute, 3 lin. long, outer margin obscurely undulate; lip narrowly unguiculate,

limb subpandurate-oblong, acute, 1½ lin. long; appendage linear,
obtuse, over twice as long as the limb; column short; rostellum
ovate-orbicular, obtuse, rather large, with deflexed somewhat
sinuate margin; arms cartilaginous, ovate-oblong, obtuse, 1¼ lin.
long.

EASTERN REGION: Natal; Oliviers Hoek, 5000 ft., *Allison*, 9!

9. **D. Tysoni** (Bolus in Journ. Linn. Soc. xxii. 79, t. 1, fig. F);
plant ½–1¼ ft. high; stem slender, somewhat flexuous; leaves
cauline, alternate, sessile, 3–5, suberect, lanceolate or ovate-
lanceolate, acute, ⅓–1 in. long; scapes ½–1¼ ft. high; spikes
somewhat secund, usually lax, 1–4½ in. long, 2–14-flowered; bracts
ovate-oblong, subacute or obtuse; pedicels 3–5 lin. long; flowers
rather small, pink; dorsal sepal galeate, with an obtuse sac as long
as the limb, ovate, subacute, 2½ lin. long; lateral sepals spreading,
obliquely elliptic-lanceolate, falcately curved and acuminate at the
apex, concave and with an obtuse sac above the middle, 2½ lin.
long; petals oblique, falcate-obovate from a narrow base, 2⅝ lin.
long; lip narrowly unguiculate; limb ovate or ovate-lanceolate,
acute or subacute, prominently bituberculate or gibbous at the base,
1¼–1½ lin. long; appendage subulate, narrowed above the middle,
then subclavate, 2–2¼ lin. long; rostellum large, suborbicular, very
obtuse; arms rounded at the base, narrowed and obliquely twisted
above. *Bolus in Journ. Linn. Soc.* xxv. 207; *Durand & Schinz,
Conspect. Fl. Afr.* v. 121; *Schlechter in Bull. Herb. Boiss.* 1ʳᵉ *sér.* vi.
918; *Kränzl. Orch. Gen. et Sp.* i. 828.

SOUTH AFRICA: without locality, *Bachmann*.
CENTRAL REGION: Somerset Div.; Bosch Berg, *MacOwan*, 521.
KALAHARI REGION: Transvaal; Belfast, 6450–6750 ft., *Burtt-Davy*, 1303!
Miss Doidge, 4814!

EASTERN REGION: Tembuland; wet places on Bazeia Mountain, 2000 ft.,
Baur, 151! Griqualand East; near Kokstad, 4500–5000 ft., *Tyson*, 1079!
summit of the Drakensberg Range, on Satsannas Peak, 9300 ft., *Galpin*, 6843!
Zuurberg Range, 4800 ft., *Schlechter*, 6592. Natal; grassy hill at Van Reenen,
5000–6000 ft., *Wood*, 5545! Mooi River Station; 4500 ft., *Kuntze*.

10. **D. anomala** (Schlechter in Engl. Jahrb. xxvi. 333); plant
6 in. high, becoming black in drying; stem straight, slender,
puberulous at the base, glabrous above; leaves cauline, amplexicaul,
3, lanceolate, acute or acuminate, glabrous, much like those of *D.
gracilis*; scapes 6 in. high; spikes 2- or few-flowered; bracts leaf-
like, lanceolate, concave, acuminate, equalling or exceeding the
ovary; flowers medium-sized, white; dorsal sepal galeate, acuminate,
as long as the lateral sepals; lateral sepals obliquely lanceolate,
acuminate, 5 lin. long, with a small sac 2 lin. long, about the
middle; petals cohering with the dorsal sepal, obliquely subfalcate-
lanceolate from a narrow base, acuminate, 4½ lin. long; lip narrowly
unguiculate, 2 lin. long; limb peltate, narrowly linear, bifid at the
apex; appendage ascending, somewhat boat-shaped, lanceolate,

verrucose at the apex, curved and acuminate; column short; rostellum large, 3-lobed in front, with small rounded side lobes and an elongate-linear front lobe; arms straight, not curved at the apex, rather short.

EASTERN REGION : Natal ; among grass near Nottingham Road, 5000 ft., *Wood,* 6078.

Said to be allied to *D. gracilis* and *D. stenoplectron,* but to be anomalous in the shape of the lip.　I have not seen a specimen of this.　　　　　　　　•

11. D. concinna (Schlechter in Engl. Jahrb. xx. Beibl. 50, 43) ; plant 5–10 in. high; stem slender, straight, glabrous; leaves cauline, alternate, sessile, 1–3, somewhat spreading, elliptic-oblong to lanceolate, subacute or obtuse, $\frac{1}{3}$–$\frac{1}{2}$ in. long; scapes 5–10 in. high ; spikes short, somewhat lax, 1–3-flowered ; bracts lanceolate or ovate-lanceolate, acute, 5–7 lin. long ; pedicels about 4 lin. long ; flowers small, rose-coloured ; dorsal sepal galeate, with an obtuse rounded sac nearly as long as the limb, acuminate or obtuse, $2\frac{1}{2}$–3 lin. long ; lateral sepals spreading, oblong below, narrowed and acuminate above, with a large conic-oblong sac about the middle, 3–$3\frac{1}{2}$ lin. long ; petals oblique, falcate-lanceolate, $2\frac{1}{2}$–3 lin. long ; lip narrowly unguiculate ; limb ovate-oblong, obtuse, conduplicate, about 1 lin. long ; appendage broadly subpandurate-oblong, apiculate, keeled, longer than the limb ; rostellum very large, ovate, acute ; arms oblong, $1\frac{1}{2}$ lin. long. *Schlechter in Bull. Herb. Boiss.* 1re *sér.* vi. 920 ; *Kränzl. Orch. Gen. et Sp.* i. 833.

KALAHARI REGION : Transvaal ; marshes near Wilge River, 4600 ft., *Schlechter.* EASTERN REGION : Natal ; Mohlamba Range, 5000–6000 ft., *Sutherland* !

12. D. gracilis (Schlechter in Engl. Jahrb. xx. Beibl. 50, 44) ; plant $\frac{1}{2}$–$1\frac{1}{4}$ ft. high ; stem slender, glabrous; leaves cauline, alternate, sessile, usually 3, suberect, oblong-lanceolate, acute or subacute, glabrous, 9–12 lin. long ; scapes $\frac{1}{2}$–$1\frac{1}{4}$ ft. high, 1–4-flowered ; bracts ovate or ovate-lanceolate, subacute, 5–6 lin. long ; pedicels about 4 lin. long ; flowers rather small, white ; dorsal sepal galeate, broadly saccate, ovate, acuminate, 4–5 lin. long, with a broad mouth ; lateral sepals spreading, falcate-oblong, acuminate, 4–$4\frac{1}{2}$ lin. long, with an oblong obtuse sac below the middle ; petals cohering with the dorsal sepal, falcate-lanceolate, acuminate, 4–$4\frac{1}{2}$ lin. long ; lip broadly unguiculate ; limb oblong-lanceolate, acuminate, $2\frac{1}{4}$–$2\frac{1}{2}$ lin. long ; appendage ovate below, narrowed upwards, subobtuse, somewhat cucullate, as long as the limb ; rostellum ovate, obtuse, somewhat 3-lobed at the apex ; arms cartilaginous, twisted below, linear, about $2\frac{1}{4}$ lin. long. *Schlechter in Bull. Herb. Boiss.* 1re *sér.* vi. 925.　*D. Wealii, Kränzl. Orch. Gen. et Sp.* i. 829, *partly, not of Reichb. f.*

KALAHARI REGION : Transvaal ; marshes on Houtbosch Mountains, 5000–6500 ft., *Rehmann,* 5853 ! *Schlechter,* 4393.

13. **D. Wealii** (Reichb. f. Otia Bot. Hamb. ii. 103) ; plant 6–8 in. high ; stem slender, glabrous ; leaves cauline, alternate, sessile, usually 3, suberect, elliptic or oblong-lanceolate, usually subacute, glabrous, 6–9 lin. long ; scapes 6–8 in. high ; spikes lax, 1–5-flowered ; bracts oblong-lanceolate, acute, 5–8 lin. long ; pedicels 6–9 lin. long ; flowers medium-sized, white with some green transverse bars on the front margin of the petals ; dorsal sepal galeate, about 5 lin. long, with a triangular very acuminate limb, and an obtuse rounded sac ; lateral sepals spreading, oblong-lanceolate, very acuminate and hooked at the apex, about 4½ lin. long, with an oblong obtuse spur above the middle ; petals cohering with the dorsal sepal, broadly falcate-lanceolate, acuminate, 4½ lin. long, front margin somewhat undulate ; lip broadly unguiculate ; limb cucullate, ovate, acuminate, 2 lin. long ; appendage about as long as the limb, and closely resembling it in shape ; column short ; rostellum broady dilated, rounded or obtuse at the apex, nearly 2 lin. broad ; arms cartilaginous, much twisted, about 3 lin. long. *Bolus in Journ. Linn. Soc.* xxii. 80, *t.* 1, *fig. A* ; xxv. 206 ; *Durand & Schinz, Conspect. Fl. Afr.* v. 121 ; *Schlechter in Bull. Herb. Boiss.* 1ʳᵉ *sér.* vi. 925 ; *Kränzl. Orch. Gen. et Sp.* i. 829, *partly, excl. syn. Schlechter.*

COAST REGION : Bedford Div. ; summit of Kaga Berg, near Bedford, 4500–5000 ft., *Weale*, 917 !
KALAHARI REGION : Orange River Colony ; Harrismith, 6500–7500 ft., *Sankey*, 258 !
EASTERN REGION : Griqualand East ; summit of Mount Ingeli, near Kokstad, 6500 ft., *Tyson*, 1077 ! Natal ; marshes near Polela, *Wood*, 4829.

14. **D. bicolor** (Rolfe) ; plant with slender pubescent stem, 4½ in. high, base not seen ; leaves 2, suberect, narrowly ovate-lanceolate, acute, ¾–1 in. long ; scapes 2-flowered at the apex ; bracts ovate-lanceolate, acute or acuminate, 6–9 lin. long ; pedicels about 6 lin. long ; flowers medium-sized, yellow with dull purple galea ; dorsal sepal galeate, 5 lin. long, with broadly triangular apiculate deflexed limb and broadly conical obtuse galea ; lateral sepals spreading, elliptic-oblong, shortly acuminate, 4½ lin. long, with a narrowly oblong sac about the middle ; petals obovate-elliptic, subobtuse, 4 lin. long, outer margin sinuate-undulate ; lip broadly unguiculate ; limb navicular, subobtuse, reflexed, 2½ lin. long ; appendage linear, obtuse, about as long as the limb ; column short ; rostellum broadly ovate, obtuse, 1 lin. long ; arms cartilaginous, slightly twisted, linear, 2½ lin. long.

EASTERN REGION : Natal ; Oliviers Hoek, 5000 ft., *Allison*, 4 !

15. **D. flava** (Rolfe) ; plant 4½ in. high ; stem moderately slender, glabrous ; leaves 2, somewhat spreading, lanceolate, acuminate, ¾–1 in. long ; scapes 4½ in. high, 2-flowered ; bracts ovate, acute, 7–9 lin. long ; pedicels 4–5 lin. long ; flowers medium-sized, yellow,

the galea tipped with orange; dorsal sepal galeate, 4 lin. long, with narrowly triangular acuminate deflexed limb and broadly saccate galea; lateral sepals spreading, obliquely ovate, very acuminate, 5 lin. long, with a large oblong obtuse sac below the middle; petals falcate-lanceolate, acuminate, 5 lin. long, outer margin slightly undulate; lip broadly unguiculate, limb ovate-lanceolate, acuminate, deflexed, cucullate, 1½ lin. long; appendage broadly oblong, obtuse, shorter and much broader than the limb; column short; rostellum broadly deltoid-ovate, obtuse, 2 lin. long; arms twisted, oblong, subobtuse, 2½ lin. long.

EASTERN REGION: Natal; Oliviers Hoek, 5000 ft., *Allison*, 3!

16. **D. oxyglossa** (Bolus in Journ. Linn. Soc. xxii. 76); plant ½–1 ft. high; stem slender, glabrous; leaves cauline, sessile, 3, suberect or somewhat spreading, elliptic or oblong-lanceolate, acute or subobtuse; scapes ½–1 ft. high; spikes lax, 2–3-flowered; bracts oblong-lanceolate, subacute, ¾–1 in. long; pedicels ½–¾ in. long; flowers medium-sized, purple with some green warts on the petals; dorsal sepal galeate, oblong-lanceolate, acuminate, 7–8 lin. long; lateral sepals spreading, falcate-lanceolate from a broader base, acuminate, 6–7 lin. long, with an oblong obtuse sac below the middle; petals cohering with the dorsal sepal, obliquely lanceolate, acuminate, 6–7 lin. long; lip linear, unguiculate at the base, 3½ lin. long, united with the base of the rostellum, the front lobe lanceolate, acuminate, suddenly reflexed, minutely tubercled at the base; appendage erect, linear, cymbiform, with a pair of broad triangular wings at the base, shorter than the limb of the lip; column short; rostellum large, convex, obtuse, with revolute margin; arms spreading, linear, truncate, spirally twisted. *Bolus in Journ. Linn. Soc.* xxv. 206; *Durand & Schinz, Conspect. Fl. Afr.* v. 120; *Schlechter in Bull. Herb. Boiss.* 1ʳᵉ sér. vi. 917; *Kränzl. Orch. Gen. et Sp.* i. 833.

COAST REGION: British Kaffraria, without precise locality, *Mrs. Barber*, 28! EASTERN REGION: Tembuland; top of Bazeia Mountain, 4000 ft., *Baur*, 813! Griqualand East; Mount Currie, above the Waterfall, 5500 ft., *Tyson*, 1603; plain at summit of Insizwa Mountains, 6500 ft., *Schlechter*. Natal; heights above Kar Kloof, *Sanderson*, 1071! Nottingham, *Buchanan*, 150! Emangweni, *Thode*.

17. **D. paludosa** (Harv. ex Lindl. in Hook. Lond. Journ. Bot. i. 14); plant ½–1½ ft. high; stems slender, flexuous, glabrous; leaves cauline, sessile, 2–4, suberect, lanceolate or linear-lanceolate, acute, ¾–1½ in. long; spikes lax, 1–6-flowered; bracts ovate-lanceolate, acute, 5–9 lin. long; pedicels 8–10 lin. long; flowers medium-sized, with dull purple sepals and pink petals, margined with green and dotted with rose near the apex; dorsal sepal galeate, broadly lanceolate, acute, 6–7 lin. long; lateral sepals spreading, falcate-lanceolate from a broad base, acuminate, 5–6 lin. long, each with a falcate-oblong sac at the base over 1 lin. long; petals

cohering with the dorsal sepal, obliquely semicordate, falcate, acute, 5–6 lin. long; lip ascending, shortly angled, lanceolate, acuminate, 3–4 lin. long, with a gland-fringed claw over the convex face of the column; appendage broadly oblong, short, gland-fringed; column short; rostellum large, convex, obtuse, with revolute margin; arms distant, divaricately spreading; capsule fusiform, peduncled, about ¾ in. long. *Sond. in Linnæa,* xix. 112, xx. 221; *Harv. Thes. Cap.* ii. 30, *t.* 148; *Bolus in Trans. S. Afr. Phil. Soc.* v. 176, *t.* 19, *Orch. Cap. Penins.* 176, *t.* 19, *and Journ. Linn. Soc.* xxv. 206; *Durand and Schinz, Conspect. Fl. Afr.* v. 120; *N. E. Br. in Gard. Chron.* 1885, xxiv. 232; *Schlechter in Bull. Herb. Boiss.* 1re *sér.* vi. 916; *Kränzl. Orch. Gen. et Sp.* i. 834, *and in Ann. Naturhist. Hofmus. Wien,* xx. 9.

COAST REGION: Paarl Div.; French Hoek Pass, 3000 ft., *Harvey!* *Schlechter*, 9306! Cape Div.; marshy places at Van Kamps Bay, *Harvey!* Table Mountain, 2400–2500 ft., *Harvey! Pappe,* 11! *Bolus,* 4449! and in *McOwan & Bolus, Herb. Norm. Austr.-Afr.* 339! Devils Mountain, 2500 ft., *Ecklon, Bolus.* Swellendam Div.; banks of the Zondereinde River, near Appels Kraal, *Zeyher,* 3935! George Div.; above Montagu Pass, 1500 ft., *Penther.* Knysna Div.; near Knysna, *Forcade.* Fort Beaufort Div.; Kat River, *Hutton!*

18. **D. capensis** (Sw. in Vet. Acad. Handl. Stockh. 1800, 220, t. 3, fig. F.); plant ½–1½ ft. high; stems erect, rather slender, somewhat flexuous, covered with long spreading hairs below the middle; leaves cauline, sessile, with sheathing base, usually 1 or 2, suberect, lanceolate or narrowly oblong-lanceolate, acute or acuminate, 1–2 in. long; scapes ½–1½ ft. high, 1- (rarely 2-) flowered; bracts ovate-lanceolate, acute or acuminate, ¾–1 in. long; pedicels ⅓–½ in. long; flowers large, yellowish or greenish, often with the petals purple or lilac and darker at the margins; dorsal sepal erect, galeate, 10–12 lin. long, obovate-oblong, broad and angled above, then suddenly contracted into a setaceous point about as long as the limb; lateral sepals spreading, lanceolate, caudate-acuminate, 12–14 lin. long, with a conical very shallow sac near the base; petals cohering with the dorsal sepal, 5 lin. long, obovate-oblong, very obtuse, with a setaceous point 2 lin. long; lip spathulate-lanceolate, very narrow at the base, acuminate, decurved and usually hooked at the apex, 6–7 lin. long; appendage linear or oblong, somewhat lacerate, short or sometimes absent; column short; rostellum broadly cordate-ovate, subobtuse, with recurved margins; arms linear from a broader base, spirally twisted, acuminate. *Ker in Quart. Journ. Sci. & Arts,* v. 104, *t.* 1, *fig.* 2; *Thunb. Fl. Cap. ed. Schult.* 25; *Lindl. Gen. & Sp. Orch.* 370; *Drège, Zwei Pfl. Documente,* 124; *Sond. in Linnæa,* xix. 112, xx. 221; *Krauss in Flora,* 1845, 306, *and Fl. Cap.- und Natal.* 158; *N. E. Br. in Gard. Chron.* 1885, xxiv. 232; *Bolus in Trans. S. Afr. Phil. Soc.* v. 175, *Orch. Cap. Penins.* 175, *Journ. Linn. Soc.* xxv. 205, *and Ic. Orch. Austr.-Afr.* i. *t.* 89; *Durand & Schinz, Conspect. Fl. Afr.* v. 118; *Schlechter in Bull. Herb. Boiss.* 1re *sér.* vi. 915; *Kränzl. Orch. Gen. et Sp.* i. 830, *and in Ann. Naturhist.*

Hofmus. Wien, xx. 9.　　*Arethusa capensis, Linn. f. Suppl.* 405 ; *Thunb.
Prodr.* 3.　*Dipera capensis, Spreng. Syst. Veg.* iii. 696.　*D. tenera,
Spreng. Syst. Veg.* iii. 696.

VAR. β, **brevicaudata** (Rolfe); flowers smaller, greenish with purple petals ;
tails of sepals reduced by about one-half ; petals much narrower than in the type.

SOUTH AFRICA : without locality, *Thunberg* ! *Masson, Lalande, Leibold, Trimen,
Forbes, Mrs. Holland,* 6 ! *Menzies* !
COAST REGION : Cape Div. ; heathy places on the Cape Flats, *Burchell,* 8527 !
Bolus, 3735 ! and in *MacOwan & Bolus, Herb. Norm. Austr.-Afr.* 177 ! *Harvey,*
122 ! *Prior* ! Table Mountain, up to 3500 ft., *Bergius, Mund, Ecklon & Zeyher,
Krauss,* 1319, *Kässner, Tyson,* 3236 ; at Orange Kloof, *Schlechter,* 1316 ! Devils
Peak, *Wilms,* 3660 a ! *Wolley-Dod,* 601 ! Lion Mountain, *Harvey* ! *Wilms,* 3660 !
Stellenbosch Div. ; Hottentots Holland Mountains, *Mund, Schlechter,* 1157 ;
Stellenbosch, *Lloyd in Herb. Sanderson,* 932 ! Caledon Div. ; Houw Hoek
Mountains, 1000-3000 ft., *Zeyher,* 3936 ! Zwart Berg, *Ecklon & Zeyher* ! Rivers-
dale Div. ; Langeberg Range, near Riversdale, 2500 ft., *Schlechter.* George Div. ;
Cradock Berg, near George, *Burchell,* 5943 ! near Montagu Pass, *Penther.* Knysna
Div. ; near Knysna, *Forcade* ; Ruigte Vallei, under 500 ft., *Drège,* 8278 ! Port
Elizabeth Div. ; bushy places at Port Elizabeth, *Kemsley* ! marshes near Algoa
Bay, *MacOwan,* 1910 ! Albany Div. ; near Grahamstown, *Williamson* ! and
without precise locality, *Hutton* ! Eastern Districts of Cape Colony, *Hallack* !
Var. β : Cape Div. ; damp rocks at Waai Vley, 3000-3200 ft., *Wolley-Dod,* 3074 !
EASTERN REGION : Transkei ; Krielis Country, *Bowker* !

The variety *brevicaudata* is quite anomalous in character, and as *D. paludosa,*
Harv., was found a few yards away, there is a suspicion that it may be a hybrid
between the two, although it is difficult to trace the character of the latter.
Major Wolley-Dod specially points out the difference in shape and colour of the
flower, and remarks that *D. capensis* is rarely seen so high up.

19. D. cardiophora (Harv. Thes. Cap. ii. 4, t. 106); plant 4–9 in.
high; stem rather stout, glabrous; leaves subradical, solitary,
sessile, spreading, suborbicular, amplexicaul, obtuse, ½–1 in. long ;
scapes 4–9 in. high ; spikes 1–3 in. long, rather dense, distichous,
secund ; bracts suborbicular, obtuse, cucullate, 2–3 lin. long;
pedicels 1–2 lin. long; flowers small, white and green with dull
purple tips to the petals; dorsal sepal galeate, 2½ lin. long, with a
short triangular subacute limb and a subglobose sac ; lateral sepals
falcate-oblong, obtuse, 2½ lin. long, with a short and broad conical
sac about the middle ; petals cohering with the dorsal sepal,
obliquely ovate-oblong, subacute, 2½ lin. long ; lip broadly unguiculate ; limb reflexed, ovate, subobtuse, under 1 lin. long ; appendage
erect, broadly oblong, subobtuse, longer than the limb ; column
short ; rostellum very broadly dilated, truncate or obscurely 3-lobed
at the apex, reflexed, 1½ lin. broad ; arms cartilaginous, twisted,
linear, obtuse, 1 lin. long ; capsule broadly elliptic-oblong, 3½ lin.
long. *Bolus, Ic. Orch. Austr.-Afr.* i. *t.* 95, *and in Journ. Linn. Soc.*
xxv. 206 ; *Durand & Schinz, Conspect. Fl. Afr.* v. 118 ; *Schlechter
in Bull. Herb. Boiss.* 1ʳᵉ *sér.* vi. 926 ; *Kränzl. Orch. Gen. et Sp.* i.
825, *and in Ann. Naturhist. Hofmus. Wien,* xx. 9.

SOUTH AFRICA : without locality, *Buchanan, Mrs. Saunders* ! *Mrs. Barber.*
COAST REGION : Stockenstrom Div. ; Kat Berg, near Stockenstrom, *Scully.*
KALAHARI REGION : Transvaal ; without precise locality, *Sanderson* !

EASTERN REGION : Tembuland ; slopes of Bazeia Range, between Umbaxa and Nxanxaza, 2500–3000 ft., *Baur*, 567 ! Natal ; Fields Hill, 1200 ft., *Sanderson*, 488 ! Dargle Farm, *Mrs. Fannin*, 10 ! 46 ! 100 ! 105 ! Durban, *Gerrard*, 1560 ! Inanda, *Wood*, 1066 ! Oliviers Hoek, 5000 ft., *Allison*, 22 ! near Charlestown, 5000–6000 ft., *Wood*, 5568 ! near Van Reenen, 5500 ft., *Schlechter*, 6957 !

20. **D. villosa** (Sw. in Vet. Acad. Handl. Stockh. 1800, 220) ; plant 1½–6 in. high ; stem moderately slender, villous ; leaves cauline, 2, petiolate or subsessile, ovate or cordate, subacute or apiculate, rarely obtuse, velvety, 4–8 (rarely 12) lin. long ; petioles 1–3 lin. long ; scapes 1½–6 in. high, 1- (rarely 2-) flowered ; bracts cucullate, ovate or elliptic-ovate, subobtuse, velvety, 4–7 lin. long ; pedicels 3–4 lin. long ; flowers medium-sized, dull canary-yellow ; dorsal sepal galeate, 3–4 lin. long, with a broadly triangular obtuse limb and a rounded obtuse sac ; lateral sepals spreading, obovate from a narrower base, 4 lin. long, with a broadly conical obtuse sac close to and exceeding the apex ; petals cohering with the dorsal sepal, broadly falcate-lanceolate, 2–2½ lin. long, outer margin undulate ; lip narrowly unguiculate ; limb very short, apiculate, cucullate ; appendage lanceolate-oblong, subobtuse, cucullate, 3 or 4 times longer than the limb ; column short ; rostellum broadly triangular-ovate, large, margin reflexed ; arms falcate-oblong, obtuse, twisted, about 2 lin. long ; capsule elliptic-oblong, velvety, 4 lin. long. *Ker in Quart. Journ. Sci. & Arts.* vi. 44, *t.* 1, *fig.* 5 ; *Thunb. Fl. Cap. ed. Schult.* 25 ; *Lindl. Gen. & Sp. Orch.* 371 ; *Drège, Zwei Pfl. Documente*, 83, 98 ; *Sond. in Linnæa*, xix. 112, xx. 221 ; *Krauss in Flora*, 1845, 306, *and Fl. Cap- und Natal.* 158 ; *Bolus in Trans. S. Afr. Phil. Soc.* v. 178, *Orch. Cap. Penins.* 178, *and in Journ. Linn. Soc.* xxv. 205 ; *Durand and Schinz, Conspect. Fl. Afr.* v. 121 ; *Kränzl. Orch. Gen. et Sp.* i. 834, *and in Ann. Naturhist. Hofmus. Wien*, xx. 10 ; *Schlechter in Bull. Herb. Boiss.* 1ᵉ sér. vi. 942. *Arethusa villosa, Linn. f. Suppl.* 405 ; *Thunb. Prodr.* 3.

SOUTH AFRICA : without locality, *Thunberg*! *Masson, Brown, Mund, Lalande, Leibold, Forbes*, 86 !

COAST REGION : Clanwilliam Div. ; near Sneeuwkop, on the Cederberg Range, *Bodkin in Herb. Bolus*! near Zwartbosch Kraal, 4000–5000 ft., *Schlechter*, 5175, *Penther, Krook*. Malmesbury Div. ; sandy places at Saldanha Bay, *Bachmann*. Tulbagh Div. ; near Tulbagh, 400 ft., *Schlechter*. Paarl Div. ; by the Berg River, near Paarl, under 500 ft., *Drège*, 481a ! between Paarl and Lady Grey Railway Bridge, *Drège*, 481b ! Drakenstein Mountains, 2000–3000 ft., *Drège*. Cape Div. ; hills and F ats of Cape Peninsula, 20–600 ft., *Bergius, Ludwig, Ecklon & Zeyher* ! *Ecklon*, 248 ! *Pappe*, 9 ! *Harvey*, 149 ! *Prior* ! *Kässner, Rehmann*, 1858, *Bolus*, 3966 ! *MacOwan in MacOwan & Bolus, Herb. Norm. Austr.-Afr.* 178 ! *Krauss*, 1321, *Hutton* ! *Schlechter*, 1338 ! *Wilms*, 3656 ! *Wolley-Dod*, 1515 ! 3084 ! *Mrs. Jameson* ! Stellenbosch Div. ; Stellenbosch, *Lloyd in Herb. Sanderson*, 934 ! Hottentots Holland Mountains, *Ecklon & Zeyher, Miss Farnham, Marloth*. Caledon Div. ; moist sandy places at the foot of mountains near Appels Kraal, Zondereinde River, *Zeyher*, 3938 ! Genadendal, *Roser* ! Albany Div. ; near Grahamstown, *Williamson* !

21. **D. cucullata** (Sw. in Vet. Acad. Handl. Stockh. 1800, 220) ; plant 3–7 in. high ; stem moderately slender, pubescent ; leaves

cauline, alternate, sessile, usually 2, spreading, ovate-oblong or lanceolate-oblong, subobtuse, pubescent, $\frac{1}{2}$–1 in. long; scapes 3–7 in. high, 1- (rarely 2-) flowered; bracts ovate or ovate-lanceolate, subobtuse or acute, cucullate, pubescent, 3–8 lin. long; pedicels 3–4 lin. long; flowers medium-sized, pale green; dorsal sepal galeate, about 6 lin. long, with a short triangular acute limb and a conical obtuse galea; lateral sepals spreading, oblong-lanceolate, acuminate, 5 lin. long, with an oblong obtuse sac about the middle; petals cohering with the dorsal sepal, falcate-lanceolate, acute, 4 lin. long; lip narrowly unguiculate; limb distinctly 3-lobed, cucullate, under 2 lin. long, with rounded side lobes and a longer triangular acute front lobe; appendage linear, obtuse, papillose at the apex, over twice as long as the limb; column short; rostellum broadly triangular, acute or subacuminate, 2 lin. long; arms cartilaginous, twisted, linear, about 2 lin. long. *Ker in Quart. Journ. Sci. & Arts*, vi. 45, *t.* 1, *f.* 4; *Lindl. Gen. & Sp. Orch.* 371; *Drège, Zwei Pfl. Documente*, 86; *Sond. in Linnæa*, xix. 112; *Bolus in Trans. S. Afr. Phil. Soc.* v. 177, *Orch. Cap. Penins.* 177, *Journ. Linn. Soc.* xxv. 206, *and Ic. Orch. Austr.-Afr.* i. *t.* 94; *Durand & Schinz, Conspect. Fl. Afr.* v. 119; *Kränzl. Orch. Gen. et Sp.* i. 842, *and in Ann. Naturhist. Hofmus.* xx. 10; *Schlechter in Bull. Herb. Boiss.* 1re *sér.* vi. 923.

SOUTH AFRICA : without locality, *Herb. Swartz*! *Mund.*

COAST REGION : Clanwilliam Div. ; Zwartbosch Kraal, 400–500 ft., *Schlechter*, 5164. Malmesbury Div. ; Hopefield, 150 ft., *Schlechter*, 5312. Tulbagh Div. ; near Tulbagh Waterfall, 600 ft., *Schlechter*, 1414. Paarl Div. ; Paarl Mountain, *Drège*, 8279 b ! *Zeyher*, 3938 ! Cape Div. ; Sea Point, *Bolus*, 4887 ! *Schlechter* ! Rondebosch, *Miss Bolus* ; Steen Berg, 800 ft., *Bolus*. Stellenbosch Div. ; near Stellenbosch, *Lloyd in Herb. Sanderson*, 933 ! *Miss Farnham in MacOwan & Bolus, Herb. Norm. Austr.-Afr.* 338 !

CENTRAL REGION : Ceres Div. ; near Ceres, 1500 ft., *Bolus*.

22. **D. bolusiana** (Schlechter in Engl. Jahrb. xxiv. 430) ; plant 4–7 in. high; stem slender, pubescent; leaves cauline, 2, amplexicaul, broadly cordate-ovate, subobtuse or apiculate, minutely puberulous and ciliolate, $\frac{1}{2}$–1 in. long, 4–9 lin. broad, with broad shallow basal lobes; scapes 4–7 in. high, 1-flowered; bracts ovate-lanceolate, acute or acuminate, 3–5 lin. long; pedicels 4–6 lin. long; flowers medium-sized, white or dull yellow, with green spots on the petals; dorsal sepal galeate, 3–3$\frac{1}{2}$ lin. long, with a triangular acuminate deflexed limb and a broadly rounded saccate galea ; lateral sepals spreading, obliquely ovate-oblong, acute, nearly 3 lin. long, with a broadly conical obtuse sac below the middle, nearly as long as the limb; petals cohering with the dorsal sepal, obliquely ovate, acuminate, 2$\frac{1}{2}$ lin. long, outer margin undulate; lip narrowly unguiculate; limb triangular, acute, cucullate, recurved; appendage ovate-lanceolate, subobtuse, cucullate, more than twice as long as the limb; column short; rostellum ovate, obtuse; arms broadly oblong, obtuse, about 1 lin. long. *Bolus, Ic. Orch. Austr.-Afr.* i. *t.* 93 ; *Kränzl. Orch. Gen. et Sp.* i. 839, *and in Ann. Naturhist. Hofmus.*

Wien, xx. 10; *Schlechter in Bull. Herb. Boiss.* 1ʳᵉ *sér.* vi. 940. *D. purpurata, var. parviflora, Bolus in Journ. Linn. Soc.* xxii. 79, xxv. 206; *Durand & Schinz, Conspect. Fl. Afr.* v. 120.

COAST REGION: Piquetberg Div.; slopes near Piquiniers Kloof, 1200 ft., *Schlechter,* 4972! Malmesbury Div.; near Groene Kloof, 300 ft., *Bolus,* 4337! hills near Malmesbury, 200 ft., *Schlechter,* 5346; among shrubs near Hopefield, 300 ft., *Bachmann, Schlechter,* 5321! Cape Div.; Tyger Berg, *Bergius.* Stellenbosch Div.; near Stellenbosch, *Lloyd in Herb. Sanderson,* 937!

23. **D. purpurata** (Reichb. f. in Linnæa, xli. 55); plant 2–6 in. high; stem sparsely pubescent; leaves cauline, 2, subamplexicaul, broadly cordate-ovate, subobtuse or apiculate, puberulous and ciliolate, ½–1 in. long, 5–9 lin. broad, with broad shallow basal lobes, the upper leaf much smaller than the lower; scapes 2–6 in. high, 1-flowered; bracts ovate-oblong or ovate-lanceolate, acute, 3–4 lin. long; pedicels 4–6 lin. long; flowers medium-sized, light rose, with a few small green dots on the petals; dorsal sepal galeate, 4–5 lin. long, with a triangular acute deflexed limb and a broadly saccate obtuse galea; lateral sepals spreading, obliquely ovate-oblong, acute, 4–5 lin. long, with a broadly conical subacute sac about the middle; petals cohering with the dorsal sepal, broadly unguiculate, obliquely ovate-oblong, acute, obtusely auricled above the base, about 4 lin. long; lip narrowly unguiculate, limb broadly ovate, acute, strongly cucullate, recurved; appendage triangular, subobtuse, cucullate, longer than the limb; column short; rostellum rhomboid-ovate, subobtuse, rather large; arms oblong, obtuse, twisted, about 2 lin. long. *Bolus, Ic. Orch. Austr.-Afr.* i. *t.* 91, *and in Journ. Linn. Soc.* xxv. 206; *Durand & Schinz, Conspect. Fl. Afr.* v. 120; *Kränzl. Orch. Gen. et Sp.* i. 843, *and in Ann. Naturhist. Hofmus. Wien,* xx. 10; *Schlechter in Bull. Herb. Boiss.* 1ʳᵉ *sér.* vi. 939. *D. namaquensis, Bolus in Journ. Linn. Soc.* xx. 486. *D. purpurata, var. namaquensis, Bolus in Trans. S. Afr. Phil. Soc.* v. 188, *in note.*

CENTRAL REGION: Calvinia Div.; Hantam Mountains, *Meyer*!
WESTERN REGION: Little Namaqualand; Spektakel Mountain, near Ookiep, *Morris in Herb. Bolus,* 5820! among stones at Kasteel Poort, near Klip Fontein, 3000 ft., *Bolus in MacOwan & Bolus, Herb. Norm. Austr.-Afr.* 179.

D. purpurata, var. *parviflora,* Bolus, is now referred to the preceding species.

24. **D. macrocorys** (Rolfe); plant 4–6 in. high; stem slender, pubescent; leaves cauline, 2, amplexicaul, cordate-ovate, apiculate, puberulous, 6–9 lin. long, 3–7 lin. broad; scapes 4–6 in. high, 1-flowered; bracts cucullate, suberect, ovate-lanceolate, shortly acuminate, 3–4 lin. long; pedicels 4–5 lin. long; flowers rather large, pale yellow; dorsal sepal galeate, 6–7 lin. long, with a short triangular acute deflexed limb and a conic-oblong galea; lateral

sepals spreading, ovate-triangular, acute, 4 lin. long, with a conical diverging sac about the middle, as long as the limb ; petals cohering with the dorsal sepal, falcate-oblong, subobtuse, 3 lin. long, outer margin sinuate-undulate ; lip narrowly unguiculate; limb deflexed, ovate-triangular, acute, cucullate, 1 lin. long ; appendage linear from a triangular cucullate base, about three times as long as the limb ; column short ; rostellum triangular-ovate, acute, with reflexed sides, 2½ lin. long; arms cartilaginous, twisted, broadly oblong, obtuse, 1 lin. long.

COAST REGION : Clanwilliam Div. ; Koudeberg Mountains, near Wupperthal, 2400 ft., *Bolus,* 9093 !

25. **D. Macowani** (Bolus in Journ. Linn. Soc. xxii. 77, t. 1, fig. 13–17) ; plant 3–6 in. high; stem slender, sparsely pubescent ; leaves cauline, 2, amplexicaul, broadly cordate-ovate, acute, minutely puberulous and ciliolate, 4–7 lin. long by nearly as broad, with broad very shallow basal lobes ; scapes 3–6 in. high, 1-flowered ; bracts broadly ovate, acute, 3–4 lin. long ; pedicels 3–4 lin. long ; flowers rather small, pale bluish with a few green dots on the petals ; dorsal sepal galeate, 2½–3 lin. long, with a triangular acute or acuminate limb and a broadly conical obtuse sac ; lateral sepals spreading, obliquely obovate-oblong, acute, about 2 lin. long, with a short broadly oblong sac near the apex ; petals cohering with the dorsal sepal, falcate-lanceolate, acute, about 2 lin. long, with an undulate margin and a rounded auricle near the base ; lip narrowly unguiculate ; limb subrhomboid, acute, reflexed, very short ; appendage linear-lanceolate, obtuse, recurved at the apex, several times longer than the limb ; column short ; rostellum large, convex, subobtuse, with reflexed margin ; arms linear-oblong, obtuse, twisted, about 1 lin. long. *Bolus in Journ. Linn. Soc.* xxv. 206 ; *Durand & Schinz, Conspect. Fl. Afr.* v. 120 ; *Schlechter in Bull. Herb. Boiss.* 1ʳᵉ *sér.* vi. 938 ; *Kränzl. Orch. Gen. et Sp.* i. 841.

COAST REGION : George Div. ; on wet rocks in woods near Silver River, 500 ft., *Schlechter,* 2458 ! Albany Div. ; rocks at Featherstone Kloof, near Grahamstown, 2000–4000 ft., *MacOwan,* 807 ! *Schlechter.*
CENTRAL REGION : Somerset East Div. ; banks of streams, on the Bosch Berg, behind Besters Hoek, near Somerset East, 4500 ft., *MacOwan,* 2626 !
EASTERN REGION : Natal ; among shrubs near Van Reenen, 7000 ft., *Schlechter,* 6937, and without precise locality, *Wood* !

26. **D. Bodkini** (Bolus, Ic. Orch. Austr.-Afr. i. t. 96) ; plant 3–6 in. high ; stem slender, puberulous ; leaves cauline, 2, amplexicaul, cordate-ovate, acute, 4–7 lin. long by nearly as broad ; scapes 3–6 in. high, 1–2-flowered ; bracts leaf-like in shape and texture, 3–5 lin. long ; pedicels 4–6 lin. long ; flowers very small, green and white, petals with a purple mid-nerve ; dorsal sepal galeate, 2–2½ lin. long, with a triangular acuminate deflexed limb and a broadly conical obtuse spur ; lateral sepals deflexed, obliquely elliptic-oblong, shortly acuminate, 1½–2 lin. long, with a very short sac above the

x 2

middle; petals obliquely rhomboid-ovate, acute, 1½–2 lin. long,
angled at the outer margin ; lip narrowly unguiculate ; limb ovate,
somewhat fleshy, reflexed, very short ; appendage linear, obtuse,
curved, shortly bifid at the apex, about thrice as long as the limb ;
column short ; rostellum very short, broadly dilated ; arms cartilaginous, linear-oblong, obtuse, under 1 lin. long ; capsule oblong,
3–4 lin. long. *Schlechter in Bull. Herb. Boiss.* 1ʳᵉ *sér.* vi. 937 ;
Kränzl. Orch. Gen. et Sp. i. 840, *excl. syn., and in Ann. Naturhist.
Hofmus. Wien,* xx. 10.

COAST REGION : Cape Div. ; Claremont Flats, near Wynberg, 86 ft., *Bodkin in
Herb. Bolus,* 7970 !

27. **D. micrantha** (Lindl. Gen. & Sp. Orch. 370) ; plant 5–7 in.
high ; stem rather slender, glabrous ; leaves cauline, 2 or 3,
amplexicaul, broadly cordate, subacute or apiculate, ¾–1½ in. long,
½–1 in. broad, with rather broad basal lobes ; scapes 5–7 in. high ;
spikes short and lax, 2–5-flowered ; bracts leaf-like in shape and
texture, 4–10 lin. long ; pedicels 3–5 lin. long ; flowers very small,
pink and green ; dorsal sepal galeate, about 1½ lin. long, with a
triangular acuminate deflexed limb and a broadly saccate galea ;
lateral sepals spreading, obliquely ovate-oblong, acuminate, 1½ lin.
long, with a short broadly conical obtuse sac above the middle ;
petals cohering with the margin of the dorsal sepal, obliquely
falcate-ovate, acute, 1½ lin. long ; lip narrowly unguiculate ; limb
reflexed, with 2 short diverging lobes ; appendage broadly oblong,
obtuse, papillose, longer than the limb ; column short ; rostellum
broadly rounded, obtuse, concave ; arms cartilaginous, spathulate-
oblong, obtuse, twisted, over ½ lin. long.   *Sond. in Linnæa,* xx. 221 ;
Bolus in Journ. Linn. Soc. xxv. 206 ; *Durand & Schinz, Conspect. Fl.
Afr.* v. 120 ; *Schlechter in Bull. Herb. Boiss.* 1ʳᵉ *sér.* vi. 934 ; *Kränzl.
Orch. Gen. et Sp.* i. 835, *excl. both syn.*

SOUTH AFRICA : without locality, *Mund* ! and the type, *without collector's
name* !
COAST REGION : Albany Div. ; shady mountains near Grahamstown, 2300 ft.,
South, 509.   Bedford Div. ; Kaga Forest near Bedford, *Hutton* ! Komgha Div. ;
Prospect Farm, near Komgha, 2000 ft., *Flanagan,* 2593 ! Maastimi Forest,
Mrs. Hutton ! Beaumont, *Hutton* ! Highlands, *Mrs. Barber,* 17 !
KALAHARI REGION : Transvaal ; near Barberton, 3600 ft., *Culver,* 83 ; Hout-
bosch Range, 5500 ft., *Schlechter,* 4739.

28. **D. disæformis** (Schlechter in Verhandl. Bot. Ver. Brandenb.
xxxv. 47) ; plant 3–7 in. high ; stem slender, glabrous ; leaves 2,
cauline, amplexicaul, cordate, subacute, 10–14 lin. long, 6–8 lin.
broad, with rather broad shallow basal lobes ; scapes 3–7 in. high,
1–2-flowered ; bracts cordate-ovate, acute, 3–5 lin. long ; pedicels
about 4 lin. long ; flowers rather small, whitish ; dorsal sepal
galeate, 2½–3½ lin. long, with a triangular acute limb and an
oblong obtuse curved spur about half as long as the limb ; lateral
sepals spreading, obliquely ovate-oblong, acuminate, 2–3 lin. long,

with a short obtuse sac above the middle ; petals cohering with the
dorsal sepal, shortly unguiculate, obliquely falcate-ovate, acute or
acuminate, 2½–3½ lin. long ; lip narrowly unguiculate ; limb sub-
hastate-auriculate, acuminate, appendage ligulate, papillose at the
apex, 1 lin. long ; column short ; rostellum suborbicular, with
reflexed margin ; arms cartilaginous, twisted, linear, obtuse, about
1 lin. long ; capsule elliptic-oblong, 5–6 lin. long. *Schlechter in
Bull. Herb. Boiss.* 1ʳᵉ *sér.* vi. 935 ; *Kränzl. Orch. Gen. et Sp.*
i. 838.

CoAST REGION : Riversdale Div. ; margins of scrub on slopes of the Langeberg
Range, near Riversdale, *Schlechter*, 2143.　Knysna Div. ; near Knysna, *Forcade.*
Albany Div. ; Oatlands Park, near Grahamstown, 1800 ft., *Galpin*, 3000.

29. **D. Thorncrofti** (Schlechter in Engl. Jahrb. xx. Beibl. 50, 19) ;
plant 6–10 in. high ; stem rather slender, glabrous ; leaves 2, the
lower situated much below the middle of the stem, broadly cordate,
amplexicaul, acute, 1¼–2 in. long, 1–1¼ in. broad, with rather
shallow basal lobes, the upper leaf narrower and 6–7 lin. long ;
scapes 6 – 10 in. high, 1 – 3 - flowered ; bracts ovate, acute,
amplexicaul, 4 – 5 lin. long ; pedicels 6–7 lin. long ; flowers
rather small, white, with many green dots on the sepals ; dorsal
sepal galeate, 4 lin. long, with a broadly ovate acuminate deflexed
limb and a broad rather shallow sac ; lateral sepals spreading,
obliquely falcate-ovate, acuminate, 3½–4 lin. long, with an oblong
obtuse somewhat curved sac about the middle ; petals obliquely
ovate, acuminate, 4 lin. long, with a crenulate outer margin ; lip
narrowly unguiculate ; limb ovate, acuminate, reflexed, about 1 lin.
long ; appendage erect, linear, slightly curved at the apex, nearly
as long as the limb ; column short ; rostellum short, broadly dilated,
obtuse ; arms cartilaginous, twisted, linear or linear-spathulate,
2–2¼ lin. long ; capsule oblong, 8–9 lin. long. *Schlechter in Bull.
Herb. Boiss.* 1ʳᵉ *sér.* vi. 933.　*D. Thornycroftii, Kränzl. Orch. Gen.
et Sp.* i. 846.

CoAST REGION : Stockenstrom Div. ; woods at base of the Kat Berg, 3000 ft.,
Galpin, 1688 b !
KALAHARI REGION : Transvaal ; near Barberton, *Thorncroft.*

30. **D. lindleyana** (Reichb. f. in Flora, 1865, 181) ; plant ½–1 ft.
high ; stem moderately slender, glabrous ; leaf solitary, situated
rather above the middle of the stem, broadly cordate, amplexicaul,
acute, 1¼–2¼ in. long, 1–1½ in. broad, with very broad ample basal
lobes ; scapes ½–1 ft. high, 1–3-flowered ; bracts leaf-like in shape,
½–1 in. long ; pedicels 7–9 lin. long ; flowers medium-sized, light
cream-yellow with some green dots and a hairy purple patch on the
petals ; dorsal sepal galeate, 4–5 lin. long, with a broadly triangular
subobtuse deflexed apex and a conical obtuse sac ; lateral sepals
spreading, falcate-ovate, acute, 4–4½ lin. long, with a very short
broad truncate sac about the middle ; petals cohering with the

dorsal sepal, falcate-oblong, subacute, 4–4½ lin. long, outer margin
undulate ; lip broadly unguiculate ; limb subpandurate-linear,
obtuse, 4 lin. long, the part beyond the appendage sharply deflexed ;
appendage obovate-oblong, subobtuse, somewhat fleshy and papillose
at the apex, shorter than the limb ; column short ; rostellum
broadly trulliform-ovate, obtuse, fleshy, 1½ lin. long ; arms cartilaginous, oblong-lanceolate, acute or somewhat toothed at the apex,
2 lin. long ; capsule oblong, 8–10 lin. long. *Bolus, Ic. Orch. Austr.-
Afr.* i. *t. 44, and in Journ. Linn.* xxv. 206 ; *Durand & Schinz,
Conspect. Fl. Afr.* v. 119 ; *Schlechter in Bull. Herb. Boiss.* 1ʳᵉ *sér.* vi.
928 ; *Kränzl. Orch. Gen. et Sp.* i. 845.

SOUTH AFRICA : without locality, *Krebs* !
COAST REGION : Albany Div.; Fern Kloof, near Grahamstown, *Glass*, 454.
Bedford Div. ; Kaga Berg, near Bedford, 4500 ft., *Bodkin.* Stockenstrom Div. ;
woods at the base of the Kat Berg, 3000 ft., *Galpin*, 1688 ! Komgha Div. ; Kloof
near Komgha, 2000 ft., *Flanagan*, 176 !
KALAHARI REGION : Transvaal ; shady mountains near Barberton, 4000 ft.,
Culver, 85 ; Houtbosch Mountains, 5000 ft., *Schlechter, Nelson*, 493.
EASTERN REGION : Natal, *Buchanan in Herb. Wood* !

31. D. Fanniniæ (Harv. Thes. Cap. ii. 46, t. 171) ; plant ½–1½ ft.
high ; stem slender or moderately stout, glabrous ; leaves cauline,
3, amplexicaul, cordate-ovate, much narrowed and acuminate above,
¾–3¼ in. long, ¼–1¼ in. broad, glabrous ; scapes ½–1½ ft. high ; spikes
lax, 1–4-flowered ; bracts leaf-like in shape and texture, ⅓–1 in.
long ; pedicels ½–¾ in. long ; flowers large, white, sometimes with a
purple tinge and some purple raised dots on the petals ; dorsal sepal
galeate, 5 lin. long, with a triangular acute deflexed limb and a
broadly saccate obtuse galea ; lateral sepals spreading, falcate-
obovate, acuminate, 3½–4½ lin. long, with a short broadly oblong
sac about the middle ; petals cohering with the dorsal sepal, broadly
falcate-obovate, acuminate, with 2 prominent rounded lobes on the
outer margin, 4½–5 lin. long ; lip linear, sharply bent above the
rostellum, dilated about the middle, acuminate ; appendage cucullate,
obtuse, shorter than the limb ; column short ; rostellum very broad,
membranous, 2-lobed at the apex, reflexed ; arms cartilaginous,
broadly oblong, twisted, about 2 lin. long. *N. E. Br. in Gard.
Chron.* 1885, xxiv. 232 ; *Bolus in Journ. Linn. Soc.* xxv. 206 ;
Durand & Schinz, Conspect. Fl. Afr. v. 119 ; *Kränzl. Orch. Gen.
et Sp.* i. 827 ; *Schlechter in Bull. Herb. Boiss.* 1ʳᵉ *sér.* vi. 944.

KALAHARI REGION : Orange River Colony ; woods on the Drakensberg Range,
Cooper, 1092 ! Basutoland ; without precise locality, *Cooper*, 3615 ! Transvaal ;
Houtbosch Range, 6650 ft., *Rehmann*, 5854 ! *Schlechter*, 4462 !
EASTERN REGION : Transkei ; Tsomo River, *Mrs. Barber*, 837 ! Griqualand
East ; shady places on the Zuurberg Range, 4300 ft., *Schlechter*, 6613. Natal ;
Drakensberg Range, *Wood*, 626 ! Dargle Farm, *Mrs. Fannin*, 1 ! 88 ! Deepdene,
Sanderson, 1050 ! Krans Kop, *McKen*, 20 ! Oliviers Hoek, 5000 ft., *Allison*, Z !
shady places near Van Reenen, 5500 ft., *Schlechter*, 6929 ! near Highlands, 6000
ft., *Schlechter*, and without precise locality, *Gerrard*, 77 ! 2172 ! *Wood*, 1789 !
Sanderson, Mrs. Saunders.

32. D. **virginalis** (Schlechter in Engl. Jahrb. xxiv. 431); plant 5–9 in. high; stem moderately slender, glabrous; leaves cauline, 2, subopposite, sessile or subsessile, broadly ovate, acute, glabrous, $\frac{3}{4}$–1$\frac{3}{4}$ in. long, $\frac{1}{2}$–1 in. broad; scapes 5–9 in. high, 1–2-flowered; bracts leaf-like in shape and texture, 4–9 lin. long; pedicels 6–7 lin. long; flowers rather small, white; dorsal sepal galeate, 3$\frac{1}{2}$–4 lin. long, with a triangular acuminate deflexed apex, and a broadly conical obtuse galea; lateral sepals spreading, obliquely obovate-oblong, subacute or apiculate, 4 lin. long, with a conical subacute sac above the middle; petals cohering with the dorsal sepal, almost 2-lobed, obliquely falcate-rhomboid, acute, 3–3$\frac{1}{2}$ lin. long, with an ample rounded lobe in front near the base; lip narrowly unguiculate, about 3 lin. long; limb minute, linear, reflexed; appendage recurved, dilated into 4 spreading lobes, or bilobed at the apex and the segments shortly 2-lobed, about four times as long as the limb; column short; rostellum broadly ovate-orbicular, shortly bifid, with reflexed margin; arms cartilaginous, narrowly linear-clavate, about 1 lin. long. *Kränzl. Orch. Gen. et Sp.* i. 849; *Schlechter in Bull. Herb. Boiss.* 1re *sér.* vi. 953. *D. Kersteni, Schlechter in Engl. Jahrb.* xx. *Beibl.* 50, 43, *not of Reichb. f.*

KALAHARI REGION : Transvaal ; woods on the Houtbosch Range, 6500–7300 ft., *Schlechter,* 4453 !

33. D. **Nelsonii** (Rolfe); plant 6–8 in. high; stem slender, glabrous; leaves cauline, 2, in an opposite pair, subsessile, broadly ovate, acute, 1$\frac{1}{2}$–1$\frac{3}{4}$ in. long, 1–1$\frac{1}{4}$ in. broad, transparent (*Nelson*); scapes 6–8 in. high, 1–2-flowered; bracts spreading, leaf-like, abruptly acuminate, 6–10 lin. long; pedicels 7–9 lin. long; flowers medium-sized, pink or rose-coloured (in the dried state); dorsal sepal galeate, 4–4$\frac{1}{2}$ lin. long, with a triangular acute deflexed limb and a broadly saccate galea; lateral sepals spreading, obliquely falcate-elliptic, shortly apiculate, 5 lin. long, with a very small saccate oblong sac below the middle; petals elliptic-oblong, obtuse, 4–4$\frac{1}{2}$ lin. long, somewhat constricted in the middle and slightly crenulate below; lip narrowly unguiculate, claw 3$\frac{1}{2}$ lin. long; limb ovate, reflexed and acute at the apex, under $\frac{1}{2}$ lin. long; appendage bipartite, with subclavate acute lobes, thrice as long as the limb; column short; rostellum broadly ovate, truncate, 1$\frac{1}{2}$ lin. long; arms oblong, obtuse, 1 lin. long.

KALAHARI REGION : Transvaal ; Houtbosch Mountains, *Nelson,* 493 !

34. D. **anthoceros** (Reichb. f. Otia Bot. Hamb. ii. 103); plant 3–8 in. high; stem rather slender, glabrous; leaves cauline, 2, sub-opposite, sessile or subsessile, broadly ovate or subcordate-ovate, acute, glabrous, $\frac{3}{4}$–1$\frac{3}{4}$ in. long, $\frac{1}{2}$–1 in. broad; scapes 3–8 in. high, short or subcorymbose, 1–5-flowered; bracts leaf-like in shape and

312 ORCHIDEÆ (Rolfe). [*Disperis.*

texture, 2–6 lin. long ; pedicels 6–8 lin. long ; flowers medium-sized,
white with lilac spots ; dorsal sepal galeate, 5–6 lin. long, with a
short broadly triangular acute deflexed limb and a narrowly conic-
oblong obtuse galea ; lateral sepals spreading, obliquely ovate, subacute, 3–3½ lin. long, with a short broad subconical sac below the
middle ; petals cohering with the dorsal sepal, broadly ovate-oblong,
subobtuse, 2½–3 lin. long, outer margin undulate ; lip narrowly
unguiculate, 4 lin. long, limb reflexed, ovate-oblong, conduplicate,
very short ; appendage recurved, bilobed, with somewhat spreading
lobes ; column short ; rostellum broadly ovate, obtuse, with reflexed
margin ; arms cartilaginous, linear-oblong, twisted, under 1 lin. long.
Bolus in Journ. Linn. Soc. xxv. 206 ; *Durand & Schinz, Conspect. Fl.
Afr.* v. 118 ; *Schlechter in Bull. Herb. Boiss.* 1ʳᵉ *sér.* vi. 951; *Kränzl.
Orch. Gen. et Sp.* i. 847.

KALAHARI REGION : Transvaal ; Rimers Creek, near Barberton, 3000 ft., *Culver*,
46 ! woods on the Houtbosch Range, 6650 ft., *Schlechter*, 4469 !
EASTERN REGION : Natal ; Inanda, *Wood*, 841 ! near Tugela River, *Wood*
4545 ; near Hilton, 4000 ft., *Pearse in Herb. Wood*, 4545 ! 5249.

35. **D. Woodii** (Bolus in Journ. Linn. Soc. xxii. 78, t. 1, fig. 18–22) ;
plant 2½–5 in. high ; stem slender, glabrous ; leaves cauline or sometimes subradical, 2, sessile, subamplexicaul, ovate or subcordate,
acute or apiculate, ½–1 in. long, dark-coloured with whitish veins ;
scapes 2½–5 in. high, 1- (rarely 2-) flowered ; bracts ovate, acuminate,
cucullate, 3–4 lin. long ; pedicels about 2½ lin. long ; flowers
medium-sized ; sepals pink ; petals white with a few green dots ;
dorsal sepal galeate, 5–6 lin. long, with a broadly triangular
deflexed acuminate limb and a narrowly conical elongated erect
galea ; lateral sepals spreading, ovate, long-acuminate, concave,
about 3 lin. long, with a very shallow sac above the middle ; petals
cohering with the dorsal sepal, broadly ovate-oblong, subacute or
apiculate, with an undulate outer margin, subauriculate near the
apex, 2½–3 lin. long ; lip narrowly unguiculate ; limb reflexed,
ovate, acute, convex, very short ; appendage linear, fleshy, shortly
2-lobed at the apex, about thrice as long as the limb ; column short ;
rostellum ovate, subobtuse, large ; arms oblong, obtuse, twisted,
about 1 lin. long ; capsule elliptic-oblong, 6 lin. long. *Bolus in
Journ. Linn. Soc.* xxv. 207, *and Ic. Orch. Austr.-Afr.* i. *t.* 92 ; *Durand
& Schinz, Conspect. Fl. Afr.* v. 121 ; *Schlechter in Bull. Herb. Boiss.*
1ʳᵉ *sér.* vi. 950 ; *Kränzl. Orch. Gen. et Sp.* i. 850. *D. McKenii,
Harv. Thes. Cap.* ii. 47, *name only.*

EASTERN REGION : Transkei ; sand dunes on the coast near Kentani, and near
the mouth of the Qolora River, *Miss Pegler*, 41 ! Natal ; Noods Berg, *Wood*, 127 !
near Durban, 50–200 ft., *Schlechter*, 2943 ; Tongaat, *Mrs. Saunders in Herb.
Sanderson*, 1010 !

36. **D. stenoglossa** (Schlechter in Engl. Jahrb. xx. Beibl. 50, 19) ;
plant 2–7 in. high ; stem straight, slender ; leaves sessile, 2, distant,

somewhat spreading, ovate, acute, glabrous, the lower 7 lin. long, 4 lin. broad, upper much smaller; scapes 2–7 in. high, 1–2-flowered; bracts leaf-like, ovate, acuminate, half as long as the pedicels; flowers medium-sized; dorsal sepal galeate, acuminate, with a narrowly conical galea 3½–4 lin. long; lateral sepals spreading, obliquely ovate-lanceolate, acuminate and deflexed at the apex, 2–2½ lin. long, with a short obtuse sac about the middle; petals cohering with the dorsal sepal, obliquely ovate, acuminate, attenuate at the base, with a rounded undulate lobe about the middle in front; lip narrowly unguiculate; limb broadly ovate, acuminate, narrowed towards the apex; appendage linear, with two spreading lobes at the apex; column short; rostellum obtusely apiculate; arms short, porrect. *Schlechter in Bull. Herb. Boiss.* 1ʳᵉ *sér.* vi. 936. *D. Bodkini, Kränzl. Orch. Gen. et Sp.* i. 840, *partly, not of Schlechter.*

EASTERN REGION: Natal; marshes between Durban and the mouth of the Umgeni River, 10 ft., *Schlechter*, 3001.

ORDER CXXXII. **SCITAMINEÆ.**

(BY C. H. WRIGHT.)

Flowers hermaphrodite or unisexual. *Calyx* superior, lobed or entire, sometimes split down one side. *Corolla* tubular below or of 3 free petals. *Stamens* 1 or 5; anthers 2-celled, dehiscing longitudinally; connective sometimes produced above and petaloid. *Staminodes* 1–6, large and petaloid or rudimentary. *Ovary* inferior, usually 3-celled; style long; stigma terminal; stylodia often present; ovules 1 or many in each cell. *Seeds* 1 or many; albumen hard or farinaceous; embryo central, straight or curved.

Herbs, usually perennial, rarely shrubs; rootstock often horizontal, sometimes tuberous; leaves lanceolate to orbicular, cauline or radical; inflorescence spicate, racemose or panicled, axillary or terminal on a leafy stem, or direct from the rootstock; flowers often large and showy.

DISTRIB. Genera 69, species about 900, in the tropics of both hemispheres.

    I. **Kæmpferia.**—*Calyx* tubular or spathaceous, lobed or entire. *Corolla* tubular below, 3-lobed. *Stamen* 1.

    II. **Strelitzia.**—*Sepals* 3, free. *Petals* 3, two connivent. *Stamens* 5. *Fruit* capsular.

I. KÆMPFERIA, Linn.

Calyx tubular or often spathaceously split down one side, membranous. *Corolla-tube* shorter or longer than the calyx ; lobes lanceolate or linear. Lateral *staminodes* conspicuous, usually longer than the corolla-lobes, sometimes unguiculate, free or (in the South African species) united to the lip ; lip more or less 2-lobed, rarely entire, usually much larger than the lateral staminodes. *Stamen* solitary ; filament broad, flat, short ; anther 2-celled, linear ; connective produced above into a petaloid entire, lobed or toothed appendage. *Ovary* 3-celled ; cells many-ovuled ; nectary consisting of 2 narrowly cylindrical or filiform glands ; style filiform ; stigma funnel-shaped.

Herbs ; rhizome subterranean, thick, often aromatic ; rootlets filiform or nodose ; stem usually very short ; leaves oblong to lanceolate, usually with long sheaths ; flowers usually large, spicate, subcapitate or solitary, pedunculate or subsessile, sometimes with the ovary subterranean, usually distinct from the leafy stem ; bracts lanceolate or oblong, membranous to almost coriaceous.

DISTRIB. Species 55, of which 8 occur in Tropical Africa, 3 in South Africa and the remainder in Tropical Asia.

All the South African species belong to the section *Cienkowskia* characterised by the lateral staminodes being united with the lip.

Corolla-lobes 1 lin. wide ...   . (1) **stenopetala**.

Corolla-lobes 6–8 lin. wide :
     Flowers all hermaphrodite   ...   . (2) **Ethelæ**.
     Flowers hermaphrodite and female   ...   ... (3) **natalensis**.

1. **K. stenopetala** (K. Schum. in Engl. Pflanzenr. Zingib. 69) ; a perennial herb ; leafy stem nearly 2 ft. high, distinct from the flowering ; leaves lanceolate, acuminate, cuneate at the base, about 9 in. long and 2 in. wide, glabrous on both surfaces ; sheath 3 in. long, 4 lin. wide, glabrous, strongly ribbed when dry ; ligule none ; flowers on short pedicels ; calyx 1¾ in. long, acute, membranous, spathaceously split in the upper third, glabrous ; corolla-tube slender, about as long as the calyx ; lobes very narrowly linear, 2 in. long, 1 lin. wide ; lip deeply 2-lobed, 2 in. long, 1½ in. wide, lobes rounded ; lateral staminodes 1¾ in. long, ½ in. wide, acute, shortly cuspidate ; filament short ; anther narrowly linear, 4½ lin. long ; connective produced into a 2-lobed appendage about 6 lin. long ; ovary ellipsoid, glabrous.

EASTERN REGION : Natal ; without precise locality, *Wood*, 1942 !

2. **K. Ethelæ** (Wood in Gard. Chron. 1898, xxiii. 94, fig. 34) ; root-stock tuberous, irregular, with a strong aromatic odour ; leafy stem 6–8 in. high at the time the flowers are produced, finally reaching 2–3 ft. ; leaves sessile, oblong-lanceolate or lanceolate, acuminate, the uppermost 10–14 in. long, 2–2½ in. wide, the lower much

shorter, glabrous on both surfaces, sheaths long ; flowers solitary
from the rootstock ; peduncle 8 in. long; calyx tubular, split to the
middle on one side, 2 in. long, glabrous; corolla white ; lobes
linear or lanceolate-oblong, acute or shortly cucullate, 2–2½ in. long,
6–8 lin. wide ; lateral staminodes oblong, erect, 3 in. long, 1¾ in.
wide, peach-coloured, darker at the outer edge, lighter towards the
centre of the flower ; lip about 3 in. long and wide, 2-lobed about
two-thirds of its length, with a broad central yellow area extending half-way down the tube; connective produced into an oblong
petaloid appendage 2–2¼ in. long, 5 lin. wide, slightly contracted
below the lacerate apex. *K. Schum. in Engl. Pflanzenr. Zingib.* 71.
K. sp., Burtt-Davy in Transv. Agric. Journ. ix. (1910) 45.

KALAHARI REGION : Transvaal ; Barberton, *Steytler in Herb. Bolus*, 6396 !
Granite Hill Farm, Nelspruit, *Wilhelm*, 4163 ! Elim, Spelunken, *Mingard*, 19 !
New Agatha, *Reckenzaan*, 6095 !
EASTERN REGION : Natal ; Riet Valley, *Mrs. Saunders* ! Swaziland, without
precise locality, *Andrews* !

Also in Tropical Africa.

According to Mr. Burtt-Davy this plant is known in the Transvaal as the
"*Sherungulu.*" Plants grown in the Cambridge Botanic Garden from tubers
supplied by Mr. Burtt-Davy as "Sherungulu" produced flowers in April, 1912,
scarcely distinguishable from those of *Curcuma æruginosa*, Roxb. A specimen
collected by Mr. Swierstra at Groot Letaba in the Zoutpansberg district of the
Transvaal (Transvaal Mus. Herb. 2497) consists of a single flower identical with
that of *K. Ethelæ*, but accompanied by a solitary detached leaf with an ovate
acuminate blade 3½ in. long, 1¾ in. wide, a rounded ligule 3 lin. high and a
sheath 5½ in. long.

3. **K. natalensis** (Schlechter & K. Schum. in Engl. Pflanzenr.
Zingib. 72, fig. 10, E–F) ; a perennial herb up to 2 ft. high ;
rhizome 8 lin. in diam. when dried ; leaves nearly all radical,
linear-lanceolate or linear, long-acuminate, tapering towards the
base, up to 1 ft. long, nearly 1 in. wide, glabrous on both surfaces ;
sheath 8 in. to 1 ft. long, 4 lin. wide ; ligule about 2 lin. long,
membranous, rotundate or 2-lobed ; scape arising outside the tuft
of leaves ; bracts, pedicels and ovary subterranean ; flowers solitary
or 3–6 racemosely arranged, hermaphrodite and female ; hermaphrodite flower : calyx tubular, white, membranous, about 1 in.
long, obtusely 3-toothed, split on one side about a third of the
way down ; corolla-tube 1 in. long; lobes 2½ in. long, 7 lin. wide,
lanceolate, acuminate, membranous, white ; lateral staminodes erect,
3 in. long, 1¼ in. wide, obovate, pink ; lip 2½ in. long, 2½ in. wide,
bifid, pink with a yellow blotch at the base, united below with the
lateral staminodes into a split tube 2½ in. long ; anther linear, 6 lin.
long ; connective produced above the cells into a toothed crest
about 1 in. long and 5–6 lin. wide ; style filiform; stigma crateri-
form, 3-lobed ; female flower : calyx and corolla as in the hermaphrodite flower ; staminodes very variable, 4–6, united into a yellow
cylindrical tube about 3 in. long ; lobes pink, sometimes one oblong,
obtuse, 1¼ in. long, 7 lin. wide, with a yellow blotch at the base,

and 5 others oblanceolate, unguiculate, obtuse, 4–5 lin. wide, sometimes 2–3 oblong, obtuse, 1–1¼ in. wide, alternating with 2–3
oblanceolate, unguiculate, about 5 lin. wide, when 5 are present 2
are broad and 3 narrow. *Siphonochilus natalensis, Wood & Franks in
Kew Bulletin*, 1911, 274, *and in Wood, Natal Plants*, vi. *tt.* 560–561.

EASTERN REGION : Natal; Inanda, 1800 ft., *Wood,* 544 ! Ungoye, *Wylie in
Herb. Wood,* 11723 ! and without precise locality, *Gerrard,* 1823 !

The hermaphrodite flowers of this do not differ from those of *K. Ethelæ,* Wood,
and further investigation in the field may result in these two species being united.
The existence of female flowers was noted by the late Prof. W. H. Harvey in a
letter to Dr. T. Thomson, written in 1865, and now preserved at Kew; the
specimens to which he refers have not been found either at Kew or Dublin. The
staminodes of the female flowers are very variable, even in Wood's 11723.

II. STRELITZIA, Ait.

Flowers hermaphrodite. *Sepals* 3, yellow or white, the anticous
more concave than the others. *Petals* 3, free, blue or white, 2
lower connivent and forming a sagittate blade with a central
channel in which the stamens and style are placed, the third much
smaller. *Stamens* 5 perfect; filaments slender; anthers linear,
2-celled, very long. *Ovary* 3-celled; style long, with 3 linear
branches; ovules many, superposed. *Capsule* oblong, triquetrous,
loculicidal. *Seeds* few, with a woolly aril.

Stemless or with a stem; leaves distichous, more or less oblong; petiole up to
6 ft. long, deeply channelled down the face; peduncle erect, variable in length;
flowers several in the axil of a coriaceous spathaceous bract.

Acaulescent ; sepals bright yellow :
    Leaves 1½ ft. long, margin undulate below   ...   ... (1) **Reginæ.**
    Leaves 9 in. long, margin flat ...   . (2) **parvifolia.**
Caulescent ; sepals white :
    Petals white   ...   . (3) **augusta.**
    Petals blue   ...   . (4) **Nicolai.**

1. **S. Reginæ** (Banks in Ait. Hort. Kew. ed. 1, i. 285, t. 2);
stemless; leaves oblong-lanceolate, acute, cuneate at the base, up to
1½ ft. long and 4 in. wide, entire, undulate especially in the lower
part, quite glabrous, bright green above, glaucescent beneath;
petiole up to 4 ft. long; peduncle as long as the petiole; bracts
tubular, oblique and acute at the mouth, uppermost one cymbiform,
acuminate, up to 8 in. long, green, edged with purple; sepals
lanceolate, 3–4 in. long, orange-yellow; petals dark blue, blade of
the two lower 2 in. long, with a rounded basal auricle, claw 1 in.
long; upper petal ovate, 1 in. long; stamens reaching to the top
of the longer petals; anthers narrowly linear, twice as long as the
filaments; style exserted, with 3 linear branches 1 in. long. *Thunb.*

Nov. Gen. 113, *Prodr.* 46, *and Fl. Cap. ed. Schult.* 216; *Bauer, Strelitzia, tt.* 6–9; *Redouté, Liliac. tt.* 77–78; *Drège, Zwei Pfl. Documente,* 141, 150; *Otto in Allg. Gartenz.* 1856, 17; *Lemaire in Ill. Hort.* x. (1863), *Misc.* 25; *Baker in Ann. Bot.* vii. 201; *K. Schum. in Engl. Pflanzenr. Musac.* 31. *S. regalis, Salisb. Prodr.* 145. *Heliconia Bihai, J. Miller, Ic. Pl. tt.* 5–6 (1780).

VAR. β, **glauca** (Baker in Ann. Bot. vii. 201); leaves oval or oval-oblong, shortly acuminate, rounded and undulate at the base, glaucous; peduncle shorter than the leaves; lower petal 8 lin. long, deeply concave, mucronate. *K. Schum. in Engl. Pflanzenr. Musac.* 33. *S. glauca, L. C. Rich. in Nova Acta Acad. Nat. Cur.* xv. *Suppl.* 17, *tt.* 2–3.

VAR. γ, **rutilans** (Morren in Ann. Soc. Bot. Gand, ii. t. 53); leaves rounded at the base, more than 3 ft. long, midrib purple; petiole not longer than the blade; sepals orange-red; petals very dark purple. *K. Schum. in Engl. Pflanzenr. Musac.* 33. *S. rutilans, K. Schum. l.c.*

VAR. δ, **farinosa** (Baker in Ann. Bot. vii. 201); leaves oblong, unequal and truncate at the base; petiole half as long again as the blade; peduncle slightly longer than the petiole, glaucous. *K. Schum. in Engl. Pflanzenr. Musac.* 33. *S. farinosa, Dryand in Ait. Hort. Kew. ed.* 2, ii. 55; *Roem. & Schult. Syst.* v. 595; *Lemaire in Ill. Hort.* x. (1863), *Misc.* 26.

VAR. ε, **ovata** (Baker in Ann. Bot. vii. 202); leaves ovate-oblong, rounded or subcordate at the base; petiole about 3 times as long as the blade; peduncle longer than the leaves. *K. Schum. in Engl. Pflanzenr. Musac.* 33. *S. ovata, Dryand. in Ait. Hort. Kew. ed.* 2, ii. 55. *S. Reginæ, Curt. Bot. Mag. tt.* 119–120; *Andr. Rep. t.* 432.

VAR. ζ, **humilis** (E. Meyer in Drège, Zwei Pfl. Documente, 141, 224); much smaller than the other varieties; leaves ovate, concave, 9 in. long, 4 in. wide; petiole twice as long as the blade; peduncle as long as the petiole. *Baker in Ann. Bot.* vii. 202; *K. Schum. in Engl. Pflanzenr. Musac.* 33. *S. humilis, Link, Enum. Pl. Hort. Berol.* i. 150. *S. pumila, Hort. ex Otto in Allg. Gartenz.* 1856, 18; *Planch. in Fl. des Serres,* xi. (1856) 53.

COAST REGION : Humansdorp Div. ; Gamtoos River, *Thunberg.* Bathurst Div. ; Glenfilling, under 1000 ft., *Drège.* Var. ζ, Uitenhage Div. ; rocky heights of Uitenhage, *Bowie.* Albany Div. ; between Blaauw Kranz to Kowie Poort, *Burchell,* 3670 ! near the Kowie River, *Bowie.* Bathurst Div. ; Fish River, under 1000 ft., *Drège* ! South-west of Cape Colony, *Masson, Nelson.* King Williamstown Div. ; banks of the Keiskamma River, nearly 7 miles from the mouth, *Gill* !
EASTERN REGION : Pondoland ; between Umtata River and St. Johns River, under 1000 ft., *Drège.*

The varieties *glauca, rutilans, farinosa* and *ovata* are known from cultivated specimens only.

Boer name "*Geele Pisang.*"

S. Reginæ, var. *Lemoinierii* (Planch. in Fl. des Serres, t. 2370. *S. Lemoinierii,* Miellez ex Planch. l.c.) is a highly coloured form. *S. prolifera,* Carr. in Rev. Hort. 1869, 159, fig. 39, is a form with several flower-clusters. *S. kewensis,* Kew Bulletin, 1910, 65, 327, and Gard. Chron. 1910, xlvii. 217, with fig., is a hybrid between *S. Reginæ* and *S. augusta,* raised at Kew in 1898, which first flowered in 1910.

2. **S. parvifolia** (Dryand. in Ait. Hort. Kew. ed. 2, ii. 56) ; stemless ; leaves oblong-lanceolate, 9 in. long, 3 in. wide, deltoid at the base, with a narrow brown scarious flat margin ; petiole

slender, 4–6 ft. long; peduncle as long as the leaves; spathe 6 in.
long, green edged with red; sepals bright orange, lanceolate, about
4 in. long, $\frac{3}{4}$ in. wide; petals blue, blade of the 2 lower 2 in. long,
with rounded basal auricles, upper broadly ovate, $\frac{1}{2}$ in. long,
mucronate; stamens about 3 in. long; style-arms $1\frac{1}{2}$ in. long,
protruding beyond the petals. *Bauer, Strelitz. tt.* 10–11; *Lemaire
in Ill. Hort.* x. (1863), *Misc.* 26; *Baker in Ann. Bot.* vii. 201;
K. Schum. in Engl. Pflanzenr. Musac. 33. *S. angustifolia, Dryand.
in Ait. Hort. Kew. ed.* 2, ii. 55. *S. Principis, Andr. ex Spreng.
Syst.* i. 833. *S. Reginæ, var. parvifolia, Smith in Rees, Cyclop.*

VAR. **juncea** (Ker in Bot. Reg. t. 516); blade of leaf very small or none.
Reichb. Fl. Exot. t. 181; *Baker in Ann. Bot.* vii. 201; *K. Schum. in Engl.
Pflanzenr. Musac.* 33. *S. juncea, Link, Enum. Hort. Berol.* i. 150; *Andr. ex
Spreng. Syst.* i. 833. *S. augustifolia, var. juncea, E. Meyer in Drège, Zwei Pfl.
Documente,* 130, 224.

SOUTH AFRICA: without locality, *Villet!* Var. β, *Villet!*
COAST REGION: Var. β, Uitenhage Div.; at Uitenhage by the Zwartkops
River, *Burchell,* 4438/3! limestone hill or kopje near the mouth of the Zwartkops
River, under 500 ft., *Drège!*

3. **S. augusta** (Thunb. Nov. Gen. Pl. 113); stem simple up to
18 ft. high, as thick as a man's arm (*Thunberg*); leaves oblong or
ovate, obtuse, deltoid or rounded at the base, glabrous, shining
green, 3–4 ft. long, $1\frac{1}{2}$–2 ft. wide; petiole 5–6 ft. long; peduncle
shorter than the petiole; spathe claret-coloured, 8–12 in. long;
sepals lanceolate, white, 5–6 in. long; petals white, 2 lower with
blades 3 in. long and small rounded auricles, basal part $1\frac{1}{2}$ in. long,
tapering upwards; upper petal ovate, 1 in. long; filaments $1\frac{1}{2}$ in.
long; anthers 3 in. long; style-arms $1\frac{1}{2}$–2 in. long. *Fl. Cap. ed.
Schult.* 216; *Bauer, Strelitzia, tt.* 1–4; *Drège, Zwei Pfl. Doc.* 125,
156; *Bot. Mag. tt.* 4167–8; *Lemaire in Ill. Hort.* x. (1863), *Misc.*
24; *Baker in Ann. Bot.* vii. 202; *K. Schum. in Engl. Pflanzenr.
Musac.* 33; *Burtt-Davy in Transv. Agric. Journ.* v. 418, 422.
S. gigantea, Kerner, Hort. Semperv. t. 589. *S. angusta, D. Dietr.
Syn. Pl.* i. 850. *Heliconia alba, Linn. f. Suppl.* 157.

COAST REGION: George Div.; Outeniqua Mountains, *Thunberg.* Knysna Div.;
Bosch River, in a wood, under 500 ft., *Drège!* Humansdorp Div.; Gamtoos
River, *Thunberg.*
EASTERN REGION: Natal; between Umzimkulu River and Umkomanzi River,
under 500 ft., *Drège*; near Durban, *Cooper,* 1225!

Boer name "*Witte Pisang.*"

4. **S. Nicolai** (Regel & Körnicke in Gartenfl. 1858, 265, t. 235);
stem up to 25 ft. high; leaves oblong, rounded (rarely cordate) at the
base, 3–4 ft. long, $1\frac{1}{2}$–2 ft. wide; petiole up to 6 ft. long; peduncle
shorter than the petiole; spathe reddish-brown, 1 ft. long, acuminate; sepals white, lanceolate, about 7 in. long, 1 in. wide;
petals blue, two lower forming a sagittate blade 4 in. long and 1 in.
wide springing from an oblong base tapering in the upper part and

2¼ in. long; upper petal rotundate, ½ in. in diam., with a cusp 6 in. long; filaments 1¾ in. long; anthers slender, 3 in. long; style 8 in. long including the 2 in.-long linear branches. *Fl. des Serres, t.* 1356; *Lemaire in Ill. Hort.* x. (1863), *Misc.* 24; *Hook. f. in Bot. Mag. t.* 7038; *Baker in Ann. Bot.* vii. 203; *K. Schum. in Engl. Pflanzenr. Musac.* 33.

SOUTH AFRICA : without locality or collector's name.

Known only from cultivated plants.

Imperfectly known species.

5. **S. lanceolata** (Hort. ex Steud. Nomencl. ed. 2, ii. 645; K. Schum. in Engl. Pflanzenr. Musac. 33, name only).

SOUTH AFRICA : without locality.

6. **S. macrophylla** (Hort. ex Steud. Nomencl. ed. 2, ii. 645; K. Schum. in Engl. Pflanzenr. Musac. 33, name only).

SOUTH AFRICA : without locality.

7. **S. Quensoni** (Lemaire in Ill. Hort. x. (1863), Misc. 20); acaulescent, but otherwise resembling in habit *S. Nicolai* and *S. augusta*; rhizome thick; leaves ovate-oblong, obtuse and cucullate at the apex, attenuate downwards, very large; petiole very robust, with a long wide sheath; inflorescence lateral; spathe much compressed, green; flowers 3–5 in each spathe, rosy violet at the base, white above; petals bright blue, auricles somewhat angled. *S. albiflos, Hort. ex Lemaire, l.c.* 21.

SOUTH AFRICA : without locality, *Villet.*

8. **S. teretifolia** (Barrow ex Steud. Nomencl. ed. 2, ii. 645; Lemaire in Ill. Hort. x. (1863), Misc. 26; K. Schum. in. Engl. Pflanzenr. Musac. 33, name only).

SOUTH AFRICA : without locality or indication of the collector.

Perhaps the same as *S. parvifolia,* var. *juncea,* Ker.

ADDENDA AND CORRIGENDA.

BURMANNIACEÆ.

1. **Burmannia capensis** (Mart.). Page 3, line 13, for *Fedde, Repert. July*, 1912, read *Fedde, Repert.* xi. 81.

2. **Burmannia inhambanensis** (Schlechter in Fedde, Repert. xi. 82); a very slender plant 6 to 9 in. high; rootlets fascicled, very thinly filiform; stem simple, very slender, subterete, glabrous; leaves scale-like, approximate on the lower part of the stem, scattered above, linear, acute, glabrous, 1–2 lin. long; cymes short, 3–5-flowered; bracts similar to the cauline leaves and nearly as large; flowers erect on short pedicels; perianth 3–4 lin. long, with 3 semi-elliptic wings decurrent from above the base of the free lobes; outer lobes 3, ovate, obtuse, $\frac{3}{4}$ lin. long, slightly thickened at the margin; inner lobes 3, oblong, obtuse, a little shorter and at the base half as wide as the outer, distinctly papillose-ciliate at the margin; stamens subsessile, inserted just below the mouth of the perianth-tube, included; connective rather fleshy, dilated, with 2 rounded lobes conspicuously papillose at their apex; cells divergent, distinctly shorter than the connective; style cylindrical with short arms; stigmas bilabiate, overtopping the anthers, the lower lip larger, but thinner, than the upper; unripe seeds narrowly and obliquely ellipsoid.

EASTERN REGION: Portuguese East Africa; in damp meadows, Inhambane, 20 ft., *Schlechter*, 12086.

ORCHIDEÆ.

22a. **Eulophia Pillansii** (Bolus, Ic. Orch. Austr.-Afr. ii. t. 27); rhizome much thickened at the nodes; leaves immature at flowering time, 6, suberect or spreading, ultimately recurved, conduplicate below, ensiform, acuminate, 4–6 in. long, 5–6 lin. broad; scape somewhat flexuous, with 1 bract-like sheath; raceme loosely 10–11-flowered; flowers spreading; bracts oblong-lanceolate, acute, half as long as the pedicel; flowers with brownish-green sepals and pale yellow petals and lip; sepals declinate towards the apex, spreading or somewhat recurved, oblong, acute, 8 lin. long; petals 7 lin. long, otherwise like the sepals; lip ovate in outline, shortly three-lobed, about 8 lin. long; side lobes erect, narrowly semi-orbicular, undulate; front lobe ovate-oblong, with a reflexed acute or apiculate apex; disc with erect plate-like keels; spur cylindrical or subclavate, obtuse, strongly incurved, $2\frac{1}{2}$–$3\frac{1}{2}$ lin. long; column oblong, produced at the base into a projecting foot, apiculate at the apex.

CENTRAL REGION : Somerset Div. ; near Cookhouse, 1,600 ft., *Pillans in Herb. Bolus*, 10479.

Bolus remarks : "The species is interesting owing to the fact of its being the solitary instance, so far as I know, of a *Eulophia* growing in that dry part of the Colony known as the Karoo."

1. **Lissochilus clitellifer** (Reichb. f.).   Delete from the synonymy :—*Eulophia platypetala, Krauss.*

7a. **Polystachya pisobulbon** (Kränzl. in Engl. Jahrb. xlviii. 397) ; pseudobulbs densely aggregated, as large as a pea, globose or slightly compressed at the base, somewhat elongated above, about 4 lin. long, with 2 or 3 short brown sheaths ; leaves broadly oblong, deeply bilobed at the apex, with obtuse lobes, $\frac{1}{2}$–$\frac{3}{4}$ in. long, 2–3 lin. broad ; scapes rather longer than the leaves, 2-flowered, with 2 or 3 rudimentary sheaths, puberulous ; bracts very minute ; pedicels puberulous, $3\frac{1}{2}$ lin. long ; flowers white, with a pale rose line on the sepals, a yellow disc to the lip, and a green column ; dorsal sepal oblong-lanceolate, acute, concave ; lateral sepals sub-oblique at the base, triangular, acute, $3\frac{1}{2}$ lin. long, 2 lin. broad, with a very short obtuse chin, puberulous at the base ; petals oblanceolate, subacute, $1\frac{1}{4}$ lin. long; lip sessile, much recurved, 3-lobed, $3\frac{1}{2}$ lin. broad ; side lobes semiobovate, rounded in front ; front lobe broadly ovate, acute or shortly acuminate ; disc granular-pilose, with a thickened elevated central keel from middle to base.

EASTERN REGION : Natal ; without precise locality. Flowered in the Berlin Botanic Garden.

A small species, most allied to *P. ottoniana*, Reichb. f.

6a. **Mystacidium Aliciæ** (Bolus, Ic. Orch. Austr.-Afr. ii. t. 6, fig. B) ; a small stemless epiphyte ; aerial roots stout ; leaves 2–4, spreading, broadly oblong, unequally and obtusely bilobed, coriaceous, $\frac{3}{4}$–1 in. long ; raceme spreading, slender, about 1 in. long, loosely 3–5-flowered ; rhachis flexuous ; pedicels very slender, about 3 lin. long ; flowers white or pale cream ; dorsal sepal erect, oblong, subacute, 2 lin. long ; lateral sepals spreading or somewhat decurved, oblong, acute, 2 lin. long ; petals somewhat spreading, ovate-oblong, acute, subfalcate, 2 lin. long ; lip arcuately decurved, ovate-lanceolate, acute, as long as the petals, with an incurved margin ; spur pendulous, cylindrical from a broader base, nearly straight, gradually attenuate towards the apex, subacute, 4–$4\frac{1}{2}$ lin. long ; rostellum projecting forwards, then deflexed, subulate, acute, furnished with a bearded appendage on each side ; pollinia sub-globose, attached by very slender filiform stipites to separate oblong glands.

EASTERN REGION : Transkei ; in a forest near Kentani, 1000 ft., *Miss Pegler*, 886.

Bolus remarks that the species is allied to *M. pusillum*, Harv., but the flowers are larger, the perianth-segments differently shaped and equal in length, the spur of the lip much shorter, and the column and rostellum also different.

10. Holothrix parvifolia (Lindl.). Page 103, line 8 from the bottom, for *O. hispida*, read *Orchis hispida*.

25. Satyrium coriifolium (Sw.). The synonym *Orchis bicornis*, Linn. *Sp. Pl. ed.* ii. 1330, should be excluded. It was based on Linn. *Amœn. Acad.* vi. 109, and *Orchis lutea, caule geniculato, Buxb. Pl. Min. Cogn. Cent.* iii. 6, *t.* 8 (misquoted by Linneus as *t.* 6), which is *Satyrium bicorne, Thunb.*, as correctly cited at p. 170 of this volume. The mistake apparently began with Loddiges, who in 1818 (*Bot. Cab. t.* 104) wrongly figured *S. coriifolium* under the name of *S. cucullatum*, remarking that it was "*Orchis bicornis* of the older authors." *S. cucullatum, Thunb.*, however, is a synonym of *S. bicorne, Thunb.* Bolus (*Orch. Cap. Penins.* 124) cited *O. bicorne*, Linn., as a synonym of *S. coriifolium*, and the error seems to have been copied into all subsequent works. Bolus also cites as *S. erectum, Lindl.*, the figures *Lodd. Bot. Cab. t.* 104 ; *Bot. Mag. t.* 2172 ; and *Bot. Reg. t.* 703, whereas the first was called *S. cucullatum* and the two others *S. coriifolium*. There is the further complication that *S. erectum, Lindl.*, is not the original *S. erectum, Sw.* Bolus cited *Orchis bicornis, Linn. Sp. Pl. ed.* ii. 1330, under *S. coriifolium*, and *O. bicornis, Linn. Amœn. Acad.* vi. 109, under *S. bicorne*, but the former includes the latter, and both were based on *Orchis lutea, caule geniculato, Buxb. Pl. Min. Cogn. Cent.* iii. *t.* 8, which is a quite characteristic figure of *S. bicorne*.

Page 161, line 27, for *O. luteo*, read *O. lutea*.

1. Penthea patens (Lindl.). Page 208, line 33, for *Disa patens, Thunb.*, read *Disa patens, Sw.* Lindley cites it as *Disa patens, Willd.* [*Sp. Pl.* iv. 53], which is antedated five years by *D. patens, Sw.*

Page 209, line 14, after *Disa patens, Sw. in Vet. Acad. Handl. Stockh.* 1800, 214, add "*partly.*"

2. Penthea filicornis (Lindl.). Page 210, line 4, after *Disa filicornis, Thunb.*, add "*Fl. Cap.* i. 87, *and*"; line 8, after *Disa patens, Sw. in Vet. Acad. Handl. Stockh.* 1800, 214, add "*partly.*" Swartz included two plants under his *Disa patens*, viz. *Serapias patens, Thunb.*, and *Orchis filicornis, Linn.*, but the former is identical with *Ophrys patens, Linn.*, on which *Disa tenuifolia, Sw.*, was based. This original mistake led to others, which were rectified in the synonymy of *Penthea patens* and *P. filicornis*.

INDEX.

[Synonyms are printed in *italics*.]

328 INDEX.

LONDON : PRINTED BY WILLIAM CLOWES AND SONS, LIMITED,
DUKE STREET, STAMFORD STREET, S.E., AND GREAT WINDMILL STREET, W.

Printed in the United States
By Bookmasters

Printed in the United States
By Bookmasters